DIGITAL SYSTEMS TESTING
AND
TESTABLE DESIGN

Revised Printing

MIRON ABRAMOVICI, *AT&T Bell Laboratories, Murray Hill*
MELVIN A. BREUER, *University of Southern California, Los Angeles*
ARTHUR D. FRIEDMAN, *George Washington University*

The Institute of Electrical and Electronics Engineers, Inc., New York

WILEY-
INTERSCIENCE
A JOHN WILEY & SONS, INC., PUBLICATION

This is the IEEE revised printing of the book previously published by W. H. Freeman and Company in 1990 under the title *Digital Systems Testing and Testable Design*.

THE INSTITUTE OF ELECTRICAL AND ELECTRONICS ENGINEERS, INC.
3 Park Avenue, 17th Floor, New York, NY 10016-5997

Published by John Wiley & Sons, Inc., Hoboken, New Jersey.

For general information on our other products and services please contact our Customer Care Department within the U.S. at 877-762-2974, outside the U.S. at 317-572-3993 or fax 317-572-4002.

ISBN 13: 978-0-7803-1062-9

10 9 8

To our families, who were not able to see their spouses and fathers for many evenings and weekends, and who grew tired of hearing "leave me alone, I'm working on the book." Thank you Gaby, Ayala, Orit, Sandy, Teri, Jeff, Barbara, Michael, and Steven. We love you.

CONTENTS

PREFACE

This book provides a comprehensive and detailed treatment of digital systems testing and testable design. These subjects are increasingly important, as the cost of testing is becoming the major component of the manufacturing cost of a new product. Today, design and test are no longer separate issues. The emphasis on the quality of the shipped products, coupled with the growing complexity of VLSI designs, require testing issues to be considered early in the design process so that the design can be modified to simplify the testing process.

This book was designed for use as a text for graduate students, as a comprehensive reference for researchers, and as a source of information for engineers interested in test technology (chip and system designers, test engineers, CAD developers, etc.). To satisfy the different needs of its intended readership the book (1) covers thoroughly both the fundamental concepts and the latest advances in this rapidly changing field, (2) presents only theoretical material that supports practical applications, (3) provides extensive discussion of testable design techniques, and (4) examines many circuit structures used to realize built-in self-test and self-checking features.

Chapter 1 introduces the main concepts and the basic terminology used in testing. Modeling techniques are the subject of Chapter 2, which discusses functional and structural models for digital circuits and systems. Chapter 3 presents the use of logic simulation as a tool for design verification testing, and describes compiled and event-driven simulation algorithms, delay models, and hardware accelerators for simulation. Chapter 4 deals with representing physical faults by logical faults and explains the concepts of fault detection, redundancy, and the fault relations of equivalence and dominance. The most important fault model — the single stuck-fault model — is analyzed in detail. Chapter 5 examines fault simulation methods, starting with general techniques — serial, parallel, deductive, and concurrent — and continuing with techniques specialized for combinational circuits — parallel-pattern single-fault propagation and critical path tracing. Finally, it considers approximate methods such as fault sampling and statistical fault analysis.

Chapter 6 addresses the problem of test generation for single stuck faults. It first introduces general concepts common to most test generation algorithms, such as implication, sensitization, justification, decision tree, implicit enumeration, and backtracking. Then it discusses in detail several algorithms — the D-algorithm, the 9V-algorithm, PODEM, FAN, and critical path test generation — and some of the techniques used in TOPS, SOCRATES, RAPS, SMART, FAST, and the subscripted D-algorithm. Other topics include random test generation, test generation for sequential circuits, test generation using high-level models, and test generation systems.

Chapter 7 looks at bridging faults caused by shorts between normally unconnected signal lines. Although bridging faults are a "nonclassical" fault model, they are dealt with by simple extensions of the techniques used for single stuck faults. Chapter 8 is concerned with functional testing and describes heuristic methods, techniques using binary decision diagrams, exhaustive and pseudoexhaustive testing, and testing methods for microprocessors.

Chapter 9 presents design for testability techniques aimed at simplifying testing by modifying a design to improve the controllability and observability of its internal signals. The techniques analyzed are general ad hoc techniques, scan design, board and system-level approaches, partial scan and boundary scan (including the proposed JTAG/IEEE 1149.1 standard).

Chapter 10 is dedicated to compression techniques, which consider a compressed representation of the response of the circuit under test. The techniques examined are ones counting, transition counting, parity checking, syndrome checking, and signature analysis. Because of its widespread use, signature analysis is discussed in detail. The main application of compression techniques is in circuits featuring built-in self-test, where both the generation of input test patterns and the compression of the output response are done by circuitry embedded in the circuit under test. Chapter 11 analyzes many built-in self-test design techniques (CSBL, BEST, RTS, LOCST, STUMPS, CBIST, CEBS, RTD, SST, CATS, CSTP, and BILBO) and discusses several advanced concepts such as test schedules and partial intrusion built-in self-test.

Chapter 12 discusses logic-level diagnosis. The covered topics include the basic concepts in fault location, fault dictionaries, guided-probe testing, expert systems for diagnosis, effect-cause analysis, and a reasoning method using artificial intelligence concepts.

Chapter 13 presents self-checking circuits where faults are detected by a subcircuit called a checker. Self-checking circuits rely on the use of coded inputs. Some basic concepts of coding theory are first reviewed, followed by a discussion of specific codes — parity-check codes, Berger codes, and residue codes — and of designs of checkers for these codes.

Chapter 14 surveys the testing of programmable logic arrays (PLAs). First it reviews the fault models specific to PLAs and test generation methods for external testing of these faults. Then it describes and compares many built-in self-test design methods for PLAs.

Chapter 15 deals with the problem of testing and diagnosis of a system composed of several independent processing elements (units), where one unit can test and diagnose other units. The focus is on the relation between the structure of the system and the levels of diagnosability that can be achieved.

In the Classroom

This book is designed as a text for graduate students in computer engineering, electrical engineering, or computer science. The book is self-contained, most topics being covered extensively, from fundamental concepts to advanced techniques. We assume that the students have had basic courses in logic design, computer science, and probability theory. Most algorithms are presented in the form of pseudocode in an easily understood format.

The progression of topics follows a logical sequence where most chapters rely on material presented in preceding chapters. The most important precedence relations among chapters are illustrated in the following diagram. For example, fault simulation (5) requires understanding of logic simulation (3) and fault modeling (4). Design for testability (9) and compression techniques (10) are prerequisites for built-in self-test (11).

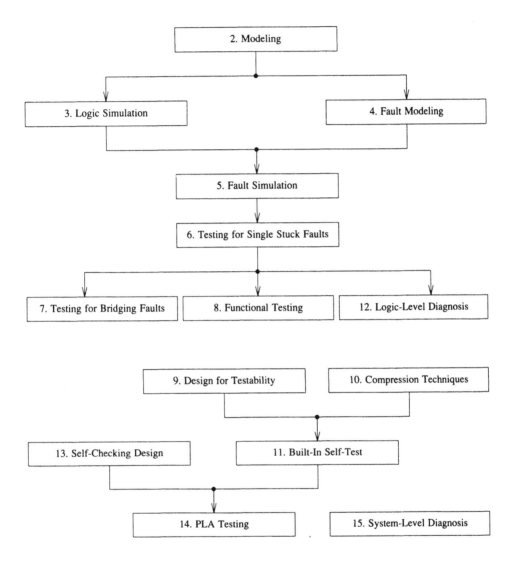

Precedence relations among chapters

The book requires a two-semester sequence, and even then some material may have to be glossed over. For a one-semester course, we suggest a "skinny path" through Chapters 1 through 6 and 9 through 11. The instructor can hope to cover only about half of this material. This "Introduction to Testing" course should emphasize the fundamental concepts, algorithms, and design techniques, and make only occasional forays into the more advanced topics. Among the subjects that could be skipped or only briefly discussed in the introductory course are Simulation Engines (Section 3.11),

The Multiple Stuck-Fault Model (4.6), Fault Sampling (5.4), Statistical Fault Analysis (5.5), Random Test Generation (6.2.3), Advanced Scan Techniques (9.9), and Advanced BIST Concepts (11.5). Most of the material in Chapter 2 and the ad hoc design for testability techniques (9.2) can be given as reading assignments.

Most of the topics not included in the introductory course can be covered in a second semester "Advanced Testing" course.

Acknowledgments

We have received considerable help in developing this book. We want to acknowledge Xi-An Zhu, who contributed to portions of the chapter on PLA testing. We are grateful for the support provided by the Bell Labs managers who made this work possible — John Bierbauer, Bill Evans, Al Fulton, Hao Nham, and Bob Taylor. Special thanks go to the Bell Labs word processing staff — Yvonne Anderson, Deborah Angell, Genevieve Przeor, and Lillian Pilz — for their superb job in producing this book, to David Hong for helping with troff and related topics, to Jim Coplien for providing the indexing software, and to John Pautler and Tim Norris for helping with the phototypesetter. We want to thank our many colleagues who have been using preprints of this book for several years and who have given us invaluable feedback, and especially S. Reddy, S. Seth, and G. Silberman. And finally, many thanks to our students at the University of Southern California and the Illinois Institute of Technology, who helped in "debugging" the preliminary versions.

Miron Abramovici
Melvin A. Breuer
Arthur D. Friedman

How This Book Was Written

Don't worry. We will not begin by saying that "because of the rapid increases in the complexity of VLSI circuitry, the issues of testing, design-for-test and built-in-self-test, are becoming increasingly more important." You have seen this type of opening a million times before. Instead, we will tell you a little of the background of this book.

The story started at the end of 1981. Miron, a young energetic researcher (at that time), noted that Breuer and Friedman's *Diagnosis & Reliable Design of Digital Systems* — known as the "yellow book" — was quickly becoming obsolete because of the rapid development of new techniques in testing. He suggested co-authoring a new book, using as much material from the yellow book as possible and updating it where necessary. He would do most of the writing, which Mel and Art would edit. It all sounded simple enough, and work began in early 1982.

Two years later, Miron had written less than one-third of the book. Most of the work turned out to be new writing rather than updating the yellow book. The subjects of modeling, simulation, fault modeling, fault simulation, and test generation were reorganized and greatly expanded, each being treated in a separate chapter. The end, however, was nowhere in sight. Late one night Miron, in a state of panic and frustration, and Mel, in a state of weakness, devised a new course of action. Miron would finalize the above topics and add new chapters on bridging faults testing, functional testing, logic-level diagnosis, delay-faults testing, and RAM testing. Mel would write the chapters dealing with new material, namely, PLA testing, MOS circuit testing, design for testability, compression techniques, and built-in self-test. And Art would update the material on self-checking circuits and system-level diagnosis.

The years went by, and so did the deadlines. Only Art completed his chapters on time. The book started to look like an encyclopedia, with the chapter on MOS testing growing into a book in itself. Trying to keep the material up to date was a continuous and endless struggle. As each year passed we came to dread the publication of another proceedings of the DAC, ITC, FTCS, or ICCAD, since we knew it would force us to go back and update many of the chapters that we had considered done.

Finally we acknowledged that our plan wasn't working and adopted a new course of action. Mel would set aside the MOS chapter and would concentrate on other, more essential chapters, leaving MOS for a future edition. Miron's chapters on delay fault testing and RAM testing would have the same fate. A new final completion date was set for January 1989.

This plan worked, though we missed our deadline by some 10 months. Out of love for our work and our profession, we have finally accomplished what we had set out to do. As this preface was being written, Miron called Mel to tell him about a paper he just read with some nice results on test generation. Yes, the book is obsolete already. If you are a young, energetic researcher — don't call us.

1. INTRODUCTION

Testing and Diagnosis

Testing of a system is an experiment in which the system is exercised and its resulting response is analyzed to ascertain whether it behaved correctly. If incorrect behavior is detected, a second goal of a testing experiment may be to *diagnose*, or locate, the cause of the misbehavior. Diagnosis assumes knowledge of the internal structure of the system under test. These concepts of testing and diagnosis have a broad applicability; consider, for example, medical tests and diagnoses, test-driving a car, or debugging a computer program.

Testing at Different Levels of Abstraction

The subject of this book is *testing and diagnosis of digital systems*. "Digital system" denotes a complex digital circuit. The *complexity* of a circuit is related to the *level of abstraction* required to describe its operation in a meaningful way. The level of abstraction can be roughly characterized by the type of information processed by the circuit (Figure 1.1). Although a digital circuit can be viewed as processing analog quantities such as voltage and current, the lowest level of abstraction we will deal with is the *logic level*. The information processed at this level is represented by discrete *logic values*. The classical representation uses *binary logic values* (0 and 1). More accurate models, however, often require more than two logic values. A further distinction can be made at this level between combinational and sequential circuits. Unlike a combinational circuit, whose output logic values depend only on its present input values, a sequential circuit can also remember past values, and hence it processes *sequences of logic values*.

Control	Data	Level of abstraction
Logic values (or sequences of logic values)		Logic level
Logic values	Words	Register level
Instructions	Words	Instruction set level
Programs	Data structures	Processor level
Messages		System level

Figure 1.1   Levels of abstraction in information processing by a digital system

We start to regard a circuit as a system when considering its operation in terms of processing logic values becomes meaningless and/or unmanageable. Usually, we view a system as consisting of a *data* part interacting with a *control* part. While the control function is still defined in terms of logic values, the information processed by the data part consists of *words*, where a word is a group (*vector*) of logic values. As data words are stored in registers, this level is referred to as the *register level*. At the next level of abstraction, the *instruction set level*, the control information is also organized as words,

1

referred to as *instructions*. A system whose operation is directed by a set of instructions is called an *instruction set processor*. At a still higher level of abstraction, the *processor level*, we can regard a digital system as processing sequences of instructions, or *programs*, that operate on blocks of data, referred to as *data structures*. A different view of a system (not necessarily a higher level of abstraction) is to consider it composed of independent subsystems, or units, which communicate via blocks of words called *messages*; this level of abstraction is usually referred to as the *system level*.

In general, the stimuli and the response defining a testing experiment correspond to the type of information processed by the system under test. Thus testing is a generic term that covers widely different activities and environments, such as

- one or more subsystems testing another by sending and receiving messages;

- a processor testing itself by executing a diagnostic program;

- automatic test equipment (ATE) checking a circuit by applying and observing binary patterns.

In this book we will not be concerned with *parametric tests*, which deal with electrical characteristics of the circuits, such as threshold and bias voltages, leakage currents, and so on.

Errors and Faults

An instance of an incorrect operation of the system being tested (or UUT for *unit under test*) is referred to as an *(observed) error*. Again, the concept of error has different meanings at different levels. For example, an error observed at the diagnostic program level may appear as an incorrect result of an arithmetic operation, while for ATE an error usually means an incorrect binary value.

The causes of the observed errors may be *design errors*, *fabrication errors*, *fabrication defects*, and *physical failures*. Examples of design errors are

- incomplete or inconsistent specifications;

- incorrect mappings between different levels of design;

- violations of design rules.

Errors occurring during fabrication include

- wrong components;

- incorrect wiring;

- shorts caused by improper soldering.

Fabrication defects are not directly attributable to a human error; rather, they result from an imperfect manufacturing process. For example, shorts and opens are common defects in manufacturing MOS Large-Scale Integrated (LSI) circuits. Other fabrication defects include improper doping profiles, mask alignment errors, and poor encapsulation. Accurate location of fabrication defects is important in improving the manufacturing yield.

Physical failures occur during the lifetime of a system due to component wear-out and/or environmental factors. For example, aluminum connectors inside an IC package thin out with time and may break because of electron migration or corrosion. Environmental factors, such as temperature, humidity, and vibrations, accelerate the aging of components. Cosmic radiation and α-particles may induce failures in chips containing high-density random-access memories (RAMs). Some physical failures, referred to as "infancy failures," appear early after fabrication.

Fabrication errors, fabrication defects, and physical failures are collectively referred to as *physical faults*. According to their stability in time, physical faults can be classified as

- *permanent*, i.e., always being present after their occurrence;

- *intermittent*, i.e., existing only during some intervals;

- *transient*, i.e., a one-time occurrence caused by a temporary change in some environmental factor.

In general, physical faults do not allow a direct mathematical treatment of testing and diagnosis. The solution is to deal with *logical faults*, which are a convenient representation of the effect of the physical faults on the operation of the system. A fault is *detected* by observing an error caused by it. The basic assumptions regarding the nature of logical faults are referred to as a *fault model*. The most widely used fault model is that of a single line (wire) being permanently "stuck" at a logic value. Fault modeling is the subject of Chapter 4.

Modeling and Simulation

As design errors precede the fabrication of the system, *design verification testing* can be performed by a testing experiment that uses a *model* of the designed system. In this context, "model" means a digital computer representation of the system in terms of data structures and/or programs. The model can be exercised by stimulating it with a representation of the input signals. This process is referred to as *logic simulation* (also called design verification simulation or true-value simulation). Logic simulation determines the evolution in time of the signals in the model in response to an applied input sequence. We will address the areas of modeling and logic simulation in Chapters 2 and 3.

Test Evaluation

An important problem in testing is *test evaluation*, which refers to determining the effectiveness, or quality, of a test. Test evaluation is usually done in the context of a fault model, and the quality of a test is measured by the ratio between the number of faults it detects and the total number of faults in the assumed fault universe; this ratio is referred to as the *fault coverage*. Test evaluation (or test grading) is carried out via a simulated testing experiment called *fault simulation*, which computes the response of the circuit in the presence of faults to the test being evaluated. A fault is detected when the response it produces differs from the expected response of the fault-free circuit. Fault simulation is discussed in Chapter 5.

Types of Testing

Testing methods can be classified according to many criteria. Figure 1.2 summarizes the most important attributes of the testing methods and the associated terminology.

Criterion	Attribute of testing method	Terminology
When is testing performed?	• Concurrently with the normal system operation • As a separate activity	On-line testing Concurrent testing Off-line testing
Where is the source of the stimuli?	• Within the system itself • Applied by an external device (tester)	Self-testing External testing
What do we test for?	• Design errors • Fabrication errors • Fabrication defects • Infancy physical failures • Physical failures	Design verification testing Acceptance testing Burn-in Quality-assurance testing Field testing Maintenance testing
What is the physical object being tested?	• IC • Board • System	Component-level testing Board-level testing System-level testing
How are the stimuli and/or the expected response produced?	• Retrieved from storage • Generated during testing	Stored-pattern testing Algorithmic testing Comparison testing
How are the stimuli applied?	• In a fixed (predetermined) order • Depending on the results obtained so far	 Adaptive testing

Figure 1.2   Types of testing

Testing by *diagnostic programs* is performed off-line, at-speed, and at the system level. The stimuli originate within the system itself, which works in a self-testing mode. In systems whose control logic is microprogrammed, the diagnostic programs can also be microprograms (microdiagnostics). Some parts of the system, referred to as *hardcore*, should be fault-free to allow the program to run. The stimuli are generated by software or

Criterion	Attribute of testing method	Terminology
How fast are the stimuli applied?	• Much slower than the normal operation speed	DC (static) testing
	• At the normal operation speed	AC testing At-speed testing
What are the observed results?	• The entire output patterns	
	• Some function of the output patterns	Compact testing
What lines are accessible for testing?	• Only the I/O lines	Edge-pin testing
	• I/O and internal lines	Guided-probe testing Bed-of-nails testing Electron-beam testing In-circuit testing In-circuit emulation
Who checks the results?	• The system itself	Self-testing Self-checking
	• An external device (tester)	External testing

Figure 1.2   (Continued)

firmware and can be adaptively applied. Diagnostic programs are usually run for field or maintenance testing.

In-circuit emulation is a testing method that eliminates the need for hardcore in running diagnostic programs. This method is used in testing microprocessor (μP)-based boards and systems, and it is based on removing the μP on the board during testing and accessing the μP connections with the rest of the UUT from an external tester. The tester can emulate the function of the removed μP (usually by using a μP of the same type). This configuration allows running of diagnostic programs using the tester's μP and memory.

In *on-line testing*, the stimuli and the response of the system are not known in advance, because the stimuli are provided by the patterns received during the normal mode of operation. The object of interest in on-line testing consists not of the response itself, but of some properties of the response, properties that should remain invariant throughout the fault-free operation. For example, only one output of a fault-free decoder should have logic value 1. The operation code (opcode) of an instruction word in an instruction set processor is restricted to a set of "legal" opcodes. In general, however, such easily definable properties do not exist or are difficult to check. The general approach to on-line testing is based on *reliable design techniques* that create invariant properties that are easy to check during the system's operation. A typical example is the use of an additional parity bit for every byte of memory. The parity bit is set to create an easy-to-check invariant property, namely it makes every extended byte (i.e., the original byte plus the parity bit) have the same parity (i.e., the number of 1 bits, taken modulo 2). The parity bit is *redundant*, in the sense that it does not carry any information useful for the normal operation of the system. This type of *information redundancy* is characteristic for systems using *error-detecting and error-correcting codes*. Another type of reliable design based on redundancy is *modular redundancy*, which is based on replicating a module several times. The replicated modules (they must have the same function, possibly with different implementations) work with the same set of inputs, and the invariant property is that all of them must produce the same response. Self-checking systems have subcircuits called *checkers*, dedicated to testing invariant properties. Self-checking design techniques are the subject of Chapter 13.

Guided-probe testing is a technique used in board-level testing. If errors are detected during the initial edge-pin testing (this phase is often referred to as a GO/NO GO test), the tester decides which internal line should be monitored and instructs the operator to place a probe on the selected line. Then the test is reapplied. The principle is to trace back the propagation of error(s) along path(s) through the circuit. After each application of the test, the tester checks the results obtained at the monitored line and determines whether the site of a fault has been reached and/or the backtrace should continue. Rather than monitoring one line at a time, some testers can monitor a group of lines, usually the pins of an IC.

Guided-probe testing is a sequential diagnosis procedure, in which a subset of the internal accessible lines is monitored at each step. Some testers use a fixture called *bed-of-nails* that allows monitoring of all the accessible internal lines in a single step.

The goal of *in-circuit testing* is to check components already mounted on a board. An external tester uses an IC clip to apply patterns directly to the inputs of one IC and to observe its outputs. The tester must be capable of electronically isolating the IC under test from its board environment; for example, it may have to overdrive the input pattern supplied by other components.

Algorithmic testing refers to the generation of the input patterns during testing. Counters and feedback shift registers are typical examples of hardware used to generate the input stimuli. *Algorithmic pattern generation* is a capability of some testers to produce combinations of several fixed patterns. The desired combination is determined by a control program written in a tester-oriented language.

The expected response can be generated during testing either from a known good copy of the UUT — the so-called *gold unit* — or by using a real-time emulation of the UUT. This type of testing is called *comparison testing*, which is somehow a misnomer, as the comparison with the expected response is inherent in many other testing methods.

Methods based on checking some function $f(R)$ derived from the response R of the UUT, rather than R itself, are said to perform *compact testing*, and $f(R)$ is said to be a *compressed* representation, or *signature*, of R. For example, one can count the number of 1 values (or the number of 0 to 1 and 1 to 0 transitions) obtained at a circuit output and compare it with the expected 1-count (or transition count) of the fault-free circuit. Such a compact testing procedure simplifies the testing process, since instead of bit-by-bit comparisons between the UUT response and the expected output, one needs only one comparison between signatures. Also the tester's memory requirements are significantly reduced, because there is no longer need to store the entire expected response. Compression techniques (to be analyzed in Chapter 10) are mainly used in self-testing circuits, where the computation of $f(R)$ is implemented by special hardware added to the circuit. Self-testing circuits also have additional hardware to generate the stimuli. Design techniques for circuits with Built-In Self-Test (BIST) features are discussed in Chapter 11.

Diagnosis and Repair

If the UUT found to behave incorrectly is to be repaired, the cause of the observed error must be diagnosed. In a broad sense, the terms diagnosis and repair apply both to physical faults and to design errors (for the latter, "repair" means "redesign"). However, while physical faults can be effectively represented by logical faults, we lack a similar mapping for the universe of design errors. Therefore, in discussing diagnosis and repair we will restrict ourselves to physical (and logical) faults.

Two types of approaches are available for fault diagnosis. The first approach is a *cause-effect analysis*, which enumerates all the possible faults (causes) existing in a fault model and determines, before the testing experiment, all their corresponding responses (effects) to a given applied test. This process, which relies on fault simulation, builds a data base called a *fault dictionary*. The diagnosis is a dictionary look-up process, in which we try to match the actual response of the UUT with one of the precomputed responses. If the match is successful, the fault dictionary indicates the possible faults (or the faulty components) in the UUT.

Other diagnosis techniques, such as guided-probe testing, use an *effect-cause analysis* approach. An effect-cause analysis processes the actual response of the UUT (the effect) and tries to determine directly only the faults (cause) that could produce that response. Logic-level diagnosis techniques are treated in Chapter 12 and system-level diagnosis is the subject of Chapter 15.

Test Generation

Test generation (TG) is the process of determining the stimuli necessary to test a digital system. TG depends primarily on the testing method employed. On-line testing methods do not require TG. Little TG effort is needed when the input patterns are provided by a feedback shift register working as a pseudorandom sequence generator. In contrast, TG for design verification testing and the development of diagnostic programs involve a large effort that, unfortunately, is still mainly a manual activity. *Automatic TG* (ATG) refers to TG algorithms that, given a model of a system, can generate tests for it. ATG has been developed mainly for edge-pin stored-pattern testing.

TG can be *fault oriented* or *function oriented*. In fault-oriented TG, one tries to generate tests that will detect (and possibly locate) specific faults. In function-oriented TG, one tries to generate a test that, if it passes, shows that the system performs its specified function. TG techniques are covered in several chapters (6, 7, 8, and 12).

Design for Testability

The *cost of testing* a system has become a major component in the cost of designing, manufacturing, and maintaining a system. The cost of testing reflects many factors such as TG cost, testing time, ATE cost, etc. It is somehow ironic that a $10 μP may need a tester thousands times more expensive.

Design for testability (DFT) techniques have been increasingly used in recent years. Their goal is to reduce the cost of testing by introducing testability criteria early in the design stage. Testability considerations have become so important that they may even dictate the overall structure of a design. DFT techniques for external testing are discussed in Chapter 9.

2. MODELING

About This Chapter

Modeling plays a central role in the design, fabrication, and testing of a digital system. The way we represent a system has important consequences for the way we simulate it to verify its correctness, the way we model faults and simulate it in the presence of faults, and the way we generate tests for it.

First we introduce the basic concepts in modeling, namely behavioral versus functional versus structural models and external versus internal models. Then we discuss modeling techniques. This subject is closely intertwined with some of the problems dealt with in Chapter 3. This chapter will focus mainly on the description and representation of models, while Chapter 3 will emphasize algorithms for their interpretation. One should bear in mind, however, that such a clear separation is not always possible.

In the review of the basic modeling techniques, we first discuss methods that describe a circuit or system as one "box"; circuits are modeled at the logic level and systems at the register level. Then we present methods of representing an interconnection of such boxes (structural models).

2.1 Basic Concepts

Behavioral, Functional, and Structural Modeling

At any level of abstraction, a digital system or circuit can be viewed as a *black box*, processing the information carried by its inputs to produce its output (Figure 2.1). The I/O mapping realized by the box defines the **behavior** of the system. Depending on the level of abstraction, the behavior can be specified as a mapping of logic values, or of data words, etc. (see Figure 1.1). This transformation occurs over time. In describing the behavior of a system, it is usually convenient to separate the value domain from the time domain. The **logic function** is the I/O mapping that deals only with the value transformation and ignores the I/O timing relations. A **functional model** of a system is a representation of its logic function. A **behavioral model** consists of a functional model coupled with a representation of the associated timing relations. Several advantages result from the separation between logic function and timing. For example, circuits that realize the same function but differ in their timing can share the same functional model. Also the function and the timing can be dealt with separately for design verification and test generation. This distinction between function and behavior is not always made in the literature, where these terms are often used interchangeably.

A **structural model** describes a box as a collection of interconnected smaller boxes called **components** or **elements**. A structural model is often *hierarchical* such that a component is in turn modeled as an interconnection of lower-level components. The bottom-level boxes are called **primitive elements**, and their functional (or behavioral) model is assumed to be known. The function of a component is shown by its **type**. A *block diagram* of a computer system is a structural model in which the types of the

9

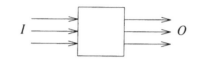

Figure 2.1   Black box view of a system

components are CPUs, RAMs, I/O devices, etc. A *schematic diagram* of a circuit is a structural model using components whose types might be AND, OR, 2-to-4 DECODER, SN7474, etc. The type of the components may be denoted graphically by special shapes, such as the shapes symbolizing the basic gates.

A structural model always carries (implicitly or explicitly) information regarding the function of its components. Also, only small circuits can be described in a strictly functional (black-box) manner. Many models characterized as "functional" in the literature also convey, to various extents, some structural information. *In practice structural and functional modeling are always intermixed.*

External and Internal Models

An **external model** of a system is the model viewed by the user, while an **internal model** consists of the data structures and/or programs that represent the system inside a computer. An external model can be graphic (e.g., a schematic diagram) or text based. A text-based model is a description in a formal language, referred to as a *Hardware Description Language* (HDL). HDLs used to describe structural models are called *connectivity languages*. HDLs used at the register and instruction set levels are generally referred to as *Register Transfer Languages* (RTLs). Like a conventional programming language, an RTL has *declarations* and *statements*. Usually, declarations convey structural information, stating the existence of hardware entities such as registers, memories, and busses, while statements convey functional information by describing the way data words are transferred and transformed among the declared entities.

Object-oriented programming languages are increasingly used for modeling digital systems [Breuer *et al.* 1988, Wolf 1989].

2.2 Functional Modeling at the Logic Level
2.2.1 Truth Tables and Primitive Cubes

The simplest way to represent a combinational circuit is by its **truth table**, such as the one shown in Figure 2.2(a). Assuming binary input values, a circuit realizing a function $Z(x_1, x_2, ..., x_n)$ of n variables requires a table with 2^n entries. The data structure representing a truth table is usually an array V of dimension 2^n. We arrange the input combinations in their increasing binary order. Then $V(0) = Z(0,0,...,0)$, $V(1) = Z(0,0,...,1)$, ..., $V(2^n-1) = Z(1,1,...,1)$. A typical procedure to determine the value of Z, given a set of values for $x_1, x_2, ..., x_n$, works as follows:

1. Concatenate the input values in proper order to form one binary word.

2. Let i be the integer value of this word.

3. The value of Z is given by $V(i)$.

x_1 x_2 x_3	Z
0 0 0	1
0 0 1	1
0 1 0	0
0 1 1	1
1 0 0	1
1 0 1	1
1 1 0	0
1 1 1	0

x_1 x_2 x_3	Z
x 1 0	0
1 1 x	0
x 0 x	1
0 x 1	1

(a)   (b)

Figure 2.2   Models for a combinational function $Z(x_1, x_2, x_3)$ (a) truth table (b) primitive cubes

For a circuit with m outputs, an entry in the V array is an m-bit vector defining the output values. Truth table models are often used for Read-Only Memories (ROMs), as the procedure outlined above corresponds exactly to the ROM addressing mechanism.

Let us examine the first two lines of the truth table in Figure 2.2(a). Note that $Z = 1$ independent of the value of x_3. Then we can compress these two lines into one whose content is $00x|1$, where x denotes an unspecified or "don't care" value. (For visual clarity we use a vertical bar to separate the input values from the output values). This type of compressed representation is conveniently described in *cubical notation*.

A **cube** associated with an ordered set of signals $(a_1\ a_2\ a_3\ \cdots)$ is a vector $(v_1\ v_2\ v_3\ \cdots)$ of corresponding signal values. A *cube of a function* $Z(x_1, x_2, x_3)$ has the form $(v_1\ v_2\ v_3\ |\ v_Z)$, where $v_Z = Z(v_1, v_2, v_3)$. Thus a cube of Z can represent an entry in its truth table. An implicant g of Z is represented by a cube constructed as follows:

1. Set $v_i = 1(0)$ if x_i $(\overline{x_i})$ appears in g.

2. Set $v_i = x$ if neither x_i nor $\overline{x_i}$ appears in g.

3. Set $v_Z = 1$.

For example, the cube $00x|1$ represents the implicant $\overline{x}_1\overline{x}_2$.

If the cube q can be obtained from the cube p by replacing one or more x values in p by 0 or 1, we say that p **covers** q. The cube $00x|1$ covers the cubes $000|1$ and $001|1$.

A cube representing a prime implicant of Z or $\bar{Z}$ is called a **primitive cube**. Figure 2.2(b) shows the primitive cubes of the function Z given by the truth table in Figure 2.2(a).

To use the primitive-cube representation of Z to determine its value for a given input combination $(v_1 \; v_2 \; \cdots \; v_n)$, where the v_i values are binary, we search the primitive cubes of Z until we find one whose input part covers v. For example, $(v_1 \; v_2 \; v_3) = 010$ matches (in the first three positions) the primitive cube $x10 \,|\, 0$; thus $Z(010) = 0$. Formally, this matching operation is implemented by the *intersection operator*, defined by the table in Figure 2.3. The intersection of the values 0 and 1 produces an *inconsistency*, symbolized by $\varnothing$. In all other cases the intersection is said to be *consistent*, and values whose intersection is consistent are said to be *compatible*.

$\cap$	0	1	x
0	0	$\varnothing$	0
1	$\varnothing$	1	1
x	0	1	x

Figure 2.3   Intersection operator

The intersection operator extends to cubes by forming the pairwise intersections between corresponding values. The intersection of two cubes is consistent iff* all the corresponding values are compatible. To determine the value of Z for a given input combination $(v_1 \; v_2 \; \cdots \; v_n)$, we apply the following procedure:

1.   Form the cube $(v_1 \; v_2 \; \cdots \; v_n \,|\, x)$.

2.   Intersect this cube with the primitive cubes of Z until a consistent intersection is obtained.

3.   The value of Z is obtained in the rightmost position.

Primitive cubes provide a compact representation of a function. But this does not necessarily mean that an internal model for primitive cubes consumes less memory than the corresponding truth-table representation, because in an internal model for a truth table we do not explicitly list the input combinations.

Keeping the primitive cubes with $Z = 0$ and the ones with $Z = 1$ in separate groups, as shown in Figure 2.2(b), aids in solving a problem often encountered in test generation. The problem is to find values for a subset of the inputs of Z to produce a specified value for Z. Clearly, the solutions are given by one of the two sets of primitive cubes of Z. For

---

*   if and only if

example, from Figure 2.2(b), we know that to set $Z = 1$ we need either $x_2 = 0$ or $x_1 x_3 = 01$.

2.2.2 State Tables and Flow Tables

A finite-state sequential function can be modeled as a *sequential machine* that receives inputs from a finite set of possible inputs and produces outputs from a finite-set of possible outputs. It has a finite number of *internal states* (or simply, states). A finite state sequential machine can be represented by a *state table* that has a row corresponding to every internal state of the machine and a column corresponding to every possible input. The entry in row q_i and column I_m represents the next state and the output produced if I_m is applied when the machine is in state q_i. This entry will be denoted by $N(q_i, I_m)$, $Z(q_i, I_m)$. N and Z are called the *next state* and *output function* of the machine. An example of a state table is shown in Figure 2.4.

	x	
	0	1
1	2,1	3,0
2	2,1	4,0
3	1,0	4,0
4	3,1	3,0

$N(q,x), Z(q,x)$

Figure 2.4   Example of state table (q = current state, x = current input)

In representing a sequential function in this way there is an inherent assumption of *synchronization* that is not explicitly represented by the state table. The inputs are synchronized in time with some timing sequence, $t(1)$, $t(2)$, ..., $t(n)$. At every time $t(i)$ the input is sampled, the next state is entered and the next output is produced. A circuit realizing such a sequential function is a *synchronous sequential circuit*.

A synchronous sequential circuit is usually considered to have a *canonical structure* of the form shown in Figure 2.5. The combinational part is fed by the primary inputs x and by the state variables y. Every state in a state table corresponds to a different combination of values of the state variables. The concept of synchronization is explicitly implemented by using an additional input, called a *clock* line. Events at times $t(1)$, $t(2)$, ..., are initiated by pulses on the clock line. The state of the circuit is stored in bistable memory elements called *clocked flip-flops* (F/Fs). A clocked F/F can change state only when it receives a clock pulse. Figure 2.6 shows the circuit symbols and the state tables of three commonly used F/Fs (C is the clock).

More general, a synchronous sequential circuit may have several clock lines, and the clock pulses may propagate through some combinational logic on their way to F/Fs.

Sequential circuits can also be designed without clocks. Such circuits are called *asynchronous*. The behavior of an asynchronous sequential circuit can be defined by a *flow table*. In a flow table, a state transition may involve a sequence of state changes,

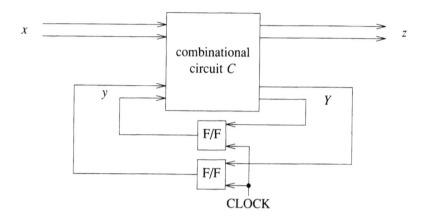

Figure 2.5   Canonical structure of a synchronous sequential circuit

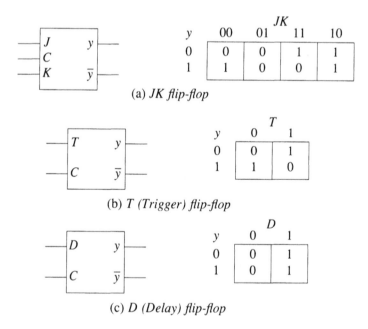

(a) JK flip-flop

(b) T (Trigger) flip-flop

(c) D (Delay) flip-flop

Figure 2.6   Three types of flip-flops

caused by a single input change to I_j, until a stable configuration is reached, denoted by the condition $N(q_i, I_j) = q_i$. Such stable configurations are shown in boldface in the flow table. Figure 2.7 shows a flow table for an asynchronous machine, and Figure 2.8 shows the canonical structure of an asynchronous sequential circuit.

$x_1 x_2$

	00	01	11	10
1	**1**,0	**5**,1	**2**,0	**1**,0
2	1,0	**2**,0	**2**,0	**5**,1
3	**3**,1	2,0	**4**,0	**3**,0
4	3,1	**5**,1	**4**,0	**4**,0
5	3,1	**5**,1	**4**,0	**5**,1

Figure 2.7   A flow table

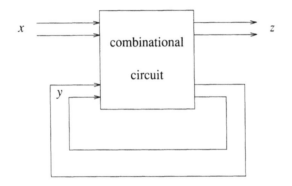

Figure 2.8   Canonical structure of an asynchronous sequential circuit

For test generation, a synchronous sequential circuit S can be modeled by a pseudocombinational iterative array as shown in Figure 2.9. This model is equivalent to the one given in Figure 2.5 in the following sense. Each cell $C(i)$ of the array is identical to the combinational circuit C of Figure 2.5. If an input sequence $x(0) \, x(1)...x(k)$ is applied to S in initial state $y(0)$, and generates the output sequence $z(0) \, z(1)...z(k)$ and state sequence $y(1) \, y(2)...y(k+1)$, then the iterative array will generate the output $z(i)$ from cell i, in response to the input $x(i)$ to cell i ($1 \leq i \leq k$). Note that the first cell also receives the values corresponding to $y(0)$ as inputs. In this transformation the clocked F/Fs are modeled as combinational elements, referred to as pseudo-F/Fs. For a JK F/F, the inputs of the combinational model are the present state q and the excitation inputs J and K, and the outputs are the next state q^+ and the device outputs y and $\bar{y}$. The present state q of the F/Fs in cell i must be equal to the q^+ output of the F/Fs in cell $i-1$. The combinational element model corresponding to a JK F/F is defined by the truth table in Figure 2.10(a). Note that here $q^+ = y$. Figure 2.10(b) shows the general F/F model.

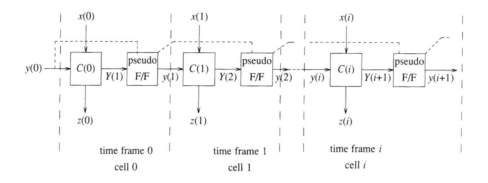

Figure 2.9   Combinational iterative array model of a synchronous sequential circuit

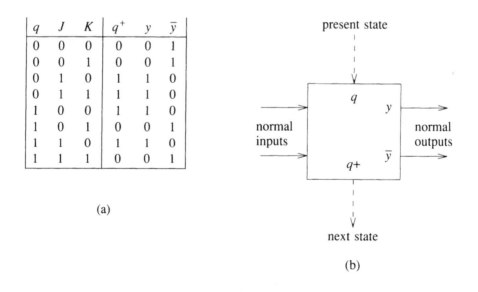

q	J	K	q^+	y	$\bar{y}$
0	0	0	0	0	1
0	0	1	0	0	1
0	1	0	1	1	0
0	1	1	1	1	0
1	0	0	1	1	0
1	0	1	0	0	1
1	1	0	1	1	0
1	1	1	0	0	1

(a)

(b)

Figure 2.10   (a) Truth table of a *JK* pseudo-F/F (b) General model of a pseudo-F/F

This modeling technique maps the time domain response of the sequential circuit into a space domain response of the iterative array. Note that the model of the combinational part need not be actually replicated. This transformation allows test generation methods developed for combinational circuits to be extended to synchronous sequential circuits. A similar technique exists for asynchronous sequential circuits.

2.2.3 Binary Decision Diagrams

A *binary decision diagram* [Lee 1959, Akers 1978] is a graph model of the function of a circuit. A simple graph traversal procedure determines the value of the output by sequentially examining values of its inputs. Figure 2.11 gives the diagram of $f = \bar{a}b\bar{c} + ac$. The traversal starts at the top. At every node, we decide to follow the left or the right branch, depending on the value (0 or 1) of the corresponding input variable. The value of the function is determined by the value encountered at the exit branch. For the diagram in Figure 2.11, let us compute f for abc=001. At node a we take the left branch, then at node b we also take the left branch and exit with value 0 (the reader may verify that when a=0 and b=0, f does not depend on the value of c). If at an exit branch we encounter a variable rather than a value, then the value of the function is the value of that variable. This occurs in our example for a=1; here f=c.

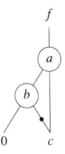

Figure 2.11   Binary decision diagram of $f = \bar{a}b\bar{c} + ac$

When one dot is encountered on a branch during the traversal of a diagram, then the final result is complemented. In our example, for a=0 and b=1, we obtain f=$\bar{c}$. If more than one dot is encountered, the final value is complemented if the number of dots is odd.

Binary decision diagrams are also applicable for modeling sequential functions. Figure 2.12 illustrates such a diagram for a *JK* F/F with asynchronous set (*S*) and reset (*R*) inputs. Here q represents the previous state of the F/F. The diagram can be entered to determine the value of the output y or $\bar{y}$. For example, when computing y for S=0, R=0, C=1, and q=1, we exit with the value of K inverted (because of the dot), i.e., y=$\bar{K}$. The outputs of the F/F are undefined (x) for the "illegal" condition S = 1 and R = 1.

The following example [Akers 1978] illustrates the construction of a binary decision diagram from a truth table.

Example 2.1: Consider the truth table of the function $f = \bar{a}b\bar{c} + ac$, given in Figure 2.13(a). This can be easily mapped into the binary decision diagram of Figure 2.13(b), which is a complete binary tree where every path corresponds to one of the eight rows of the truth table. This diagram can be simplified as follows. Because both branches from the leftmost node c result in the same value 0, we remove this

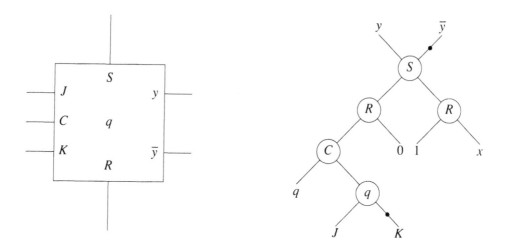

Figure 2.12   Binary decision diagram for a *JK* F/F

node and replace it by an exit branch with value 0. Because the left and right branches from the other *c* nodes lead to 0 and 1 (or 1 and 0) values, we remove these nodes and replace them with exit branches labeled with *c* (or $\bar{c}$). Figure 2.13(c) shows the resulting diagram. Here we can remove the rightmost *b* node (Figure 2.13(d)). Finally, we merge branches leading to *c* and $\bar{c}$ and introduce a dot to account for inversion (Figure 2.13(e)).   □

2.2.4 Programs as Functional Models

A common feature of the modeling techniques presented in the previous sections is that the model consists of a data structure (truth table, state table, or binary decision diagram) that is interpreted by a model-independent program. A different approach is to model the function of a circuit directly by a program. This type of code-based modeling is always employed for the primitive elements used in a structural model. In general, models based on data structures are easier to develop, while code-based models can be more efficient because they avoid one level of interpretation. (Modeling options will be discussed in more detail in a later section).

In some applications, a program providing a functional model of a circuit is automatically generated from a structural model. If only binary values are considered, one can directly map the logic gates in the structural model into the corresponding logic operators available in the target programming language. For example, the following assembly code can be used as a functional model of the circuit shown in Figure 2.14 (assume that the variables *A,B,...,Z* store the binary values of the signals *A,B,...,Z*):

a	b	c	f
0	0	0	0
0	0	1	0
0	1	0	1
0	1	1	0
1	0	0	0
1	0	1	1
1	1	0	0
1	1	1	1

(a)

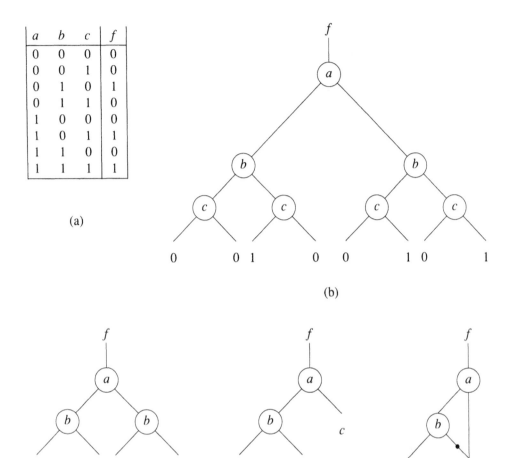

(b)

(c)   (d)   (e)

Figure 2.13   Constructing a binary decision diagram

```
LDA A     /* load accumulator with value of A */
AND B     /* compute A.B */
AND C     /* compute A.B.C */
STA E     /* store partial result */
LDA D     /* load accumulator with value of D */
INV       /* compute D̄ */
OR E      /* compute A.B.C + D̄ */
STA Z     /* store result */
```

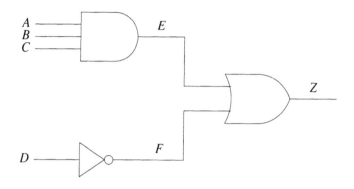

Figure 2.14

Using the C programming language, the same circuit can be modeled by the following:

$$E = A \ \& \ B \ \& \ C$$
$$F = \sim D$$
$$Z = E \mid F$$

As the resulting code is compiled into machine code, this type of model is also referred to as a *compiled-code* model.

2.3 Functional Modeling at the Register Level

2.3.1 Basic RTL Constructs

RTLs provide models for systems at the register and the instruction set levels. In this section we will discuss only the main concepts of RTL models; the reader may refer to [Dietmeyer and Duley 1975, Barbacci 1975, Shahdat *et al.* 1985] for more details. Data (and control) words are stored in registers and memories organized as arrays of registers. Declaring the existence of registers inside the modeled box defines a *skeleton structure* of the box. For example:

register *IR* [0→7]

defines an 8-bit register *IR*, and

memory *ABC* [0→255; 0→15]

denotes a 256-word memory *ABC* with 16-bit/word.

The data paths are implicitly defined by describing the processing and the transfer of data words among registers. RTL models are characterized as functional, because they emphasize functional description while providing only summary structural information.

The processing and the transfer of data words are described by reference to *primitive operators*. For example, if A, B, and C are registers, then the statement

$$C = A + B$$

denotes the addition of the values of A and B, followed by the transfer of the result into C, by using the primitive operators "+" for addition and "=" for transfer. This notation does not specify the hardware that implements the addition, but only implies its existence.

A reference to a primitive operator defines an *operation*. The control of data transformations is described by *conditional operations*, in which the execution of operations is made dependent on the truth value of a control condition. For example,

if X **then** $C = A + B$

means that $C = A + B$ should occur when the control signal X is 1. More complex control conditions use Boolean expressions and relational operators; for example:

if $(CLOCK$ **and** $(AREG < BREG))$ **then** $AREG = BREG$

states that the transfer $AREG = BREG$ occurs when $CLOCK = 1$ *and* the value contained in $AREG$ is smaller than that in $BREG$.

Other forms of control are represented by constructs similar to the ones encountered in programming languages, such as 2-way decisions (**if...then...else**) and multiway decisions. The statement below selects the operation to be performed depending on the value contained in the bits 0 through 3 of the register *IR*.

> **test** $(IR[0 \rightarrow 3])$
> **case** 0: operation_0
> **case** 1: operation_1
> .
> .
> .
> **case** 15: operation_{15}
> **testend**

This construct implies the existence of a hardware decoder.

RTLs provide compact constructs to describe hardware addressing mechanisms. For example, for the previously defined memory *ABC*,

$$ABC[3]$$

denotes the word at the address 3, and, assuming that *BASEREG* and *PC* are registers of proper dimensions, then

$$ABC[BASEREG + PC]$$

denotes the word at the address obtained by adding the current values of the two registers.

Combinational functions can be directly described by their equations using Boolean operators, as in

$$Z = (A \textbf{ and } B) \textbf{ or } C$$

When Z, A, B, and C are n-bit vectors, the above statement implies n copies of a circuit that performs the Boolean operations between corresponding bits.

Other primitive operators often used in RTLs are shift and count. For example, to shift $AREG$ right by two positions and to fill the left positions with a value 0, one may specify

$$\textbf{shift\_right} \ (AREG, 2, 0)$$

and incrementing PC may be denoted by

$$\textbf{incr} \ (PC)$$

Some RTLs allow references to past values of variables. If time is measured in "time units," the value of X two time units ago would be denoted by $X(-2)$ [Chappell *et al.* 1976]. An action caused by a positive edge (0 to 1 transition) of X can be activated by

$$\textbf{if} \ (X(-1)=0 \textbf{ and } X=1) \textbf{ then } ...$$

where X is equivalent to $X(0)$ and represents the current value of X. This concept introduces an additional level of abstraction, because it implies the existence of some memory that stores the "history" of X (up to a certain depth).

Finite-state machines can be modeled in two ways. The *direct approach* is to have a *state register* composed of the state variables of the system. State transitions are implemented by changing the value of the state register. A more *abstract approach* is to partition the operation of the system into disjoint blocks representing the states and identified by *state names*. Here state transitions are described by using a **go to** operator, as illustrated below:

$$\textbf{state} \quad S1, S2, S3$$

```
S1:  if X then
        begin
            P = Q + R
            go to S2
        end
     else
            P = Q - R
            go to S3
S2: ...
```

Being in state *S1* is an implicit condition for executing the operations specified in the *S1* block. Only one state is active (current) at any time. This more abstract model allows a finite state machine to be described before the state assignment (the mapping between state names and state variables) has been done.

2.3.2 Timing Modeling in RTLs

According to their treatment of the concept of time, RTLs are divided into two categories, namely *procedural languages* and *nonprocedural languages*. A procedural RTL is similar to a conventional programming language where statements are sequentially executed such that the result of a statement is immediately available for the following statements. Thus, in

$$A = B$$
$$C = A$$

the value transferred into C is the new value of A (i.e., the contents of B). Many procedural RTLs directly use (or provide extensions to) conventional programming languages, such as Pascal [Hill and vanCleemput 1979] or C [Frey 1984].

By contrast, the statements of a nonprocedural RTL are (conceptually) executed in parallel, so in the above example the old value of A (i.e., the one before the transfer $A = B$) is loaded into C. Thus, in a nonprocedural RTL, the statements

$$A = B$$
$$B = A$$

accomplish an exchange between the contents of A and B.

Procedural RTLs are usually employed to describe a system at the instruction set level of abstraction. An implicit cycle-based timing model is often used at this level. Reflecting the instruction cycle of the modeled processor, during which an instruction is fetched, decoded, and executed, a cycle-based timing model assures that the state of the model accurately reflects the state of the processor at the end of a cycle.

More detailed behavioral models can be obtained by specifying the delay associated with an operation, thus defining when the result of an operation becomes available. For example:

$$C = A + B, \textbf{ delay} = 100$$

shows that C gets its new value 100 time units after the initiation of the transfer.

Another way of defining delays in an RTL is to specify delays for variables, rather than for operations. In this way, if a declaration

$$\textbf{delay } C \text{ } 100$$

has been issued, then this applies to every transfer into C.

Some RTLs allow a mixture of procedural and nonprocedural interpretations. Thus, an RTL that is mostly procedural may have special constructs to denote concurrent operations or may accept delay specifications. An RTL that is mostly nonprocedural may allow for a special type of variable with the property that new values of these variables become immediately available for subsequent statements.

2.3.3 Internal RTL Models

An RTL model can be internally represented by data structures or by compiled code. A model based on data structures can be interpreted for different applications by model-independent programs. A compiled-code model is usually applicable only for simulation.

Compiled-code models are usually generated from procedural RTLs. The code generation is a 2-step process. First, the RTL description is translated to a high-level language, such as C or Pascal. Then the produced high-level code is compiled into executable machine code. This process is especially easy for an RTL based on a conventional programming language.

Internal RTL models based on data structures are examined in [Hemming and Szygenda 1975].

2.4 Structural Models

2.4.1 External Representation

A typical structural model of a system in a connectivity language specifies the I/O lines of the system, its components, and the I/O signals of each component. For example, the following model conveys the same information as the schematic diagram in Figure 2.15 (using the language described in [Chang *et al.* 1974]).

CIRCUIT XOR
INPUTS = *A,B*
OUTPUTS = *Z*

NOT *D, A*
NOT *C, B*
AND *E, (A,C)*
AND *F, (B,D)*
OR *Z, (E,F)*

The definition of a gate includes its type and the specification of its I/O terminals in the form *output, input_list*. The interconnections are implicitly described by the signal names attached to these terminals. For example, the primary input *A* is connected to the inverter *D* and to the AND gate *E*. A signal line is also referred to as a *net*.

Gates are components characterized by single output and functionally equivalent inputs. In general, components may have multiple outputs, both the inputs and the outputs may be functionally different, and the component itself may have a distinct name. In such a case, the I/O terminals of a component are distinguished by predefined names. For example, to use a *JK* F/F in the circuit shown in Figure 2.16, the model would include a statement such as:

U1 JKFF Q=ERR, QB=NOTERR, J=ERRDET, K=A, C=CLOCK1

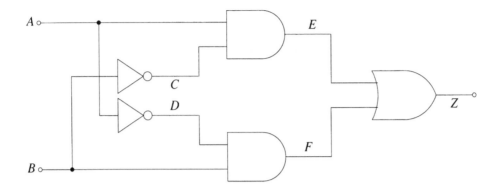

Figure 2.15   Schematic diagram of an XOR circuit

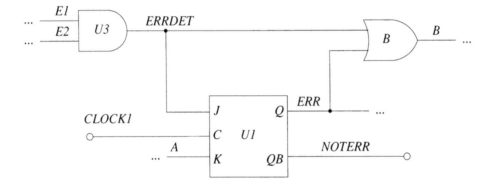

Figure 2.16

which defines the name of the component, its type, and the name of the nets connected to its terminals.

The compiler of an external model consisting of such statements must understand references to types (such as *JKFF*) and to terminal names (such as *Q, J, K*). For this, the model can refer to primitive components of the modeling system (*system primitive components*) or components previously defined in a *library of components* (also called *part library* or *catalog*).

The use of a library of components implies a bottom-up hierarchical approach to describing the external model. First the components are modeled and the library is built. The model of a component may be structural or an RTL model. Once defined, the library components can be regarded as *user primitive components*, since they become the building blocks of the system. The library components are said to be *generic*, since they define new types. The components used in a circuit are *instances* of generic components.

Usually, structural models are built using a *macro approach* based on the macro processing capability of the connectivity language. This approach is similar to those used in programming languages. Basically, a macro capability is a text-processing technique, by means of which a certain text, called the macro body, is assigned a name. A reference to the macro name results in its replacement by the macro body; this process is called *expansion*. Certain items in the macro body, defined as macro parameters, may differ from one expansion to another, being replaced by actual parameters specified at the expansion time. When used for modeling, the macro name is the name of the generic component, the macro parameters are the names of its terminals, the actual parameters are the names of the signals connected to the terminals, and the macro body is the text describing its structure. The nesting of macros allows for a powerful hierarchical modeling capability for the external model.

Timing information in a structural model is usually defined by specifying the delays associated with every instance of a primitive component, as exemplified by

AND X, (A,B) delay 10

When most of the components of the same type have the same delay, this delay is associated with the type by a declaration such as

delay AND 10

and this will be the default delay value of any AND that does not have an explicit delay specification.

Chapter 3 will provide more details about timing models.

2.4.2 Structural Properties

A structural model of a system can be represented by a graph. This is useful since many concepts and algorithms of the graph theory can be applied. Initially we will consider only circuits in which the information is processed by components and is unidirectionally transmitted along signal lines. In cases when every net connects only two components, a natural representation of a structural model is a *directed graph* in which the components and nets are mapped, respectively, into nodes and edges of the graph (Figure 2.17). The primary inputs and outputs are mapped into nodes of special types. In general, however, a net may propagate a signal from one source to more than one destination. Such a signal is said to have **fanout**. A circuit in which no signal has fanout is said to be **fanout-free**. The graph model of a fanout-free combinational circuit is always a tree. A general way to represent a circuit with fanout is by a *bipartite directed graph*, in which both the components and the nets are mapped into nodes such that any edge connects a node representing a component to a node representing a net, as illustrated in Figure 2.18. (The nodes marked with *X* correspond to nets.) This model also allows multioutput components.

Reconvergent fanout refers to different paths from the same signal reconverging at the same component. This is the case of the two paths from *ERRDET* to *B* in Figure 2.18.

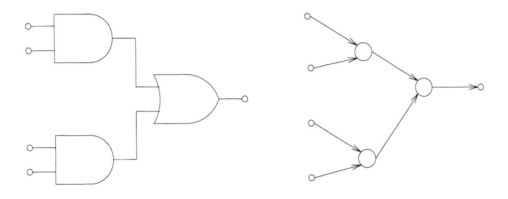

Figure 2.17   Fanout-free circuit and its graph (tree) representation

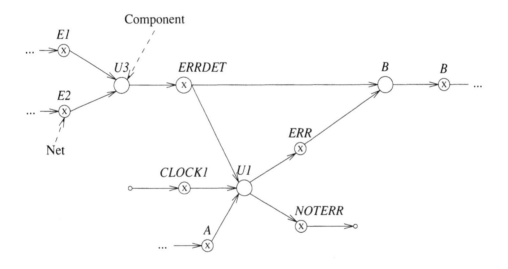

Figure 2.18   Bipartite graph model for the circuit of Figure 2.16

The next chapters will show that the presence of reconvergent fanout in a circuit greatly complicates the problems of test generation and diagnosis.

In circuits composed only of AND, OR, NAND, NOR, and NOT gates, we can define the **inversion parity** of a path as being the number, taken modulo 2, of the inverting gates (NAND, NOR, and NOT) along that path.

In circuits in which every component has only one output it is redundant to use separate nodes to represent nets that connect only two components. In such cases, the graph model used introduces additional nodes, called *fanout nodes*, only to represent nets with fanout. This is illustrated in Figure 2.19. A net connecting one component (the source) with k other components (destinations or loads) is represented by one fanout node with one edge (the *stem*) from the node corresponding to the source and with k edges (*fanout branches*) going to the nodes corresponding to the destinations.

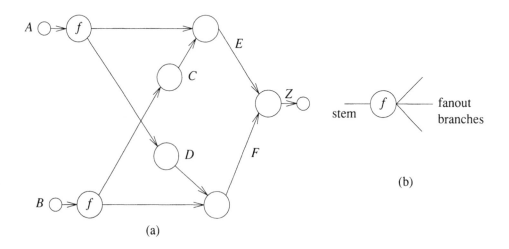

(a)

(b)

Figure 2.19   (a) Graph model for the circuit of Figure 2.15 (b) A fanout node

The (logic) **level** of an element in a combinational circuit is a measure of its "distance" from the primary inputs. The level of the primary inputs is defined to be 0. The level of an element i, $l(i)$, whose inputs come from elements $k_1, k_2, ..., k_p$, is given by

$$l(i) = 1 + \max_{j} l(k_j) \tag{2.1}$$

The following procedure computes levels via a breadth-first traversal of the circuit:

1.   For every primary input i set $l(i) = 0$.

2.   For every element i not yet assigned a level, such that all the elements feeding i have been assigned a level, compute $l(i)$ by equation (2.1).

The concept of level is extended to sequential circuits by considering the feedback lines to be pseudo primary inputs and hence at level 0. For example, for the circuit of Figure 2.20, the following levels are determined:

> level 1: E, D, J
> level 2: F, K
> level 3: G
> level 4: W

Instead of explicit levels, we can use an implicit leveling of the circuit by ordering the elements such that for any two elements numbered i_1 and i_2, $i_1 < i_2$ if $l(i_1) \le l(i_2)$. These numbers can then be used as internal names of the elements (see next section).

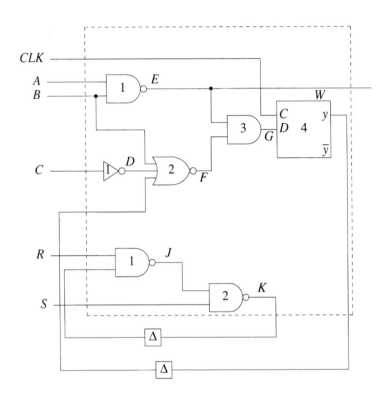

Figure 2.20 Logic levels in a circuit

2.4.3 Internal Representation

Figure 2.21 illustrates a portion of typical data structures used to represent a structural model of the circuit shown in Figure 2.16. The data structure consists of two sets of "parallel" tables, the ELEMENT TABLE and the SIGNAL TABLE, and also of a FANIN TABLE and a FANOUT TABLE. An element is internally identified by its position (index) in the ELEMENT TABLE, such that all the information associated with the element i can be retrieved by accessing the i-th entry in different columns of that table. Similarly, a signal is identified by its index in the SIGNAL TABLE. For the element i, the ELEMENT TABLE contains the following data:

1. NAME(i) is the external name of i.

2. TYPE(i) defines the type of i. (PI denotes primary input; PO means primary output.)

3. NOUT(i) is the number of output signals of i.

4. OUT(i) is the position (index) in the SIGNAL TABLE of the first output signal of i. The remaining NOUT(i)−1 output signals are found in the next NOUT(i)−1 positions.

ELEMENT TABLE

	NAME	TYPE	NOUT	OUT	NFI	FIN
1	U1	JKFF	2	1	3	1
2	B	OR	1	8	2	4
3	U3	AND	1	3	2	6
4	CLOCK1	PI	1	4	0	—
5	NOTERR	PO	0	—	1	8

	SIGNAL TABLE				FANIN TABLE		FANOUT TABLE
	NAME	SOURCE	NFO	FOUT	INPUTS		FANOUTS
1	ERR	1	2	1	1	3	1	2
2	NOTERR	1	1	3	2	4	2	...
3	ERRDET	3	2	4	3	7	3	5
4	CLOCK1	4	1	6	4	1	4	1
5	E1	...	1	7	5	3	5	2
6	E2	...	1	8	6	5	6	1
7	A	...	1	9	7	6	7	3
8	B	2	...	...	8	2	8	3
9							9	1

Figure 2.21 Typical data structures for the circuit of Figure 2.16

5. NFI(i) is the fanin count (the number of inputs) of i.

6. FIN(i) is the position in the FANIN TABLE of the first input signal of i. The remaining NFI(i)−1 input signals are found in the next NFI(i)−1 positions.

For a signal j, the SIGNAL TABLE contains the following data:

1. NAME(j) is the external name of j.

2. SOURCE(j) is the index of the element where j originates.

3. NFO(j) is the fanout count of j (the number of elements fed by j).

4. FOUT(j) is the position in the FANOUT TABLE of the first element fed by j. The remaining NFO(j)−1 elements are found in the next NFO(j)−1 positions.

If every element has only one output, the data structure can be simplified by merging the ELEMENT and SIGNAL tables.

The described data structure is somewhat redundant, since the FANOUT TABLE can be constructed from the FANIN TABLE and vice versa, but it is versatile as it allows the graph representing the circuit to be easily traversed in both directions.

This data structure is extended for different applications by adding columns for items such as delays, physical design information, signal values, or internal states.

Although the macro approach allows hierarchical modeling for a structural external model, it does not carry this hierarchy to the internal model. This is because every reference to a defined macro (library component) results in its expansion, such that the final model interconnects only primitive elements.

An alternative approach, referred to as the *subsystem approach*, allows a hierarchical internal structural model. We will consider only two levels of hierarchy. The components of the top level are called subsystems, and the internal model describing their interconnections is similar to that described in Figure 2.21. The ELEMENT TABLE has an additional column pointing to a set of tables describing the internal structure of the subsystem as an interconnection of primitive elements.

Thus in the subsystem approach all the instances of the same type of a subsystem point to the same data structures representing the subsystem. This is helpful for systems containing many replicas of the same type of subsystem, such as bit-slice architectures. By contrast, the use of a macro approach would result in replicating the same structure as many times as necessary. Hence the subsystem approach can realize important memory savings for systems exhibiting repetitive structure.

On the negative side, the subsystem approach is more complicated to implement and many applications based on following connectivity incur a slight overhead by alternating between the two levels of hierarchy. This is why more than two levels appear to be impractical.

2.4.4 Wired Logic and Bidirectionality

The modeling techniques described so far rely on two basic assumptions:

- The information is processed by components and transmitted by signals.

- The information flow is unidirectional, that is, from input(s) to output(s) within a component and from source to destination(s) along a signal line.

These assumptions, however, are not generally valid. For example, in many technologies it is possible to directly connect several outputs such that the connecting net introduces a new logic function. This mechanism is referred to as *wired logic*, to denote that the logic function is generated by a wire rather than a component. An AND function realized in this way is called a wired-AND. Figure 2.22(a) shows a schematic representation of a wired-AND. A simple modeling technique for wired logic is to insert a "dummy" gate to perform the function of the wire, as illustrated in Figure 2.22(b). Although its output values are logically correct, the unidirectionality implied by the dummy gate may lead to incorrect values for its inputs. In the model in Figure 2.22(b), A and B may have different values, say $A=0$ and $B=1$. This, however, cannot occur in the real circuit (where A and B are tied together) and will cause further errors if B has additional fanout. A correct modeling technique should always force the value of C back onto A and B. One way to achieve this is to make A and B appear as special fanouts of C, as suggested by the dashed arrows in Figure 2.22(b). Then for every wired signal X we should maintain two values:

- the *driven value*, which is the value computed by the element driving X;
- the *forced value*, which is the value resulting from the wired logic.

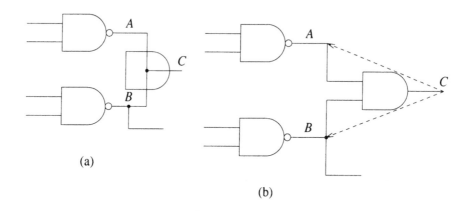

(a)

(b)

Figure 2.22   Wired-AND (a) schematic representation (b) modeling by a
                 "dummy" gate

Note that the arguments for the wired logic function (which produces the forced value) are the driven values of the wired signals. Modeling of a tristate bus is similar and will be discussed in Chapter 3.

The insertion of dummy components for wired logic is a "work-around" introduced to maintain the basic modeling assumptions. The insertion of dummy components for modeling complicates the correspondence between the model and the real circuit and thus adversely affects fault modeling and diagnosis. A direct modeling technique that allows logic functions to be performed by bidirectional wires is therefore preferable. This is especially important in MOS circuits where wired logic is commonly employed. MOS circuits deviate even more from the basic modeling assumptions. The MOS transistor is a bidirectional component that acts like a voltage-controlled switch. The capacitance of a wire can be used to store information as charge (dynamic memory). The behavior of a MOS circuit may also be influenced by the relative resistances of transistors (ratioed logic). Although several work-around techniques [Wadsack 1978, Flake *et al.* 1980, Sherwood 1981, Levendel *et al.* 1981, McDermott 1982] have been devised to extend a conventional modeling framework to accommodate MOS technology, these methods offer only limited accuracy in their ability to model MOS circuits. Accurate solutions are based on *switch-level models* (see [Hayes 1987] for a review), which use transistors as primitive components.

2.5 Level of Modeling

The level of modeling refers to the types of primitive components used in a model. A model that (after expansion) contains only gates is called a *gate-level model*. Similarly, there is a *transistor-level model*. The lowest-level primitive components in a modeling system are those that cannot be structurally decomposed into simpler components; usually at the lowest level we have either gates or transistors. Any other primitive component is said to be *functional* (or a *high-level*) primitive component. Hence, in a modeling system that uses transistors as the lowest-level primitive components, a gate can be considered a functional primitive component.

Nonprimitive components whose function is described using RTL primitive operators are also characterized as functional. In the literature, a component modeled at the functional level is also called a *module*, or a *functional block*.

Let us review what alternatives one may have in modeling a system. First, one can have a specialized program to represent the system. This type of approach is used, for example, to provide a model of a processor to support its software development. In this way the entire system is viewed as a primitive (top-level) component. Second, one may develop an RTL model of the system. Third, one can structurally model the system as an interconnection of lower-level components. The same three alternatives exist for modeling every component (except those at the lowest level): primitive functional model, RTL model, or further structural decomposition.

Note that a primitive functional model of a component is usually provided by the developer of the modeling system, while an RTL model of a component is provided by the user. In general, the effort involved in developing primitive functional models for complex components is larger than that involved in developing the corresponding RTL models. In addition, an RTL model is more adequate for fault simulation and test generation than is a procedure invisible to the user. This is why primitive functional models are usually provided only for components of low to moderate complexity that can be used as building blocks of many other different components. The ability to define and interconnect user-specified functional blocks is very important since low-level models may not be available, either because they are not provided by the IC manufacturers or because they have not yet been designed.

Some simulation systems allow *physical models* to replace functional models of existing ICs [Widdoes and Stump 1988]. A physical model uses a hardware device as a generic model for all its instances in a circuit. The use of physical models eliminates the effort involved in developing software models for complex devices.

Figure 2.23 illustrates different levels of modeling used in describing a 2-bit counter. The highest level is an RTL model. The next lower level is a structural model composed of F/Fs and gates. The *D* F/F can be a functional primitive, or a functional block described by an RTL model, or it can be expanded to its gate-level model. Finally, a gate can be regarded as a primitive component, or it can be expanded to its transistor-level model.

Different levels of modeling are needed during the design process. Following a top-down approach, a design starts as a high-level model representing the specification and progresses through several stages of refinement toward its final implementation. The transition from specification (*what* is to be done) toward implementation (*how* it is done) corresponds to a transition from a higher to a lower level of modeling.

A high-level model provides a more abstract view of the system than a corresponding low-level model (gate or transistor-level), but it involves a loss of detail and accuracy, especially in the area of timing modeling (this subject will be further discussed in the next chapter). At the same time, the size and amount of detail of a low-level model

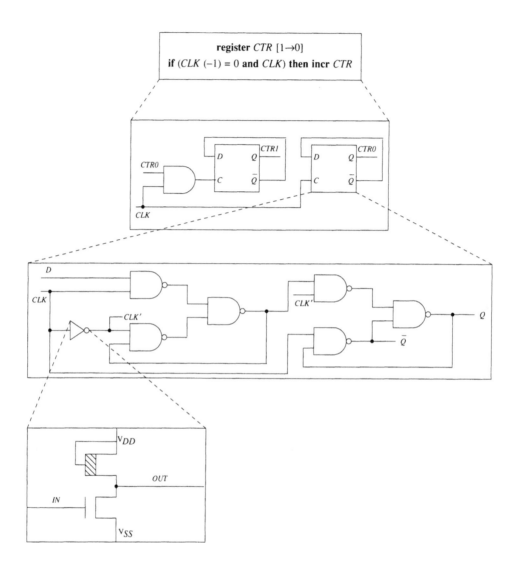

Figure 2.23   Levels of modeling for a 2-bit counter

require large computing resources, in terms of both memory capacity and processing time, for any application using that model. The need for large computing resources has become critical because of the increasing complexity of VLSI circuits.

We can observe that the level of modeling controls the trade-off between accuracy and complexity in a model. An efficient solution of this problem is offered by *hierarchical mixed-level modeling* different, which allows different levels of modeling to be mixed in the same description of a system. Mixed-level modeling is usually coupled with *multimodeling*, which consists of having more than one model for the same component and being able to switch from one model to another.

With mixed-level modeling, different parts of the system can be modeled at different levels. It is also possible to mix analog electrical models with digital ones. Thus, parts in different stages of development can be combined allowing members of design teams to proceed at different rates. Also, mixed-level modeling allows one to focus on low-level details only in the area of interest, while maintaining high-level models for the rest of the system for the sake of efficiency.

REFERENCES

[Akers 1978] S. B. Akers, "Binary Decision Diagrams," *IEEE Trans. on Computers*, Vol. C-27, No. 6, pp. 509-516, June, 1978.

[Barbacci 1975] M. R. Barbacci, "A Comparison of Register Transfer Languages for Describing Computers and Digital Systems," *IEEE Trans. on Computers*, Vol. C-24, No. 2, pp. 137-150, February, 1975.

[Breuer *et al.* 1988] M. A. Breuer, W. Cheng, R. Gupta, I. Hardonag, E. Horowitz, and S. Y. Lin, "Cbase 1.0: A CAD Database for VLSI Circuits Using Object Oriented Technology," *Proc. Intn'l. Conf. on Computer-Aided Design*, pp. 392-395, November, 1988.

[Bryant 1986] R. E. Bryant, "Graph-Based Algorithms for Boolean Function Manipulation," *IEEE Trans. on Computers*, Vol. C-35, No. 8, pp. 677-691, August, 1986.

[Chang *et el.* 1974] H. Y. Chang, G. W. Smith, and R. B. Walford, "LAMP: System Description," *Bell System Technical Journal*, Vol. 53, No. 8, pp. 1431-1449, October, 1974.

[Chappell *et al.* 1976] S. G. Chappell, P. R. Menon, J. F. Pellegrin, and A. M. Schowe, "Functional Simulation in the LAMP System," *Proc. 13th Design Automation Conf.*, pp. 42-47, June, 1976.

[Dietmeyer and Duley 1975] D. L. Dietmeyer and J. R. Duley, "Register Transfer Languages and Their Translation," in *"Digital System Design Automation: Languages, Simulation, and Data Base"* (M. A. Breuer, ed.), Computer Science Press, Woodland Hills, California, 1975.

[Flake *et al.* 1980] P. L. Flake, P. R. Moorby, and G. Musgrave, "Logic Simulation of Bi-directional Tri-state Gates," *Proc. 1980 IEEE Conf. on Circuits and Computers*, pp. 594-600, October, 1980.

[Frey 1984] E. J. Frey, "*ESIM*: A Functional Level Simulation Tool," *Proc. Intn'l. Conf. on Computer-Aided Design*, pp. 48-50, November, 1984.

[Hayes 1987] J. P. Hayes, "An Introduction to Switch-Level Modeling," *IEEE Design & Test of Computers*, Vol. 4, No. 4, pp. 18-25, August, 1987.

[Hemming and Szygenda 1975] C. W. Hemming, Jr., and S. A. Szygenda, "Register Transfer Language Simulation," in *"Digital System Automation: Languages, Simulation, and Data Base"* (M. A. Breuer, ed.), Computer Science Press, Woodland Hills, California, 1975.

[Hill and vanCleemput 1979] D. Hill and V. vanCleemput, "SABLE: A Tool for Generating Structured, Multi-level Simulations," *Proc. 16th Design Automation Conf.*, pp. 272-279, June, 1979.

[Johnson *et al.* 1980] W. Johnson, J. Crowley, M. Steger, and E. Woosley, "Mixed-Level Simulation from a Hierarchical Language," *Journal of Digital Systems*, Vol. 4, No. 3, pp. 305-335, Fall, 1980.

[Lee 1959] C. Lee, "Representation of Switching Circuits by Binary Decision Diagrams," *Bell System Technical Journal*, Vol. 38, No. 6, pp. 985-999, July, 1959.

[Levendel *et al.* 1981] Y. H. Levendel, P. R. Menon, and C. E. Miller, "Accurate Logic Simulation for TTL Totempole and MOS Gates and Tristate Devices," *Bell System Technical Journal*, Vol. 60, No. 7, pp. 1271-1287, September, 1981.

[McDermott 1982] R. M. McDermott, "Transmission Gate Modeling in an Existing Three-Value Simulator," *Proc. 19th Design Automation Conf.*, pp. 678-681, June, 1982.

[Shahdad *et al.* 1985] M. Shahdad, R. Lipsett, E. Marschner, K. Sheehan, H. Cohen, R. Waxman, and D. Ackley, "VHSIC Hardware Description Language," *Computer*, Vol. 18, No. 2, pp. 94-103, February, 1985.

[Sherwood 1981] W. Sherwood, "A MOS Modeling Technique for 4-State True-Value Hierarchical Logic Simulation," *Proc. 18th Design Automation Conf.*, pp. 775-785, June, 1981.

[Wadsack 1978] R. L. Wadsack, "Fault Modeling and Logic Simulation of CMOS and MOS Integrated Circuits," *Bell System Technical Journal*, Vol. 57, No. 5, pp. 1449-1474, May-June, 1978.

[Widdoes and Stump 1988] L. C. Widdoes and H. Stump, "Hardware Modeling," *VLSI Systems Design*, Vol. 9, No. 7, pp. 30-38, July, 1988.

[Wolf 1989] W. H. Wolf, "How to Build a Hardware Description and Measurement System on an Object-Oriented Programming Language," *IEEE Trans. on Computer-Aided Design*, Vol. 8, No. 3, pp. 288-301, March, 1989.

PROBLEMS

2.1 Determine whether the following cubes can be cubes of a function $Z(x_1,x_2,x_3,x_4)$.

$$
\begin{array}{cccc|c}
0 & x & 1 & x & 0 \\
x & 1 & 0 & 1 & 0 \\
1 & x & 1 & 1 & 0 \\
0 & 1 & 0 & x & 1 \\
x & 0 & 1 & 0 & 1 \\
\end{array}
$$

2.2 Construct a binary decision diagram for the exclusive-OR function of two variables.

2.3 Derive a binary decision diagram for the function given in Figure 2.2(a).

2.4 Write two RTL models for a positive edge-triggered D flip-flop (the output takes the value of D after a 0 to 1 transition of the clock). First assume an RTL that does not allow accessing past values of signals. Remove this restriction for the second model.

3. LOGIC SIMULATION

About This Chapter

First we review different aspects of using logic simulation as a tool for design verification testing, examining both its usefulness and its limitations. Then we introduce the main types of simulators, compiled and event-driven. We analyze several problems affecting the accuracy and the performance of simulators, such as treatment of unknown logic values, delay modeling, hazard detection, oscillation control, and values needed for tristate logic and MOS circuits. We discuss in detail techniques for element evaluation and gate-level event-driven simulation algorithms. The last section describes special-purpose hardware for simulation.

3.1 Applications

Logic simulation is a form of *design verification testing* that uses a model of the designed system. Figure 3.1 shows a schematic view of the simulation process. The simulation program processes a representation of the input stimuli and determines the evolution in time of the signals in the model.

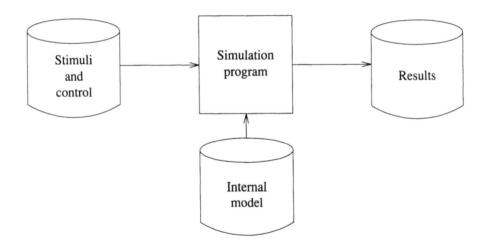

Figure 3.1   Simulation process

The verification of a logic design attempts to ascertain that the design performs its specified behavior, which includes both function and timing. The verification is done by comparing the results obtained by simulation with the expected results provided by the specification. In addition, logic simulation may be used to verify that the operation of the system is

- correct independent of the initial (power-on) state;

- not sensitive to some variations in the delays of the components;

- free of critical races, oscillations, "illegal" input conditions, and "hang-up" states.

Other applications of simulation in the design process are

- *evaluation of design alternatives* ("what-if" analysis), to improve performance/cost trade-offs;

- *evaluation of proposed changes* of an existing design, to verify that the intended modifications will produce the desired change without undesired side effects;

- *documentation* (generation of *timing diagrams*, etc.).

Traditionally, designers have used a prototype for the verification of a new design. The main advantage of a prototype is that it can run at operating speed, but building a prototype is costly and time-consuming. Prototypes built with discrete components lack accuracy as models of complex ICs. Simulation *replaces the prototype with a software model*, which is easily analyzed and modified. Simulation-based design verification benefits from additional features not available to a prototype-based process, such as

- checking error conditions (e.g., bus conflicts);

- ability to change delays in the model to check worst-case timing conditions;

- checking user-specified expected values during simulation;

- starting the simulated circuit in any desired state;

- precise control of the timing of asynchronous events (e.g., interrupts);

- the ability to provide an automated testing environment for the simulated circuit, by coupling it with an RTL model that drives and/or observes the circuit during simulation.

Interestingly, simulation has also been used during the *debugging of the prototype* of a new system. As reported in [Butler *et al.* 1974], designers found that tracing some problems on the simulated model was easier than tracing them on the prototype hardware. Although simulation runs much slower than the hardware, debugging using simulation provides the user with the ability to suspend the simulation on the occurrence of user-specified conditions and to display the value of any desired signal, including lines not directly observable on the hardware. Such a feature becomes more valuable in LSI-based designs, where the number of signals not directly observable is large.

Debugging software or microcode to run on hardware still under development can be started using a simulation model before a prototype is available.

Another use of logic simulation is to *prepare the data base for guided-probe testing*, consisting of the expected values of the lines accessible for probing.

3.2 Problems in Simulation-Based Design Verification

Three interrelated problems in simulation-based design verification testing (and in any type of testing experiment as well) are

- How does one generate the input stimuli? (test generation)

- How does one know the results are correct?

- How "good" are the applied input stimuli, i.e., how "complete" is the testing experiment? (test evaluation)

The input stimuli are usually organized as a sequence of *test cases*, where a test case is intended to verify a certain aspect of the behavior of the model. The results are considered correct when they match the expected results, which are provided by the specification of the design. Initially, the specification consists of an informal (mental or written) model of the intended behavior of the system. During a top-down design process, the highest-level formal model (usually an RTL model) is checked against the initial informal model. After that, any higher-level model defines the specification for its implementation at the next lower level. Checking that the implementation satisfies the specification is reduced to applying the same test cases to the two models and using the results obtained from the higher level as the expected results for the lower level. An example of such a hierarchical approach is the design verification process described in [Sasaki *et al.* 1981].

It is important to understand the difference between test generation for design verification, where the objective is to find design errors, and test generation for detecting physical faults in a manufactured system. Most types of physical faults can be represented by logical faults, whose effects on the behavior of the system are well defined. Based on the difference between the behavior in the presence of a fault and the fault-free behavior, one can derive a test for that fault. Many logical fault models allow the possible faults to be enumerated. The existence of a set of enumerable faults allows one to determine the quality of a test by computing the *fault coverage* as the ratio between the number of faults detected by that test and the total number of faults in the model. In contrast, the space of design errors is not well defined and the set of design errors is not enumerable. Consequently, it is impossible to develop test generation algorithms or rigorous quality measures for design verification tests.

Although the set of design errors is not enumerable, experience shows that most design errors are related to the sequencing of the data transfers and transformations rather than to the data operations themselves. The reason is that data (i.e., arithmetic and logic) operations are more regular and "local," while their control and timing in a complex system may depend on other operations occurring concurrently. Therefore, the usual strategy in design verification is to emphasize exercising the control. For example, minimal test cases for an instruction set processor consist of executing every instruction in the repertoire. In addition, the test designer must consider sequences of instructions that are relevant to the operation of the processor, interactions between these sequences and interrupts, and so on.

Design verification via simulation suffers from several limitations. As there are no formal procedures to generate tests, producing the stimuli is generally a heuristic process that relies heavily on the designer's intuition and understanding of the system under test. A system that passes the test is shown to be correct only with respect to the applied test cases, and hence only a partial correctness can be proved. Moreover, the completeness of the tests cannot be rigorously determined. (Some heuristic measures used to determine the quality of a test with respect to the control flow of the system will be presented in Chapter 8.)

In spite of these limitations, simulation is an effective technique for design verification, and experience (for example, [Monachino 1982]) has shown that it helps discover most design errors early in the design process. For LSI/VLSI circuits, where design errors are costly and prototypes are impractical, logic simulation is an invaluable aid.

3.3 Types of Simulation

Simulators can be classified according to the type of internal model they process. A simulator that executes a compiled-code model is referred to as a *compiler-driven simulator*, or a *compiled simulator*. The compiled code is generated from an RTL model, from a functional model written in a conventional programming language, or from a structural model. A simulator that interprets a model based on data structures is said to be *table-driven*. The data structures are produced from an RTL model or a structural model. The interpretation of the model is controlled by the applied stimuli, and results in a series of calls to the routines implementing primitive operators (for an RTL model) or primitive components (for a structural model).

Let us consider a circuit in operation and look at the signals changing value at some arbitrary time. These are called *active* signals. The ratio between the number of active signals and the total number of signals in the circuit is referred to as *activity*. In general, the activity of a circuit is between 1 and 5 percent. This fact forms the basis of *activity-directed simulation* [Ulrich 1965, 1969], which simulates only the active part of the circuit.

An **event** represents a change in the value of a signal line. When such an event on line i occurs, the elements having i as input are said to be *activated*. The process of determining the output values of an element is called **evaluation**. Activity-directed simulation evaluates only the activated elements. Some of the activated elements may in turn change their output values, thus generating new events. As activity is caused by events, activity-directed simulation is also referred to as *event-driven simulation*. To propagate events along the interconnections among elements, an event-driven simulator needs a structural model of a circuit. Hence event-driven simulation is usually table driven.

Compiled simulation is mainly oriented toward functional verification and is not concerned with the timing of the circuit. This makes it applicable mostly to synchronous circuits, for which timing can be separately verified [Hitchcock 1982]. In contrast, the passage of time is central to event-driven simulation, which can work with accurate timing models. Thus event-driven simulation is more general in scope, being also applicable to asynchronous circuits.

Event-driven simulation can process *real-time inputs*, that is, inputs whose times of change are independent of the activity in the simulated circuit. This is an important feature for design verification testing, as it allows accurate simulation of nonsynchronized events, such as interrupts or competing requests for use of a bus.

Compiled simulation allows inputs to be changed only when the circuit is stable. This is adequate when the input stimuli are *vectors* applied at a fixed rate. Note that real-time inputs include the fixed-rate vectors as a particular case.

Often, the two simulation types are combined, such that an event-driven algorithm propagates events among components, and the activated components are evaluated by compiled-code models.

The *level of simulation* corresponds to the level of modeling employed to represent the simulated system. Thus, we can have

- *register-level simulation*, for systems modeled entirely in RTL or as an interconnection of components modeled in RTL;

- *functional-level simulation*, for systems modeled as an interconnection of primitive functional blocks (sometimes this term is also used when the components are modeled in RTL);   .

- *gate-level simulation*;

- *transistor-level simulation* (we consider only logic-level and not circuit-level analog simulation);

- *mixed-level simulation*.

3.4 The Unknown Logic Value

In general, the response of a sequential circuit to an input sequence depends on its initial state. However, when a circuit is powered-up, the initial state of memory elements such as F/Fs and RAMs is usually unpredictable. This is why, before the normal operation begins, an initialization sequence is applied with the purpose of bringing the circuit into a known "reset" state. To process the unknown initial state, simulation algorithms use a separate logic value, denoted by u, to indicate an **unknown logic value**. The u logic value is processed together with the binary logic values during simulation. The extension of Boolean operators to 3-valued logic is based on the following reasoning. The value u represents one value in the set $\{0,1\}$. Similarly, we can treat the values 0 and 1 as the sets $\{0\}$ and $\{1\}$, respectively. A Boolean operation B between p and q, where $p,q \in \{0,1,u\}$, is considered an operation between the sets of values representing p and q, and is defined as the union set of the results of all possible B operations between the components of the two sets. For example,

$$AND(0,u)=AND(\{0\},\{0,1\})=\{AND(0,0),AND(0,1)\}=\{0,0\}=\{0\}=0$$

$$OR(0,u)=OR(\{0\},\{0,1\})=\{OR(0,0),OR(0,1)\}=\{0,1\}=u$$

Similarly, the result of $NOT(q)$, where $q \in \{0,1,u\}$, is defined as the union set of the results of NOT operations applied to every component of the set corresponding to q. Hence

$$NOT(u)=NOT(\{0,1\})=\{NOT(0),NOT(1)\}=\{1,0\}=\{0,1\}=u$$

Figure 3.2 shows the truth tables for AND, OR, and NOT for 3-valued logic. A general procedure to determine the value of a combinational function $f(x_1, x_2, ...,x_n)$ for a given input combination $(v_1 v_2 \cdots v_n)$ of 0, 1, and u values, works as follows:

1. Form the cube $(v_1 v_2 \cdots v_n \mid x)$.

2. Using the modified intersection operator given in Figure 3.3, intersect this cube with the primitive cubes of f. If a consistent intersection is found, then the value of f is obtained in the right-most position; otherwise set $f=u$.

AND	0	1	u		OR	0	1	u		NOT	0	1	u
0	0	0	0		0	0	1	u			1	0	u
1	0	1	u		1	1	1	1					
u	0	u	u		u	u	1	u					

Figure 3.2   Truth tables for 3-valued logic

$\cap$	0	1	x	u
0	0	$\varnothing$	0	$\varnothing$
1	$\varnothing$	1	1	$\varnothing$
x	0	1	x	u
u	$\varnothing$	$\varnothing$	u	u

Figure 3.3   Modified intersection operator

To understand this procedure, recall that an x in a primitive cube denotes a "don't care" value; hence an unknown value is consistent with an x in a primitive cube. However, a binary input value specified in a primitive cube is a required one, so a u on that input cannot generate the corresponding output value. For example, for an AND gate with two inputs, the cube $u0 \mid x$ matches a primitive cube, but $u1 \mid x$ does not.

There is a loss of information associated with the use of 3-valued logic [Breuer 1972]. This can be seen from the NOT truth table: in the case where both the input and the output have value u, we lose the complementary relation between them. The same effect occurs between complementary outputs of a F/F whose state is u. In the presence of reconvergent fanout with unequal inversion parities, this loss of information may lead to pessimistic results. This is illustrated in Figure 3.4, where the output of NAND gate is actually 1, but is computed as u according to the rules of 3-valued logic.

It may appear that the use of complementary unknown values u and $\bar{u}$, along with the rules $u.\bar{u}=0$ and $u+\bar{u}=1$, would solve the above problems. This is true, however, only

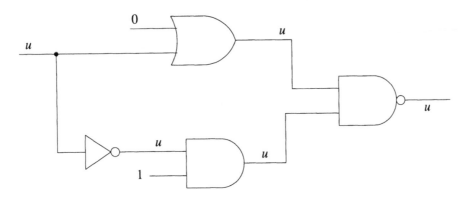

Figure 3.4 Pessimistic result in 3-valued simulation

when we have only one state variable set to u. Figure 3.5 illustrates how the use of u and $\bar{u}$ may lead to incorrect results. Since it is better to be pessimistic than incorrect, using u and $\bar{u}$ is not a satisfactory solution. A correct solution would be to use several distinct unknown signals u_1, u_2, ..., u_k (one for every state variable) and the rules $u_i.\bar{u}_i=0$ and $u_i+\bar{u}_i=1$. Unfortunately, this technique becomes cumbersome for large circuits, since the values of some lines would be represented by large Boolean expressions of u_i variables.

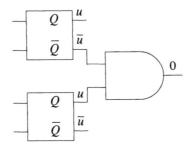

Figure 3.5 Incorrect result from using u and $\bar{u}$

Often the operation of a functional element is determined by decoding the values of a group of control lines. A problem arises in simulation when a functional element needs to be evaluated and some of its control lines have u values. In general, if k control lines have u values, the element may execute one of 2^k possible operations. An accurate (but potentially costly) solution is to perform all 2^k operations and to take as result the union set of their individual results. Thus if a variable is set to 0 in some operations and to 1 in others, its resulting value will be $\{0,1\} = u$. Of course, this solution is practical if 2^k is a small number. For example, assume that two bits in the address of a ROM have u values. This leads to four evaluations, each accessing a different word. The resulting output will have a binary value b in those bit positions where every accessed word has value b, and u values wherever the accessed words do not match.

In an asynchronous circuit, the presence of u values may indicate an oscillation, as shown in Figure 3.6. Signals involved in a high-frequency oscillation often assume a voltage between the levels corresponding to logic 0 and 1. Thus, in addition to a static unknown value, u may also represent a dynamic unknown value or an indeterminate logic value.

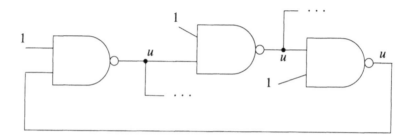

Figure 3.6   Oscillation indicated by u values

3.5 Compiled Simulation

In compiled simulation, the compiled-code model becomes part of the simulator. In the extreme case, the simulator is nothing but the compiled-code model. Then this code also reads the input vectors and outputs the results. In general, the compiled-code model is linked with the simulator's core, whose tasks include reading the input vectors, executing the model for every vector, and displaying the results.

We will illustrate the operation of a compiled simulator using the circuit of Figure 3.7. It is a synchronous circuit controlled by the periodic clock signal *CLK*. We assume that, after a new input vector is applied, there is enough time for the data input of the F/F to become stable at least t_{setup} before the F/F is clocked, where t_{setup} is the setup time of the F/F. This assumption can be independently verified by a timing verification program [Hitchcock 1982]. If this assumption is satisfied, the simulation can ignore the individual gate delays, as the exact times when signals in the combinational logic change are not important. Then, for every vector, the simulation needs only to compute the static value of F and to transfer this value to Q.

The code model is generated such that the computation of values proceeds level by level. This assures that whenever the code evaluates a gate, the gates feeding it have already been evaluated. The values of the primary inputs A and B (which have level 0) are read from a stimulus file. The only other signal with level 0 is the state variable Q; first we assume that its initial value is known. Then the simulator must process

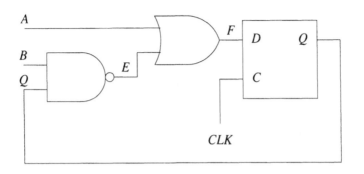

Figure 3.7   Synchronous circuit

only binary values, which are stored in the variables $A, B, Q, E,$ and D. The following is the assembly code model of the circuit:

```
LDA    B
AND    Q
INV
STA    E
OR     A
STA    F
STA    Q
```

Note that a compiled simulator evaluates all the elements in the circuit for every input vector.

If the initial value of Q is unknown, the simulator must process the values 0, 1, and u. These values are coded with 2-bit vectors as follows:

$$0 - 00$$
$$1 - 11$$
$$u - 01$$

We can easily observe that an AND (OR) operation between 2-bit vectors is correctly done by ANDing (ORing) the individual bits. However a NOT operation cannot be done only by complementing the bits, because the complement of u would result in the illegal code 10. The solution for NOT is to swap the two bits after complementation.

Let us consider the evaluation of an AND gate with inputs A and B and output C by an operation $C = A.B$, where "." represents the AND instruction of the host computer. If we restrict ourselves to 2-valued logic, we need only one bit to represent the value of a signal. In *parallel-pattern evaluation* [Barzilai *et al*. 1987], we use a W-bit memory location of the host computer to store the values of the same signal in W different vectors. If the "." instruction works on W-bit operands, then $C = A.B$ simultaneously computes the values of C in W vectors. Of course, this is valid only in combinational circuits, where the order in which vectors are applied is not relevant.

This method speeds up 2-valued compiled simulation by a factor of W (typically, $W = 32$). For three logic values, only $W/2$ vectors can be simulated concurrently.

Compiled simulation can also be used for asynchronous circuits, using the model shown in Figure 3.8, which assumes that delays are present only on the feedback lines. In response to an input vector x, the circuit may go through a series of state transitions, represented by changes of the state variables y. We assume that an input vector is applied only when the circuit is stable, i.e., $y=Y$. Feedback lines, which have level 0, must be identified before the code model for the combinational circuit C is generated. Figure 3.9 outlines the general procedure for simulating an asynchronous circuit. The execution of the model computes the values of z and Y based on x and y.

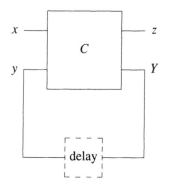

Figure 3.8 Asynchronous circuit model

This type of simulation is not accurate for asynchronous circuits whose operation is based on certain delay values. For example, the circuit of Figure 3.10(a) could be used as a pulse generator: when A has a $0\rightarrow1$ transition, the delay of the inverter B creates an interval during which both inputs of C have value 1, thus causing a $0\rightarrow1\rightarrow0$ pulse on C. Without careful modeling, this pulse cannot be predicted by a compiled simulator, which deals only with the static behavior of the circuit (statically, C is always 0). To take into account the delay of B, which is essential for the intended operation of the circuit, B should be treated as a feedback line, as shown in Figure 3.10(b). With this model, the circuit can be correctly simulated. In general, however, deriving such a "correct" model cannot be done automatically and requires input from the user.

Even when the structure of the circuit allows the identification of the feedback loops, different models can be derived from the same circuit (see Figure 3.11). Because the different models have different assumptions about the location of the delays, they may respond differently to the same stimuli. For example, consider the simulation of the latch given in Figure 3.11(a) for the vector 00 followed by 11. Using the model with Q as the feedback line, we determine that QN changes from 1 to 0, while $Q=1$ in both vectors. However, using the model with QN as the feedback line, we would compute that Q changes from 1 to 0, while $QN=1$. The reason for the different results is that

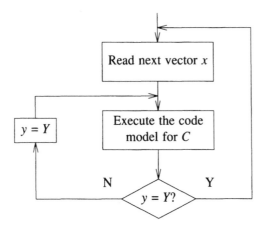

Figure 3.9 Asynchronous circuit simulation with compiled-code model

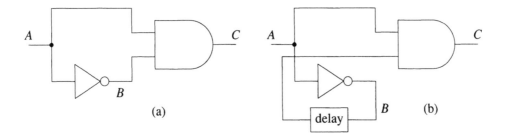

Figure 3.10 (a) Circuit used as pulse generator (b) Correct model for compiled simulation

the two vectors cause a critical race, whose result depends on the actual delays in the circuit. This example shows that a compiled simulator following the procedure outlined in Figure 3.9, cannot deal with races and hazards which often affect the operation of an asynchronous circuit. Techniques for detecting hazards will be discussed in Section 3.9.

3.6 Event-Driven Simulation

An event-driven simulator uses a structural model of a circuit to propagate events. The changes in the values of the primary inputs are defined in the stimulus file. Events on the other lines are produced by the evaluations of the activated elements.

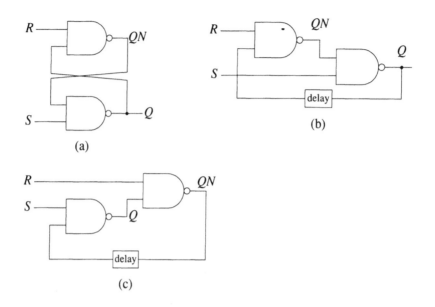

Figure 3.11   (a) Latch (b)&(c) Possible models for compiled simulation

An event occurs at a certain (simulated) time. The simulation *time-flow mechanism* manipulates the events such that they will occur in a correct temporal order. The applied stimuli are represented by sequences of events whose times are predefined. The events scheduled to occur in the future (relative to the current simulated time) are said to be *pending* and are maintained in a data structure called an *event list*.

Figure 3.12 shows the main (conceptual) flow of event-directed simulation. The simulation time is advanced to the next time for which events are pending; this becomes the current simulation time. Next, the simulator retrieves from the event list the events scheduled to occur at the current time and updates the values of the active signals. The fanout list of the active signals is then followed to determine the activated elements; this process parallels the propagation of changes in the real circuit. The evaluation of the activated elements may result in new events. These are scheduled to occur in the future according to the delays associated with the operation of the elements. The simulator inserts the newly generated events in the event list. The simulation continues as long as there is logic activity in the circuit; that is, until the event list becomes empty.

The evaluation of an element M modeled by a nonprocedural RTL may generate a *state event* that denotes a change in the value of an internal state variable of M. (In procedural RTL, changes of state variables occur immediately, without processing via the event list.) When such a state event occurs, it activates only the element M that has generated it; that is, it causes M to be reevaluated.

For simplicity, in the above description we have assumed that all the events defining the applied stimuli were inserted in the event list before simulation. In fact, the

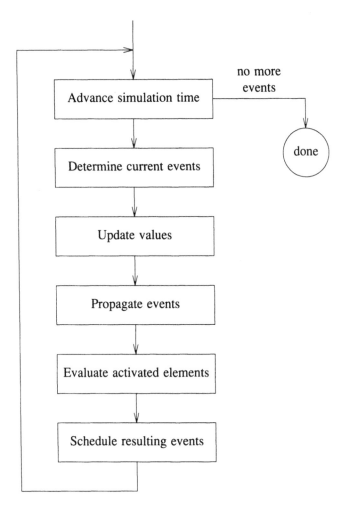

Figure 3.12   Main flow of event-driven simulation

simulator has periodically to read the stimulus file and to merge the events on the primary inputs with the events internally generated.

In addition to the events that convey updates in signal values, an event-driven simulator may also process *control events*, which provide a convenient means to initiate different activities at certain times. The following are some typical actions requested by control events:

• Display values of certain signals.

• Check expected values of certain signals (and, possibly, stop the simulation if a mismatch is detected).

• Stop the simulation.

3.7 Delay Models

Many different variants of the general flow shown in Figure 3.12 exist. The differences among them arise mainly from different *delay models* associated with the behavior of the components in the model. Delay modeling is a key element controlling the trade-off between the accuracy and the complexity of the simulation algorithm.

3.7.1 Delay Modeling for Gates

Every gate introduces a delay to the signals propagating through it. In modeling the behavior of a gate, we separate its function and its timing as indicated in Figure 3.13. Thus in simulation an activated element is first evaluated, then the delay computation is performed.

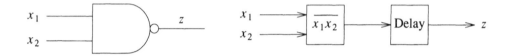

Figure 3.13 Separation between function and delay in modeling a gate

Transport Delays

The basic delay model is that of a *transport delay,* which specifies the interval d separating an output change from the input change(s) which caused it.

To simplify the simulation algorithm, delay values used in simulation are usually integers. Typically they are multiples of some common unit. For example, if we are dealing with gate delays of 15, 20, and 30 ns, for simulation we can scale them respectively to 3, 4, and 6 units, where a unit of delay represents the greatest common divisor (5 ns) of the individual delays. (Then the times of the changes at the primary inputs should be similarly scaled.) If all transport delays in a circuit are considered equal, then we can scale them to 1 unit; this model is called a *unit-delay model.*

The answers to the following two questions determine the nature of the delay computation.

• Does the delay depend on the direction of the resulting output transition?

• Are delays precisely known?

For some devices the times required for the output signal to rise (0 to 1 transition*) and to fall (1 to 0) are greatly different. For some MOS devices these delays may

---

*   We assume a *positive logic* convention, where the logic level 1 represents the highest voltage level.

differ by a ratio of 3 to 1. To reflect this phenomenon in simulation, we associate different *rise and fall delays*, d_r and d_f, with every gate. The delay computation selects the appropriate value based on the event generated for the gate output. If the gate delays are not a function of the direction of the output change, then we can use a *transition-independent delay model*. Figures 3.14(a) and (b) illustrate the differences between these two models. Note that the result of having different rise and fall delays is to change the width of the pulse propagating through the gate.

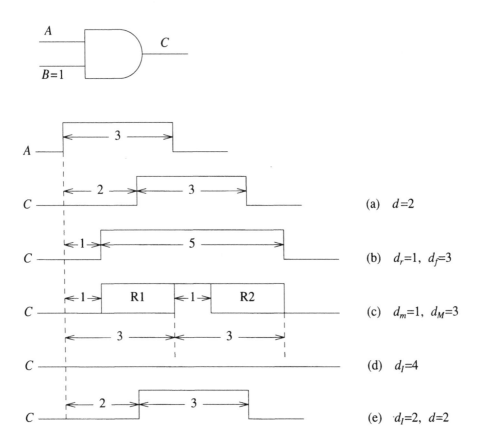

Figure 3.14   Delay models (a) Nominal transition-independent transport delay
            (b) Rise and fall delays (c) Ambiguous delay (d) Inertial delay (pulse
            suppression) (e) Inertial delay

Often the exact transport delay of a gate is not known. For example, the delay of a certain type of NAND gate may be specified by its manufacturer as varying from 5 ns to 10 ns. To reflect this uncertainty in simulation, we associate an *ambiguity interval*, defined by the minimum (d_m) and maximum (d_M) delays, with every gate. This model, referred to as an *ambiguous delay model*, results in intervals (R1 and R2 in Figure 3.14(c)) during which the value of a signal is not precisely known. Under the assumption that the gate delays are known, we have a *nominal delay model*. The rise and fall delay model and the ambiguous delay model can be combined such that we have different ambiguity intervals for the rise (d_{rm}, d_{rM}) and the fall (d_{fm}, d_{fM}) delays [Chappel and Yau 1971].

Inertial Delays

All circuits require energy to switch states. The energy in a signal is a function of its amplitude and duration. If its duration is too short, the signal will not force the device to switch. The minimum duration of an input change necessary for the gate output to switch states is called the *input inertial delay* of the gate, denoted by d_I. An input pulse whose duration is less than d_I is said to be a *spike,* which is *filtered* (or *suppressed*) by the gate (Figure 3.14(d)). If the pulse width is at least d_I, then its propagation through the gate is determined by the transport delay(s) of the gate. If the gate has a transition-independent nominal transport delay d, then the two delays must satisfy the relation $d_I \le d$ (Problem 3.6). Figure 3.14(e) shows a case with $d_I = d$.

A slightly different way of modeling inertial delays is to associate them with the gate outputs. This *output inertial delay* model specifies that the gate output cannot generate a pulse whose duration is less than d_I. An output pulse may be caused by an input pulse (and hence we get the same results as for the input inertial delay model), but it may also be caused by "close" input transitions, as shown in Figure 3.15(a). Another case in which the two inertial delay models would differ is illustrated in Figure 3.15(b). With the input inertial delay model, the input pulses are considered separately and they cannot switch the gate output. However, under the output inertial delay model, the combined effect of the input pulses forces the output to switch.

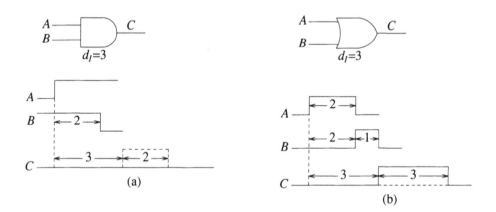

Figure 3.15   Output inertial delay

3.7.2 Delay Modeling for Functional Elements

Both the logic function and the timing characteristics of functional elements are more complex than those of the gates. Consider, for example, the behavior of an edge-triggered D F/F with asynchronous set (S) and reset (R), described by the table given in Figure 3.16. The symbol $\uparrow$ denotes a 0-to-1 transition. The notation $d_{I/O}$

represents the delay in the response of the output O to the change of the input I. The superscript (r or f) distinguishes between rise and fall delays, where necessary. For example, the third row says that if the F/F is in initial state $q = 1$ and S and R are inactive (1), a 0-to-1 transition of the clock C causes the output Q to change to the value of D (0) after a delay $d^f_{C/Q} = 8$. Similarly, the output QN changes to 1 after a delay $d^r_{C/QN} = 6$. The last row specifies that the illegal input condition $SR = 00$ causes both outputs to be set to u.

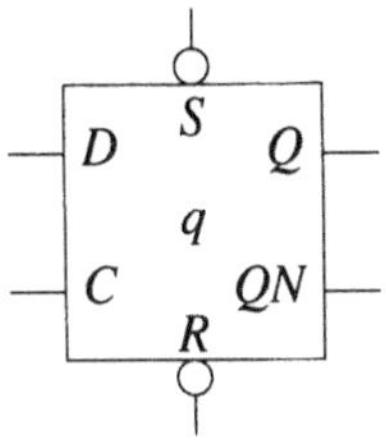

q	S	R	C	D	Q	QN	Delays
0	0	1	x	x	1	0	$d_{S/Q} = 4 \quad d_{S/QN} = 3$
1	1	0	x	x	0	1	$d_{R/Q} = 3 \quad d_{R/QN} = 4$
1	1	1	↑	0	0	1	$d^f_{C/Q} = 8 \quad d^r_{C/QN} = 6$
0	1	1	↑	1	1	0	$d^r_{C/Q} = 6 \quad d^f_{C/QN} = 8$
x	0	0	x	x	u	u	

Figure 3.16   I/O delays for a D F/F

Similar to the input inertial delay model for gates, the specifications of the F/F may include the *minimum pulse widths* for C, S, and R required to change the state of the F/F.

Additional timing specifications deal with requirements for avoiding race conditions between C and D. The *setup time* and the *hold time* are the minimum intervals preceding and following an active C transition during which D should remain stable in order to have a correct (i.e., predictable) F/F operation. Some simulation systems can detect improper operation with respect to setup and hold conditions [Evans 1978, Tokoro *et al.* 1978]. Rather than having these checks done by the simulator, a better approach is to include them in the functional models developed for F/Fs.

3.7.3 Delay Modeling in RTLs

Different ways of specifying delays in RTL models have been discussed in Chapter 2. In general, RTLs offer a more abstract view of the system, and hence the delay modeling is less detailed. Many RTLs use a cycle timing model. If delays are assigned to individual operations, they have the meaning of nominal transport delays.

3.7.4 Other Aspects of Delay Modeling

In high-speed circuits, the delays introduced by the propagation of signals along wires become as significant as the delays of components. Since these delays depend on wire lengths, they are known only after the routing of the circuit is performed (at least tentatively). Many design automation systems are able to extract information automatically from the layout data and to update the simulation data base. The delay introduced by a signal line i with only one fanout can be assimilated in the delay of the gate that generates i. However, if i has several fanouts, each one of its fanout branches may involve a different propagation delay. These are usually modeled by inserting *delay elements* in the appropriate places (see Figure 3.17). A delay element realizes an identity logic function, and its purpose is only to delay the signal propagation.

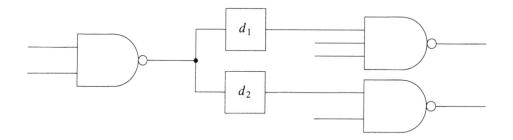

Figure 3.17   Wire delays modeled by delay elements

Another factor that affects delay modeling is the *loading* of a signal, where the apparent delay of a gate grows with the fanout count of its output signal.

3.8 Element Evaluation

The evaluation of a combinational element is the process of computing its output values given its current input values. The evaluation of a sequential element is also based on its current state and computes its next state as well. Evaluation techniques depend on many interrelated factors, such as the system of logic values used in simulation, the way values are stored, the type of the elements, and the way they are modeled.

As the evaluation of a combinational element G must analyze the input values of G, a first question is how are these values made available to the evaluation routine. First let us assume that signal values are stored in a table parallel to the signal tables (see Figure 2.21). Then finding the input values of G is an indirect process, which first determines the inputs by following the fanin list of G and then accesses their values. A second way is to maintain the input values of G in a contiguous area of memory associated with G (see Figure 3.18). Although this scheme may appear wasteful, since it replicates the value of a signal with k fanouts in the value area of every one of its fanout elements, it presents several advantages:

• The evaluation routines are faster, because they can directly access the needed values.

• Values can be "packed" together, which allows more efficient evaluation techniques.

- Since the evaluation routines no longer need to determine the inputs of the evaluated elements, storage can be saved by not loading the fanin data in memory during simulation.

- The separation between the value of a signal and the values of its fanout branches is useful in fault simulation, as these values can be different in the presence of faults (in logic simulation they are always the same).

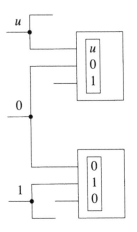

Figure 3.18   Input values maintained per element

In the following, we assume that input values are maintained per element (as illustrated in Figure 3.18) and that values of the state variables of sequential elements are similarly stored.

Truth Tables

Let n be the number of inputs and state variables of an element. Assuming only binary values, a truth table of the element has 2^n entries, which are stored as an array V, whose elements are vectors of output and state variable values. For evaluation, the n values of the inputs and state variables are packed in the same word, and the evaluation uses the value of this word — say, i — as an index to retrieve the corresponding output and next state stored in $V[i]$.

Truth tables can also be generalized for multivalued logic. Let k be the number of logic values, and let q be the number of bits needed to code the k values. That is, q is the smallest integer such that $k \leq 2^q$. Then the size of the array needed to store a truth table of a function of n k-valued variables is 2^{qn}. For example, a truth table for an element that depends on five binary variables requires $2^5 = 32$ entries. For 3-valued logic, the truth table has $3^5 = 243$ entries, and the size of the array needed to store it is $2^{10} = 1024$.

Evaluation techniques based on truth tables are fast, but because of the exponential increase in the amount of memory they require, they are limited to elements that depend on a small number of variables.

A trade-off between speed and storage can be achieved by using a one-bit flag in the value area of an element to indicate whether any variable has a nonbinary value. If all values are binary, then the evaluation is done by accessing the truth table; otherwise a special routine is used. This technique requires less memory, since truth tables are now defined only for binary values. The loss in speed for nonbinary values is not significant, because, in general, most of the evaluations done during a simulation run involve only binary values.

Zoom Tables

An evaluation technique based on truth tables must first use the type of the evaluated element to determine the truth table to access. Thus checking the type and accessing the truth table are separate steps. These two consecutive steps can be combined into a single step as follows. Let t be the number of types and let S be the size of the largest truth table. We build a *zoom table* of size tS, in which we store the t individual truth tables, starting at locations 0, S, ..., $(t-1)S$. To evaluate an element, we pack its type code (in the range 0 to $t-1$) in the same word with its values, such that we can use the value of this word as an index into the zoom table (see Figure 3.19).

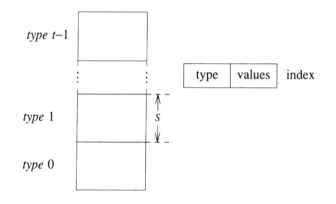

Figure 3.19   Zoom table structure

This type of zoom table is an instance of a general speed-up technique that reduces a sequence of k decisions to a single step. Suppose that every decision step i is based on a variable x_i which can take one of a possible set of m_i values. If all x_i's are known beforehand, we can combine them into one cross-product variable $x_1 \times x_2 \times ... \times x_k$ which can take one of the possible $m_1 m_2 ... m_k$ values. In this way the k variables are examined simultaneously and the decision sequence is reduced to one step. More complex zoom tables used for evaluation are described in [Ulrich *et al.* 1972].

Input Scanning

The set of primitive elements used in most simulation systems includes the basic gates — AND, OR, NAND, and NOR. These gates can be characterized by two parameters, the *controlling value c* and the *inversion i*. The value of an input is said to be controlling if it determines the value of the gate output regardless of the values of the other inputs; then the output value is $c \oplus i$. Figure 3.20 shows the general form of the primitive cubes of any gate with three inputs.

$$
\begin{array}{ccc|c}
c & x & x & c \oplus i \\
x & c & x & c \oplus i \\
x & x & c & c \oplus i \\
\bar{c} & \bar{c} & \bar{c} & \bar{c} \oplus i \\
\end{array}
\qquad
\begin{array}{l|c|c}
      & c & i \\
\hline
\text{AND} & 0 & 0 \\
\text{OR}   & 1 & 0 \\
\text{NAND} & 0 & 1 \\
\text{NOR} & 1 & 1 \\
\end{array}
$$

Figure 3.20   Primitive cubes for a gate with controlling value c and inversion i

Figure 3.21 outlines a typical gate evaluation routine for 3-valued logic, based on scanning the input values. Note that the scanning is terminated when a controlling value is encountered.

```
evaluate (G, c, i)
begin
     u_values = FALSE
     for every input value v of G
          begin
               if v = c then return c⊕i
               if v = u then u_values = TRUE
          end
     if u_values return u
     return c̄⊕i
end
```

Figure 3.21   Gate evaluation by scanning input values

Input Counting

Examining Figure 3.21, we can observe that to evaluate a gate using 3-valued logic, it is sufficient to know whether the gate has any input with c value, and, if not, whether it has any input with u value. This suggests that, instead of storing the input values for every gate, we can maintain a compressed representation of the values, in the form of two counters — c_count and u_count — which store, respectively, the number of inputs with c and u values [Schuler 1972]. The step of updating the input values (done before evaluation) is now replaced by updating of the two counters. For example, a $1 \rightarrow 0$ change at an input of an AND gate causes the c_count to be incremented, while a $0 \rightarrow u$ change results in decrementing the c_count and incrementing the u_count. The evaluation of a gate involves a simple check of the counters (Figure 3.22). This technique is faster than the input scanning method and is independent of the number of inputs.

> *evaluate* (G, c, i)
> **begin**
>   **if** $c\_count > 0$ **then return** $c \oplus i$
>   **if** $u\_count > 0$ **then return** u
>   **return** $\bar{c} \oplus i$
> **end**

Figure 3.22   Gate evaluation based on input counting

3.9 Hazard Detection

Static Hazards

In the circuit of Figure 3.23, assume that $Q = 1$ and A changes from 0 to 1, while B changes from 1 to 0. If these two changes are such that there exists a short interval during which $A=B=1$, then Z may have a spurious $1 \rightarrow 0 \rightarrow 1$ pulse, which may reset the latch. The possible occurrence of a transient pulse on a signal line whose static value does not change is called a *static hazard*.

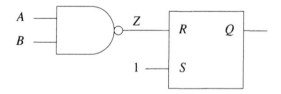

Figure 3.23

To detect hazards, a simulator must analyze the transient behavior of signals. Let $S(t)$ and $S(t+1)$ be the values of a signal S at two consecutive time units. If these values are different, the exact time when S changes in the real circuit is uncertain. To reflect this uncertainty in simulation, we will introduce a "pseudo time unit" t' between t and $t+1$ during which the value of S is unknown, i.e., $S(t') = u$ [Yoeli and Rinon 1964, Eichelberger 1965]. This is consistent with the meaning of u as one of the values in the set $\{0,1\}$, because during the transition period the value of S can be independently observed by each of its fanouts as either 0 or 1. Then the sequence $S(t)\ S(t')\ S(t+1) = 0u1$ represents one of the sequences in the set $\{001,011\}$. These two sequences can be interpreted, respectively, as a "slow" and a "fast" $0{\rightarrow}1$ transition. Returning now to the example of Figure 3.23, the corresponding sequences are $A = 0u1$ and $B = 1u0$. The resulting sequence for Z, computed by bitwise NAND operations, is $1u1$. This result shows that a possible output sequence is 101, and thus it detects the unintended pulse.

The general procedure for detecting static hazards in a combinational circuit C works as follows. Assume that C has been simulated for time t and now it is simulated for time $t+1$. Let E be the set of inputs changing between t and $t+1$.

Procedure 3.1

1. Set every input in E to value u and simulate C. (The other inputs keep their values.) Let $Z(t')$ denote the value obtained for a signal Z.

2. Set every input in E to its value at $t+1$ and simulate C.   □

Theorem 3.1: In a combinational circuit, a static hazard exists on line Z between the times t and $t+1$ if and only if the sequence of values $Z(t)\ Z(t')\ Z(t+1)$ (computed by Procedure 3.1) is $1u1$ or $0u0$.

Proof: Clearly, a sequence $0u0$ or $1u1$ is a sufficient condition for the existence of a static hazard. To prove that it also a necessary one, we rely on the following two facts, easily derived from the 3-valued truth tables of the basic gates.

1. If one or more gate inputs change from binary values to u, then the gate output either remains unchanged or changes from a binary value to u.

2. If one or more gate inputs change from u to binary values, then the gate output either remains unchanged or changes from u to a binary value.

Hence any gate whose value in t' is not u has the same binary value in t, t', and $t+1$, and therefore cannot have a static hazard.   □

Because Procedure 3.1 ignores the delays in the circuit, it performs a worst-case analysis whose results are independent of the delay model. Hence, we say that it uses an *arbitrary delay model*.

The next example illustrates how different delay models (0-delay, unit-delay, and arbitrary delays) affect hazard detection.

Example 3.1: Consider again the circuit of Figure 3.10(a) and the input sequence $A=010$. For the 0-delay model we obtain $B=101$ and $C=000$. Thus no hazard is predicted. This is because a 0-delay model deals only with the static behavior of a circuit and ignores its dynamic behavior.

For the unit-delay model, the signal sequences are $B=1101$ and $C=0010$. Thus a pulse is predicted in response to the $0{\rightarrow}1$ transition of A.

For the arbitrary delay model, the signal sequences (obtained with Procedure 3.1) are $A=0u1u0$, $B=1u0u1$, and $C=0u0u0$. Thus a hazard is predicted for both the rise and fall input transitions. This is an overly pessimistic result, since in general either the path through the inverter or the direct path from A to C has the most delay, and then only one of the input transitions should cause a hazard. However, under an arbitrary delay model it is not known which path has the most delay, and hazards are predicted for both transitions.   □

The analysis based on sequences of consecutive values underlies the hazard detection mechanism of most simulators [Hayes 1986]. Many simulators use multivalued logic systems that represent (implicitly or explicitly) different sets of sequences. Figure 3.24 shows such a set of values and their corresponding sequences. The result of a logic operation between these values (Figure 3.25) can be obtained by performing the same operation bitwise between the corresponding sequences.

Value	Sequence(s)	Meaning
0	000	Static 0
1	111	Static 1
0/1, R	$\{001,011\} = 0u1$	Rise (0 to 1) transition
1/0, F	$\{110,100\} = 1u0$	Fall (1 to 0) transition
0*	$\{000,010\} = 0u0$	Static 0-hazard
1*	$\{111,101\} = 1u1$	Static 1-hazard

Figure 3.24   6-valued logic for static hazard analysis

AND	0	1	R	F	0*	1*
0	0	0	0	0	0	0
1	0	1	R	F	0*	1*
R	0	R	R	0*	0*	R
F	0	F	0*	F	0*	F
0*	0	0*	0*	0*	0*	0*
1*	0	1*	R	F	0*	1*

Figure 3.25   AND truth table for 6-valued logic

Some simulators combine the values 0* and 1* into a single value that denotes a hazard. Sometimes this value is also combined with the unknown (u) value [Lewis 1972].

Dynamic Hazards

A *dynamic hazard* is the possible occurrence of a transient pulse during a 0→1 or 1→0 signal transition. The analysis for detecting dynamic hazards requires 4-bit sequences. For example, the sequence 0101 describes a 1-pulse during a 0→1 transition. The "clean" 0→1 transition corresponds to the set {0001,0011,0111}. Figure 3.26 shows an 8-valued logic system used for static and dynamic hazard analysis [Breuer and Harrison 1974]. We can observe that it includes the six values of Figure 3.24, to which it adds the values $R*$ and $F*$ to represent dynamic hazards.

Value	Sequence(s)	Meaning
0	0000	Static 0
1	1111	Static 1
0/1, R	{0001,0011,0111}	Rise transition
1/0, F	{1110,1100,1000}	Fall transition
0*	{0000,0100,0010,0110}	Static 0-hazard
1*	{1111,1011,1101,1001}	Static 1-hazard
R*	{0001,0011,0111,0101}	Dynamic 1-hazard
F*	{1110,1100,1000,1010}	Dynamic 0-hazard

Figure 3.26   8-valued logic for static and dynamic hazard analysis

A Boolean operation B between p and q, where p and q are among the eight values shown in Figure 3.26, is considered an operation between the sets of sequences corresponding to p and q and is defined as the union set of the results of all possible B operations between the sequences in the two sets. For example:

AND(R,1*) = AND({0001,0011,0111},{1111,1011,1101,1001}) =
= {0001,0011,0111,0101} = $R*$

Hazard Detection in Asynchronous Circuits

We will analyze hazards in an asynchronous circuit of the form shown in Figure 3.8 using an arbitrary delay model. Assume that all values at time t are known (and stable); now we apply a new vector x at time $t+1$. Let E be the set of primary inputs changing between t and $t+1$.

Procedure 3.2

1. Set every input in E to value u and simulate C. For every feedback line Y_i that changes to u, set the corresponding state variable y_i to u and resimulate C. Repeat until no more Y_i changes to u.

2. Set every input in E to its value at $t+1$ and simulate C. For every Y_i that changes to a binary value b_i, set the corresponding y_i to b_i and resimulate. Repeat until no more Y_i changes to a binary value.   □

Theorem 3.2: If the final value of Y_i computed by Procedure 3.2 is binary, then the feedback line Y_i stabilizes in this state (under the given input transition), regardless of the delays in the circuit.

Proof: Exercise.   □

The following is an important consequence of Theorem 3.2.

Corollary 3.1: If the final value of Y_i computed by Procedure 3.2 is u, then the given input transition may cause a critical race or an oscillation.   □

Procedure 3.2 does not require the feedback lines to be identified. It can work with an event-driven simulation mechanism by propagating first the changes that occur between t and t' until values stabilize, then the changes occurring between t' and $t+1$. In each simulation pass the values are guaranteed to stabilize (see Problem 3.17). Because of the arbitrary delay model, the order in which elements are evaluated is not important. Note that while executing step 1, any element whose value is currently u need not be reevaluated. Also, in step 2, any element whose current value is binary need not be reevaluated.

Example 3.2: Consider again the latch given in Figure 3.11(a). Assume that at time t, $R=S=0$ and $Q=QN=1$, and that at time $t+1$ both R and S change to 1. Following Procedure 3.2, in step 1 we set $R=S=u$, and as a result we obtain $Q=QN=u$. In step 2 we set $R=S=1$, and Q and QN remain stable at u. This shows that under an arbitrary delay model the operation of the circuit is unpredictable. Depending on the actual delays in the circuit, the final state may be $Q=0$, $QN=1$ or $Q=1$, $QN=0$ or the circuit may oscillate.   □

An important assumption of Procedure 3.2 is that the circuit is operated in *fundamental mode*, that is, stimuli are applied only when the values in the circuit are stable. This assumption precludes the simulation of real-time inputs (see Problem 3.18).

3.10 Gate-Level Event-Driven Simulation

3.10.1 Transition-Independent Nominal Transport Delays

Now we will particularize the general simulation algorithm given in Figure 3.12 for gate-level simulation using a transition-independent nominal transport delay model.

First we consider that the event list is organized as shown in Figure 3.27, where events scheduled to occur at the same time in the future are stored in the same list. The time order is maintained by chaining the list headers in appropriate order, i.e., $t_p < t_q < t_r$. An entry (i, v_i') in the list associated with t_q indicates that at time t_q the value of line i is scheduled to be set to v_i'.

We assume that values and delays are kept in tables similar to those depicted in Figure 2.21; $v(i)$ denotes the current value of gate i and $d(i)$ denotes the nominal delay of i.

Figure 3.28 shows the general structure of the event-driven simulation algorithms discussed in this section. Algorithm 3.1, given in Figure 3.29, provides a first implementation of the line "process entries for time t" of Figure 3.28. Algorithm 3.1 employs a *two-pass strategy*. In the first pass it retrieves the entries from the event list associated with the current time t and determines the activated gates. In the second pass it evaluates the activated gates and schedules their computed values. This strategy assures that gates activated by more than one event are evaluated only once.

Note that in Algorithm 3.1, an entry (i, v_i') in the event list does not always represent a change in the value of i. The problem of determining when a new output value of an activated gate is indeed an event is illustrated in Figure 3.30. In response to the

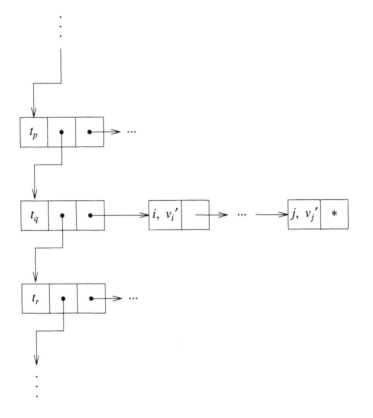

Figure 3.27   Event list implemented as a linked list structure

while (event list not empty)
    begin
        t = next time in list
        process entries for time t
    end

Figure 3.28   General structure for event-driven simulation

event $(a,1)$ at time 0, $(z,1)$ is scheduled for time 8. Since now the gate is in a transition state (i.e., the scheduled output event has not yet occurred), there is a temporary logic inconsistency between the current output value (0) and the input values (both 1). Thus when the next input event $(b,0)$ occurs at time 2, Algorithm 3.1 cannot determine whether setting $z = 0$ at time 10 will represent a change. The strategy employed is to enter in the event list all the new values of the activated gates

Activated = ∅   /* set of activated gates */
for every entry (i,v_i') pending at the current time t
    if $v_i' \neq v(i)$ **then**
        begin   /* it is indeed an event */
            $v(i) = v_i'$ /* update value */
            for every j on the fanout list of i
                begin
                    update input values of j
                    add j to *Activated*
                end
        end
for every $j \in$ *Activated*
    begin
        $v_j' =$ evaluate (j)
        schedule (j,v_j') for time $t+d(j)$
    end

Figure 3.29   Algorithm 3.1

and to do the check for activity when the entries are retrieved. For the example in Figure 3.30, $(z,0)$ will be scheduled for time 12 as a result of $(a,0)$ at time 4, but z will already be set to 0 at time 10.

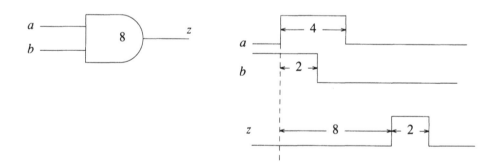

Figure 3.30   Example of activation of an already scheduled gate

Algorithm 3.2, given in Figure 3.31, is guaranteed to schedule only true events. This is done by comparing the new value v_j', determined by evaluation, with the *last scheduled value* of j, denoted by $lsv(j)$. Algorithm 3.2 does fewer schedule operations than Algorithm 3.1, but requires more memory and more bookkeeping to maintain the last scheduled values. To determine which one is more efficient, we have to estimate how many unnecessary entries in the event list are made by Algorithm 3.1. Our analysis focuses on the flow of events in and out of the event list.

$$Activated = \varnothing$$

for every event (i, v_i') pending at the current time t
   begin
      $v(i) = v_i'$
      for every j on the fanout list of i
         begin
            update input values of j
            add j to $Activated$
         end
   end
for every $j \in Activated$
   begin
      $v_j' = $ evaluate (j)
      if $v_j' \neq lsv(j)$ **then**
         begin
            schedule (j, v_j') for time $t+d(j)$
            $lsv(j) = v_j'$
         end
   end

Figure 3.31   Algorithm 3.2 — guaranteed to schedule only events

Let N be the total number of events occurring during a simulation run. If we denote by f the average fanout count of a signal, the number of gate activations is Nf. A gate that has $k > 1$ simultaneously active inputs is activated k times, but it is evaluated only once. (The entries in the set $Activated$ are unique.) Hence the number of gate evaluations is bounded by Nf. Because most gates are evaluated as a result of only one input change, we can approximate the number of evaluations by Nf. Let q be the fraction of gate evaluations generating output events. Then

$$N_1 = qNf \tag{3.1}$$

represents the total number of events scheduled during simulation. To these N_1 events, we should add N_2 events corresponding to the changes of the primary inputs entered in the event list before simulation. The equation

$$N_1 + N_2 = N \tag{3.2}$$

states that all the events entering the event list are eventually retrieved. Since usually $N_2 \ll N_1$, we obtain

$$qNf \cong N \tag{3.3}$$

Hence, $qf \cong 1$. This result shows that on the average, only one out of f evaluated gates generates an output event. Thus Algorithm 3.1 schedules and retrieves f times more items than Algorithm 3.2. Typically the average fanout count f is in the range 1.5 to 3. As the event list is a dynamic data structure, these operations involve some form of free-space management. Also scheduling requires finding the proper time slot where an item should be entered. We can conclude that the cost of the unnecessary event list operations done by Algorithm 3.1 is greater than that involved in maintaining the lsv values by Algorithm 3.2.

One-Pass Versus Two-Pass Strategy

The reason the *two-pass strategy* performs the evaluations only after all the concurrent events have been retrieved is to avoid repeated evaluations of gates that have multiple input changes. Experience shows, however, that most gates are evaluated as a result of only one input change. Hence a *one-pass strategy* [Ulrich 1969], which evaluates a gate as soon as it is activated, would be more efficient, since it avoids the overhead of constructing the *Activated* set. Algorithm 3.3, shown in Figure 3.32, implements the one-pass strategy.

> **for every** event (i, v_i') pending at the current time t
>   **begin**
> $v(i) = v_i'$
>   **for every** j on the fanout list of i
>   **begin**
>   update input values of j
> $v_j' = $ evaluate (j)
>   **if** $v_j' \neq lsv(j)$ **then**
>   **begin**
>   schedule (j, v_j') for time $t+d(j)$
> $lsv(j) = v_j'$
>   **end**
>   **end**
>   **end**

Figure 3.32   Algorithm 3.3 — one-pass strategy

Figure 3.33 illustrates the problem associated with Algorithm 3.3. Inputs a and b are scheduled to change at the same time. If the events are retrieved in the sequence $(b,0)$, $(a,1)$, then z is never scheduled. But if $(a,1)$ is processed first, this results in $(z,1)$ scheduled for time 4. Next, $(b,0)$ causes $(z,0)$ also to be scheduled for time 4. Hence at time 4, z will undergo both a 0-to-1 and a 1-to-0 transition, which will create a zero-width "spike." Although the propagation of this spike may not pose a problem, it is unacceptable to have the results depending on the order of processing of the concurrent events. Algorithm 3.4, shown in Figure 3.34, overcomes this problem by detecting the case in which the gate output is repeatedly scheduled for the same time and by canceling the previously scheduled event. Identifying this situation is helped by maintaining the *last scheduled time*, $lst(j)$, for every gate j.

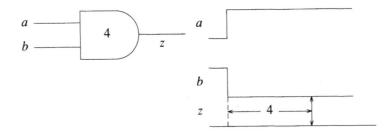

Figure 3.33 Processing of multiple input changes by Algorithm 3.3

for every event (i,v_i') pending at the current time t
    begin
        $v(i) = v_i'$
        for every j on the fanout list of i
            begin
                update input values of j
                $v_j' = evaluate\ (j)$
                if $v_j' \neq lsv(j)$ **then**
                    begin
                        $t' = t+d(j)$
                        if $t' = lst(j)$
                              then cancel event $(j,lsv(j))$ at time t'
                        schedule (j,v_j') for time t'
                        $lsv(j) = v_j'$
                        $lst(j) = t'$
                    end
            end
    end

Figure 3.34 Algorithm 3.4 — one-pass strategy with suppression of zero-width
            spikes

Let us check whether with this extra bookkeeping, the one-pass strategy is still more
efficient than the two-pass strategy. The number of insertions in the *Activated* set
performed during a simulation run by Algorithm 3.2 is about Nf; these are eliminated
in Algorithm 3.4. Algorithm 3.4 performs the check for zero-width spikes only for
gates with output events, hence only N times. Canceling a previously scheduled event
is an expensive operation, because it involves a search for the event. However, in
most instances $t + d(j) > lst(j)$, thus event canceling will occur infrequently, and the

associated cost can be ignored. Therefore we can conclude that the one-pass strategy is more efficient.

The Timing Wheel

With the event list implemented by the data structure shown in Figure 3.27, scheduling an event for time t requires first a search of the ordered list of headers to find the position for time t. The search time is proportional to the length of the header list, which, in general, increases with the size of the circuit and its activity. The header list is usually dense, that is, the differences between consecutive times for which events are scheduled are small. This suggests that employing an array of headers (see Figure 3.35), rather than a linked list, would be more efficient, since the search is avoided by using the time t to provide an index into the array. The disadvantage is that the process of advancing the simulation time has now to scan all headers (which follow the current time) until the next one with activity is encountered. But if the headers are dense, this represents only a minor overhead.

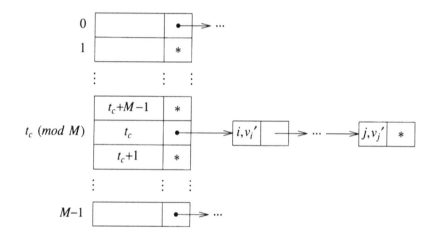

Figure 3.35 Timing wheel

The array of headers and their associated lists store events scheduled to occur between the current simulation time t_c and t_c+M-1, where M is the size of the array. The header for the events scheduled for a time t in this range appears in the array in position t modulo M. Because of the circular structure induced by the modulo-M operation, the array is referred to as a *timing wheel* [Ulrich 1969]. Any event scheduled for time t_c+d, where $d<M$, can be inserted in the event list without search. Events beyond the range of the wheel ($d\geq M$) are stored in an overflow "remote" event list of the type shown in Figure 3.27. Insertions in the remote list require searching, but in general, most events will be directly processed via the timing wheel. To keep the number of headers in the remote list at a minimum, events from that list should be brought into the wheel as soon as their time becomes included in the wheel range.

3.10.2 Other Logic Values

3.10.2.1 Tristate Logic

Tristate logic allows several devices to time-share a common wire, called a *bus*. A bus connects the outputs of several *bus drivers* (see Figure 3.36). Each bus driver is controlled by an "enable" signal E. When $E=1$, the driver is enabled and its output O takes the value of the input I; when $E=0$, the driver is disabled and O assumes a high-impedance state, denoted by an additional logic value Z. $O=Z$ means that the output is in effect disconnected from the bus, thus allowing other bus drivers to control the bus. In normal operation, at most one driver is enabled at any time. The situation when two bus drivers drive the bus to opposite binary values is called a *conflict* or a *clash* and may result in physical damage to the devices involved. A pull-up (pull-down) function, realized by connecting the bus to power (ground) via a resistor, provides a "default" 1 (0) logic value on the bus when all the bus drivers are disabled.

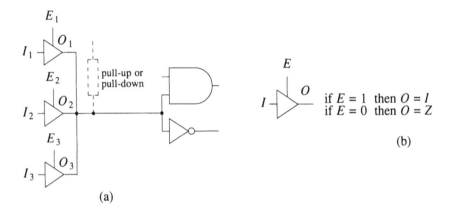

Figure 3.36   (a) Bus (b) bus driver

A bus represents another form of wired logic (see Section 2.4.4) and can be similarly modeled by a "dummy" component, whose function is given in Figure 3.37. Note that while the value observed on the bus is the forced value computed by the bus function, for every bus driver the simulator should also maintain its own driven value, obtained by evaluating the driver. One task of the bus evaluation routine is to report conflicts or potential conflicts to the user, because they usually indicate design errors. Sometimes, it may also be useful to report situations when multiple drivers are enabled, even if their values do not clash. When all the drivers are disabled, the resulting Z value is converted to a binary value if a pull-up or a pull-down is present; otherwise the devices driven by the bus will interpret the Z value as u (assuming TTL technology).

Recall that the unknown logic value u represents a value in the set $\{0,1\}$. In simulation, both the input I and the enable signal E of a bus driver can have any of the values $\{0,1,u\}$. Then what should be the output value of a bus driver with $I=1$ and $E=u$? A case-by-case analysis shows that one needs a new "uncertain" logic value to

	0	1	Z	u
0	0^1	u^2	0	u^3
1	u^2	1^1	1	u^3
Z	0	1	Z^4	u
u	u^3	u^3	u	u^3

Figure 3.37 Logic function of a bus with two inputs (1 — report multiple drivers enabled; 2 — report conflict; 3 — report potential conflict; 4 — transform to 1(0) if pull-up (pull-down) present)

represent a value in the set $\{1,Z\}$. Figure 3.38 gives a complete truth table of a bus driver; the results for $E=u$ are given as sets of values.

		E		
		0	1	u
	0	Z	0	$\{0,Z\}$
I	1	Z	1	$\{1,Z\}$
	u	Z	u	$\{u,Z\}$

Figure 3.38 Truth table for a bus driver

3.10.2.2 MOS Logic

In this section we describe several techniques that allow traditional simulation methods to be extended to MOS components. The basic component in MOS logic is the *transmission gate* (Figure 3.39(a)), which works as a switch controlled by the gate signal G. When $G=1$, the transistor is on and the switch is closed, thus connecting its source (S) and drain (D) terminals; when $G=0$ the transistor is off and the switch is open. Although a transmission gate is intrinsically bidirectional, it can be used as a unidirectional or as a bidirectional element. As a unidirectional element (Figure 3.39(b)), its function is similar to that of a bus driver, except that when the transmission gate is open, the wire connected to its output retains its previous value. This is caused by the charge stored on the stray capacitance C associated with the output wire. (After a long *decay time*, C will be discharged, but usually circuits are operated sufficiently fast so that decay times can be considered infinite.) If the output of the transmission gate is connected to a bus without a pull-up or pull-down (Figure 3.39(c)), the stored value is the last forced value on the bus. This stored value, however, is overridden by the value driven by any gate (connected to the same bus) that is turned on and thus provides an escape path for the stored charge.

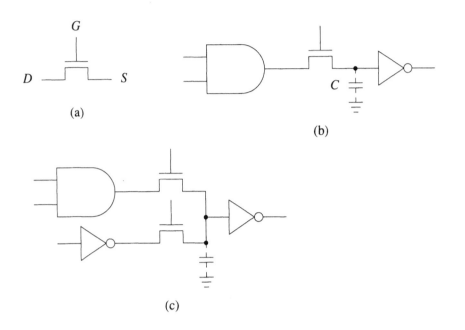

Figure 3.39   (a) Transmission gate (b) Transmission gate used as a unidirectional
component (c) Transmission gates connected to a bus

This behavior illustrates the concept of *logical strength*, which is a digital measure of
the relative current drive capability of signal lines. The values stored on wires — also
called *floating* or *weak values* and denoted by Z_0 and Z_1 — correspond to low-current
drive capability. They can be overridden by *strong values* (0 and 1) supplied by
outputs of turned-on gates (or power and ground leads, or primary input leads), which
have high-current drive capability.

The diagram given in Figure 3.40 shows the strength relations of a typical set of values
used in MOS simulation. The level of a value in the diagram corresponds to its
strength. The unknown value u is the strongest, and the high-impedance Z is the
weakest. The weak unknown value Z_u denotes an unknown stored charge associated
with a wire. The function B of a bus driven with the values v_1, v_2, ..., v_n can be
defined as producing a unique value $v_c = B (v_1, v_2, ...v_n)$, where v_c is the weakest value
whose strength is greater than or equal to the strength of every v_i $(1 \le i \le n)$. Thus
$B(0,1) = u$, $B(Z_0,1) = 1$, $B(Z_0,Z_1) = Z_u$, $B(1,Z) = 1$.

The concept of strength allows a pull-up load connected to power to be treated as a
logic component, called an *attenuator* [Hayes 1982], whose function is to transform the
strong 1 provided by the power line into a Z_1 value supplied to the bus. This value is
overridden by a strong value driven by any gate, and it overrides the Z value provided
by the turned-off gates. Similarly, a pull-down load connected to ground transforms a
0 value into a Z_0.

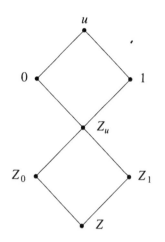

Figure 3.40 Typical values used in MOS simulation and their relative strengths

Figure 3.41(a) illustrates a transmission gate used as a bidirectional switch. When $G=0$, the switch is open, and C and F are independent. But when $G=1$, C and F become connected by a bus. One way of modeling the bidirectionality of the gate controlled by G is with two unidirectional gates connected back to back, as shown in Figure 3.41(b) [Sherwood 1981].

3.10.3 Other Delay Models

3.10.3.1 Rise and Fall Delays

Conceptually, the rise/fall delay model is a simple extension of the transition-independent delay model. Instead of using one delay value for scheduling events, one selects the rise or the fall delay of an element, depending on the result of its evaluation. Since rise and fall delays are associated (respectively) with $0{\rightarrow}1$ and $1{\rightarrow}0$ transitions, a first question is what delays should be used for transitions involving u values. For example, a $0{\rightarrow}u$ transition is either a $0{\rightarrow}1$ transition or not an event at all. Thus if an event occurs, its associated delay is the rise delay. Similarly $u{\rightarrow}1$ will use the rise delay, and $1{\rightarrow}u$ and $u{\rightarrow}0$ will use the fall delay.

The following example illustrates some of the complications arising from the rise/fall delay model.

Example 3.3: Consider an inverter with $d_r = 12$ and $d_f = 7$. Assume that a $1{\rightarrow}0{\rightarrow}1$ pulse of width 4 arrives at its input. The first input transition (occurring at time 0) results in scheduling the output to go to 1 at time $0 + t_r = 12$. The second input transition at time 4 would cause the output to be scheduled to go to 0 at time $4 + t_f = 11$. This "impossible" sequence of events appears because the effect of the second input transition propagates to the output faster than the effect of the first one. Here the correct result should be that the output does not change at all; thus the first

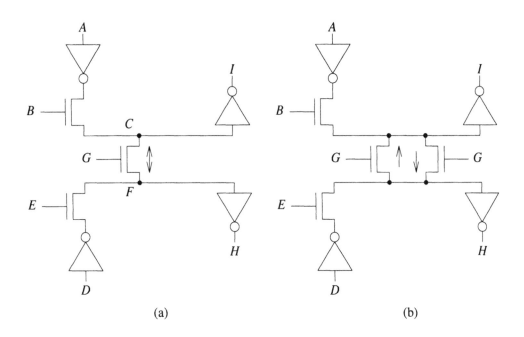

Figure 3.41 (a) Transmission gate used as bidirectional component (b) Model by unidirectional gates

event should be canceled and the second one not scheduled. Note that the input pulse is suppressed, even without an inertial delay model.   □

As this example shows, the main difficulty associated with the rise/fall delay model arises when previously scheduled events have to be canceled. Let v_j' be the value computed by the evaluation of an activated gate j at the current time t. The procedure *Process_result*, outlined in Figure 3.42, processes the result of the evaluation and determines whether event canceling is needed. For this analysis, we assume (as in Algorithm 3.4) that for every signal j, one maintains its last scheduled value — $lsv(j)$ — and its last scheduled time — $lst(j)$. In the "normal" case, the new event of j occurs after the last scheduled event of j. When the order of these two events is reversed, the value v_j' reaches the output before $lsv(j)$, and the last scheduled event is canceled. After that, $lst(j)$ and $lsv(j)$ are updated to correspond to the new last scheduled event of j and the entire process is repeated. If after canceling its last scheduled event, j has more pending events, finding the last one requires a search of the event list. Otherwise, $lsv(j)$ can be set to the current value of j, and $lst(t)$ can be set to the current simulation time. (This analysis is helped by maintaining the count of pending events of every signal.)

Let us apply this procedure to handle the result of the second evaluation of the inverter of Example 3.3. Let j be the output of the inverter. At time $t=4$, the last scheduled event is a transition to $lsv(j)=1$ at time $lst(j)=12$. The result of the second evaluation is $v_j'=0$. The delay d for the transition $lsv(j) \rightarrow v_j'$ is $d_f=7$, so this event would occur at time $t'=11$. Because $t' < lst(j)$, the event scheduled for time 12 is canceled. As there

Process_result (j, v_j')
begin
    while $v_j' \neq lsv(j)$
        begin
            d = delay for the transition $lsv(j) \to v_j'$
            $t' = t+d$
            if $t' > lst(j)$ **then**
                begin
                    schedule *(j, v_j')* for time t'
                    $lst(j) = t'$
                    $lsv(j) = v_j'$
                    return
                end
            cancel event *(j, lsv(j))* at time *lst(j)*
            update *lst(j)* and *lsv(j)*
        end
end

Figure 3.42   Processing an evaluation result with the rise/fall delay model

are no more pending events for j, $lsv(j)$ is set to its current value (0), and $lst(j)$ is set to $t=4$. Now $v_j'=lsv(j)$ and no event is scheduled.

3.10.3.2 Inertial Delays

The inertial delay model leads to problems similar to those of the rise/fall delay model. Here event canceling is required to implement spike filtering. For example, consider a gate j with an output inertial delay d_I. Assume that j has been scheduled to go to $lsv(j)$ at time $lst(j)$, and that a second evaluation of j has computed a new value $v_j' \neq lsv(j)$, which would occur at time t'. If $t'-lst(j) < t_I$, then the output pulse is suppressed by canceling the last scheduled event of j.

3.10.3.3 Ambiguous Delays

In the ambiguous delay model, the transport delay of a gate is within a range $[d_m, d_M]$. (We assume transition-independent delays). Thus when the gate responds to a transition occurring at an input at time t, its output will change at some time during the interval $[t+d_m, t+d_M]$. (See Figure 3.14(c).) To reflect this uncertainty in simulation, we will use the 6-valued logic of Figure 3.24, where 0/1 (1/0) is the value of a signal during the interval in which it changes from 0 to 1 (1 to 0). Hence a gate output changing from 0 to 1 goes first from 0 to 0/1, holds this value for an interval d_M-d_m, then goes from 0/1 to 1. For uniformity, we will treat a $0 \to 1$ transition on a primary input as a $0 \to 0/1$ event, followed (at the same time) by a $0/1 \to 1$ event. When scheduling a gate output, its minimum delay d_m will be used for changes to the 0/1 and 1/0 values, and its maximum delay d_M for changes to binary values.

Example 3.4: In the circuit shown in Figure 3.43 the delay of C is between 2 and 4, and that of D between 2 and 5. Assume that at time 0, A changes from 0 to 1. C is evaluated twice, first with $A=0/1$, then with $A=1$. The first evaluation causes C to change to 0/1 after a delay $d_m(C)=2$; as a result of the second evaluation, C is scheduled to take value 1 after a delay $d_M(C)=4$. At time $t=2$, D is evaluated with $C=0/1$, and its new 1/0 value is scheduled for time $t+d_m(D)=4$. The next evaluation of D (at time 4) produces a binary value (0), which is scheduled using the delay $d_M(D)=5$.

□

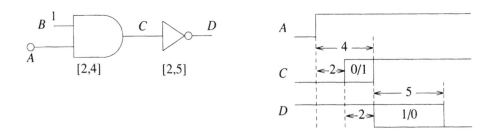

Figure 3.43 Simulation with ambiguous delay model

A simulator using the ambiguous delay model computes the earliest and the latest times signal changes may occur. In the presence of reconvergent fanout, this analysis may lead to overly pessimistic results. For example, in the circuit shown in Figure 3.44, assume that A changes from 1 to 0 during the interval [20, 30]. If the delay of B is between 6 and 10, the simulator predicts that B changes from 0 to 1 during the interval [26, 40]. Thus it appears that in the interval [26, 30], $A=1/0$ and $B=0/1$, which creates a static hazard at C. This result, however, is incorrect because the transitions of B and A are not independent; the transition of B occurs between 6 and 10 time units after the transition of A, so that no pulse can appear at C. Additional processing is required to remove the *common ambiguity* among the inputs of the same gate [Bowden 1982].

A less pessimistic way of treating ambiguous delays is to perform several simulation runs, each one preceded by a random selection of the delay value of every element from its specified range.

When a tester applies input vectors to a circuit, the inputs will not change at the same time because of skews introduced by the tester. These skews can be taken into account in simulation by adding a gate with ambiguous delays to every primary input of the model.

3.10.4 Oscillation Control

Although Procedure 3.2 can detect potential oscillations, the arbitrary delay model it uses often leads to pessimistic results. That is, the problems reported will not occur under a more realistic delay model. Another drawback of Procedure 3.2 is that it is not applicable for real-time inputs. In this section we discuss other methods for detecting oscillations in event-driven simulation.

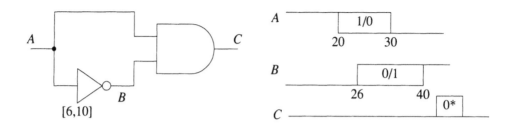

Figure 3.44 Pessimistic result in ambiguous delay simulation

Why is this problem important? An accurate simulation of a circuit that oscillates will result in repeated scheduling and processing of the same sequence of events, with the simulation program apparently caught in an "endless loop." This may consume a lot of CPU time. Figure 3.45 illustrates a simple example of oscillation. Assume that every gate has a delay of 3 nsec, and that initially $S=R=1$, $y=0$, and $\bar{y}=1$. If S goes to 0 for 4 nsec, y and $\bar{y}$ will oscillate as a result of a pulse continuously traveling around the loop composed of the two gates.

The process of detecting oscillations during simulation and taking appropriate corrective action is called *oscillation control*. The corrective action is to set the oscillating signals to u. In addition, it is desirable to inform the user about the oscillation and to provide the option of aborting the simulation run, which may have become useless. Oscillation can be controlled at two levels, referred to as local control and global control.

Local oscillation control is based on identifying conditions causing oscillations in specific subcircuits, such as latches or flip-flops. For the example in Figure 3.45, $y=\bar{y}=0$ is an oscillating condition. Note that this implies $R=S=1$, which is the condition enabling propagation along the loop. Also, the states $y=0$, $\bar{y}=u$ and $y=u$, $\bar{y}=0$ can cause oscillation. The appropriate corrective action is to set $y=\bar{y}=u$.

Local oscillation control can be implemented in several ways. One way is to have the simulator monitor user-specified conditions on certain gates [Chappel *et al.* 1974]; these are typically cross-coupled gates used in latches. An easier way to implement local oscillation control is via modeling techniques. If latches are modeled as primitive or user-defined functional elements, checking for conditions causing oscillations (and for other "illegal" states as well) can be part of the model. Gate-level models can also be used to detect and correct oscillations [Ulrich *et al.* 1972]. Figure 3.45(c) shows such a model. In the normal operation of the latch ($y=0$ and $\bar{y}=1$, or $y=1$ and $\bar{y}=0$, or $y=\bar{y}=1$), $G=1$ and this value does affect the rest of the circuit. However, any of the "illegal" states ($y=\bar{y}=0$, or $y=0$ and $\bar{y}=u$, or $\bar{y}=0$ and $y=u$) will cause $G=u$, which in turn (with $R=S=1$) will stop the oscillation by setting $y=\bar{y}=u$.

Because local oscillation control applies only for specific subcircuits, it may fail to detect oscillations involving feedback loops that are not contained in those subcircuits. Correct identification of this type of global oscillation is not computationally feasible,

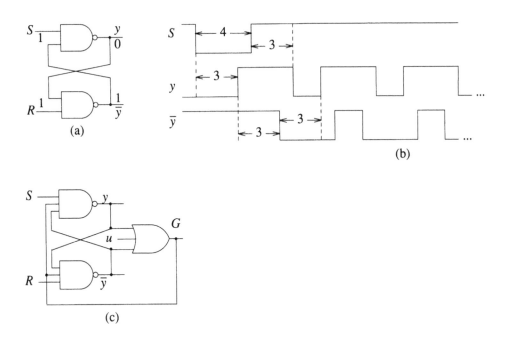

Figure 3.45 (a) Latch (b) Oscillation waveforms (c) Model for oscillation control

since it would require detecting cyclic sequences of values for any signal in the circuit. Most simulators implement *global oscillation control* by identifying unusually high activity during simulation. A typical procedure is to count the number of events occurring after any primary input change. When this number exceeds a predefined limit, the simulator assumes that an oscillation has been detected and sets all currently active signals to u to terminate the activity caused by oscillations. Of course, this procedure may erroneously label legitimate high activity as an oscillation, but the user can correct this mislabeling by adjusting the activity limit.

3.11 Simulation Engines

Logic simulation of complex VLSI circuits and systems is a time-consuming process. This problem is made more acute by the extensive use of simulation in the design cycle to validate the numerous design changes a project usually undergoes. One solution to this problem is provided by hardware specially designed to speed up simulation by using parallel and/or distributed processing architectures. Such special-purpose hardware, called a *simulation engine* or a simulator hardware accelerator, is usually attached to a general-purpose host computer. Typically, the host prepares and downloads the model and the applied stimuli into the simulation engine, which performs the simulation and returns the results to the host.

Simulation engines achieve their speed-up based on two different strategies, referred to as model partitioning and algorithm partitioning. *Model partitioning* consists of dividing the model of the simulated circuit into disjoint subcircuits, concurrently simulated by identical processing units (PUs), interconnected by a communication medium (see Figure 3.46). *Algorithm partitioning* consists of dividing the simulation algorithm into tasks distributed among different PUs working concurrently as stages of a pipeline. Many simulation engines combine both model partitioning and algorithm partitioning.

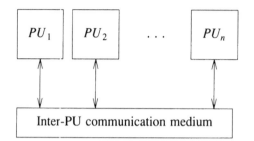

Figure 3.46   Simulation engine based on model partitioning (connections with host not shown)

The Yorktown Simulation Engine (YSE) [Denneau 1982] is based on partitioning the model among up to 256 PUs, interconnected by a crossbar switch. Each PU simulates a subcircuit consisting of up to 4K gates, with some specialized PUs dedicated for simulating RAMs and ROMs. The YSE implements a compiled simulation algorithm, in which every gate is evaluated for every input vector. Internally, every PU employs algorithm partitioning, with pipeline stages corresponding to the following structure of a compiled simulator:

> **for** every gate G
> **begin**
> > determine inputs of G
> > get the values of the inputs
> > evaluate G
> > store value of G
> **end**

Figure 3.47 shows a block diagram of a PU. The "program counter" provides the index of the next gate to be evaluated (recall that in compiled simulation a gate is evaluated only after all its input values are known). This index is used to access the "instruction memory," which provides the inputs of the gate, its type and other information needed for evaluation, and the address where the computed output value should be stored. Signal values are stored in the "data memory" block. The YSE uses four logic values (0, 1, u, and Z), which are coded with two bits. Evaluations are done by the "function unit," whose structure is shown in Figure 3.48. A gate in the YSE model can have up to four inputs, whose values are provided by the data memory. These values are first passed through "generalized DeMorgan" (GDM) memory blocks, which contain truth tables for 16 functions of one 4-valued variable. The function of a

GDM is selected by a 4-bit GDM code provided by the instruction memory. Typical GDM functions are identity (in which the input value is passed unchanged), inversion, and constant functions. (Constant functions are primarily used to supply noncontrolling values to unused gate inputs.) Evaluations are done with a zoom table technique. The tables are stored in the "function memory," whose address is obtained by concatenating the gate type with the input values transformed by the GDM codes. The output value is similarly passed through another GDM.

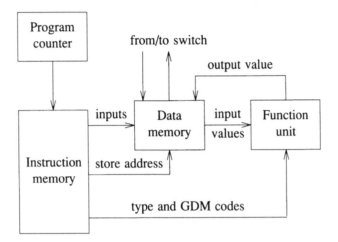

Figure 3.47   YSE processing unit

Model partitioning is done by the host as a preprocessing step. Because of partitioning, inputs of a subcircuit assigned to one PU may be outputs of subcircuits assigned to different PUs. The values of such signals crossing the partition boundaries must be transmitted across the inter-PU switch. All PUs are synchronized by a common clock, and each PU can evaluate a gate during every clock cycle. The newly computed output values are sent out to the switch, which can route them further to the other PUs that use them. The scheduling of the inter-PU communications is done together with the model partitioning [Kronstadt and Pfister 1982]. The evaluation of a gate by one PU may be delayed until all its input values computed by other PUs arrive through the switch. The goal of the partitioning is to minimize the total waiting time during simulation.

Figure 3.49 depicts the architecture of an event-driven simulation engine, similar to that described in [Abramovici *et al.* 1983]. The tasks of the algorithm are partitioned among several PUs, configured as stages of a circular pipeline. The PUs may be hardwired (i.e., their tasks are directly implemented in hardware) or programmable. Programmable PUs offer more flexibility at a cost of some loss in performance. Another advantage of programmable PUs is that they can be identically built and their different tasks can be realized by different code. (Microprogrammable PUs, whose tasks are directly implemented in microcode, offer a good trade-off between

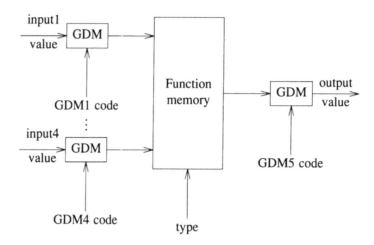

Figure 3.48   YSE function unit

performance and flexibility.) Every PU has a local memory that can be loaded by the host (connections with the host — usually a shared bus — are not shown in Figure 3.49).

A brief outline of the main tasks of the PUs of Figure 3.49 follows. The Event List Manager maintains the event list in its local memory. It advances the simulation time to the next time for which events are scheduled, retrieves these concurrent events, and sends them to the Current Event PU. This PU updates the signal values (which are maintained in its local memory) and sends the incoming events to the Event Router. In addition, the Current Event PU performs auxiliary tasks such as global oscillation control and sending results to the host. The Event Router stores the fanout list in its local memory. It determines the activated gates and sends them (together with their input events) to the Evaluation Unit. The Evaluation Unit maintains the input values of every gate in its local memory. It updates the input values of the activated gates and evaluates them. The resulting output events are sent to the Scheduler, whose local memory stores delay data. The Scheduler determines the appropriate delay to be used for each event and sends the events and their times of occurrence to the Event List Manager, which inserts them in the event list. The Event List Manager also processes the events scheduled for the primary inputs, which are received from the host.

The efficiency of a pipeline depends on the distribution of the workload among its stages. If the average workload is not uniformly distributed, the stage that is most heavily loaded becomes the bottleneck of the pipeline. Such an imbalance can be corrected by replacing the bottleneck stage with several parallel processors. Even a pipeline that is balanced on the average may have temporary disparities between the workloads of different stages. These variations can be smoothed out by buffering the data flow between PUs using first-in first-out (FIFO) buffers.

A simulation engine implementing event-driven simulation can also be based on model partitioning among parallel PUs, as shown in Figure 3.46. The communication

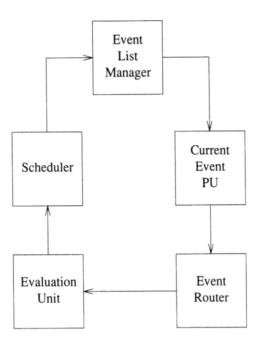

Figure 3.49 Simulation engine based on algorithm partitioning for event-driven
simulation

medium is primarily used to transmit events that propagate between subcircuits
assigned to different PUs. The architecture described in [VanBrunt 1983] combines
model partitioning and algorithm partitioning, by connecting up to 16 PUs of the type
illustrated in Figure 3.49 via a shared bus.

In an architecture based on model partitioning, one PU, called the master, assumes the
task of coordinating the work of the other PUs, referred to as slaves [Levendel *et
al.* 1982]. Each slave has its own event list, but the simulation time is maintained by
the master. In every simulation cycle, the master advances the simulation time by one
time unit and initiates all the slaves. Then each slave processes the events pending for
that time (if any) in its local event list. Events generated for signals internal to a
subcircuit are inserted into the local event list, while events crossing the partition
boundaries are sent to the appropriate PUs via the communication medium. Every PU
reports to the master when it is done. The master synchronizes the slaves by waiting
for all of them to complete processing before advancing the time and initiating the new
cycle.

The partitioning of the simulated circuit into subcircuits assigned to different PUs is
guided by two objectives: (1) maximize the concurrency achieved by the ensemble of
PUs, by trying to uniformly distribute the simulation activity among PUs, and
(2) minimize the inter-PU communication. Partitioning algorithms are discussed in
[Levendel *et al.* 1982, Agrawal 1986].

REFERENCES

[Abramovici *et al.* 1983] M. Abramovici, Y. H. Levendel, and P. R. Menon, "A Logic Simulation Machine," *IEEE Trans. on Computer-Aided Design*, Vol. CAD-2, No. 2, pp. 82-94, April, 1983.

[Agrawal *et al.* 1980] V. D. Agrawal, A. K. Bose, P. Kozak, H. N. Nham, and E. Pacas-Skewes, "A Mixed-Mode Simulator," *Proc. 17th Design Automation Conf.*, pp. 618-625, June, 1980.

[Agrawal 1986] P. Agrawal, "Concurrency and Communication in Hardware Simulators," *IEEE Trans. on Computer-Aided Design*, Vol. CAD-5, No. 4, pp. 617-623, October, 1986.

[Agrawal *et al.* 1987] P. Agrawal, W. T. Dally, W. C. Fisher, H. V. Jagadish, A. S. Krishnakumar, and R. Tutundjian, "MARS: A Multiprocessor-Based Programmable Accelerator," *IEEE Design & Test of Computers*, Vol. 4, No. 5, pp. 28-36, October, 1987.

[Barzilai *et al.* 1987] Z. Barzilai, J. L. Carter, B. K. Rosen, and J. D. Rutledge, "HSS — A High-Speed Simulator," *IEEE Trans. on Computer-Aided Design*, Vol. CAD-6, No. 4, pp. 601-617, July, 1987.

[Blank 1984] T. Blank, "A Survey of Hardware Accelerators Used in Computer-Aided Design," *IEEE Design & Test of Computers*, Vol. 1, No. 3, pp. 21-39, August, 1984.

[Bowden 1982] K. R. Bowden, "Design Goals and Implementation Techniques for Time-Based Digital Simulation and Hazard Detection," *Digest of Papers 1982 Intn'l. Test Conf.*, pp. 147-152, November, 1982.

[Breuer 1972] M. A. Breuer, "A Note on Three Valued Logic Simulation," *IEEE Trans. on Computers*, Vol. C-21, No. 4, pp. 399-402, April, 1972.

[Breuer and Harrison 1974] M. A. Breuer and L. Harrison," Procedures for Eliminating Static and Dynamic Hazards in Test Generation," *IEEE Trans. on Computers*, Vol. C-23, No. 10, pp. 1069-1078, October, 1974.

[Butler *et al.* 1974] T. T. Butler, T. G. Hallin, K. W. Johnson, and J. J. Kulzer, "LAMP: Application to Switching System Development," *Bell System Technical Journal*, Vol. 53, No. 8, pp. 1535-1555, October, 1974.

[Chappell and Yau 1971] S. G. Chappell and S. S. Yau, "Simulation of Large Asynchronous Logic Circuits Using an Ambiguous Gate Model," *Proc. Fall Joint Computer Conf.*, pp. 651-661, 1971.

[Chappell *et al.* 1974] S. G. Chappell, C. H. Elmendorf, and L. D. Schmidt, "LAMP: Logic-Circuit Simulators," *Bell System Technical Journal*, Vol. 53, No. 8, pp. 1451-1476, October, 1974.

[Chen *et al.* 1984] C. F. Chen, C-Y. Lo, H. N. Nham, and P. Subramaniam, "The Second Generation MOTIS Mixed-Mode Simulator," *Proc. 21st Design Automation Conf.*, pp. 10-17, June, 1984.

[Denneau 1982] M. M. Denneau, "The Yorktown Simulation Engine," *Proc. 19th Design Automation Conf.*, pp. 55-59, June, 1982.

[Eichelberger 1965] E. B. Eichelberger, "Hazard Detection in Combinational and Sequential Circuits," *IBM Journal of Research and Development*, Vol. 9, pp. 90-99, March, 1965.

[Evans 1978] D. J. Evans, "Accurate Simulation of Flip-Flop Timing Characteristics," *Proc. 15th Design Automation Conf.*, pp. 398-404, June, 1978.

[Hayes 1982] J. P. Hayes, "A Unified Switching Theory with Applications to VLSI Design," *Proc. of the IEEE*, Vol. 70, No. 10, pp. 1140-1151, October, 1982.

[Hayes 1986] J. P. Hayes, "Uncertainty, Energy, and Multiple-Valued Logics," *IEEE Trans. on Computers*, Vol. C-35, No. 2, pp. 107-114, February, 1986.

[Hitchcock 1982] R. B. Hitchcock, "Timing Verification and Timing Analysis Program," *Proc. 19th Design Automation Conf.*, pp. 594-604, June, 1982.

[Ishiura *et al.* 1987] N. Ishiura, H. Yasuura, and S. Yajma, "High-Speed Logic Simulation on Vector Processors," *IEEE Trans. on Computer-Aided Design*, Vol. CAD-6, No. 3, pp. 305-321, May, 1987.

[Kronstadt and Pfister 1982] E. Kronstadt and G. Pfister, "Software Support for the Yorktown Simulation Engine," *Proc. 19th Design Automation Conf.*, pp. 60-64, June, 1982.

[Levendel *et al.* 1982] Y. H. Levendel, P. R. Menon, and S. H. Patel, "Special-Purpose Logic Simulator Using Distributed Processing," *Bell System Technical Journal*, Vol. 61, No. 10, pp. 2873-2909, December, 1982.

[Lewis 1972] D. W. Lewis, "Hazard Detection by a Quinary Simulation of Logic Devices with Bounded Propagation Delays," *Proc. 9th Design Automation Workshop*, pp. 157-164, June, 1972.

[Monachino 1982] M. Monachino, "Design Verification System for Large-Scale LSI Designs," *Proc. 19th Design Automation Conf.*, pp. 83-90, June, 1982.

[Saab *et al.* 1988] D. G. Saab, R. B. Mueller-Thuns, D. Blaauw, J. A. Abraham, and J. T. Rahmeh, "CHAMP: Concurrent Hierarchical and Multilevel Program for Simulation of VLSI Circuits," *Proc. Intn'l. Conf. on Computer-Aided Design*, pp. 246-249, November, 1988.

[Sasaki *et al.* 1981] T. Sasaki, A. Yamada, and T. Aoyama, "Hierarchical Design Verification for Large Digital Systems," *Proc. 18th Design Automation Conf.*, pp. 105-112, June, 1981.

[Schuler 1972] D. M. Schuler, "Simulation of NAND Logic," *Proc. COMPCON 72*, pp. 243-245, September, 1972.

[Sherwood 1981] W. Sherwood, "A MOS Modeling Technique for 4-State True-Value Hierarchical Logic Simulation," *Proc. 18th Design Automation Conf.*, pp. 775-785, June, 1981.

[Szygenda *et al.* 1970] S. A. Szygenda, D. W. Rouse, and E. W. Thompson, "A Model and Implementation of a Universal Time Delay Simulator for Digital Nets," *Proc. AFIPS Spring Joint Computer Conf.*, pp. 207-216, 1970.

[Szygenda and Lekkos 1973] S. A. Szygenda and A. A. Lekkos, "Integrated Techniques for Functional and Gate-Level Digital Logic Simulation," *Proc. 10th Design Automation Conf.*, pp. 159-172, June, 1973.

[Tokoro *et al.* 1978] M. Tokoro, M. Sato, M. Ishigami, E. Tamura, T. Ishimitsu, and H. Ohara, "A Module Level Simulation Technique for Systems Composed of LSIs and MSIs," *Proc. 15th Design Automation Conf.*, pp. 418-427, June, 1978.

[Ulrich 1965] E. G. Ulrich, "Time-Sequenced Logic Simulation Based on Circuit Delay and Selective Tracing of Active Network Paths," *Proc. 20th ACM Natl. Conf.*, pp. 437-448, 1965.

[Ulrich 1969] E. G. Ulrich, "Exclusive Simulation of Activity in Digital Networks," *Communications of the ACM*, Vol. 13, pp. 102-110, February, 1969.

[Ulrich *et al.* 1972] E. G. Ulrich, T. Baker, and L. R. Williams, "Fault-Test Analysis Techniques Based on Logic Simulation," *Proc. 9th Design Automation Workshop*, pp. 111-115, June, 1972.

[VanBrunt 1983] N. VanBrunt, "The Zycad Logic Evaluator and Its Application to Modern System Design," *Proc. Intn'l. Conf. on Computer Design*, pp. 232-233, October, 1983.

[Yoeli and Rinon 1964] M. Yoeli and S. Rinon, "Applications of Ternary Algebra to the Study of Static Hazards," *Journal of the ACM*, Vol. 11, No. 1, pp. 84-97, January, 1964.

PROBLEMS

3.1 Consider a function $f(x_1,x_2,x_3)$ defined by the following primitive cubes: $x10 \mid 0$, $11x \mid 0$, $x0x \mid 1$, and $011 \mid 1$. Determine the value of f for the following input vectors: $10u$, $01u$, $u1u$, and $u01$.

3.2 The D F/F used in the circuit shown in Figure 3.50 is activated by a $0{\to}1$ transition on its C input. Assume that a sequence of three $0{\to}1{\to}0$ pulses is applied to CLOCK.

    a. Show that simulation with 3-valued logic fails to initialize this circuit.

    b. Show that simulation with multiple unknown values does initialize this circuit (denote the initial values of the F/Fs by u_1 and u_2).

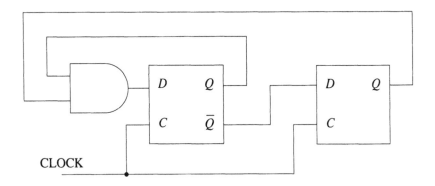

Figure 3.50

3.3 Use 3-valued compiled simulation to simulate the circuit of Figure 3.7 starting with $Q = u$ for the input vectors 01 and 11.

3.4 For 3-valued compiled simulation, someone may suggest using both 01 and 10 to represent the value u. The benefit of this dual encoding would be that the NOT(a) operation could be done just by complementing the 2-bit vector a (without bit swapping). Show that this encoding may lead to incorrect results.

3.5 Using the compiled simulation procedure outlined in Figure 3.9, simulate the latch given in Figure 3.11(a) for the vectors 00 and 11. Assume a model in which both Q and QN are treated as feedback lines (with equal delays).

3.6 Analyze the propagation of the pulse at A shown in Figure 3.14 for the following rise and fall delays of gate C: (1) $d_r = 3$, $d_f = 1$; (2) $d_r = 4$, $d_f = 1$; (3) $d_r = 5$, $d_f = 1$.

3.7 Explain why the (input or output) inertial delay of a gate cannot be greater than the smallest transport delay associated with the gate, i.e., $d_I \le \min \{d_r,d_f\}$.

3.8 Determine the response of the F/F of Figure 3.16 to the two sets of stimuli given in Figure 3.51. Assume $q = 0$, $D = 1$, $S = 1$.

3.9 How can a truth table of an OR gate with four inputs be used to evaluate an OR gate with a different number of inputs?

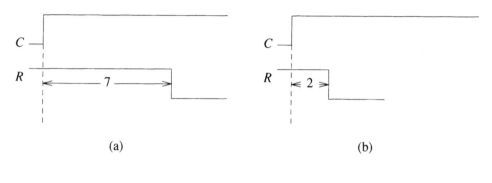

Figure 3.51

3.10 Determine a formula to compute the *utilization factor* of an array used to store the truth table of a function of n k-valued variables, where the value of a variable is coded with q bits (that is, the proportion of the array actually used).

3.11 Construct a zoom table for evaluating AND, OR, NAND, and NOR gates with two inputs with binary values.

3.12 Let us model an OR gate with two inputs as a sequential machine M whose state is given by the two counters, *c_count* and *u_count*, used to evaluate the gate via the input counting method. The input of M represents an input event of the OR gate, that is, a variable whose values are in the set $E = \{0{\to}1,\ 1{\to}0,\ u{\to}0,\ u{\to}1,\ 0{\to}u,\ 1{\to}u\}$. The output of M represents the result of evaluating the OR gate by a variable whose values are in the set $E \cup \Phi$, where Φ denotes no output event. Construct a state diagram for M, showing all possible state transitions during simulation. What is the initial state of M?

3.13 Outline evaluation routines for an XOR gate, based on: (1) input scanning, and (2) (a technique similar to) input counting.

3.14 Construct a truth table for a 2-input NAND gate for the 8-valued logic system of Figure 3.26.

3.15 Show that the existence of a static hazard is a necessary condition for the creation of a dynamic hazard.

3.16 Consider the circuit shown in Figure 3.52 and the input sequence shown. Calculate the output sequence at J and K using the 6-valued logic of Figure 3.24 and the 8-valued logic of Figure 3.26. Verify that the 8-valued logic system correctly identifies a dynamic hazard at J, while the 6-valued logic system produces the same result for lines J and K.

3.17 Show that for an asynchronous circuit with n feedback lines, Procedure 3.2 will simulate the circuit C at most $2n$ times.

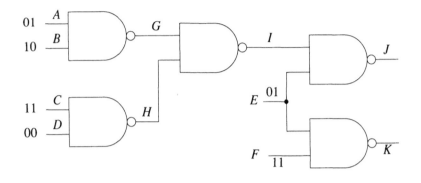

Figure 3.52

3.18 Consider the latch of Figure 3.53 and assume that, starting with the shown values, A changes to 1, then goes back to 0.

a. Simulate the circuit assuming an arbitrary delay model (apply Procedure 3.2).

b. Simulate the circuit assuming that all gates have 8 nsec delay and that A is 1 for 12 nsec.

c. If the results for a. and b. are different, explain why.

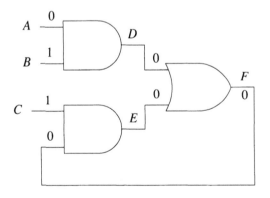

Figure 3.53

3.19 Regarding Algorithm 3.2 (Figure 3.31), it may appear that if $v(i) = lsv(i)$, i.e., the current and the last scheduled values of i are the same, then the gate i is "stable." Give a counterexample showing how i can have pending events when $v(i) = lsv(i)$. Will Algorithm 3.2 correctly handle such a case?

3.20 Assume that the largest possible gate delay is D. Discuss the advantages and disadvantages of using a timing wheel of size $M = D + 1$.

3.21 Extend the truth table of a bus with two inputs (given in Figure 3.37) to process all the output values of a bus driver shown in Figure 3.38.

3.22 Simulate the circuit of Figure 3.36(a), with $I_1 = u$, $E_1 = 0$, $I_2 = 0$, $E_2 = u$, $I_3 = 0$, $E_3 = 1$.

3.23 Simulate the circuit of Figure 3.41(a) for the following input vectors.

A	B	D	E	G
1	1	0	1	0
1	0	0	1	1
0	1	1	0	1
1	0	1	0	0

3.24 Simulate the latch shown in Figure 3.11(a) for the input sequence shown in Figure 3.54, assuming the following delays for the two gates:

a. $d = 2$

b. $d_l = 2$

c. $d_l = 1$

d. $d_r = 3$, $d_f = 1$

e. $d_r = 1$, $d_f = 3$

f. $d_m = 1$, $d_M = 2$

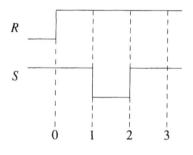

Figure 3.54

3.25 Determine the function of the circuit shown in Figure 3.55.

3.26 For the circuit of Figure 3.45(a), determine the relation between the width of a $1\rightarrow0\rightarrow1$ pulse at S and the transport delays of the gates in the circuit, necessary for the oscillation to occur.

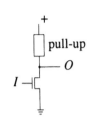

Figure 3.55

4. FAULT MODELING

About This Chapter

First we discuss the representation of physical faults by logical faults and the major concepts in logical fault modeling — namely explicit versus implicit faults, structural versus functional faults, permanent versus intermittent faults, and single versus multiple faults. We introduce the main types of structural faults — shorts and opens — and show how they are mapped into stuck and bridging faults. Next we deal with fault detection and redundancy, and the fault relations of equivalence and dominance. Then we discuss the single and the multiple stuck-fault models, and we present extensions of the stuck-fault model for modeling faults in RTL models. Bridging faults, functional faults, and technology-specific faults are treated in separate chapters.

4.1 Logical Fault Models

Logical faults represent the effect of physical faults on the behavior of the modeled system. (For studies of physical faults in integrated circuits, see [Case 1976, Partridge 1980, Timoc *et al.* 1983, Banerjee and Abraham 1984, Shen *et al.* 1985].) Since in modeling elements we differentiate between the logic function and timing, we also distinguish between **faults that affect the logic function** and **delay faults** that affect the operating speed of the system. In this chapter we will be mainly concerned with the former category.

What do we gain by modeling physical faults as logical faults? First, the problem of fault analysis becomes a logical rather than a physical problem; also its complexity is greatly reduced since many different physical faults may be modeled by the same logical fault. Second, some logical fault models are technology-independent in the sense that the same fault model is applicable to many technologies. Hence, testing and diagnosis methods developed for such a fault model remain valid despite changes in technology. And third, tests derived for logical faults may be used for physical faults whose effect on circuit behavior is not completely understood or is too complex to be analyzed [Hayes 1977].

A logical fault model can be explicit or implicit. An **explicit fault model** defines a fault universe in which every fault is individually identified and hence the faults to be analyzed can be explicitly enumerated. An explicit fault model is practical to the extent that the size of its fault universe is not prohibitively large. An **implicit fault model** defines a fault universe by collectively identifying the faults of interest — typically by defining their characterizing properties.

Given a logical fault and a model of a system, we should be able in principle to determine the logic function of the system in the presence of the fault. Thus, fault modeling is closely related to the type of modeling used for the system. Faults defined in conjunction with a structural model are referred to as **structural faults**; their effect is to modify the interconnections among components. **Functional faults** are defined in conjunction with a functional model; for example, the effect of a functional fault may be to change the truth table of a component or to inhibit an RTL operation.

Although intermittent and transient faults occur often, their modeling [Breuer 1973, Savir 1980] requires statistical data on their probability of occurrence. These data are needed to determine how many times an off-line testing experiment should be repeated to maximize the probability of detecting a fault that is only sometimes present in the circuit under test. Unfortunately, this type of data is usually not available. Intermittent and transient faults are better dealt with by on-line testing. In this chapter we will discuss only **permanent faults**.

Unless explicitly stated otherwise, we will always assume that we have at most one logical fault in the system. This simplifying **single-fault assumption** is justified by the *frequent testing strategy*, which states that we should test a system often enough so that the probability of more than one fault developing between two consecutive testing experiments is sufficiently small. Thus if maintenance intervals for a working system are too long, we are likely to encounter multiple faults. There are situations, however, in which frequent testing is not sufficient to avoid the occurrence of multiple faults. First, what may appear in the real system between two consecutive testing experiments is a physical fault, and some physical faults manifest themselves as multiple logical faults. This is especially true in high-density circuits, where many physical faults can affect an area containing several components. Second, in newly manufactured systems prior to their first testing, multiple faults are likely to exist. And third, if the testing experiment does not detect every single fault (which is usually the case), then the circuit may contain one of the undetected faults at any time, and the occurrence of a second single fault between two testing experiments creates a multiple fault. But even when multiple faults are present, the tests derived under the single-fault assumption are usually applicable for the detection of multiple faults, because, *in most cases, a multiple fault can be detected by the tests designed for the individual single faults that compose the multiple one.*

In general, *structural fault models assume that components are fault-free and only their interconnections are affected.* Typical faults affecting interconnections are shorts and opens. A **short** is formed by connecting points not intended to be connected, while an **open** results from the breaking of a connection. For example, in many technologies, a short between ground or power and a signal line can make the signal remain at a fixed voltage level. The corresponding logical fault consists of the signal being **stuck at** a fixed logic value v ($v \in \{0,1\}$), and it is denoted by $s\text{-}a\text{-}v$. A short between two signal lines usually creates a new logic function. The logical fault representing such a short is referred to as a **bridging fault**. According to the function introduced by a short we distinguish between AND bridging faults and OR bridging faults.

In many technologies, the effect of an open on a unidirectional signal line with only one fanout is to make the input that has become unconnected due to the open assume a constant logic value and hence appear as a stuck fault (see Figure 4.1(a)). This effect may also result from a physical fault internal to the component driving the line, and without probing the two endpoints of the line we cannot distinguish between the two cases. This distinction, however, is not needed in edge-pin testing, where we can assume that the entire signal line is stuck. Note how a single logical fault, namely the line i stuck at $a \in \{0,1\}$, can represent many totally different physical faults: i open, i shorted to power or ground, and any internal fault in the component driving i that keeps i at the logic value a.

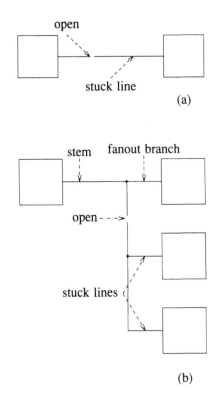

Figure 4.1   Stuck faults caused by opens (a) Single stuck fault (b) Multiple stuck fault

An open in a signal line with fanout may result in a multiple stuck fault involving a subset of its fanout branches, as illustrated in Figure 4.1(b). If we restrict ourselves to the single stuck-fault model, then we have to consider any single fanout branch stuck fault separately from the stem fault.

In the macro approach for hierarchical modeling, every component is expanded to its internal structural model. However, if components are individually tested before their assembly, then it may be enough to test only for faults affecting their interconnections. Then we do not want to consider the faults internal to components but only faults associated with their I/O pins. This restricted fault assumption is referred to as the *pin-fault model*.

4.2 Fault Detection and Redundancy

4.2.1 Combinational Circuits

Let $Z(x)$ be the logic function of a combinational circuit N, where x represents an arbitrary input vector and $Z(x)$ denotes the mapping realized by N. We will denote by t a specific input vector, and by $Z(t)$ the response of N to t. For a multiple-output circuit $Z(t)$ is also a vector. The presence of a fault f transforms N into a new circuit N_f. Here we assume that N_f is a combinational circuit with function $Z_f(x)$. The circuit is tested by applying a sequence T of test vectors $t_1, t_2, ..., t_m$, and by comparing the obtained output response with the (expected) output response of N, $Z(t_1), Z(t_2), ..., Z(t_m)$.

Definition 4.1: A test (vector) t **detects** a fault f iff $Z_f(t) \neq Z(t)$.

Note that the tests in the sequence T may be applied in any order; therefore, for a combinational circuit we will refer to T as a *set of tests*. Note that Definition 4.1 does not apply if N_f is a sequential circuit. We also emphasize that this definition of detection assumes edge-pin testing with full comparison of the results; other definitions apply for other forms of testing, such as compact testing techniques, which will be described in a separate chapter.

Example 4.1: In the circuit of Figure 4.2, let f be the OR bridging fault between x_1 and x_2. This fault changes the functions realized by the two outputs to $Z_{1f} = x_1 + x_2$ (instead of $Z_1 = x_1 x_2$) and to $Z_{2f} = (x_1 + x_2)x_3$ (instead of $Z_2 = x_2 x_3$). The test 011* detects f because $Z(011) = 01$ while $Z_f(011) = 11$.   □

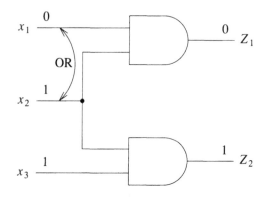

Figure 4.2

For a single-output circuit, a test t that detects a fault f makes $Z(t)=0$ and $Z_f(t)=1$ or vice versa. Thus, the set of all tests that detect f is given by the solutions of the equation

$$Z(x) \oplus Z_f(x) = 1 \qquad (4.1)$$

where the symbol $\oplus$ denotes the exclusive-OR operation.

---

*   The order of the bits is always $x_1, x_2, ..., x_n$ (or alphabetical order if the names $A, B, C...$ are used).

Example 4.2: The function realized by the circuit of Figure 4.3(a) is $Z = (x_2+x_3)x_1 + \bar{x}_1x_4$. Let f be x_4 s-a-0. In the presence of f the function becomes $Z_f = (x_2+x_3)x_1$, and equation (4.1) reduces to $\bar{x}_1x_4 = 1$. Thus any test in which $x_1 = 0$ and $x_4 = 1$ is a test for f. The expression $\bar{x}_1x_4$ represents, in compact form, any of the four tests (0001, 0011, 0101, 0111) that detect f.   □

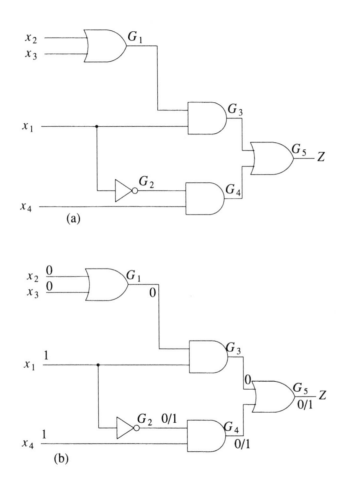

Figure 4.3

Sensitization

Let us simulate the circuit of Figure 4.3(a) for the test $t = 1001$, both without and with the fault G_2 s-a-1 present. The results of these two simulations are shown in Figure 4.3(b). The results that are different in the two cases have the form v/v_f, where v and v_f are corresponding signal values in the fault-free and in the faulty circuit. The fault is detected since the output values in the two cases are different.

Figure 4.3(b) illustrates two basic concepts in fault detection. First, a test t that detects a fault f **activates** f, i.e., **generates an error** (or a **fault effect**) by creating different v and v_f values at the site of the fault. Second, t **propagates the error** to a primary output w, that is, makes all the lines along at least one path between the fault site and w have different v and v_f values. In Figure 4.3(b) the error propagates along the path (G_2, G_4, G_5). (Sometimes the term "fault propagation" is used instead of "error propagation" or "fault-effect propagation.") A line whose value in the test t changes in the presence of the fault f is said to be **sensitized** *to the fault f by the test t*. A path composed of sensitized lines is called a **sensitized path**.

A gate whose output is sensitized to a fault f has at least one of its inputs sensitized to f as well. The following lemma summarizes the properties of such a gate.

Lemma 4.1: Let G be a gate with inversion i and controlling value c, whose output is sensitized to a fault f (by a test t).

1. All inputs of G sensitized to f have the same value (say, a).

2. All inputs of G not sensitized to f (if any) have value $\bar{c}$.

3. The output of G has value $a \oplus i$.

Proof

1. Let us assume, by contradiction, that there are two sensitized inputs of G, k, and l, which have different values. Then one of them (say, k) has the controlling value of the gate. In the presence of the fault f, both k and l change value and then l has the controlling value. But this means that G has the same value independent of f, which contradicts the assumption that the gate output is sensitized to f. Therefore all the sensitized inputs of G must have the same value (say, a).

2. The output of G cannot change in the presence of f if one of its inputs that is not sensitized to f has value c. Hence all inputs of G not sensitized to f (if any) must have value $\bar{c}$.

3. If $a = c$, the gate output has value $c \oplus i$. If $a = \bar{c}$, then all inputs of G have value $\bar{c}$, and the gate output has value $\bar{c} \oplus i$. Hence, in both cases the gate output has value $a \oplus i$.   □

The value $\bar{c}$ is referred to as an *enabling value*, since it enables error propagation. Thus a NAND gate that satisfies Lemma 4.1 has either all inputs (sensitized or not) set to 1 or all sensitized inputs set to 0 and the remaining inputs (if any) set to the enabling value 1 (see Figure 4.4).

By repeatedly applying Part 3 of the above lemma, we can derive the following properties of sensitized paths.

Corollary 4.1: Let j be a line sensitized to the fault l *s-a-v* (by a test t), and let p be the inversion parity of a sensitized path between l and j.

1. The value of j in t is $\bar{v} \oplus p$.

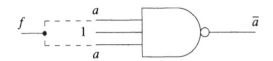

Figure 4.4   NAND gate satisfying Lemma 4.1 ($c=0$, $i=1$)

2. If there are several sensitized paths between l and j, then all of them have the same inversion parity.   □

Detectability

A fault f is said to be **detectable** if there exists a test t that detects f; otherwise, f is an *undetectable* fault. For an undetectable fault f, $Z_f(x) = Z(x)$ and no test can simultaneously activate f and create a sensitized path to a primary output. In the circuit of Figure 4.5(a) the fault a *s-a*-1 is undetectable. Since undetectable faults do not change the function of the circuit, it may appear that they are harmless and hence can be ignored. However, a circuit with an undetectable fault may invalidate the single-fault assumption. Recall that based on the frequent testing strategy, we assume that we can detect one fault before a second one occurs. But this may not be possible if the first fault is undetectable.

When generating a set of tests for a circuit, a typical goal is to produce a *complete detection test set*, that is, a set of tests that detect any detectable fault. However, a complete test set may not be sufficient to detect all detectable faults if an undetectable one is present in the circuit [Friedman 1967].

Example 4.3: Figure 4.5(a) shows how the fault b *s-a*-0 is detected by $t = 1101$. Figure 4.5(b) shows that b *s-a*-0 is no longer detected by the test t if the undetectable fault a *s-a*-1 is also present. Thus, if t is the only test that detects b *s-a*-0 in a complete detection test set T, then T is no longer complete in the presence of a *s-a*-1.□

The situation in which the presence of an undetectable fault prevents the detection of another fault by a certain test is not limited to faults of the same category; for example, an undetectable bridging fault can similarly invalidate a complete test set for stuck faults.

Example 4.4: [Kodandapani and Pradhan 1980]. Consider the circuit of Figure 4.6(a) that realizes the function $xy + \bar{x}z$. The OR bridging fault between y and $\bar{x}$ is undetectable, since the function realized in the presence of the fault is $xy + yz + z\bar{x} = xy + z\bar{x}$. Figure 4.6 shows how the test 111 detects the fault q *s-a*-0 but no longer does so in the presence of the bridging fault. The test set $T = \{111, 010, 001, 101\}$ is a complete detection test set for single stuck faults, and 111 is the only test in T that detects q *s-a*-0. Hence T is no longer complete in the presence of the undetectable bridging fault.   □

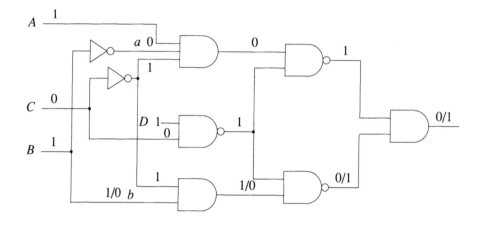

(a)

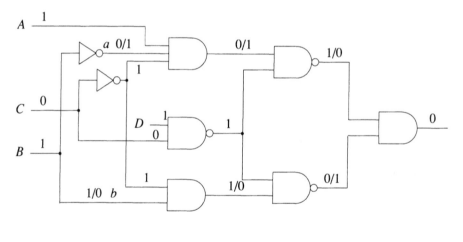

(b)

Figure 4.5

Redundancy

A combinational circuit that contains an undetectable stuck fault is said to be
redundant, since such a circuit can always be simplified by removing at least one gate
or gate input. For instance, suppose that a $s\text{-}a\text{-}1$ fault on an input of an AND gate G is
undetectable. Since the function of the circuit does not change in the presence of the
fault, we can permanently place a 1 value on that input. But an n-input AND with a
constant 1 value on one input is logically equivalent to the $(n-1)$-input AND obtained
by removing the gate input with the constant signal. Similarly, if an AND input $s\text{-}a\text{-}0$
is undetectable, the AND gate can be removed and replaced by a 0 signal. Since now

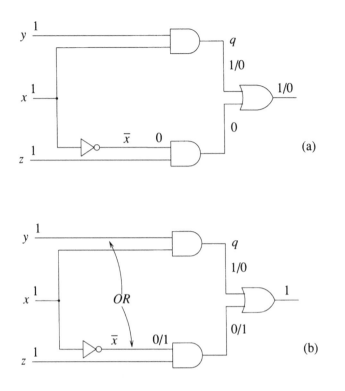

Figure 4.6

we have a new signal with a constant value, the simplification process may continue. The simplification rules are summarized in the following table.

Undetectable fault	Simplification rule
AND(NAND) input s-a-1	Remove input
AND(NAND) input s-a-0	Remove gate, replace by 0(1)
OR(NOR) input s-a-0	Remove input
OR(NOR) input s-a-1	Remove gate, replace by 1(0)

The general concept of redundancy is broader than the particular one related to the existence of undetectable stuck faults, and it denotes a circuit that can be simplified. One possible simplification, not covered by the rules given above, is to replace a string of two inverters by a single line. A general type of redundancy [Hayes 1976] exists in a circuit when it is possible to cut a set of r lines and to connect $q \leq r$ of the cut lines to some other signals in the circuit without changing its function. In the following, we will restrict ourselves to the definition of redundancy related to undetectable stuck faults. The term "redundant" is also applied to undetectable stuck faults and to the lines that can be removed. A combinational circuit in which all stuck faults are detectable is said to be *irredundant*.

Redundancy does not necessarily denote an inefficient or undesirable implementation of a function. For example, triple modular redundancy (TMR) is a basic technique used in fault-tolerant design. For the TMR configuration shown in Figure 4.7, any fault occurring in one of the identical modules A will be masked by the two other fault-free modules because of the majority voter circuit M. But this also makes the TMR circuit almost untestable by off-line testing. The solution is to remove the redundancy for off-line testing.

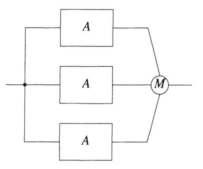

Figure 4.7   TMR configuration

Redundancy may also be introduced in a circuit to avoid hazards. This design technique is illustrated by the circuit of Figure 4.8, which implements the function $z = ab + bc + \bar{a}c = ab + \bar{a}c$. Thus the gate Y that produces the term bc is not logically required. Without this gate, however, the circuit would have a static hazard, as a spurious 0-pulse may appear at the output when the input vector changes from 111 to 011. The role of gate Y is to keep the output constant at 1 during this transition. But the fault Y s-a-0 is undetectable.*

We noted how the presence of a redundant fault may invalidate a complete test set. Other problems that can arise in redundant circuits include the following [Friedman 1967]:

1. If f is a detectable fault and g is an undetectable fault, then f may become undetectable in the presence of g (see Problem 4.6). Such a fault f is called a *second-generation redundant fault*.

2. Two undetectable single faults f and g may become detectable if simultaneously present in the circuit (see Problem 4.7). In other words, the multiple fault $\{f,g\}$ may be detectable even if its single-fault components are not.

---

*   The fault Y s-a-0 could be detected by observing the spurious 0-pulse that the term bc is used to prevent. However, we assume a testing experiment in which only the static output values are observed. Moreover, the occurrence of the spurious pulse depends on the delays of the circuit.

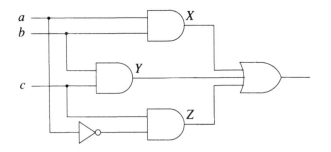

Figure 4.8

Note that in practice, when we deal with large combinational circuits, even irredundant circuits may not be tested with complete detection test sets. The reason is that generating tests for some faults may consume too much time, and all practical test generation programs are set up to stop the test generation process for a fault when it becomes too costly. *In a practical sense, there is no difference between an undetectable fault f and a detectable one g that is not detected by an applied test set.* Clearly, g could be present in the circuit and hence invalidate the single-fault assumption.

Identifying redundancy is closely related to the problem of test generation. To show that a line is redundant means to prove that no test exists for the corresponding fault. The test generation problem belongs to a class of computationally difficult problems, referred to as *NP-complete* [Ibarra and Sahni 1975]. The traveling salesman problem is one famous member of this class [Horowitz and Sahni 1978]. Let n be the "size" of the problem. For the traveling salesman problem n is the number of cities to be visited; for test generation n is the number of gates in a circuit. An important question is whether there exists an algorithm that can solve any instance of a problem of size n using a number of operations proportional to n^r, where r is a finite constant. At present, no such polynomial-time algorithm is known for any *NP*-complete problem. These problems are related in the sense that either all of them or none of them can be solved by polynomial-time algorithms.

Although test generation (and identification of redundancy) is a computationally difficult problem, practical test generation algorithms usually run in polynomial time. The fact that the test generation problem is *NP*-complete means that polynomial time cannot be achieved in all instances, that is, any test generation algorithm may encounter a circuit with a fault that cannot be solved in polynomial time. Experience has shown that redundant faults are usually the ones that cause test generation algorithms to exhibit their worst-case behavior.

4.2.2 Sequential Circuits

Testing sequential circuits is considerably more difficult than testing combinational circuits. To detect a fault a test sequence is usually required, rather than a single input vector, and the response of a sequential circuit is a function of its initial state.

We will first illustrate a general form of fault detection for a sequential circuit. Let T be a test sequence and $R(q,T)$ be the response to T of a sequential circuit N starting in the initial state q. Now consider the circuit N_f obtained in the presence of the fault f. Similarly we denote by $R_f(q_f,T)$ the response of N_f to T starting in the initial state q_f.

Definition 4.2: A test sequence T **strongly detects** the fault f iff the output sequences $R(q,T)$ and $R_f(q_f,T)$ are different for every possible pair of initial states q and q_f.

Example 4.5: Figure 4.9 shows a synchronous circuit, its state table, and the output sequences obtained in response to the input sequence $T = 10111$, from the fault-free circuit and the circuits with the single faults α (line a s-a-1) and β (line b s-a-0).

Since all sequences generated in the presence of β are different from those of the fault-free circuit, T strongly detects β. Since the sequences of the fault-free circuit in initial state B and the circuit with fault α in initial state B are identical, T does not strongly detect α. □

In this example, although T strongly detects β, the error symptom cannot be simply specified. That is, we cannot say that at some point in the output sequence the normal circuit will have a 1 output and the faulty circuit a 0 output or vice versa. Instead, we must list all possible responses of the normal and faulty machines. This is obviously not practical, as a circuit with n memory elements may have 2^n possible initial states. Moreover, we have to consider how the tester operates. A tester performing edge-pin testing with full comparison of the output results compares the obtained and the expected output sequences on a vector-by-vector basis. Hence, the expected output response must be known in advance. Practically feasible decisions on fault location and practical test generation algorithms cannot use ambiguity of the type "detection in either vector i or j." Thus the response R_f must also be predictable. Therefore, we will use the following (less general) concept of detection.

Definition 4.3: A test sequence T **detects** the fault f iff, for every possible pair of initial states q and q_f, the output sequences $R(q,T)$ and $R_f(q_f,T)$ are different for some specified vector $t_i \in T$.

To determine the vector t_i when an error caused by f can be observed at the primary outputs, independent of the initial states q and q_f, the testing experiment is usually divided into two distinct phases. In the first phase we apply an *initialization sequence* T_I such that at the end of T_I both N and N_f are brought to known states q_I and q_{If}. The output responses are ignored during T_I since they are not predictable. In the second phase we apply a sequence T'. Now both the expected response $R(q_I,T')$ and the faulty response $R_f(q_{If},T')$ are predictable. Usually t_i is taken as the first vector of T' for which an error is observed.

This type of testing is based on the fundamental assumption that such an initialization sequence T_I exists. Almost any circuit used in practice has an initialization sequence, which is employed to start its operation from a known state. Circuits are usually designed so that they can be easily initialized. An often used technique is to provide a common reset (or preset) line to every F/F. Then a single vector is needed to initialize

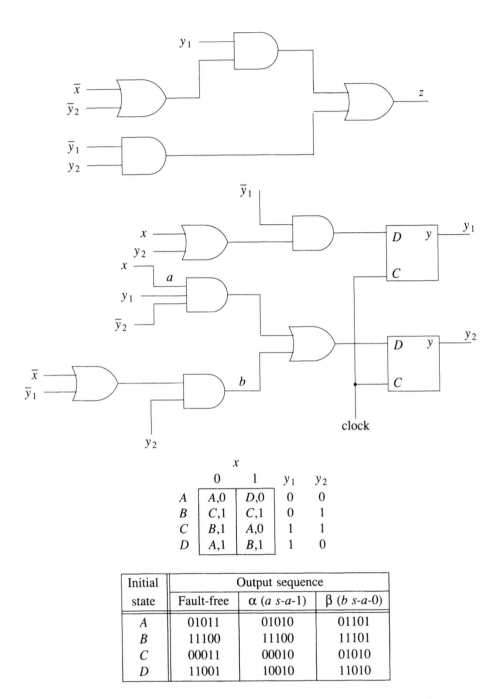

Figure 4.9 Output sequences as a function of initial state and fault

the circuit. However, an initialization sequence for the fault-free circuit N may fail to initialize some faulty circuit N_f. Such a fault f is said to *prevent initialization*.

Example 4.6: Consider the D F/F of Figure 4.10 (this is a typical configuration for a 1-bit counter) and the fault R s-a-1. While the fault-free circuit can be simply initialized by $R=0$, there is no way the faulty circuit can be initialized to a known state.

$\square$

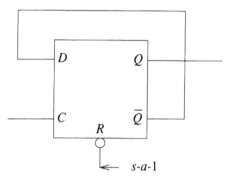

Figure 4.10   Example of a fault preventing initialization

One may derive a test sequence for a fault that prevents initialization by separately analyzing every possible initial state in the presence of that fault, but, in general, such an approach is impractical. Therefore faults that prevent initialization should be considered undetectable by edge-pin testing with full output comparison. However, unlike in a combinational circuit, this does not mean that the circuit is redundant [Abramovici and Breuer 1979]. Also it should be pointed out that a fault that prevents initialization may be detected by other types of testing, such as compact testing (to be discussed in a separate chapter).

4.3 Fault Equivalence and Fault Location

4.3.1 Combinational Circuits

Definition 4.4: Two faults f and g are said to be **functionally equivalent** iff $Z_f(x) = Z_g(x)$.

A test t is said to **distinguish** between two faults f and g if $Z_f(t) \neq Z_g(t)$; such faults are *distinguishable*. There is no test that can distinguish between two functionally equivalent faults. The relation of functional equivalence partitions the set of all possible faults into *functional equivalence classes*. For fault analysis it is sufficient to consider only one representative fault from every equivalence class.

For a single-output circuit, a test t that distinguishes between f and g makes $Z_f(t) = 0$ and $Z_g(t) = 1$ or vice versa. Thus, the set of all tests that distinguish between f and g is given by the solutions of the equation

$$Z_f(x) \oplus Z_g(x) = 1$$

Note that the faults f and g in Definition 4.4 are not restricted to the same fault universe. For example, in the circuit in Figure 4.11 the AND bridging fault between x and y is functionally equivalent to the stuck fault z s-a-0. Also an AND bridging fault between two complementary signals q and $\bar{q}$ is functionally equivalent to the multiple stuck fault $\{q$ s-a-0, $\bar{q}$ s-a-$0\}$. In general, however, we will limit the analysis of equivalence relations to those among faults of the same type.

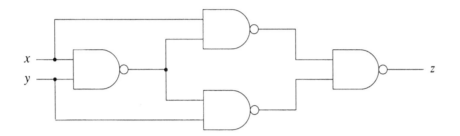

Figure 4.11

With any n-input gate we can associate $2(n+1)$ single stuck faults. For a NAND gate all the input s-a-0 faults and the output s-a-1 are functionally equivalent. In general, for a gate with controlling value c and inversion i, all the input s-a-c faults and the output s-a-$(c\oplus i)$ are functionally equivalent. Thus for an n-input gate ($n>1$) we need to consider only $n+2$ single stuck faults. This type of reduction of the set of faults to be analyzed based on equivalence relations is called **equivalence fault collapsing**. An example is shown in Figure 4.12(a) and (b), where a black (white) dot denotes a s-a-1 (s-a-0) fault.

If in addition to fault detection, the goal of testing is *fault location* as well, we need to apply a test that not only detects the detectable faults but also distinguishes among them as much as possible. A *complete location test set* distinguishes between every pair of distinguishable faults in a circuit.

It is convenient to consider that a fault-free circuit contains an *empty fault* denoted by Φ. Then $Z_\Phi(x) = Z(x)$. This artifact helps in understanding the relation between fault location and fault detection. Namely, fault detection is just a particular case of fault location, since a test that detects a fault f distinguishes between f and Φ. Undetectable faults are in the same equivalence class with Φ. Hence, *a complete location test set must include a complete detection test set*.

The presence of an undetectable fault may invalidate a complete location test set. If f and g are two distinguishable faults, they may become functionally equivalent in the presence of an undetectable fault (see Problem 4.8).

A complete location test set can diagnose a fault to within a functional equivalence class. This represents the maximal diagnostic resolution that can be achieved by edge-pin testing. In practice, large circuits are tested with test sets that are not

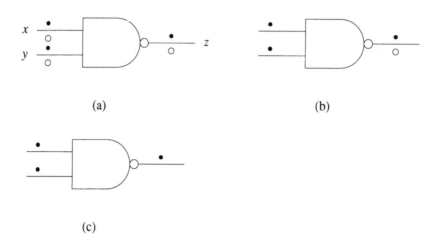

(a)   (b)

(c)

Figure 4.12   (a) NAND gate with uncollapsed fault set (b) Equivalence fault
collapsing (c) Dominance fault collapsing

complete.   Another equivalence relation can be used to characterize the resolution
achieved by an arbitrary test set.

Definition 4.5: Two faults f and g are **functionally equivalent under a test set** T iff
$Z_f(t) = Z_g(t)$ for every test $t \in T$.

Functional equivalence implies equivalence under any test set, but equivalence under a
given test set does not imply functional equivalence (see Problem 4.9).

4.3.2 Sequential Circuits

Definition 4.6: Two faults f and g are said to be **strongly functionally equivalent** iff
the corresponding sequential circuits N_f and N_g have equivalent state tables. (For state
table equivalence refer to [Friedman and Menon 1975].)

Like the concept of strong detection, strong equivalence cannot be used in practice.
The practical definition of functional equivalence is based on the responses of N_f and
N_g to a test sequence T. We assume the same type of testing experiment as discussed
for detection; namely, T contains first an initialization sequence T_I that brings N_f and
N_g to known states q_{If} and q_{Ig}. The output responses are not monitored during the
application of T_I since they are not predictable. Let T' be the sequence applied after
T_I.

Definition 4.7: Two faults f and g are said to be **functionally equivalent** iff
$R_f(q_{If}, T') = R_g(q_{Ig}, T')$ for any T'.

Again, this definition does not apply to faults preventing initialization.

Similarly, functional equivalence under a test sequence $T = \{T_I, T'\}$ means that
$R_f(q_{If}, T') = R_g(q_{Ig}, T')$.

4.4 Fault Dominance

4.4.1 Combinational Circuits

If the objective of a testing experiment is limited to fault detection only, then, in addition to fault equivalence, another fault relation can be used to reduce the number of faults that must be considered.

Definition 4.8: Let T_g be the set of all tests that detect a fault g. We say that a fault f **dominates** the fault g iff f and g are functionally equivalent under T_g.

In other words, if f dominates g, then any test t that detects g, i.e., $Z_g(t) \neq Z(t)$, will also detect f (on the same primary outputs) because $Z_f(t) = Z_g(t)$. Therefore, for fault detection it is unnecessary to consider the dominating fault f, since by deriving a test to detect g we automatically obtain a test that detects f as well (see Figure 4.13).

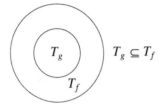

Figure 4.13   The sets T_f and T_g when f dominates g

For example, consider the NAND gate of Figure 4.12 and let g be y s-a-1 and f be z s-a-0. The set T_g consists of only one test, namely 10, and this also detects f. Then z s-a-0 dominates y s-a-1. In general, for a gate with controlling value c and inversion i, the output $s-a-\overline{(c \oplus i)}$ fault dominates any input $s-a-\overline{c}$ fault. Then the output fault can be removed from the set of faults we consider for test generation (see Figure 4.12(c)). This type of reduction of the set of faults to be analyzed based on dominance relations is called **dominance fault collapsing**.

It is interesting to observe that we can have two faults, f and g, such that any test that detects g also detects f (i.e., $T_g \subseteq T_f$), without f dominating g. Consider, for example, the circuit of Figure 4.14. Let f be z_2 s-a-0 and g be y_1 s-a-1. The set T_g consists only of the test 10, which also detects f (on a different primary output). But according to Definition 4.8, f does not dominate g. Although for fault detection it is not necessary to consider f, it would be difficult to determine this fact from an analysis of the circuit.

When choosing a fault model it is important to select one whose faults are generally dominated by faults of other fault models, because a test set detecting the faults of the chosen model will also detect many other faults that are not even explicitly considered. The best fault model with such a property appears to be the single stuck-fault model.

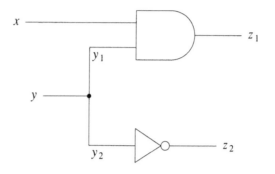

Figure 4.14

4.4.2 Sequential Circuits

Although the concept of dominance can be extended to sequential circuits, its applicability in practice is difficult, since dominance relations in a combinational circuit N may not remain valid when N is embedded in a sequential circuit. The following example illustrates this phenomenon.

Example 4.7: Consider the circuit of Figure 4.15. Assume that initially $y=1$ and consider the test sequence shown in the figure. For the fault-free circuit the output sequence generated is 0000. For the fault α (x_2 *s-a*-1) the first input fails to reset y and the fourth input propagates this error to z. Thus the generated output sequence is 0001 and x_2 *s-a*-1 is detected. Now consider the same test sequence and the fault β (G_1 *s-a*-0), which dominates α in a combinational circuit sense. The first input again fails to reset y. However, the fourth input generates an erroneous 0 at G_2 and the two effects of this single fault — the one stored in the F/F and the one propagating along the path (G_1, G_2) — cancel each other, and the fault is not detected. $\quad\square$

While equivalence fault-collapsing techniques for combinational circuits remain valid for sequential circuits, dominance fault-collapsing techniques are no longer applicable.

4.5 The Single Stuck-Fault Model

The single-stuck fault (SSF) model is also referred to as the *classical* or *standard* fault model because it has been the first and the most widely studied and used. Although its validity is not universal, its usefulness results from the following attributes:

- It represents many different physical faults (see, for example, [Timoc *et al.* 1983]).

- It is independent of technology, as the concept of a signal line being stuck at a logic value can be applied to any structural model.

- Experience has shown that tests that detect SSFs detect many nonclassical faults as well.

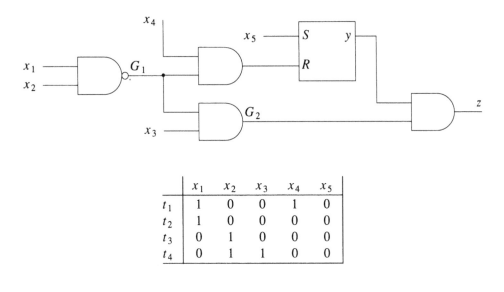

	x_1	x_2	x_3	x_4	x_5
t_1	1	0	0	1	0
t_2	1	0	0	0	0
t_3	0	1	0	0	0
t_4	0	1	1	0	0

Figure 4.15

- Compared to other fault models, the number of SSFs in a circuit is small. Moreover, the number of faults to be explicitly analyzed can be reduced by fault-collapsing techniques.

- SSFs can be used to model other type of faults.

The last point is illustrated in Figure 4.16. To model a fault that changes the behavior of the signal line z, we add to the original circuit a multiplexor (selector) that realizes the function

$$z' = z \quad \text{if } f = 0$$
$$z' = z_f \quad \text{if } f = 1$$

The new circuit operates identically to the original circuit for $f=0$ and can realize any faulty function z_f by inserting the fault f s-a-1. For example, connecting x to z_f would create the effect of a functional fault that changes the function of the inverter from $z=\bar{x}$ to $z=x$. Connecting x to z_f via an inverter with a different delay would create the effect of a delay fault. Although flexible, this approach to nonclassical fault modeling is limited by the need to increase significantly the size of the model.

When considering an explicit fault model it is important to know how large is the fault universe it defines. If there are n lines on which SSFs can be defined, the number of possible faults is $2n$. To determine n we have to consider every fanout branch as a separate line. We will do this computation for a gate-level model. Every signal source i (gate output or primary input) with a fanout count of f_i contributes k_i lines to the number of possible fault sites, where k_i is defined as follows:

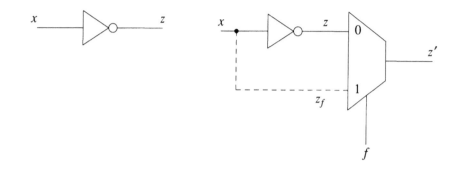

Figure 4.16 Model modification for nonclassical fault modeling

$$k_i = \begin{cases} 1 & \text{if } f_i = 1 \\ 1+f_i & \text{if } f_i > 1 \end{cases} \tag{4.2}$$

The number of possible fault sites is

$$n = \sum_i k_i \tag{4.3}$$

where the summation is over all the signal sources in the circuit. Let us define a variable q_i by

$$q_i = \begin{cases} 1 & \text{if } f_i = 1 \\ 0 & \text{if } f_i > 1 \end{cases} \tag{4.4}$$

Then (4.3) becomes

$$n = \sum_i (1 + f_i - q_i) \tag{4.5}$$

Let us denote the number of gates by G and the number of primary inputs by I. The average fanout count in the circuit is given by

$$f = \frac{\sum\limits_i f_i}{G + I} \tag{4.6}$$

The fraction of signal sources with only one fanout can be expressed as

$$q = \frac{\sum\limits_i q_i}{G + I} \tag{4.7}$$

With this notation we obtain from (4.5)

$$n = (G + I)(1 + f - q) \tag{4.8}$$

(Just as a check, apply (4.8) to a fanout-free circuit). In a large circuit, usually $G \gg I$ and $q > 0.5$, so the dominating factor in (4.8) is Gf. Thus we can conclude that the number of SSFs is slightly larger than $2Gf$. It is important to observe that it *depends on both the gate count and on the average fanout count.*

In the previous section we noted that the number of faults to be analyzed can be reduced by fault collapsing based on equivalence and dominance relations. Functional equivalence relations, however, cannot be directly applied for this purpose, because determining whether two arbitrary faults are functionally equivalent is an *NP*-complete problem [Goundan 1978]. For example, there is no simple way to determine that the faults c *s-a-*1 and d *s-a-*1 in Figure 4.17 are functionally equivalent. (We can do this by computing the two faulty functions and showing that they are identical.)

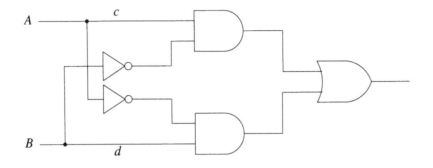

Figure 4.17

What we can do is to determine equivalent faults that are structurally related. The relation on which this analysis relies is called **structural equivalence**, and it is defined as follows. In the circuit N_f, the presence of the stuck fault f creates a set of lines with constant values. By removing all these lines (except primary outputs) we obtain a simplified circuit $S(N_f)$, which realizes the same function as N_f. Two faults f and g are said to be structurally equivalent if the corresponding simplified circuits $S(N_f)$ and $S(N_g)$ are identical. An example is shown in Figure 4.18. Obviously, structurally equivalent faults are also functionally equivalent. But, as illustrated by c *s-a-*1 and d *s-a-*1 in Figure 4.17, there exist functionally equivalent faults that are not structurally equivalent. The existence of such faults is related to the presence of reconvergent fanout [McCluskey and Clegg 1971] (see Problem 4.13).

The advantage of structural equivalence relations is that they allow fault collapsing to be done as a simple *local analysis* based on the structure of the circuit, while functional equivalence relations imply a global analysis based on the function of the circuit. Figure 4.19 illustrates the process of fault collapsing based on structural equivalence relations. Conceptually, we start by inserting *s-a-*1 and *s-a-*0 faults on every signal source (gate output or primary input) and destination (gate input). This is shown in Figure 4.19(a), where a black (white) dot denotes a *s-a-*1 (*s-a-*0) fault. Then we traverse the circuit and construct structural equivalence classes along the way. For a signal line with a fanout count of 1, the faults inserted at its source are structurally equivalent to the corresponding faults at its destination. For a gate with controlling value c and inversion i, any *s-a-c* input fault is structurally equivalent to the output *s-a-*$(c \oplus i)$. A *s-a-*0 (*s-a-*1) input fault of an inverter is structurally equivalent to the

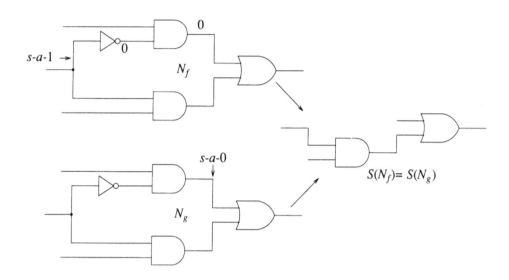

Figure 4.18 Illustration for structural fault equivalence

output $s\text{-}a\text{-}1$ ($s\text{-}a\text{-}0$). Finally, from every equivalence class we retain one fault as representative (Figure 4.19(b)).

A structural equivalence class determined in this way is confined to a fanout-free region of the circuit. The reason is that in general, a stem $s\text{-}a\text{-}v$ fault is not functionally equivalent with a $s\text{-}a\text{-}v$ fault on any of its fanout branches. Reconvergent fanout, however, may create structurally equivalent faults in different fanout-free regions; Figure 4.20 illustrates such a case. Thus our method of equivalence fault collapsing will not identify faults such as b $s\text{-}a\text{-}0$ and f $s\text{-}a\text{-}0$ as structurally equivalent. In such a case, the obtained structural equivalence classes are not maximal; i.e., there are at least two classes that can be further merged. However, the potential gain achievable by extending the fault collapsing across fanout-free regions is small and does not justify the cost of the additional analysis required.

Although the equivalence classes obtained by structural equivalence fault collapsing are not maximal, this process nevertheless achieves a substantial reduction of the initial set of $2n$ faults, where n is given by (4.8). For every gate j with g_j inputs, it eliminates m_j input faults, where

$$m_j = \begin{cases} g_j & \text{if } g_j > 1 \\ 2 & \text{if } g_j = 1 \end{cases} \tag{4.9}$$

The number of faults eliminated is

$$m = \sum_j m_j \tag{4.10}$$

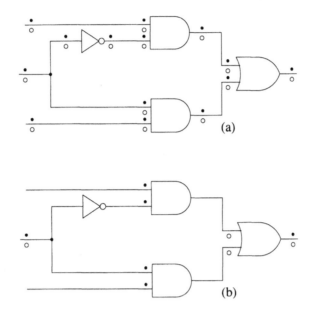

Figure 4.19   Example of structural equivalence fault collapsing

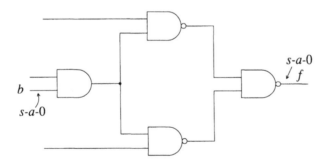

Figure 4.20

where the summation is over all the gates in the circuit. Let us define a variable p_j by

$$p_j = \begin{cases} 0 & \text{if } g_j > 1 \\ 1 & \text{if } g_j = 1 \end{cases} \qquad (4.11)$$

Then (4.10) becomes

$$m = \sum_j (g_j + p_j) \qquad (4.12)$$

The average input count in the circuit is given by

$$g = \frac{\sum_j g_j}{G} \qquad (4.13)$$

The fraction of the gates with only one input can be expressed as

$$p = \frac{\sum_j p_j}{G} \qquad (4.14)$$

With this notation we obtain from (4.12)

$$m = G(g + p) \qquad (4.15)$$

Since the average input count g is nearly equal to the average fanout count f, it follows that more than Gf faults are eliminated. Thus structural equivalence fault collapsing reduces the initial set of faults by about 50 percent.

If we are interested only in fault detection, we can also do dominance fault collapsing. This process is based on the following two theorems.

Theorem 4.1: In a fanout-free combinational circuit C, any test set that detects all SSFs on the primary inputs of C detects all SSFs in C.

Proof: Assume, by contradiction, that a test set T detects all SSFs on the primary inputs of C but does not detect all internal SSFs. Because every line in C has only one fanout, it is sufficient to consider internal faults only on gate outputs. Since T detects all SSFs on the primary inputs, there must be some gate G in C such that T detects all faults in the circuit feeding G but does not detect some output fault f. First assume G is an AND gate. Then f cannot be the s-a-0 fault, since this is equivalent to any input s-a-0. Also f cannot be the s-a-1 fault, since this dominates any input s-a-1. Hence such a fault f does not exist. A similar argument holds if G is an OR, NAND, or NOR gate. Therefore all the internal faults in C are detected. □

A similar proof mechanism can be used for the following theorem.

Theorem 4.2: In a combinational circuit C any test set that detects all SSFs on the primary inputs and the fanout branches of C detects all SSFs in C. □

The primary inputs and the fanout branches are called *checkpoints*. The number of checkpoints is

$$r = I + \sum_i (f_i - q_i) = I + (G+I)(f - q) \qquad (4.16)$$

Thus dominance fault collapsing reduces the number of faults dealt with from $2n$ to $2r$. Usually the number of fanout branches — $(G+I)(f-q)$ — is much larger than the number I of primary inputs, so we can approximate

$$r \simeq (G+I)(f - q) \qquad (4.17)$$

The fraction of faults eliminated is

$$1 - \frac{2r}{2n} = 1 - \frac{f - q}{1 + f - q} = \frac{1}{1 + f - q} \tag{4.18}$$

For example, for typical values such as $f = 2.5$ and $q = 0.7$, we eliminate about 40 percent of the faults. The set of checkpoint faults can be further collapsed by using structural equivalence and dominance relations, as illustrated in the following example.

Example 4.8: The circuit of Figure 4.21 has 24 SSFs and 14 checkpoint faults (10 on the primary inputs — a, b, c, d, e — and 4 on the fanout branches g and h). Since a s-a-0 and b s-a-0 are equivalent, we can delete the latter. Similarly, we can delete d s-a-0, which is equivalent to h s-a-0. The fault g s-a-1 is equivalent to f s-a-1, which dominates a s-a-1. Therefore, g s-a-1 can be eliminated. Similarly, e s-a-1 is equivalent to i s-a-1, which dominates h s-a-1; hence e s-a-1 can be eliminated. The original set of 24 faults has thus been reduced to 10.   □

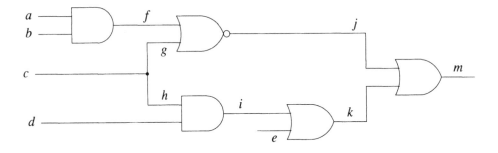

Figure 4.21

Based on Theorem 4.2, many test generation systems generate tests explicitly only for the checkpoint faults of a circuit. But Theorem 4.2 is meaningful only for an irredundant circuit. In a redundant circuit, some of the checkpoint faults are undetectable. If we consider only checkpoint faults and we generate a test set that detects all detectable checkpoint faults, this test set is not guaranteed to detect all detectable SSFs of the circuit; in such a case, additional tests may be needed to obtain a complete detection test set [Abramovici et al. 1986].

Understanding the relations between faults on a stem line and on its fanout branches is important in several problems that we will discuss later. Clearly a stem s-a-v is equivalent to the multiple fault composed of all its fanout branches s-a-v. But, in general, neither equivalence nor dominance relations exist between a stem s-a-v and an individual fanout branch s-a-v. Figure 4.22 shows an example [Hayes 1979] in which a stem fault (j s-a-0) is detected, but the (single) faults on its fanout branches (k s-a-0 and m s-a-0) are not. Figure 4.23 illustrates the opposite situation [Abramovici et al. 1984], in which faults on all fanout branches (x_1 s-a-0 and x_2 s-a-0) are detected, but the stem fault (x s-a-0) is not.

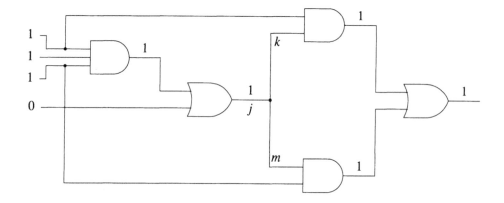

Figure 4.22

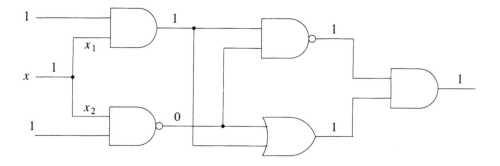

Figure 4.23

4.6 The Multiple Stuck-Fault Model

The multiple stuck-fault (MSF) model is a straightforward extension of the SSF model in which several lines can be simultaneously stuck. If we denote by n the number of possible SSF sites (given by formula (4.8)), there are $2n$ possible SSFs, but there are $3^n - 1$ possible MSFs (which include the SSFs). This figure assumes that any MSF can occur, including the one consisting of all lines simultaneously stuck. If we assume that the multiplicity of a fault, i.e., the number of lines simultaneously stuck, is no greater than a constant k, then the number of possible MSFs is $\sum_{i=1}^{k} \binom{n}{i} 2^i$. This is usually too large a number to allow us to deal explicitly with all multiple faults. For example, the number of double faults ($k=2$) in a circuit with $n=1000$ possible fault sites is about half a million.

Let us first consider the question of why we need to consider MSFs altogether. Since a multiple fault F is just a set $\{f_1, f_2, ..., f_k\}$ of single faults f_i, why isn't F detected by the tests that detect the single faults f_i? The explanation is provided by the *masking relations* among faults.

Definition 4.9: Let T_g be the set of all tests that detect a fault g. We say that a fault f **functionally masks** the fault g iff the multiple fault $\{f,g\}$ is not detected by any test in T_g.

Example 4.9: In the circuit of Figure 4.24 the test 011 is the only test that detects the fault c s-a-0. The same test does not detect the multiple fault $\{c$ s-a-0, a s-a-1$\}$. Thus a s-a-1 masks c s-a-0. □

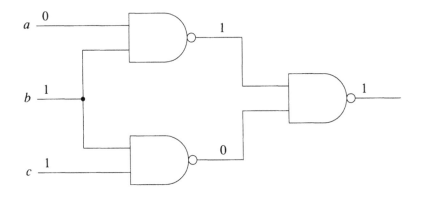

Figure 4.24

Masking can also be defined with respect to a test set T.

Definition 4.10: Let $T_g{'} \subseteq T$ be the set of all tests in T that detect a fault g. We say that a fault f **masks** the fault g **under a test set** T iff the multiple fault $\{f,g\}$ is not detected by any test in $T_g{'}$.

Functional masking implies masking under any test set, but the converse statement is not always true.

Masking relations can also be defined among different type of faults. In Example 4.4 we have an undetectable bridging fault that masks a detectable SSF under a complete test set for SSFs.

If f masks g, then the fault $\{f,g\}$ is not detected by the tests that detect g alone. But $\{f,g\}$ may be detected by other tests. This is the case in Example 4.9, where the fault $\{c$ s-a-0, a s-a-1$\}$ is detected by the test 010.

An important problem is, given a complete test set T for single faults, can there exist a multiple fault $F = \{f_1, f_2, ..., f_k\}$ such that F is not detected by T? (Remember that T detects every f_i alone.) The answer is provided by the following example.

Example 4.10: The test set $T = \{1111, 0111, 1110, 1001, 1010, 0101\}$ detects every SSF in the circuit of Figure 4.25. Let f be B s-a-1 and g be C s-a-1. The only test in T that detects the single faults f and g is 1001. However, the multiple fault $\{f,g\}$ is not detected because under the test 1001, f masks g and g masks f. $\square$

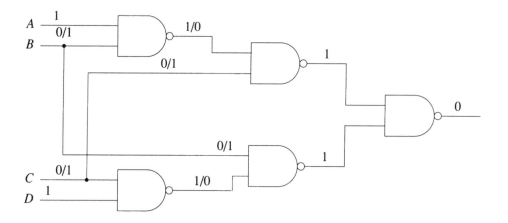

Figure 4.25

In the above example, a multiple fault F is not detected by a complete test set T for single faults because of *circular masking relations* under T among the components of F. Circular functional masking relations may result in an undetectable multiple fault F, as shown in the following example [Smith 1978].

Example 4.11: In the circuit of Figure 4.26, all SSFs are detectable. Let f be D s-a-1 and g be E s-a-1. One can verify that f functionally masks g and vice versa and that the multiple fault $\{f,g\}$ is undetectable. (This represents another type of redundancy, called *multiple-line redundancy*.) $\square$

Note that the existence of circular masking relations among the SSF components f_i of a MSF F is a necessary but not a sufficient condition for F to be undetectable [Smith 1979].

An important practical question is: What percentage of the MSFs can escape detection by a test set designed to detect SSFs? The answer depends on the structure (but not on the size) of the circuit. Namely, the following results have been obtained for combinational circuits:

1. In an irredundant two-level circuit, *any* complete test set for SSFs also detects all MSFs [Kohavi and Kohavi 1972].

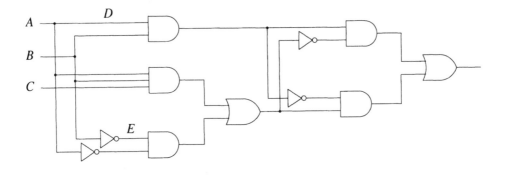

Figure 4.26

2.   In a fanout-free circuit, any complete test set for SSFs detects all double and triple faults, and there exists a complete test set for SSFs that detects all MSFs [Hayes 1971].

3.   In an internal fanout-free circuit (i.e., a circuit in which only the primary inputs may be stems), any complete test set for SSFs detects at least 98 percent of all MSFs of multiplicity less than 6 [Agarwal and Fung 1981].

4.   In an internal fanout-free circuit C, any complete test set for SSFs detects all MSFs unless C contains a subcircuit with the same interconnection pattern as the 5-gate circuit shown in Figure 4.25 [Schertz and Metze 1972].

5.   A test set that detects all MSFs defined on all primary inputs without fanout and all fanout branches of a circuit C detects all multiple faults in C [Bossen and Hong 1971].

Since a test set T designed for detecting SSFs also detects most of the MSFs, for MSF analysis we would like to deal directly only with those MSFs *not* detected by T. Algorithms to identify MSFs not detected by a test set are presented in [Cha 1979, Abramovici and Breuer 1980] for combinational circuits and in [Abramovici and Breuer 1982] for sequential circuits.

Unless circuits are used in extremely high-reliability applications, the MSF model is mainly of theoretical interest. Although the MSF model is generally more realistic than the SSF model, the use of the latter is justified since circular masking conditions are seldom encountered in practice.

Jacob and Biswas [1987] identified SSFs that are *guaranteed to be detected* (GTBD), that is, they can be detected regardless of the presence of other faults in the circuit. For example, any fault on a primary output is GTBD. All the MSFs containing a SSF f that is GTBD will also be detected by the tests for f. It can be shown that a very large fraction of the MSFs (at least 99.6 percent for circuits with three or more primary outputs) contain GTBD faults, which means that circular masking cannot occur. This result takes into account MSFs of any multiplicity, and it reflects the contribution of the large number of MSFs of higher multiplicity, which are more likely to contain

GTBD faults. In general, MSFs of lower multiplicity are more difficult to detect. However, experimental results obtained by Hughes and McCluskey [1984], simulating the 74LS181 4-bit ALU, show that complete detection test sets for SSFs detect all (or more than 99.9 percent) of the double faults and most of the sampled triple and quadruple faults.

4.7 Stuck RTL Variables

A straightforward extension of the stuck-fault model is to replace the concept of a signal line that is stuck with that of an internal RTL variable being stuck. (Stuck I/O variables of functional models are covered by the pin stuck-fault model.) We can further differentiate between *data faults* and *control faults*, depending on the type of the stuck variable. Typical data faults are register or memory bits that are stuck. Control faults are defined on variables that control conditional operations. Thus in the statement

$$\textbf{if } (X \textbf{ and } CLK) \textbf{ then } A = B$$

if X is s-a-1 the transfer $A = B$ will always occur when $CLK = 1$, and it will never occur if X is s-a-0. More generally, we can allow the result of any expression that is part of a condition (or the entire condition itself) to have a stuck fault. In the above example, if the expression $(X \text{ and } CLK)$ is s-a-1, the transfer $A=B$ becomes unconditional. This functional fault model is often used because it is easy to integrate with gate-level stuck-fault modeling.

4.8 Fault Variables

The effect of a fault f on the behavior of a system can be explicitly described using RTL constructs. To do this we introduce a *fault variable* f to reflect the existence of the fault f [Menon and Chappel 1978]. Then we can use f as a condition to switch-in the behavior in the presence of f, as illustrated in

$$\textbf{if } f \textbf{ then } A = B{-}C$$
$$\textbf{else } A = B{+}C$$

where $A = B+C$ and $A = B-C$ represent, respectively, the fault-free and the faulty operations. The fault variable f is treated as a state variable whose value is always 0 for normal operation. Thus the fault f can be modeled as a s-a-1 fault on the internal variable f. This is similar to the technique illustrated in Figure 4.16.

This general method can explicitly model any functional fault. The next example describes a (pattern-sensitive) fault f whose effect — complementing Z — occurs only in the presence of a certain pattern ($DREG$=1111):

$$\textbf{if } (f \textbf{ and } DREG\text{=}1111) \textbf{ then } Z = \bar{Z}$$

We can also apply this technique to model delay faults, as illustrated in

$$\textbf{if } f \textbf{ then } C = A{+}B \textbf{ delay } 40$$
$$\textbf{else } C = A{+}B \textbf{ delay } 10$$

Functional fault modeling based on fault variables is limited by the need to define explicitly the faulty behavior, and it is applicable only to a small number of specific faults of interest.

REFERENCES

[Abraham and Fuchs 1986] J. A. Abraham and W. K. Fuchs, "Fault and Error Models for VLSI," *Proc. of the IEEE*, Vol. 75, No. 5, pp. 639-654, May, 1986.

[Abramovici *et al.* 1984] M. Abramovici, P. R. Menon, and D. T. Miller, "Critical Path Tracing: An Alternative to Fault Simulation," *IEEE Design & Test of Computers*, Vol. 1, No. 1, pp. 83-93, February, 1984.

[Abramovici *et al.* 1986] M. Abramovici, P. R. Menon, and D. T. Miller, "Checkpoint Faults Are Not Sufficient Target Faults for Test Generation," *IEEE Trans. on Computers*, Vol. C-35, No. 8, pp. 769-771, August, 1986.

[Abramovici and Breuer 1979] M. Abramovici and M. A. Breuer, "On Redundancy and Fault Detection in Sequential Circuits," *IEEE Trans. on Computers*, Vol. C-28, No. 11, pp. 864-865, November, 1979.

[Abramovici and Breuer 1980] M. Abramovici and M. A. Breuer, "Multiple Fault Diagnosis in Combinational Circuits Based on an Effect-Cause Analysis," *IEEE Trans. on Computers*, Vol. C-29, No. 6, pp. 451-460, June, 1980.

[Abramovici and Breuer 1982] M. Abramovici and M. A. Breuer, "Fault Diagnosis in Synchronous Sequential Circuits Based on an Effect-Cause Analysis," *IEEE Trans. on Computers*, Vol. C-31, No. 12, pp. 1165-1172, December, 1982.

[Agarwal and Fung 1981] V. K. Agarwal and A. S. F. Fung, "Multiple Fault Testing of Large Circuits by Single Fault Test Sets," *IEEE Trans. on Computers*, Vol. C-30, No. 11, pp. 855-865, November, 1981.

[Banerjee and Abraham 1984] P. Banerjee and J. A. Abraham, "Characterization and Testing of Physical Failures in MOS Logic Circuits," *IEEE Design & Test of Computers*, Vol. 1, No. 4, pp. 76-86, August, 1984.

[Bossen and Hong 1971] D. C. Bossen and S. J. Hong, "Cause-Effect Analysis for Multiple Fault Detection in Combination Networks," *IEEE Trans. on Computers*, Vol. C-20, No. 11, pp. 1252-1257, November, 1971.

[Breuer 1973] M. A. Breuer, "Testing for Intermittent Faults in Digital Circuits," *IEEE Trans. on Computers*, Vol. C-22, No. 3, pp. 241-246, March, 1973.

[Case 1976] G. R. Case, "Analysis of Actual Fault Mechanisms in CMOS Logic Gates," *Proc. 13th Design Automation Conf.*, pp. 265-270, June, 1976.

[Cha 1979] C. W. Cha, "Multiple Fault Diagnosis in Combinational Networks," *Proc. 16th Design Automation Conf.*, pp. 149-155, June, 1979.

[Ferguson and Shen 1988] F. J. Ferguson and J. P. Shen, "A CMOS Fault Extractor for Inductive Fault Analysis," *IEEE Trans. on Computer-Aided Design*, Vol. 7, No. 11, pp. 1181-1194, November, 1988.

[Friedman 1967] A. D. Friedman, "Fault Detection in Redundant Circuits," *IEEE Trans. on Electronic Computers*, Vol. EC-16, pp. 99-100, February, 1967.

[Friedman and Menon 1975] A. D. Friedman and P. R. Menon, *Theory and Design of Switching Circuits*, Computer Science Press, Woodland Hills, California, 1975.

[Galiay *et al.* 1980] J. Galiay, Y. Crouzet, and M. Vergniault, "Physical Versus Logical Fault Models in MOS LSI Circuits: Impact on Their Testability," *IEEE Trans. on Computers*, Vol. C-29, No. 6, pp. 527-531, June, 1980.

[Goundan 1978] A. Goundan, *Fault Equivalence in Logic Networks*, Ph.D. Thesis, University of Southern California, March, 1978.

[Hayes 1971] J. P. Hayes, "A NAND Model for Fault Diagnosis in Combinational Logic Networks," *IEEE Trans. on Computers*, Vol. C-20, No. 12, pp. 1496-1506, December, 1971.

[Hayes 1976] J. P. Hayes, "On the Properties of Irredundant Logic Networks," *IEEE Trans. on Computers*, Vol. C-25, No. 9, pp. 884-892, September, 1976.

[Hayes 1977] J. P. Hayes, "Modeling Faults in Digital Logic Circuits," in *Rational Fault Analysis*, R. Saeks and S. R. Liberty, eds., Marcel Dekker, New York, pp. 78-95, 1977.

[Hayes 1979] J. P. Hayes, "Test Generation Using Equivalent Normal Forms," *Journal of Design Automation and Fault-Tolerant Computing*, Vol. 3, No. 3-4, pp. 131-154, Winter, 1979.

[Horowitz and Sahni 1978] E. Horowitz and S. Sahni, *Fundamentals of Computer Algorithms*, Computer Science Press, Rockville, Maryland, 1978.

[Hughes and McCluskey 1984] J. L. A. Hughes and E. J. McCluskey, "An Analysis of the Multiple Fault Detection Capabilities of Single Stuck-at Fault Test Sets," *Proc. Intn'l. Test Conf.*, pp. 52-58, October, 1984.

[Ibarra and Sahni 1975] O. H. Ibarra and S. Sahni, "Polynomially Complete Fault Detection Problems," *IEEE Trans. on Computers*, Vol. C-24, No. 3, pp. 242-249, March, 1975.

[Jacob and Biswas 1987] J. Jacob and N. N. Biswas, "GTBD Faults and Lower Bounds on Multiple Fault Coverage of Single Fault Test Sets," *Proc. Intn'l. Test Conf.*, pp. 849-855, September, 1987.

[Kohavi and Kohavi 1972] I. Kohavi and Z. Kohavi, "Detection of Multiple Faults in Combinational Logic Networks," *IEEE Trans. on Computers*, Vol. C-21, No. 6, pp. 556-668, June, 1972.

[Kodandapani and Pradhan 1980] K. L. Kodandapani and D. K. Pradhan, "Undetectability of Bridging Faults and Validity of Stuck-At Fault Test Sets," *IEEE Trans. on Computers*, Vol. C-29, No. 1, pp. 55-59, January, 1980.

[McCluskey and Clegg 1971] E. J. McCluskey and F. W. Clegg, "Fault Equivalence in Combinational Logic Networks," *IEEE Trans. on Computers*, Vol. C-20, No. 11, pp. 1286-1293, November, 1971.

[Menon and Chappell 1978] P. R. Menon and S. G. Chappell, "Deductive Fault Simulation with Functional Blocks," *IEEE Trans. on Computers*, Vol. C-27, No. 8, pp. 689-695, August, 1978.

[Partridge 1980] J. Partridge, "Testing for Bipolar Integrated Circuit Failure Modes," *Digest of Papers 1980 Test Conf.*, pp. 397-406, November, 1980.

[Savir 1980] J. Savir, "Detection of Single Intermittent Faults in Sequential Circuits," *IEEE Trans. on Computers*, Vol. C-29, No. 7, pp. 673-678, July, 1980.

[Schertz and Metze 1972] D. R. Schertz and G. Metze, "A New Representation of Faults in Combinational Digital Circuits," *IEEE Trans. on Computers*, Vol. C-21, No. 8, pp. 858-866, August, 1972.

[Shen *et al.* 1985] J. P. Shen, W. Maly, and F. J. Ferguson, "Inductive Fault Analysis of MOS Integrated Circuits," *IEEE Design & Test of Computers*, Vol. 2, No. 6, pp. 13-26, December, 1985.

[Smith 1978] J. E. Smith, "On the Existence of Combinational Logic Circuits Exhibiting Multiple Redundancy," *IEEE Trans. on Computers*, Vol. C-27, No. 12, pp. 1221-1225, December, 1978.

[Smith 1979] J. E. Smith, "On Necessary and Sufficient Conditions for Multiple Fault Undetectability," *IEEE Trans. on Computers*, Vol. C-28, No. 10, pp. 801-802, October, 1979.

[Timoc *et al.* 1983] C. Timoc, M. Buehler, T. Griswold, C. Pina, F. Scott, and L. Hess, "Logical Models of Physical Failures," *Proc. Intn'l Test Conf.*, pp. 546-553, October, 1983.

[Wadsack 1978] R. L. Wadsack, "Fault Modeling and Logic Simulation of CMOS and MOS Integrated Circuits," *Bell System Technical Journal*, Vol. 57, pp. 1449-1474, May-June, 1978.

PROBLEMS

4.1 Find a circuit that has an undetectable stuck fault.

4.2 Is it possible to have a combinational circuit C with some signal S and test t such that t detects both S s-a-1 and S s-a-0? Give an example or prove it is impossible.

4.3 Determine the output function of the circuit of Figure 4.27 for the following faults:

   a. AND bridge between inputs of gate G_1

   b. The multiple fault $\{x_3 \ s\text{-}a\text{-}1, \ x_2 \ s\text{-}a\text{-}0\}$

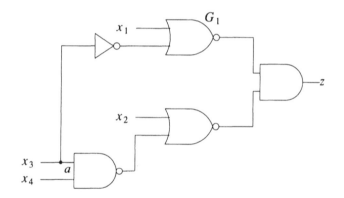

Figure 4.27

4.4 In the circuit of Figure 4.27 which if any of the following tests detect the fault x_1 s-a-0?

   a. (0,1,1,1)

   b. (1,1,1,1)

   c. (1,1,0,1)

   d. (1,0,1,0)

4.5 For the circuit of Figure 4.27 find a Boolean expression for the set of all tests that detect the fault:

   a. $x_3 \ s\text{-}a\text{-}0$

    b. x_2 s-a-0

    c. x_2 s-a-1

4.6 For the circuit of Figure 4.28

    a.   Find the set of all tests that detect the fault c s-a-1.

    b.   Find the set of all tests that detect the fault a s-a-0.

    c.   Find the set of all tests that detect the multiple fault {c s-a-1, a s-a-0}.

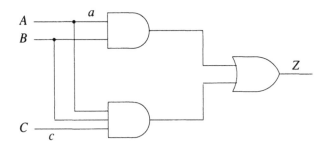

Figure 4.28

4.7 For the circuit of Figure 4.29

    a.   Find the set of all tests that detect the fault a s-a-0.

    b.   Find the set of all tests that detect the fault b s-a-0.

    c.   Find the set of all tests that detect the multiple fault {a s-a-0, b s-a-0}.

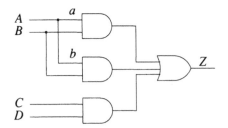

Figure 4.29

4.8 For the circuit of Figure 4.30

    a.   Find the set of all tests that detect the fault b s-a-1.

    b.   Find the set of all tests that distinguish the faults a s-a-0 and c s-a-0.

    c.   Find the set of all tests that distinguish the multiple faults $\{a$ s-a-0, b s-a-1$\}$ and $\{c$ s-a-0, b s-a-1$\}$.

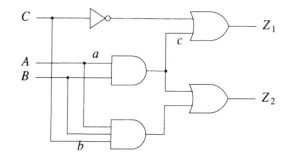

Figure 4.30

4.9

    a.   Find an example of a combinational circuit with two faults that are functionally equivalent under a test set but not functionally equivalent.

    b.   Prove that in a combinational circuit two faults functionally equivalent under a complete location test set are functionally equivalent.

4.10

    a.   Find a counterexample to the following statement:

       "In a combinational circuit two faults f and g are functionally equivalent iff they are always detected by the same tests."

    b.   Find a class of combinational circuits for which the above statement is true and prove it.

4.11   Prove that in a combinational circuit, if two faults dominate each other, then they are functionally equivalent.

4.12   Prove that for combinational circuits fault dominance is a transitive relation; i.e., if f dominates g and g dominates h, then f dominates h.

4.13   Prove that in a fanout-free circuit, any pair of functionally equivalent faults are also structurally equivalent.

4.14   Analyze the relation between the detection of a stem line s-a-v and the detection of (single) s-a-v faults on its fanout branches for the case of nonreconvergent fanout.

4.15

    a. For the circuit and the test in Figure 4.22, show that j s-a-0 is detected, but k s-a-0 and m s-a-0 are not.

    b. For the circuit and the test in Figure 4.23, show that both x_1 s-a-0 and x_2 s-a-0 are detected, but x s-a-0 is not.

4.16 Let N be a combinational circuit composed only of NAND gates. Assume that every primary input has only one fanout. Show that a test set that detects all s-a-1 faults in N detects all s-a-0 faults as well.

4.17 Consider the circuit of Figure 4.31. Let f be the fault b s-a-0 and g be a s-a-1.

    a. Does f mask g under the test 0110? Does f mask g under the test 0111?

    b. Are the faults f and $\{f,g\}$ distinguishable?

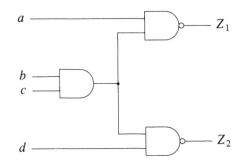

Figure 4.31

4.18 Prove that in an irredundant two-level combinational circuit, any complete test set for SSFs also detects all MSFs.

4.19 Let $Z(x)$ be the function of a single-output irredundant circuit N. Show that none of the SSFs in N can change its function from $Z(x)$ to $\overline{Z}(x)$.

5. FAULT SIMULATION

About This Chapter

First we review applications of fault simulation. Then we examine fault simulation techniques for SSFs. We describe both general methods — serial, parallel, deductive, and concurrent — and techniques specialized for combinational circuits. We also discuss fault simulation with fault sampling and statistical fault analysis. This chapter emphasizes gate-level and functional-level models and SSFs. Fault simulation for technology-specific faults models and fault simulation for bridging faults are dealt with in separate chapters.

5.1 Applications

Fault simulation consists of simulating a circuit in the presence of faults. Comparing the fault simulation results with those of the fault-free simulation of the same circuit simulated with the same applied test T, we can determine the faults detected by T.

One use of fault simulation is to **evaluate (grade) a test** T. Usually the grade of T is given by its *fault coverage*, which is the ratio of the number of faults detected by T to the total number of simulated faults. This figure is directly relevant only to the faults processed by the simulator, as even a test with 100 percent fault coverage may still fail to detect faults outside the considered fault model. Thus the fault coverage represents only a lower bound on the *defect coverage*, which is the probability that T detects any physical fault in the circuit. Experience has shown that a test with high coverage for SSFs also achieves a high defect coverage. Test evaluation based on fault simulation has been applied mainly to the SSF model, both for external testing and for self-testing (i.e., evaluation of self-test programs).

The quality of the test greatly influences the quality of the shipped product. Let Y be the manufacturing *yield*, that is, the probability that a manufactured circuit is defect-free. Let DL denote the *defect level*, which is the probability of shipping a defective product, and let d be the defect coverage of the test used to check for manufacturing defects. The relation between these variables is given by

$$DL = 1 - Y^{1-d}$$

[Williams and Brown 1981]. Assuming that the fault coverage is close to the defect coverage, we can use this relation, illustrated in Figure 5.1, to determine the fault coverage required for a given defect level. For example, consider a process with 0.5 yield. Then to achieve a 0.01 defect level — that is, 1 percent of the shipped products are likely to be defective — we need 99 percent fault coverage. A test with only 95 percent fault coverage will result in a defect level of 0.035. If, however, the yield is 0.8, then 95 percent fault coverage is sufficient to achieve a defect level of 0.01. Other aspects of the relation between product quality and fault coverage are analyzed in [Agrawal *et al.* 1981], [Seth and Agrawal 1984], [Daniels and Bruce 1985], and [McCluskey and Buelow 1988].

Fault simulation plays an important role in **test generation**. Many test generation systems use a fault simulator to evaluate a proposed test T (see Figure 5.2), then

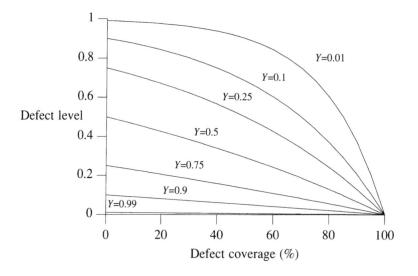

Figure 5.1   Defect level as a function of yield and defect coverage

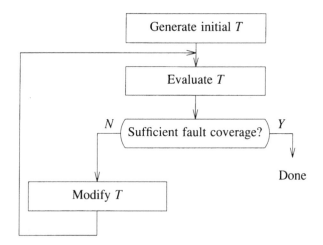

Figure 5.2   General use of fault simulation in test generation

change T according to the results of the fault simulation until the obtained coverage is considered satisfactory. The test T is modified by adding new vectors and/or by discarding some of its vectors that did not contribute to achieving good coverage. These changes may be made by a program or by a test designer in an interactive mode.

Another use of fault simulation in test generation is illustrated in Figure 5.3. Many test generation algorithms are fault-oriented; that is, they generate a test for one specified fault, referred to as the *target fault*. Often the same test also detects many other faults that can be determined by fault simulation. Then all the detected faults are discarded from the set of simulated faults and a new target fault is selected from the remaining ones.

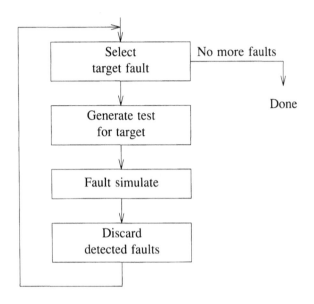

Figure 5.3   Fault simulation used in the selection of target faults for test generation

Fault simulation is also used to **construct fault dictionaries**. Conceptually, a fault dictionary stores the output response to T of every faulty circuit N_f corresponding to a simulated fault f. In fact, a fault dictionary does not store the complete output response R_f of every N_f, but some function $S(R_f)$, called the *signature* of the fault f. The fault location process relies on comparing the signature obtained from the response of the circuit under test with the entries in the precomputed fault dictionary.

Constructing a fault dictionary by fault simulation means, in principle, computing the response in the presence of every possible fault *before testing*. A different approach, referred to as **post-test diagnosis**, consists of first isolating a reduced set of "plausible" faults (i.e., faults that may be consistent with the response obtained from the circuit under test), then simulating only these faults to identify the actual fault. Diagnosis techniques will be discussed in more detail in a separate chapter.

Another application of fault simulation is to **analyze the operation of a circuit in the presence of faults**. This is especially important in high-reliability systems, since some faults may drastically affect the operation. For example:

- A fault can induce races and hazards not present in the fault-free circuit.

- A faulty circuit may oscillate or enter a deadlock (hang-up) state.

- A fault can inhibit the proper initialization of a sequential circuit.

- A fault can transform a combinational circuit into a sequential one or a synchronous circuit into an asynchronous one.

Fault simulation of self-checking circuits is used to verify the correct operation of their error-detecting and error-correcting logic.

5.2 General Fault Simulation Techniques

5.2.1 Serial Fault Simulation

Serial fault simulation is the simplest method of simulating faults. It consists of transforming the model of the fault-free circuit N so that it models the circuit N_f created by the fault f. Then N_f is simulated. The entire process is repeated for each fault of interest. Thus faults are simulated one at a time. The main advantage of this method is that no special fault simulator is required, as N_f is simulated by a fault-free simulator. Another advantage is that it can handle any type of fault, provided that the model of N_f is known. However, the serial method is impractical for simulating a large number of faults because it would consume an excessive amount of CPU time.

The other general fault simulation techniques — parallel, deductive, and concurrent — differ from the serial method in two fundamental aspects:

- They determine the behavior of the circuit N in the presence of faults without explicitly changing the model of N.

- They are capable of simultaneously simulating a set of faults.

These three techniques will be described in the following sections.

5.2.2 Common Concepts and Terminology

All three techniques — parallel, deductive, and concurrent — simultaneously simulate the fault-free circuit N (also called the *good circuit*) and a set of faulty (or *bad*) circuits $\{N_f\}$. Thus any fault simulation always includes a fault-free simulation run. If all the faults of interest are simulated simultaneously, then fault simulation is said to be done in *one pass*. Otherwise the set of faults is partitioned and fault simulation is done as a *multipass* process, in which one subset of the total set of faults is dealt with in one pass. In general, large circuits require multipass fault simulation.

In addition to the activities required for fault-free simulation (described in Chapter 3), fault simulation involves the following tasks:

- fault specification,

- fault insertion,

- fault-effect generation and propagation,

- fault detection and discarding.

Fault specification consists of defining the set of modeled faults and performing fault collapsing. **Fault insertion** consists of selecting a subset of faults to be simulated in one pass and creating the data structures that indicate the presence of faults to the simulation algorithm. These data structures are used to **generate effects of the inserted faults** during simulation. For example, let f be a s-a-1 fault inserted on line i. Then whenever a value 0 would propagate on line i (in the circuit N_f), the simulator changes it to 1. Most of the work in fault simulation is related to the **propagation of fault effects**. Whenever an effect of a fault f propagates to a primary output j (such that the values of j in the good and the faulty circuit are both binary), the simulator marks f as detected. The user may specify that a fault detected k times should be discarded from the set of simulated faults (usually $k=1$). **Fault discarding** (also called *fault dropping*) is the inverse process of fault insertion.

5.2.3 Parallel Fault Simulation

In parallel fault simulation [Seshu 1965] the good circuit and a fixed number, say W, of faulty circuits are simultaneously simulated. A set of F faults requires $\lceil F/W \rceil$ passes.[1] The values of a signal in the good circuit and the values of the corresponding signals in the W faulty circuits are packed together in the same memory location of the host computer. Depending on the implementation, a "location" consists of one or more words.

For example, if we use 2-valued logic and a 32-bit word, then $W=31$ (see Figure 5.4). Consider an AND gate with inputs A and B. Each bit in the word associated with a signal represents the value of that signal in a different circuit. Traditionally, bit 0 represents a value in the good circuit. Then using a logical AND instruction between the words associated with A and B, we evaluate (in parallel) the AND gate in the good circuit and in each of the 31 faulty circuits. Similarly we use an OR operation to evaluate an OR gate, and a bit complement operation for a NOT. A NAND gate requires an AND followed by a complement, and so on. A sequential element is represented by a Boolean expression; for example, for a JK F/F

$$Q^+ = J\overline{Q} + \overline{K}Q$$

where Q^+ and Q are, respectively, the new and the current state of the F/F. Thus the evaluation can be carried out using AND, OR, and complement operators.

Let v_i be the value propagating onto line i in the faulty circuit N_f, where f is the fault j s-a-c. Every line $i \neq j$ takes the value v_i, but the value of j should always be c. The new value of line i, v_i', can be expressed as

$$v_i' = v_i \overline{\delta_{ij}} + \delta_{ij} c$$

where

---

1. $\lceil x \rceil$ denotes the smallest integer greater than or equal to x.

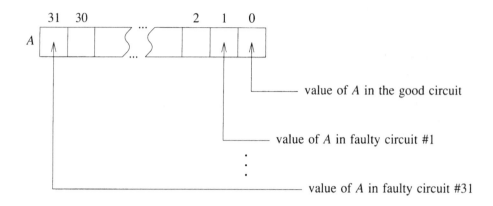

Figure 5.4   Value representation in parallel simulation

$$\delta_{ij} = \begin{cases} 0 & i \neq j \\ 1 & i = j \end{cases}$$

The above equations represent the process of fault insertion for one fault f (j s-a-c). For W faults, this process is carried on in parallel using two *mask words* storing the values δ_{ij} and c in the bit position corresponding to fault f. Figure 5.5 shows a portion of a circuit, the masks used for fault insertion on line Z, and the values of Z before and after fault insertion. The first mask — I — associated with a line indicates whether faults should be inserted on that line and in what bit positions, and the second — S — defines the stuck values of these faults. Thus after evaluating gate Z by $Z = X1.Y$, the effect of inserting faults on Z is obtained by

$$Z' = Z.\overline{I}_Z + I_Z.S_Z$$

Fault insertion for $X1$ and Y is similarly done before evaluating Z.

The above technique has several possible implementations [Thompson and Szygenda 1975].

For 3-valued logic $(0,1,u)$, one bit is not sufficient to represent a signal value. The coding scheme shown in Figure 5.6 uses two words, $A1$ and $A2$, to store the W values associated with signal A. Since the codes for the values 0 and 1 are, respectively, 00 and 11, the logical AND and OR operations can be applied directly to the words $A1$ and $A2$. Hence, to evaluate an AND gate with inputs A and B and output C, instead of $C = A.B$ we use

$$C1 = A1.B1$$
$$C2 = A2.B2$$

The complement operator, however, cannot be applied directly, as the code 01 will generate the illegal code 10. An inversion $B = \text{NOT}\,(A)$ is realized by

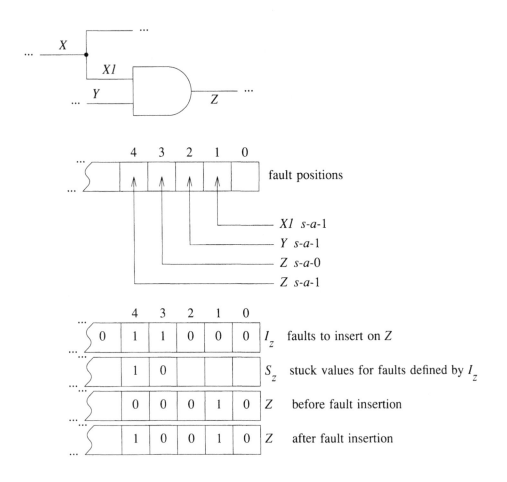

Figure 5.5   Fault insertion on Z

	Value of A		
	0	1	u
$A1$	0	1	0
$A2$	0	1	1

Figure 5.6   Coding for 3-valued logic

$$B1 = \overline{A2}$$
$$B2 = \overline{A1}$$

that is, we complement $A1$ and $A2$ and interchange them. Figure 5.7 shows a sample computation for $Z = \overline{X}.Y$ (with a 3-bit word).

	Values	Encoding	
X	0 1 u	$X1$	0 1 0
		$X2$	0 1 1
Y	0 1 1	$Y1$	0 1 1
		$Y2$	0 1 1
$X.Y$	0 1 u	$X1.Y1$	0 1 0
		$X2.Y2$	0 1 1
$Z = \overline{X.Y}$	1 0 u	$Z1$	1 0 0
		$Z2$	1 0 1

Figure 5.7   Example of NAND operation in 3-valued parallel simulation

Other coding techniques for parallel fault simulation with three or more logic values are described in [Thompson and Szygenda 1975] and [Levendel and Menon 1980].

It is possible to reduce the number of passes by simulating several *independent faults* simultaneously. Let S_i denote the set of lines in the circuit that can be affected by the value of line i. Faults defined on lines i and j, such that $S_i \cap S_j = \emptyset$, are said to be independent. Independent faults cannot affect the same part of the circuit, and they can be simulated in the same bit position. The least upper bound on the number of independent faults that can be processed simultaneously in the same bit position is the number of primary outputs — p — of the circuit. Hence the minimum number of passes is $\lceil F/(W.p) \rceil$. The potential reduction in the number of passes should be weighed against the cost of identifying independent faults. Algorithms for determining subsets of independent faults are presented in [Iyengar and Tang 1988].

To discard a fault f, first we have to stop inserting it. This is simply done by turning off the corresponding bit position in the mask word used for fault insertion. However, the effects of the fault f may have been propagated to many other lines, and additional processing is needed to stop all the activity in the circuit N_f [Thompson and Szygenda 1975].

Limitations

In parallel fault simulation several faulty circuits can be simulated in parallel, provided that for evaluation we use only operations that process each bit independently of all others. This requires that we model elements by Boolean equations, and therefore we cannot directly use evaluation routines that examine the input values of the evaluated elements, or routines based on arithmetic operations. But these types of evaluations are convenient for evaluating functional elements, such as memories and counters. To integrate these techniques into a parallel fault simulator, the individual bits of the faulty circuits are extracted from the packed-value words, the functional elements are individually evaluated, and then the resulting bits are repacked. Thus *parallel simulation is compatible only in part with functional-level modeling.*

Evaluation techniques based on Boolean equations are adequate for binary values, but they become increasingly complex as the number of logic values used in modeling increases. Hence, *parallel simulation becomes impractical for multivalued logic.*

In parallel fault simulation, an event occurs when the new value of a line differs from its old value in at least one bit position. Such an event always causes W evaluations, even if only one of the W evaluated elements has input events. Although it may appear that the unnecessary evaluations do not take extra time, because they are done in parallel with the needed ones, they do represent wasted computations. W evaluations are done even after all the faults but one have been detected and discarded. Thus *parallel fault simulation cannot take full advantage of the concept of selective trace simulation, or of the reduction in the number of faults caused by fault dropping.*

5.2.4 Deductive Fault Simulation

The deductive technique [Armstrong 1972, Godoy and Vogelsberg 1971] simulates the good circuit and deduces the behavior of *all* faulty circuits. "All" denotes a theoretical capability, subject in practice to the size of the available memory. The data structure used for representing fault effects is the **fault list**. A fault list L_i is associated with every signal line i. During simulation, L_i is the set of all faults f that cause the values of i in N and N_f to be different at the current simulated time. If i is a primary output and all values are binary, then L_i is the set of faults detected at i.

Figure 5.8 illustrates the difference between the value representation in parallel and in deductive simulation. Suppose that we have F faults and a machine word with $W \geq F + 1$, hence we can simulate all the faults in one pass. Then in parallel simulation the word associated with a line i stores the value of i in every faulty circuit. During simulation, however, the value of i in most faulty circuits is the same as in the good circuit. This waste is avoided in deductive simulation by keeping only the bit positions (used as fault names) that are different from the good value.

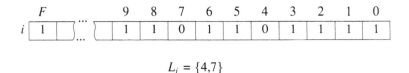

$$L_i = \{4,7\}$$

Figure 5.8 Fault-effects representation in parallel and deductive fault simulation

Given the fault-free values and the fault lists of the inputs of an element, the basic step in deductive simulation is to compute the fault-free output value and the output fault list. The computation of fault lists is called *fault-list propagation*. Thus in addition to the logic events which denote changes in signal values, a deductive fault simulator also propagates *list events* which occur when a fault list changes, i.e., when a fault is either added to or deleted from a list.

5.2.4.1 Two-Valued Deductive Simulation

In this section we assume that all values are binary. Consider, for example, an AND gate with inputs A and B and output Z (Figure 5.9). Suppose $A=B=1$. Then $Z=1$ and any fault that causes a 0 on A or B will cause Z to be erroneously 0. Hence

$$L_Z = L_A \cup L_B \cup \{Z \ s\text{-}a\text{-}0\}$$

Figure 5.9

Now suppose that $A=0$ and $B=1$. Then $Z=0$ and any fault that causes A to be 1 without changing B, will cause Z to be in error; i.e., $Z=1$. These are the faults in L_A that are not in L_B:

$$L_Z = (L_A \cap \overline{L_B}) \cup \{Z \ s\text{-}a\text{-}1\} = (L_A - L_B) \cup \{Z \ s\text{-}a\text{-}1\}$$

where $\overline{L_B}$ is the set of all faults *not* in L_B. Note that a fault whose effect propagates to both A and B does not affect Z.

Let I be the set of inputs of a gate Z with controlling value c and inversion i. Let C be the set of inputs with value c. The fault list of Z is computed as follows:

$$\textbf{if } C = \varnothing \textbf{ then } L_Z = \{\bigcup_{j \in I} L_j\} \cup \{Z \ s\text{-}a\text{-}(c \oplus i)\}$$

$$\textbf{else } L_Z = \{\bigcap_{j \in C} L_j\} - \{\bigcup_{j \in I-C} L_j\} \cup \{Z \ s\text{-}a\text{-}(\overline{c} \oplus i)\}$$

In other words, if no input has value c, any fault effect on an input propagates to the output. If some inputs have value c, only a fault effect that affects all the inputs at c without affecting any of the inputs at $\overline{c}$ propagates to the output. In both cases we add the local fault of the output.

Example 5.1: Consider the circuit in Figure 5.10. After fault collapsing (see Example 4.8), the set of faults we simulate is $\{a_0, a_1, b_1, c_0, c_1, d_1, e_0, g_0, h_0, h_1\}$ (α_v denotes α s-a-v). Assume the first applied test vector is 00110. The computation of fault lists proceeds as follows:

$$L_a = \{a_1\}, \ L_b = \{b_1\}, \ L_c = \{c_0\}, \ L_d = \varnothing, \ L_e = \varnothing$$

$$L_f = L_a \cap L_b = \varnothing, \ L_g = L_c \cup \{g_0\} = \{c_0, g_0\}, \ L_h = L_c \cup \{h_0\} = \{c_0, h_0\}$$

$$L_j = L_g - L_f = \{c_0, g_0\}, \quad L_i = L_d \cup L_h = \{c_0, h_0\}$$

$$L_k = L_i - L_e = \{c_0, h_0\}$$

$$L_m = L_k - L_j = \{h_0\}$$

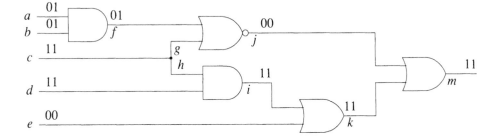

Figure 5.10

Since h_0 is detected, we drop it from the set of simulated faults by deleting h_0 from every fault list where it appears, namely L_h, L_i, L_k, and L_m. (Note that c_0 is not detected.)

Now assume that both inputs a and b change to 1. Then $L_a = \{a_0\}$, $L_b = \varnothing$, $f = 1$, and $L_f = \{a_0\}$. The evaluation of gate j generates no logic event, but now $L_j = L_f \cap L_g = \varnothing$. This shows that a list event may occur even without a corresponding logic event. Propagating this list event to gate m, we obtain $L_m = L_k - L_j = \{c_0\}$. Hence c_0 is now detected.   □

Note that when L_α is computed, to determine whether a list event has occurred, the new L_α must be compared with the old L_α (denoted by $\tilde{L}_\alpha$), before the latter is destroyed; i.e., we must determine whether $L_\alpha = \tilde{L}_\alpha$.

Fault propagation becomes more involved when there is feedback. Care must be exercised when the effect of a fault, say α_0 or α_1, feeds back onto the line α itself. If the fault list propagating to line α contains an α_0 and if the good value of α is 0, then α_0 should be deleted from L_α because the values of α in the fault-free circuit and the circuit with the fault α_0 are the same. Similarly α_1 should be deleted from L_α if the good value of α is 1.

Additional complexities occur in propagation of fault lists through memory elements. Consider the *SR* latch shown in Figure 5.11. Let the state at time t_1 be $(y_1, y_2) = (0,1)$, and the input be $(S,R) = (1,1)$. If at time t_2 R changes to 0, the outputs should remain the same. Let L_S and L_R be the input fault lists at time t_2 associated with lines S and R, and let $\tilde{L}_1$ and $\tilde{L}_2$ be the fault lists associated with lines y_1 and y_2 at time t_1. The new fault lists L_1 and L_2 associated with lines y_1 and y_2 resulting from the input logic event at t_2 can be computed as follows (faults internal to the latch will be ignored).

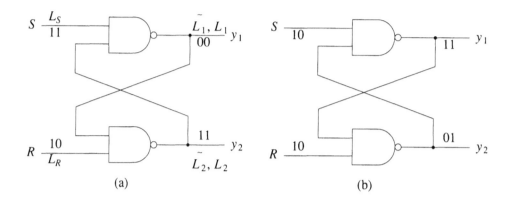

Figure 5.11 Propagation of fault lists through an SR latch (a) Good circuit
(b) Faulty circuit for some fault f in the set $\{L_S \cap \bar{L}_R \cap \tilde{L}_1 \cap \tilde{L}_2\}$

Initially set $L_1 = \tilde{L}_1$ and $L_2 = \tilde{L}_2$. The new values for L_1 and L_2 are given by the expressions $L_1 = L_S \cup L_2$ and $L_2 = L_R \cap L_1$. Since a change in the fault lists for y_1 and y_2 may have occurred, this calculation must be repeated until the fault lists stabilize. Under some conditions, such as a critical race, the lists may not stabilize and special processing is required. This problem is dealt with later in this section.

A second and faster approach to propagation of fault lists through a latch is to consider this element to be a primitive and calculate the rules for determining the steady-state fault lists for y_1 and y_2 in terms of L_S, L_R, $\tilde{L}_1$, and $\tilde{L}_2$. These rules can be derived by inspecting all 16 minterms over the variables L_S, L_R, $\tilde{L}_1$, and $\tilde{L}_2$, i.e., $(L_S \cap L_R \cap \tilde{L}_1 \cap \tilde{L}_2)$, $(L_S \cap L_R \cap \tilde{L}_1 \cap \bar{\tilde{L}}_2)$, ..., $(\bar{L}_S \cap \bar{L}_R \cap \bar{\tilde{L}}_1 \cap \bar{\tilde{L}}_2)$.

For example, consider again the initial state condition $(S,R,y_1,y_2) = (1,1,0,1)$. Any fault f in the set (minterm) $L_S \cap \bar{L}_R \cap \tilde{L}_1 \cap \tilde{L}_2$ causes both outputs to be incorrect at time t_1 (due to $\tilde{L}_1$ and $\tilde{L}_2$), and input S to be incorrect at time t_2 (see Figure 5.11(b)). Hence at t_2 we have that $(S,R) = (0,0)$ in the faulty circuit, and this fault produces $(y_1,y_2) = (1,1)$ and hence is an element in L_1 but not L_2 (note that $y_1 \neq \bar{y}_2$, hence the outputs are not labeled y_1 and $\bar{y}_1$). Similarly, each of the remaining 15 minterms can be processed and the rules for calculating L_1 and L_2 developed. These rules become more complex if internal faults are considered or if an internal fault propagates around a loop and appears in L_S or L_R. Finally, the rules for L_1 and L_2 must be developed for each initial state vector (S,R,y_1,y_2).

Let z be the output of a combinational block implementing the function $f(a,b,c,...)$. A general formula for computing the fault list L_z as a function of the variables $a,b,c,...$ and their fault lists $L_a,L_b,L_c,...$ was obtained by Levendel [1980]. Let us define an exclusive-OR operation between a variable x and its fault list L_x by

$$x \oplus L_x = \begin{cases} L_x & \text{if } x = 0 \\ \bar{L}_x & \text{if } x = 1 \end{cases}$$

Since L_x is the list of faults that cause x to have a value different from its fault-free value, it follows that $x{\oplus}L_x$ is the list of faults that cause x to take value 1. Let us denote by $F(A,B,C,...)$ the set function obtained by replacing all AND and OR operations in $f(a,b,c,...)$ by $\cap$ and $\cup$, respectively. Then the list of faults that cause z to take value 1 (ignoring z s-a-1) is given by

$$z{\oplus}L_z = F(a{\oplus}L_a,\ b{\oplus}L_b,\ c{\oplus}L_c,...)$$

(see Problem 5.6). Hence

$$L_z = f(a,b,c,...){\oplus}F(a{\oplus}L_a,\ b{\oplus}L_b,\ c{\oplus}L_c,...).$$

For example, let us apply the above equation to compute the output fault list of a *JK* flip-flop modeled by its characteristic equation $Y = J\bar{y} + \bar{K}y$, where Y and y are the new and the current output values. The new output fault list L_y is given by

$$L_y = (J\bar{y} + \bar{K}y){\oplus}[(J{\oplus}L_J)\cap(\overline{Y{\oplus}L_y})\cup(\overline{K{\oplus}L_K})\cap(y{\oplus}L_y)]$$

For $J = 0$, $K = 0$, and $y = 0$, L_Y becomes

$$L_Y(J = 0,\ K = 0,\ y = 0) = (L_J{\cap}\bar{L}_y)\cup(\bar{L}_K{\cap}L_y).$$

The fault list of the complementary output $\bar{Y}$ is $L_{\bar{Y}} = L_Y$.

We shall now consider several aspects related to the efficiency of deductive simulators.

Fault Storage and Processing

Fault lists can be stored as linked lists, sequential tables, or characteristic vectors. These three structures are illustrated in Figure 5.12. For a list structure, the faults are stored in an ordered sequence to simplify the computation of set union and intersection. List structures require the overhead of an available space list, as well as a pointer to find the next element in a list. However, insertion and deletion of elements is a simple task. Lists can be easily destroyed by assigning them to the available space list. Using a sequential table organization leads to faster processing and eliminates the need for pointers. However, repacking of storage is required occasionally to make a section of memory reusable.

The computation of set union and intersection for unordered and ordered sequential tables is about of the same complexity. It is easier to add a single element to an unordered set, since it can be placed at the end of the table. In an ordered table it must be inserted in its correct position, and all entries below it repositioned in memory. It is easier to delete an arbitrary element in an ordered table than in an unordered one, since its position can be found by a binary search process.

Using the characteristic-vector structure, all list operations are simple and fast. A fault is inserted or deleted by storing a 1 or 0 in the appropriate bit. Set union and intersection are carried out by simple OR and AND word operations. However, this structure typically requires more storage than the preceding two, though it is of fixed size. An exact comparison of the storage requirements requires knowledge of the average size of a fault list.

For large circuits, it is possible to run out of memory while processing the fault lists, which are dynamic and unpredictable in size. When this occurs, the set of faults must

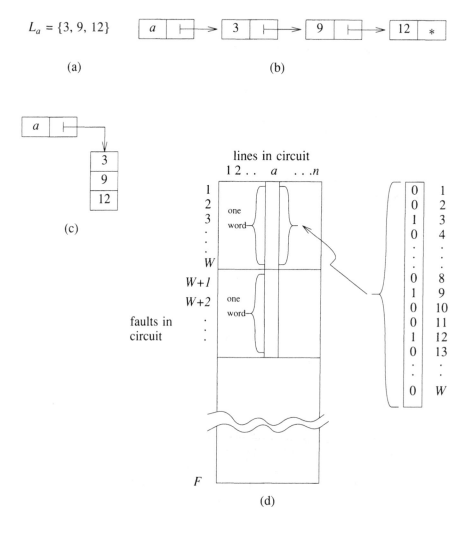

Figure 5.12   Three storage structures for lists (a) Fault list (b) Linked list
(c) Sequential table (d) Characteristic vector

be partitioned and each subset processed separately.   This process can be done
dynamically and leads to a multipass simulation process.

Oscillation and Active Faults

It is possible that the fault-free circuit will stabilize while the circuit with some fault f
oscillates.   For this case there will be an arbitrarily long sequence of list events.
Unfortunately, even though only one fault may be causing the oscillation, repeated
complex processing of long lists may be required, which is time-consuming.   Faults
that produce circuit oscillation, as well as those that achieve stability only after a great
amount of logical activity, should be purged from fault lists whenever possible.

5.2.4.2 Three-Valued Deductive Simulation

When 3-valued logic is employed, the complexity of deductive simulation greatly increases. In this section we will briefly outline two approaches of varying degree of accuracy (pessimism). We refer to these as third-order analysis (least pessimistic) and second-order analysis (most pessimistic).

Third-Order Analysis

Since each line α in the fault-free and faulty circuits can take on the logic values 0,1 and u, two lists L_α^δ and L_α^ε will be associated with a line α whose normal value is v, where $\{\delta,\varepsilon\} = \{0,1,u\} - \{v\}$. For example, if $v = 1$, then the lists L_α^0 and L_α^u are associated with line α; L_α^0 (L_α^u) represents the set of all faults that cause α to have value $0(u)$. Since the set of all faults currently processed equals the set $L_\alpha^0 \cup L_\alpha^1 \cup L_\alpha^u$, we have that $L_\alpha^v = \overline{L_\alpha^\varepsilon \cup L_\alpha^\delta}$.

Example 5.2: Consider the gate and input values shown in Figure 5.13. A fault that changes d to 0 must change a and b to 1, and it must not change c. Hence

$$L_d^0 = (L_a^1 \cap L_b^1 \cap (\overline{L_c^0 \cup L_c^u})) \cup d_0$$

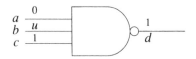

Figure 5.13

A fault that changes d to u must cause a, b, and c to have any combination of u and 1 values, except $a = b = c = 1$. Then

$$L_d^u = (L_a^u \cup L_a^1) \cap \overline{L_b^0} \cap \overline{L_c^0} - L_a^1 \cap L_b^1 \cap \overline{L_c^0 \cup L_c^u}$$ □

The disadvantage of this approach is that two lists exist for each line, and the complexity of the processing required to propagate the fault lists through an element is more than doubled.

Second-Order Analysis

To reduce computational complexity, we can associate a single fault list L_α with each line α. If the value of line α in the fault-free circuit is u, then $L_\alpha = \varnothing$. This means that if the value of the line is u, we will make no predictions on the value of this line in any of the faulty circuits. Hence some information is lost, and incorrect initialization of some faulty circuits is possible. If f is a fault that produces a u on a line α whose fault-free value is 0 or 1, then the entry in L_α corresponding to fault f is flagged (denoted by *f) and is called a *star fault*. This means that we do not know whether f is in L_α. If α is a primary output, then f is a *potentially detected* fault.

The rules for set operations for star faults are given in the table of Figure 5.14; here λ denotes an entry different from f or *f.

A	B	$A \cup B$	$A \cap B$	$A - B$	$B - A$
$*f$	λ	$\{*f,\lambda\}$	$\varnothing$	$*f$	λ
$*f$	f	f	$*f$	$\varnothing$	$*f$
$*f$	$*f$	$*f$	$*f$	$*f$	$*f$

Figure 5.14   Operations with star faults

Whenever a race or oscillation condition caused by fault f is identified on line α, f is entered as a star fault on the associated list. If it is decided that line α is oscillating, then those faults causing oscillation are precisely those entries in the set $(\tilde{L}_\alpha - L_\alpha) \cup (L_\alpha - \tilde{L}_\alpha)$ where $\tilde{L}_\alpha$ and L_α are the old and new fault lists. By changing these entries to star faults, the simulation can be continued and the simulation oscillation should cease [Chappell *et al.* 1974].

An interesting open problem is: What should be the initial contents of the fault lists? It is often not known how the initial state of the circuit was arrived at, and therefore it is not apparent whether a fault f would have influenced the initializing of line α. It is erroneous to set all fault lists initially to the empty set, and it is too pessimistic to place all faults into the set L_α^u. The former approach, however, is usually taken.

A more detailed discussion of multiple logic values in deductive simulation can be found in [Levendel and Menon 1980].

Limitations

The propagation of fault lists through an element is based on its Boolean equations. Thus *deductive fault simulation is compatible only in part with functional-level modeling*; namely it is applicable only to models using Boolean equations [Menon and Chappell 1978]. In principle, it can be extended to multiple logic values by increasing the number of fault lists associated with a line [Levendel and Menon 1980], but the corresponding increase in the complexity of the algorithm renders this approach impractical. Hence *deductive simulation is limited in practice to two or three logic values.*

The fault-list propagation mechanism cannot take full advantage of the concept of activity-directed simulation. For example, suppose that at a certain time we have activity only in one faulty circuit. Propagating this list event may generate many long fault-list computations. But most of the fault-list entries involved in this computation correspond to faulty circuits without activity at that time. This is especially time-consuming when a faulty circuit oscillates.

5.2.5 Concurrent Fault Simulation

Concurrent fault simulation [Ulrich and Baker 1974] is based on the observation that most of the time during simulation, most of the values in most of the faulty circuits agree with their corresponding values in the good circuit. The concurrent method simulates the good circuit N, and for every faulty circuit N_f, it simulates only those elements in N_f that are different from the corresponding ones in N. These differences are maintained for every element x in N in the form of a **concurrent fault list**, denoted by CL_x. Let x_f denote the replica of x in the circuit N_f. Let V_x (V_{x_f}) denote the ensemble of input, output, and (possibly) internal state values of x (x_f). During simulation, CL_x represents the set of all elements x_f that are different from x at the current simulated time. Elements x and x_f may differ in two ways. First, we may have $V_{x_f} \neq V_x$; this occurs when a fault effect caused by f has propagated to an input/output line or state variable of x_f. Second, f can be a *local fault* of x_f, that is, a fault inserted on an input/output line or state variable of x_f. A local fault f makes x_f different from x, even if $V_{x_f} = V_x$; this occurs when the input sequence applied so far does not activate f.

An entry in CL_x has the form (f, V_{x_f}). Figure 5.15(a) illustrates a concurrent fault list in pictorial form. The gates "hanging" from the good gate c are replicas of c in the faulty circuits with the faults α, β, a_1, and b_1. Here α and β are faults whose effects propagate to gate c; they cause, respectively, $a=1$ and $b=0$. Faults a_1 and b_1 are local faults of gate c. Note that b_1 appears in CL_c even if the values of a, b, and c in the presence of b_1 are the same as in the good circuit. Figure 5.15(b) shows CL_c in tabular form. By contrast, the fault list of c in deductive simulation is $L_c = \{\alpha, a_1\}$.

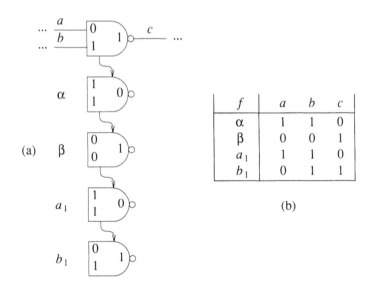

	f	a	b	c
	α	1	1	0
	β	0	0	1
	a_1	1	1	0
	b_1	0	1	1

(b)

Figure 5.15   Concurrent fault list for gate c (a) Pictorial representation (b) Tabular representation

A fault f is said to be *visible* on a line i when the values of i in N and N_f are different. Among the entries in the concurrent fault list of a gate x, only those corresponding to faults visible on its output appear also in the deductive fault list L_x. (In Figure 5.15, α and a_1 are visible faults.) In this sense, a deductive fault list is a subset of the corresponding concurrent fault list. Thus concurrent simulation requires more storage than deductive simulation.

Most of the work in concurrent simulation involves updating the dynamic data structures that represent the concurrent fault lists. Fault insertion takes place during the initialization phase (after fault collapsing); the initial content of every list CL_x consists of entries corresponding to the local faults of x. A local fault of x remains in CL_x until it is dropped. During simulation, new entries in CL_x represent elements x_f, whose values become different from the values of x; these are said to *diverge from x.* Conversely, entries removed from CL_x represent elements x_f, whose values become identical to those of x; these are said to *converge to x.* A fault is dropped by removing its entries from every list where it appears.

The following example illustrates how the concurrent fault lists change during simulation. For simplicity, we will refer to the circuit with the fault f as "circuit f."

Example 5.3: Figure 5.16(a) shows logic values and concurrent fault lists for a circuit in a stable state. Now assume that the primary input a changes from 1 to 0. This event occurs not only in the good circuit but (implicitly) also in all the faulty circuits that do not have entries in CL_c; the entries in CL_c have to be separately analyzed. In the good circuit, c is scheduled to change from 0 to 1. The event on a does not propagate in the circuit a_1; hence that gate is not evaluated. Evaluating gate c_α does not result in any event.

When the value of c is updated, the fault a_1 becomes visible. In the good circuit, the change of c propagates to the gate e. At the same time, we propagate a "list event" to indicate that a_1 is a newly visible fault on line c (see Figure 5.16(b)).

The evaluation of gate e in the good circuit does not produce an output event. An entry for the newly visible fault a_1 is added to CL_e. The other entries in CL_e are analyzed as follows (see Figure 5.16(c)):

- c does not change in the circuit c_1; the entry for c_1 remains in the list because c_1 is a local fault of gate e.

- Gate e in the circuit d_1 is evaluated and generates a 1/0 event; the same event occurs in the circuit β.

- c does not change in the circuit α; the entry for α is deleted from the list. □

An important feature of the concurrent fault simulation mechanism is that it individually evaluates elements in both the good and the faulty circuits. For evaluation, an entry in a fault list denoting a replica of a gate in the good circuit is just a gate with a different set of input/output values.

A line i in a faulty circuit may change even if i is stable in the good circuit. This is illustrated in Figure 5.16(c), where line e changes to 0 in circuits d_1 and β but remains stable at 1 in the good circuit. Figure 5.17 shows that a line i in the good circuit and some faulty circuits may also have simultaneous but different events. An event generated by an element in a faulty circuit propagates only inside that circuit.

We shall describe the concurrent simulation mechanism by reference to Figure 5.18. Line i is an output of A, and B is one of its fanouts. At a certain simulation time we may have an event on a line i in the good circuit and/or in several faulty circuits. The set of the simultaneous events occurring on the same line i is called a *composed event* and has the form (i,L), where L is a list of pairs (f,v_f'), in which f is a fault name

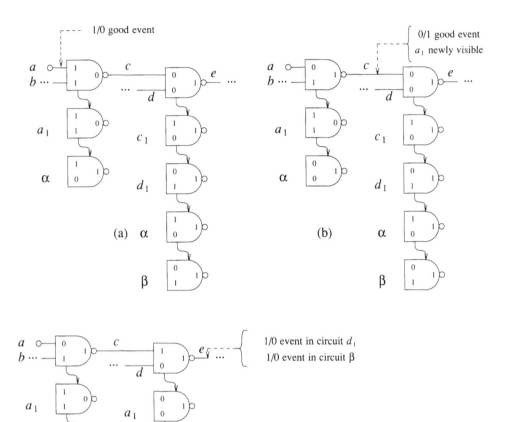

Figure 5.16   Changes in fault lists during simulation

(index) and v_f' is the scheduled value of line i in the circuit f. (The good circuit has $f=0$). The good-circuit event also occurs (implicitly) in all the faulty circuits that do *not* have an entry in CL_A.

The overall flow of event-directed logic simulation shown in Figure 3.12 is valid for concurrent fault simulation with the understanding that the events processed are composed events. Suppose that a composed event (i,L) has just been retrieved from the event list. First we update the values and the concurrent fault list of the source element A where the event originated. Then we update the values and the concurrent

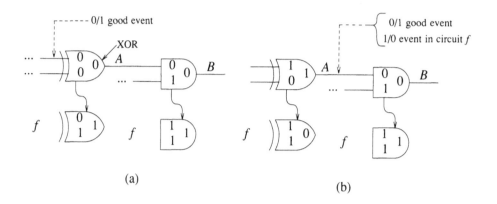

(a)   (b)

Figure 5.17

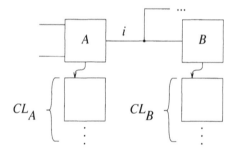

Figure 5.18

fault list of every element B on the fanout list of i and evaluate the activated elements (in the good and faulty circuits).

Figure 5.19 outlines the processing of the composed event at the source element A. Let v (v_f) be the current value of line i in the good circuit (in the circuit f). If an event v/v' occurs in the good circuit, then we have to analyze every entry in CL_A; otherwise we analyze only those entries with independent events. In the former case, if in a circuit f line i stays at value $v_f = v$ (i.e., the value of i in the circuit f is the same as the value of i in the good circuit before the change), then f is a newly visible fault on line i; these faults are collected into a list NV. The values of every analyzed entry f in CL_A, except for the newly visible ones, are compared to the values of the good element A, and if they agree, f is deleted from CL_A. (Practical implementation of the processing described in Figure 5.19 is helped by maintaining entries in concurrent fault lists and in lists of events ordered by their fault index.)

$NV = \varnothing$
if i changes in the good circuit **then**
  begin
    set i to v' in the good circuit
    for every $f \in CL_A$
      begin
        if $f \in L$ **then**
          begin
            set i to v_f' in circuit f
            if $V_{A_f} = V_A$ **then** delete f from CL_A
          end
        else /* no event in circuit f */
          if $v_f = v$ **then** add newly visible fault f to NV
          else if $V_{A_f} = V_A$ **then** delete f from CL_A
      end
  end
  else /* no good event for i */
    for every $f \in L$
      begin
        set i to v_f' in circuit f
        if $V_{A_f} = V_A$ **then** delete f from CL_A
      end

Figure 5.19   Processing of a composed event (i, L) at the source element A

Next, the composed event (i,L), together with the list NV of newly visible faults on line i, is propagated to every fanout element B. If a good event exists, then it activates B (for simplicity, we assume a two-pass strategy; thus evaluations are done after all the activated elements are determined). The processing of an element B_f depends on which lists (CL_B, L, NV) contain f. The way NV is constructed (see Figure 5.19) implies that f cannot belong to both NV and L. The different possible cases are labeled 1 through 5 in the Karnaugh map shown in Figure 5.20. The corresponding actions are as follows:

Case 1: (B_f exists in CL_B and no independent event on i occurs in N_f.) If a good event exists and it can propagate in N_f, then activate B_f. The good event on line i can propagate in the circuit f if $v_f = v$ and f is not the local s-a-v fault on the input i of B_f. For example, in Figure 5.16(b) the change of c from 0 to 1 propagates in the circuits d_1 and β but not in the circuits c_1 and α.

Case 2: (B_f exists in CL_B and an independent event on i occurs in N_f.) Activate B_f. Here we have independent activity in a faulty circuit; this is illustrated in Figure 5.17(b), where the event 1/0 activates the gate B_f.

Case 3: (An independent event on i occurs in N_f, but f does not appear in CL_B.) Add an entry for f to CL_B and activate B_f. This is shown in Figure 5.21. Here A and A_f

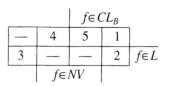

		$f \in CL_B$		
—	4	5	1	
3	—	—	2	$f \in L$
	$f \in NV$			

Figure 5.20 Possible cases in processing a composed event propagated to a fanout element B

have been evaluated because the input b changed from 0 to 1, and now i changes from 0 to 1 in the good circuit and from 0 to u in the circuit f. An entry for f is added to CL_B (with the same values as B) and activated.

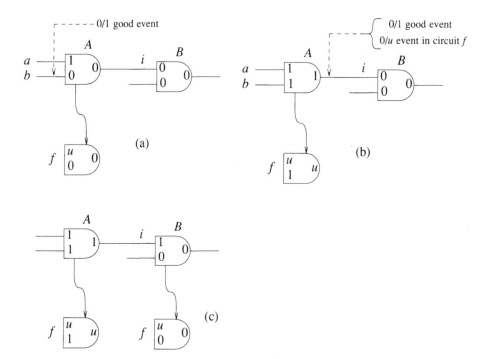

Figure 5.21

Case 4: (f is newly visible on line i and does not appear in CL_B.) Add an entry for f to CL_B. This is illustrated in Figure 5.16 by the addition of a_1 to CL_e.

Case 5: (f is a newly visible fault on line i, but an entry for f is already present in CL_B.) No action. In a combinational circuit this may occur only when there is reconvergent fanout from the origin of the fault f to the element B. Figure 5.22 provides such an example.

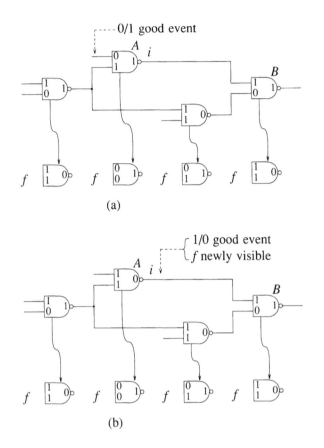

Figure 5.22

The activated elements are individually evaluated and resulting events for their outputs are merged into composed events. If a good event is generated, then all the comparisons for deletions from the fault list will be done when the composed event will have been retrieved from the event list (see Figure 5.19). If no good event is generated, however, any activated element B_f that does *not* generate an event should be immediately compared to the good element (they are both stable), and if their values agree, f should be deleted from CL_B.

Since the concurrent fault simulation is based on evaluation of individual elements, it is directly applicable to functional-level modeling [Abramovici *et al.* 1977]. In functional simulation, an event may also denote a change of an internal state variable of a functional element. Such a state event has the form (M, i, v'), where M is the element whose state variable i is scheduled to get value v'. In concurrent simulation we propagate *composed state events* of the form (M, i, L), where L has the same form as for line events. A composed state event is processed like a composed line event, with the only difference that both the source (A) and the fanout (B) of the event are the same element M.

Our definition of a concurrent fault list CL_x implies that for any entry x_f we store all the values of its inputs, outputs, and state variables. But this would waste large amounts of memory for a functional element x with a large number of state variables, because most of the time during simulation, most of the values of the elements in CL_x would be the same as the good values. Then it is preferable to maintain only the differences between the values of every x_f in the fault list and the good element values. (This represents a second-level application of the principle of concurrent simulation!) Memories are typical examples of elements of this category. We can view a memory as being "composed" of addressable words, and we can keep associated with every good word a concurrent list of faults that affect the contents of that word [Schuler and Cleghorn 1977].

5.2.6 Comparison

In this section we compare the parallel, the deductive, and the concurrent techniques with respect to the following criteria:

- capability of handling multiple logic values,

- compatibility with functional-level modeling,

- ability to process different delay models,

- speed,

- storage requirements.

While parallel and deductive simulation are reasonably efficient for two logic values, the use of three logic values $(0, 1, u)$ significantly increases their computational requirements, and they do not appear to be practical for more than three logic values. Both methods are compatible only in part with functional-level modeling, as they can process only components that can be entirely described by Boolean equations.

In both parallel and deductive simulation, the basic data structures and algorithms are strongly dependent on the number of logic values used in modeling. By contrast, the concurrent method provides only a mechanism to represent and maintain the differences between the good circuit and a set of faulty circuits, and this mechanism is independent of the way circuits are simulated. Thus concurrent simulation is transparent to the system of logic values used and does not impose any limitation on their number. Being based on evaluation of individual elements, concurrent simulation does not constrain the modeling of elements, and hence it is totally compatible with functional-level modeling and can support mixed-level and hierarchical modeling [Rogers *et al.* 1987]. It has also been applied to transistor-level circuits [Bose *et al.* 1982, Bryant and Schuster 1983, Saab and Hajj 1984, Kawai and Hayes 1984, Lo *et al.* 1987].

We have described the three methods by assuming a transition-independent nominal delay model. The use of more accurate timing models — such as rise and fall delays, inertial delays, etc. — leads to severe complications in the parallel and deductive methods. In contrast, because of its clean separation between fault processing and element evaluation, a concurrent simulator can be as sophisticated as the underlying good-circuit simulator. This means that detailed delay models can be included in concurrent simulation without complications caused by the fault-processing mechanism.

Theoretical analysis done by Goel [1980] shows that for a large combinational circuit of n gates, the run times of the deductive and parallel methods are proportional to n^2 and n^3 respectively. It is unlikely that any fault simulation algorithm can have worst-case linear complexity [Harel and Krishnamurthy 1987]. Experimental results presented in [Chang *et al.* 1974] show that (using 2-valued logic) deductive simulation is faster than parallel simulation for most sequential circuits, except small ($n<500$) ones.

Although no direct comparisons between the deductive and concurrent methods have been reported, qualitative arguments suggest that concurrent simulation is faster. One significant difference between the two methods is that a concurrent simulator processes only the active faulty circuits. This difference becomes more apparent when changes occur only in faulty circuits and especially when a faulty circuit oscillates. The reason is that a deductive simulator always recomputes a complete fault list (even to add or to delete a single entry), while a concurrent simulator propagates changes only in the active circuit. Another difference is that a concurrent simulator, being based on evaluation of individual elements, can make use of fast evaluation techniques, such as table look-up or input counting.

The main disadvantage of concurrent simulation is that it requires more memory than the other methods. The reason is that the values V_{x_f} of every element x_f in a concurrent fault list must be available for the individual evaluation of x_f. However, this disadvantage can be overcome by partitioning the set of faults for multipass simulation. Even if unlimited memory is available, partitioning the set of faults improves the efficiency of concurrent simulation by limiting the average size of the concurrent fault list [Henckels *et al.* 1980]. The most important advantages of concurrent simulation are its compatibility with different levels of modeling and its ability to process multiple logic values; these factors make it suitable for increasingly complex circuits and evolving technology.

The *parallel-value list* (PVL) is a fault simulation method [Moorby 1983, Son 1985] which combines several features of the parallel, deductive, and concurrent techniques. Consider the fault list representation based on a characteristic-vector structure, illustrated in Figure 5.12(d). Assume that the faulty values of every line are stored in another vector parallel to the characteristic vector (or in several parallel vectors for multivalued logic). The fault set is partitioned into groups of W faults. When all the characteristic-vector values in a group are 0, all the faulty values are the same as the good value and are not explicitly represented. The remaining groups, in which at least one faulty value is different, are maintained in a linked list, similar to a concurrent fault list. Given the groups in fault lists of the inputs of a device, the groups of the output fault list are determined by set-union and set-intersection operations similar to those used in deductive simulation. The computation of faulty values for the output groups proceeds as in parallel simulation. The PVL method requires less storage than concurrent simulation.

5.3 Fault Simulation for Combinational Circuits

Specialized fault simulation methods for combinational circuits are justified by the widespread use of design for testability techniques (to be described in a separate chapter) that transform a sequential circuit into a combinational one for testing purposes. For static testing, in which the basic step consists of applying an input vector and observing the results after the circuit has stabilized, we are interested only in the final (stable) values. In a combinational circuit the final values of the outputs do not depend on the order in which inputs or internal lines change. Thus a fault simulation method for combinational circuits can use a simplified (zero or unit) delay model.

5.3.1 Parallel-Pattern Single-Fault Propagation

The *Parallel-Pattern Single-Fault Propagation* (PPSFP) method [Waicukauski *et al.* 1985] combines two separate concepts — single-fault propagation and parallel-pattern evaluation.

Single-fault propagation is a serial fault simulation method specialized for combinational circuits. After a vector has been simulated using a fault-free simulator, SSFs are inserted one at a time. The values in every faulty circuit are computed by the same fault-free simulation algorithm and compared to their corresponding good values. The computation of faulty values starts at the site of the fault and continues until all faulty values become identical to the good values or the fault is detected.

Parallel-pattern evaluation is a simulation technique, introduced in Section 3.5, which simulates W vectors concurrently. For 2-valued logic, the values of a signal in W vectors are stored in a W-bit memory location. Evaluating gates by Boolean instructions operating on W-bit operands generates output values for W vectors in parallel. Of course, this is valid only in combinational circuits, where the order in which vectors are applied is not relevant.

A simulator working in this way cannot be event-driven, because events may occur only in some of the W vectors simulated in parallel. This implies that all the gates in the circuit should be evaluated in every vector, in the order of their level (for this we can use a compiled simulation technique). Let $a < 1$ denote the average activity in the circuit, that is, the average fraction of the gates that have events on their inputs in one vector. Then parallel-pattern evaluation will simulate $1/a$ more gates than an event-driven simulator. However, because W vectors are simultaneously simulated, parallel-pattern evaluation is more efficient if $W > 1/a$. For example, $W \geq 20$ will compensate for a value a of 5 percent. The overall speed-up is Wa. The implementation described in [Waicukauski *et al.* 1985] uses $W = 256$.

The PPSFP method combines single-fault propagation with parallel-pattern evaluation. First it does a parallel fault-free simulation of a group of W vectors. Then the remaining undetected faults are serially injected and faulty values are computed in parallel for the same set of vectors. Comparisons between good and faulty values involve W bits. The propagation of fault effects continues as long as faulty values are different from the good values in at least one vector. Detected faults are dropped and the above steps are repeated until all vectors are simulated or all faults are detected.

The PPSFP method has been successfully used to evaluate large sets (containing 1/2 million) of random vectors. However, it cannot support an algorithmic test generation process of the type illustrated in Figure 5.3, in which we need to know the faults detected by one vector before we generate the next one.

5.3.2 Critical Path Tracing

In this section we present the main concepts of the method called *critical path tracing* [Abramovici *et al.* 1984], which includes and extends features of earlier fault simulation techniques for combinational circuits [Roth *et al.* 1967, Ozguner *et al.* 1979, Su and Cho 1972, Hong 1978].

For every input vector, critical path tracing first simulates the fault-free circuit, then it determines the detected faults by ascertaining which signal values are *critical*.

Definition 5.1: A line *l* has a *critical value v* in the test (vector) *t* iff *t* detects the fault *l* s-a-$\overline{v}$. A line with a critical value in *t* is said to be *critical* in *t*.

Finding the lines critical in a test *t*, we immediately know the faults detected by *t*. Clearly, the primary outputs are critical in any test. (We assume completely specified tests; hence all values are binary.) The other critical lines are found by a backtracing process starting at the primary outputs. This process determines paths composed of critical lines, called *critical paths*. It uses the concept of *sensitive inputs*.

Definition 5.2: A gate input is *sensitive* (in a test *t*) if complementing its value changes the value of the gate output.

The sensitive inputs of a gate with two or more inputs are easily determined as follows:

1. If only one input *j* has the controlling value of the gate (*c*), then *j* is sensitive.

2. If all inputs have value $\overline{c}$, then all inputs are sensitive.

3. Otherwise no input is sensitive.

The sensitive inputs of a gate can be easily identified during the fault-free simulation of the circuit, as scanning for inputs with controlling value is an inherent part of gate evaluation.

The following lemma provides the basis of the critical path tracing algorithm.

Lemma 5.1: If a gate output is critical, then its sensitive inputs, if any, are also critical.

The next example illustrates critical path tracing in a fanout-free circuit.

Example 5.4: Figure 5.23(a) shows a circuit, its line values for the given test, and the sensitive gate inputs (marked by dots). Critical path tracing starts by marking the primary output as critical. The other critical lines are identified by recursive applications of Lemma 5.1. Figure 5.23(b) shows the critical paths as heavy lines. □

It is important to observe that critical path tracing has completely avoided the areas of the circuit bordered by *B* and *C*, because by working backward from the output, it first determined that *B* and *C* are not critical. In contrast, a conventional fault simulator would propagate the effects of all faults in these areas (to *B* and *C*), before discovering (at the output gate) that these faults are not detected.

For a fanout-free circuit, critical path tracing can be done by a simple depth-first tree-traversal procedure that marks as critical and recursively follows in turn every sensitive input of a gate with critical output. The next example illustrates the problem

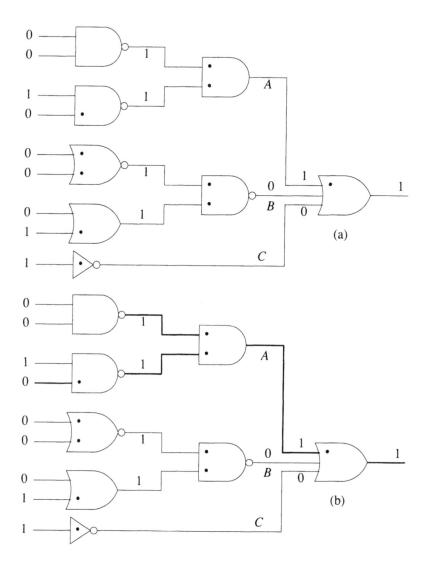

Figure 5.23   Example of critical path tracing in a fanout-free circuit

that appears in extending critical path tracing to the general case of circuits with reconvergent fanout.

Example 5.5:   For the circuit and the test given in Figure 5.24(a), we start at the primary output, and by repeatedly using Lemma 5.1, we identify F, D, A, and $B1$ as critical. We cannot, however, determine whether the stem B is critical without additional analysis. Indeed, the effects of the fault B s-a-0 propagate on two paths with different inversion parities such that they cancel each other when they reconverge at gate F. This phenomenon, referred to as *self-masking*, does not occur for the test shown in Figure 5.24(b), because the propagation of the fault effect along the path starting at $B2$ stops at gate E. Here B is critical.   □

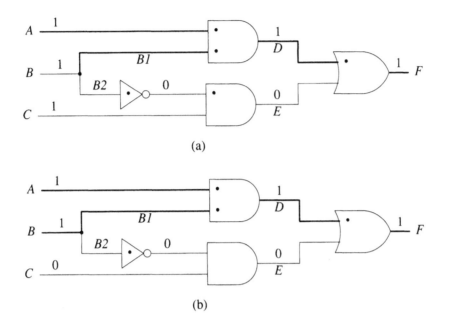

Figure 5.24   Example of stem analysis (a) B is self-masking (b) B is critical

Thus the main problem is to determine whether a stem x is critical, given that some of its fanout branches are critical. Let v be the value of x in the analyzed test t. One obvious solution is to explicitly simulate the fault x s-a-$\bar{v}$, and if t detects this fault, then mark x as critical. Critical path tracing solves this problem by a different type of analysis, which we describe next.

Definition 5.3: Let t be a test that activates fault f in a single-output combinational circuit. Let y be a line with level l_y, sensitized to f by t. If every path sensitized to f either goes through y or does not reach any line with level greater than l_y, then y is said to be a *capture line* of f in test t.

A capture line, if one exists, is a bottleneck for the propagation of fault effects, because it is common to all the paths on which the effects of f can propagate to the primary output in test t. If t detects f, then there exists at least one capture line of f, namely the primary output itself. If the effect of f propagates on a single path, then every line on that path is a capture line of f.

Example 5.6: Consider the circuit in Figure 5.25 and let f be b s-a-0. The capture lines of f in the test shown are e, j, and q. Lines n and p are sensitized to f but are not capture lines of f.   □

Let y be a capture line of fault f in test t and assume that y has value v in t. It is easy to show that any capture line of y s-a-$\bar{v}$ is also a capture line of f. In other words, the capture lines of a fault form a "transitive chain." For example, in Figure 5.25, j and q are also capture lines of e s-a-0, and q is also a capture line of j s-a-1.

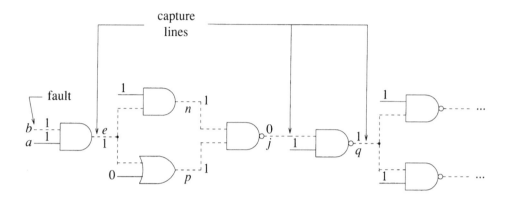

- - - sensitized path

Figure 5.25   Example of capture lines

Theorem 5.1: A test t detects the fault f iff all the capture lines of f in t are critical in t.

Proof: First note that if no capture line exists, then t does not detect f (because otherwise there exists at least one capture line). Let y be an arbitrary capture line of f in t and let v be its value. The value of y in the presence of f is $\bar{v}$, and no other effect of f but $y=\bar{v}$ can reach any line whose level is greater than the level of y.

1. Let us assume that all the capture lines of f in t are critical. Hence, the error $y=\bar{v}$ caused by y s-a-$\bar{v}$ propagates to the primary output. Then the same is true for the error $y=\bar{v}$ caused by f. Therefore t detects f.

2. Let us assume that t detects f. Hence, the error $y=\bar{v}$ caused by f at y propagates to the primary output. Then the same is true for the error $y=\bar{v}$ caused by y s-a-$\bar{v}$. Thus t detects y s-a-$\bar{v}$ and therefore all the capture lines of f are critical in t.   □

Theorem 5.1, together with the transitivity property of a set of capture lines, shows that to determine whether a stem is critical, we may not need to propagate the effects of the stem fault "all the way" to a primary output as done in explicit fault simulation. Rather, it is sufficient to propagate the fault effects only to the first capture line of the stem fault (i.e., the one closest to the stem). Then the stem is critical iff the capture line is critical.

As capture lines are defined for a single-output circuit, we partition a circuit with m primary outputs into m single-output circuits called *cones*. A cone contains all the logic feeding one primary output. To take advantage of the simplicity of critical path tracing in fanout-free circuits, within each cone we identify *fanout-free regions* (FFRs). Figure 5.26 shows these structures for an adder with two outputs. The inputs of a FFR are checkpoints of the circuit, namely fanout branches and/or primary inputs. The output of a FFR is either a stem or a primary output. Constructing cones and FFRs is a preprocessing step of the algorithm.

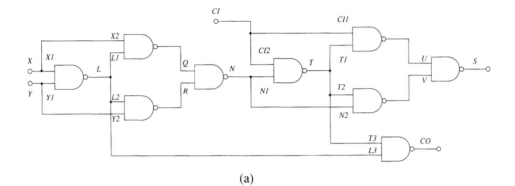

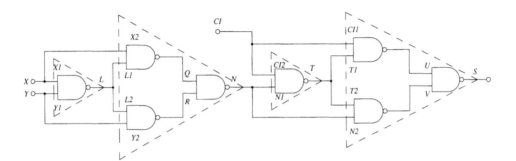

(a)

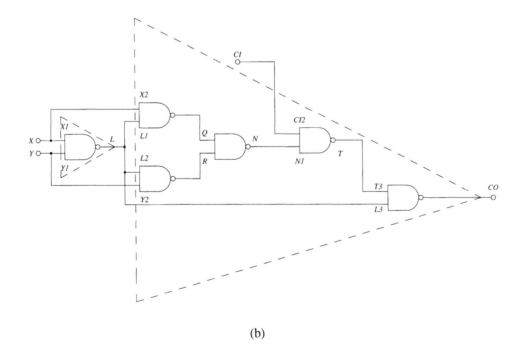

(b)

Figure 5.26　(a) Full-adder circuit (b) Cones for S and CO partitioned into FFRs

Figure 5.27 outlines the critical path tracing algorithm for evaluating a given test. It assumes that fault-free simulation, including the marking of the sensitive gate inputs, has been performed. The algorithm processes every cone starting at its primary output and alternates between two main operations: critical path tracing inside a FFR, represented by the procedure *Extend*, and checking a stem for criticality, done by the function *Critical*. Once a stem j is found to be critical, critical path tracing continues from j.

```
for every primary output z
    begin
        Stems_to_check = ∅
        Extend(z)
        while (Stems_to_check ≠ ∅)
            begin
                j = the highest level stem in Stems_to_check
                remove j from Stems_to_check
                if Critical(j) then Extend(j)
            end
    end
```

Figure 5.27   Outline of critical path tracing

Figure 5.28 shows the recursive procedure *Extend(i)*, which backtraces all the critical paths inside a FFR starting from a given critical line i by following lines marked as *sensitive*. *Extend* stops at FFR inputs and collects all the stems reached in the set *Stems_to_check*.

```
Extend(i)
begin
    mark i as critical
    if i is a fanout branch then
        add stem(i) to Stems_to_check
    else
        for every input j of i
            if sensitive(j) then Extend(j)
end
```

Figure 5.28   Critical path tracing in a fanout-free region

Figure 5.29 outlines the function *Critical(j)*, which determines whether the stem j is critical by a breadth-first propagation of the effects of the stem fault. *Frontier* is the set of all gates currently on the frontier of this propagation. A gate in *Frontier* has been reached by one or more fault effects from j, but we have not yet determined whether these propagate through the gate. This analysis is represented by the function *Propagates*. Because fault effects are propagated only in one cone, *Frontier* is bound to become empty. If the last gate removed from *Frontier* propagates fault effects, then its output is the first capture line of the stem fault and the result depends on the status (critical or not) of the capture line.

```
Critical(j)
begin
    Frontier = {fanouts of j}
    repeat
      begin
        i = lowest level gate in Frontier
        remove i from Frontier
        if (Frontier ≠ ∅) then
          begin
            if Propagates(i) then add fanouts of i to Frontier
          end
        else
          begin
            if Propagates(i) and i is critical then return TRUE
            return FALSE
          end
      end
end
```

Figure 5.29 Stem analysis

The function *Propagates(i)* determines whether gate i propagates the fault effects reaching its inputs based on the following rule.

Lemma 5.2: A gate i propagates fault effects iff:

1. either fault effects arrive only on sensitive inputs of i,

2. or fault effects arrive on all the nonsensitive inputs of i with controlling value and only on these inputs.

Proof: Based on Lemma 4.1 and the definition of sensitive inputs. □

Figure 5.30 shows different cases of propagation of fault effects through an AND gate. Fault effects are indicated by arrows.

Example 5.7: The table in Figure 5.31 shows the execution trace of critical path tracing in the circuit of Figure 5.26(a) for the test 111. Figure 5.32 shows the obtained critical paths. Note that the stem L is critical in the cone of S, but it is self-masking in the cone of CO. □

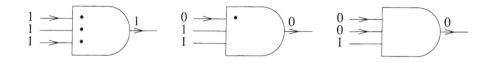

Figure 5.30 Propagation of fault effects through an AND gate

FFR traced	Critical lines	*Stems_to_check*	Stem checked	Capture line
S	S, U, CI1, T1	T, CI		
		CI	T	U
T	T, N1	CI, N		
		CI	N	U
N	N, Q, R, L1, L2	CI, L		
		CI	L	N
L	L, X1, Y1	CI, X, Y		
		X, Y	CI	U
CI	CI	X, Y		
		Y	X	R
X	X	Y		
		Ø	Y	Q
Y	Y	Ø		
CO	CO, L3	L		
		Ø	L	—

Figure 5.31 Execution trace of critical path tracing in the circuit of Figure 5.26 for the test 111

When we analyze a stem, we have already backtraced a critical path between a primary output and one of the fanout branches of the stem. The effect of the stem fault would propagate on the same critical path, unless self-masking occurs. This may take place only when effects of the stem fault reach a reconvergence gate on paths with different inversion parities (see Figure 5.24). If we examine the circuit of Figure 5.26, we can observe that this situation can never occur for the stems T and L in the cone of S. A simple preprocessing technique to identify this type of stem is described in [Abramovici *et al.* 1984]. The resulting benefit is that whenever such a stem is reached by backtracing, it can be immediately identified as critical without any additional analysis.

Other techniques to speed up critical path tracing are presented in [Abramovici *et al.* 1984].

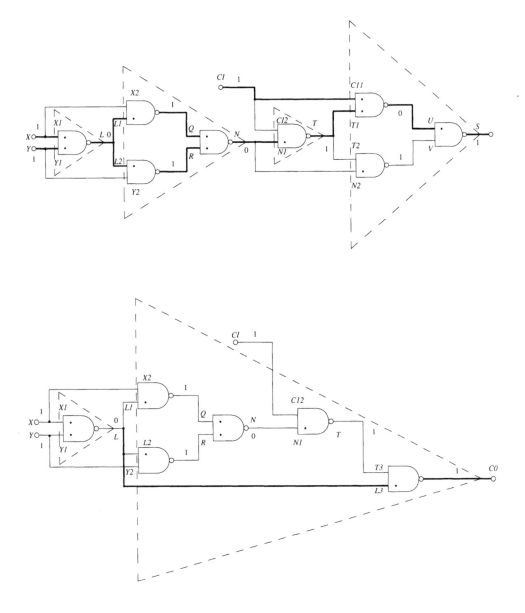

Figure 5.32   Critical paths in the circuit of Figure 5.26 for the test 111

Sometimes critical path tracing may not identify all the faults detected by a test. This may occur in a test t that propagates the effect of a fault on multiple paths that reconverge at a gate without sensitive inputs in t. For example, in Figure 5.33 the effects of j s-a-0 propagate on two paths and reach the reconvergence gate on nonsensitive inputs. In this case critical path tracing would not reach j. This situation occurs seldom in practical circuits. In the following we shall show that even when it does occur, its impact is negligible.

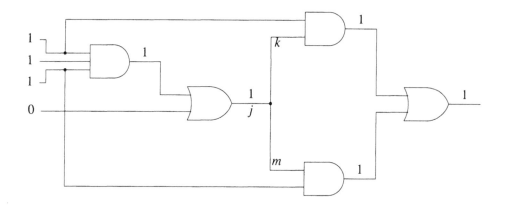

Figure 5.33

For *test grading*, it does not matter which test detects a fault but whether the fault is detected by any of the tests in the evaluated test set. Even if a fault is not correctly recognized as detected in one test, it is likely that it will be detected in other tests. Eventually the only faults incorrectly flagged as not detected are those that are detected *only* by multiple-path sensitization with reconvergence at a gate with nonsensitive inputs. If this unlikely situation occurs, then the computed fault coverage will be slightly pessimistic.

In the context of test generation, the function of critical path tracing is to aid in the selection of the next target fault (see Figure 5.3). Again, for the overall process it does not matter which test detects a fault but whether the fault is detected by any of the tests generated so far. The only adverse effect on the test generation process will occur if critical path tracing incorrectly identifies a fault f as not detected by any of the currently generated tests *and* f is selected as the next target fault. In this unlikely situation the test generator will needlessly generate a test for the already detected fault f. As we will show in the next chapter, usually f will be detected by sensitizing a single path, and then critical path tracing will mark f as detected. If the test generator has obtained a test for f, but f is not marked as detected by critical path tracing, then we directly mark the corresponding line as critical and restart critical path tracing from that line. Thus in practical terms there is no impact on the test generation process.

Compared with conventional fault simulation, the distinctive features of critical path tracing are as follows:

- *It directly identifies the faults detected by a test*, without simulating the set of all possible faults. Hence the work involved in propagating the faults that are not detected by a test towards the primary outputs is avoided.

- *It deals with faults only implicitly*. Therefore fault enumeration, fault collapsing, fault partitioning (for multipass simulation), fault insertion, and fault dropping are no longer needed.

- *It is based on a path tracing algorithm* that does not require computing values in the faulty circuits by gate evaluations or fault list processing.

- *It is an approximate method.* The approximation seldom occurs and results in not marking as detected some faults that are actually detected by the evaluated test. We have shown that even if the approximation occurs, its impact on the applications of critical path tracing is negligible.

Consequently, critical path tracing is faster and requires less memory than conventional fault simulation. Experimental results presented in [Abramovici *et al.* 1984] show that critical path tracing is faster than concurrent fault simulation.

Additional features of critical path tracing and their use in test generation are described in Chapter 6.

Antreich and Schulz [1987] developed a method that combines features similar to those used in critical path tracing with parallel-pattern evaluation.

5.4 Fault Sampling

The computational requirements (i.e., run-time and memory) of the general fault simulation methods increase with the number of simulated faults. Let M be the number of (collapsed) SSFs in the analyzed circuit and let K be the number of faults detected by the evaluated test sequence T. Then the fault coverage of T is $F=K/M$. *Fault sampling* [Butler *et al.* 1974, Agrawal 1981, Wadsack 1984] is a technique that reduces the cost of fault simulation by simulating only a random sample of $m<M$ faults.

Let k be the number of faults detected by T when simulating m faults. There exists an obvious trade-off between the cost of fault simulation and the accuracy of the estimated fault coverage $f=k/m$. This trade-off is controlled by the size m of the sample. The problem is to determine a sample size m such that we can be confident (with a specified confidence level c) that the error in the estimated fault coverage is bounded by a given e_{max}. In other words, we want to determine m such that the estimated fault coverage f belongs to the interval $[F-e_{max}, F+e_{max}]$ with a probability c.

Since the m faults in the sample are randomly selected, we can regard the number of detected faults k as a random variable. The probability that T will detect k faults from a random sample of size m, given that it detects K faults from the entire set of M faults, is

$$P_k(m,M,K) = \frac{\begin{pmatrix} K \\ k \end{pmatrix} \begin{pmatrix} M-K \\ m-k \end{pmatrix}}{\begin{pmatrix} M \\ m \end{pmatrix}}$$

This is a hypergeometric distribution with mean

$$\mu_k = m \, \frac{K}{M} = mF$$

and variance

$$\sigma_k^2 = m\frac{K}{M}\left(1 - \frac{K}{M}\right)\frac{M-m}{M-1} \approx mF(1-F)(1 - m/M)$$

where σ_k denotes the standard deviation. For large M, this distribution can be approximated by a normal distribution with mean μ_k and standard deviation σ_k. Hence the estimated fault coverage f can be considered to be a random variable with normal distribution and mean

$$\mu_f = \mu_k/m = F$$

and variance

$$\sigma_f^2 = \sigma_k^2/m^2 = \frac{1}{m}F(1-F)(1 - m/M)$$

With a confidence level of 99.7 percent, we can claim that the estimated fault coverage f is in the interval $[F-3\sigma_f, F+3\sigma_f]$. Therefore, it is almost certain that the estimation error is bounded by $3\sigma_f$; i.e., the maximum error e_{max} is given by

$$e_{max} = 3\sqrt{F(1-F)(1-m/M)}\,1/m$$

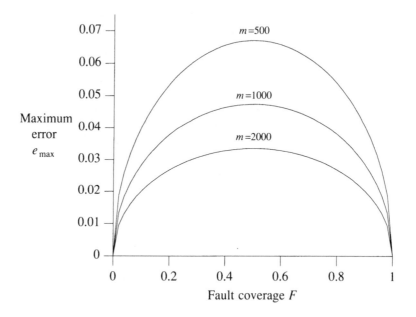

Figure 5.34   The maximum error in the estimated fault coverage

To reduce significantly the cost of fault simulation, the sample size m should be much smaller than the total number of faults M. With $m \ll M$, we can approximate $(1-m/M) \approx 1$. Then the error becomes independent of the total number of faults M. Figure 5.34 illustrates how e_{max} depends on F for several values of m. Note that the worst case occurs when the true fault coverage F is 50 percent. This analysis shows that a sample size of 1000 ensures that the estimation error is less than 0.05.

Fault sampling can provide an accurate estimate of the fault coverage at a cost substantially lower than that of simulating the entire set of faults. However, no information is available about the detection status of the faults not included in the sample; hence it may be difficult to improve the fault coverage.

5.5 Statistical Fault Analysis

Because exact fault simulation is an expensive computational process, approximate fault simulation techniques have been developed with the goal of trading off some loss of accuracy in results for a substantial reduction in the computational cost. In this section we examine the principles used in STAFAN, a *Sta*tistical *F*ault *An*alysis method which provides a low-cost alternative to exact fault simulation [Jain and Agrawal 1985].

Like a conventional fault simulator, STAFAN includes a fault-free simulation of the analyzed circuit N for a test sequence T. STAFAN processes the results of the fault-free simulation to estimate, for every SSF under consideration, its probability of being detected by T. The overall fault coverage is then estimated based on the detection probabilities of the individual faults.

First assume that N is a combinational circuit. STAFAN treats T as a set of n independent random vectors. Let d_f be the probability that a randomly selected vector of T detects the fault f. Because the vectors are assumed independent, the probability of not detecting f with n vectors is $(1-d_f)^n$. Then the probability d_f^n that a set of n vectors detects f is

$$d_f^n = 1 - (1-d_f)^n$$

Let Φ be the set of faults of interest. The expected number of faults detected by n vectors is $D_n = \sum_{f \in \Phi} d_f^n$, and the corresponding expected fault coverage is $D_n / |\Phi|$ (which is also the average detection probability).

The basis for the above computations is given by the detection probabilities d_f. Let f be the *s-a-v* fault on line l. To detect f, a vector must activate f by setting l to value $\bar{v}$ and must propagate the resulting fault effect to a primary output. STAFAN processes the results of the fault-free simulation to estimate separately the probability of activating f and the probability of propagating a fault effect from l to a primary output. These probabilities are defined as follows:

- $C1(l)$, called the *1-controllability* of l, is the probability that a randomly selected vector of T sets line l to value 1. The *0-controllability*, $C0(l)$, is similarly defined for value 0.

- $O(l)$, called the *observability* of l, is the probability that a randomly selected vector of T propagates a fault effect from l to a primary output (our treatment of observabilities is slightly different from that in [Jain and Agrawal 1985]).

STAFAN uses the simplifying assumption that activating the fault and propagating its effects are independent events. Then the detection probabilities of the *s-a-0* and *s-a-1* faults on l can be computed by $C1(l)O(l)$ and $C0(l)O(l)$. (Some pitfalls of the independence assumption are discussed in [Savir 1983]).

To estimate controllabilities and observabilities, STAFAN counts different events that occur during the fault-free simulation of n vectors. A *0-count* and a *1-count* are maintained for every line l; they are incremented, respectively, in every test in which l has value 0 or 1. After simulating n vectors, the 0- and the 1- controllability of l are computed as

$$C0(l) = \frac{0\text{-}count}{n}$$

and

$$C1(l) = \frac{1\text{-}count}{n}$$

An additional *sensitization-count* is maintained for every gate input l. This count is incremented in every test in which l is a sensitive input of the gate (see Definition 5.2). After simulating n vectors, the probability $S(l)$ that a randomly selected vector propagates a fault effect from l to the gate output is computed as

$$S(l) = \frac{sensitization\text{-}count}{n}$$

The computation of observabilities starts by setting $O(i)=1$ for every primary output i. Observabilities for the other lines are computed via a backward traversal of the circuit, in which $O(l)$ is computed based on the observability of its immediate successor(s). Let l be an input of a gate with output m. To propagate a fault effect from l to a primary output, we need to propagate it from l to m and from m to a primary output. Using the simplifying assumption that these two problems are independent, we obtain $O(l)$ as

$$O(l) = S(l)O(m)$$

Now let l be a stem with k fanout branches $l_1, l_2, ..., l_k$ (see Figure 5.35). Assume that we have computed all the $O(l_i)$ values, and now we want to determine $O(l)$. Let L_i (L) denote the event whose probability is $O(l_i)$ ($O(l)$), that is, the propagation of a fault effect from l_i (l) to a primary output. Here we use two more simplifying assumptions: (1) the events $\{L_i\}$ are independent, and (2) the event L occurs if and only if any subset of $\{L_i\}$ occurs. Then

$$L = \bigcup_{i=1}^{k} L_i$$

Because of reconvergent fanout, L may occur even when none of the L_i events happens, and the occurrence of a subset of $\{L_i\}$ does not guarantee that L will take place. STAFAN treats the probability of $\bigcup_{i=1}^{k} L_i$ as an upper bound on the value of $O(l)$. The lower bound is obtained by assuming that L can be caused by the most probable L_i. Thus

$$\max_{1 \leq i \leq k} O(l_i) \leq O(l) \leq P\left(\bigcup_{i=1}^{k} L_i\right)$$

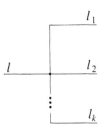

Figure 5.35

For example, for $k=2$, $P(L_1 \cup L_2) = O(l_1) + O(l_2) - O(l_1)O(l_2)$. Based on these bounds, STAFAN computes the observability of the stem l by

$$O(l) = (1-\alpha)\max_{1 \le i \le k} O(l_i) + \alpha P(\bigcup_{i=1}^{k} L_i)$$

where α is a constant in the range [0,1]. The lower bound for $O(l)$ is obtained with $\alpha=0$, and the upper bound corresponds to $\alpha=1$. The experiments reported in [Jain and Agrawal 1985] show that the results are insensitive to the value of α.

STAFAN applies the same computation rules to estimate detection probabilities in a sequential circuit (some additional techniques are used to handle feedback loops). This results in more approximations, because both fault activation and error propagation are, in general, achieved by sequences of related vectors, rather than by single independent vectors. In spite of its approximations, the fault coverage obtained by STAFAN was found to be within 6 percent of that computed by a conventional fault simulator [Jain and Singer 1986].

Based on their detection probabilities computed by STAFAN, faults are grouped in the following three ranges:

- the high range, containing faults with detection probabilities greater than 0.9;

- the low range, containing faults whose detection probabilities are smaller than 0.1;

- the middle range, grouping faults with detection probabilities between 0.1 and 0.9.

The faults in the high range are considered likely to be detected by the applied test; those in the low range are assumed not detected, and no prediction is made about the detectability of the faults in the middle range. The results for sequential circuits reported in [Jain and Singer 1986] show that the accuracy of the predictions made by STAFAN is as follows:

- Between 91 and 98 percent of the faults in the high range are indeed detected.

- Between 78 and 98 percent of the faults in the low range are indeed undetected.

- The middle range contains between 2 and 25 percent of the faults (in most cases less than 10 percent).

The results presented in [Huisman 1988] also show that lack of accuracy in fault labeling is the main problem in applying STAFAN.

STAFAN requires the following additional operations to be performed during a fault-free simulation run. First, the 0-count and the 1-count of every gate output, and the sensitization-count of every gate input, must be updated in every vector. Second, controllabilities, observabilities, and detection probabilities are computed after simulating a group of n vectors. Since n can be large, the complexity of STAFAN is dominated by the first type of operations (updating counters); their number is proportional to Gn, where G is the gate count of the circuit.

5.6 Concluding Remarks

Fault simulation plays an important role in ensuring a high quality for digital circuits. Its high computational cost motivates new research in this area. The main research directions are hardware support, new algorithms, and the use of hierarchical models.

One way to use hardware support to speed up fault simulation is by special-purpose accelerators attached to a general-purpose host computer. The architectures for fault simulation are similar to those described in Chapter 3 for logic simulation. Although any hardware accelerator for logic simulation can be easily adapted for serial fault simulation, the serial nature of this process precludes achieving a significant speed-up. Hardware and microprogrammed implementations of the concurrent algorithm (for example, [Chan and Law 1986], [Stein et al. 1986]) obtain much better performance.

Interconnected general-purpose processors can also provide support for fault simulation by partitioning the set of faults among processors. Each processor executes the same concurrent algorithm working only on a subset of faults. The processors can be parallel processors connected by a shared bus or independent processors connected by a network [Goel et al. 1986, Duba et al. 1988]. With proper fault-partitioning procedures, the speed-up obtained grows almost linearly with the number of processors used.

Two new algorithms have been developed for synchronous sequential circuits using a 0-delay model. One is based on extending critical path tracing [Menon et al. 1988]. The other, called *differential fault simulation* (DSIM), combines concepts of concurrent simulation and single fault propagation [Cheng and Yu 1989]. DSIM simulates in turn every faulty circuit, by keeping track of the differences between its values and those of the previously simulated circuit.

Hierarchical fault simulation [Rogers et al. 1987] relies on a hierarchical model of the simulated circuit, whose components have both a functional model (a C routine) and a structural one (an interconnection of primitive elements). The simulator uses the concurrent method and is able to switch between the functional and the structural models of components. Faults are inserted in the components modeled at the lower level. Efficiency is gained by using the higher-level models for fault-effect propagation.

REFERENCES

[Abramovici *et al.* 1977] M. Abramovici, M. A. Breuer, and K. Kumar, "Concurrent Fault Simulation and Functional Level Modeling," *Proc. 14th Design Automation Conf.*, pp. 128-137, June, 1977.

[Abramovici *et al.* 1984] M. Abramovici, P. R. Menon, and D. T. Miller, "Critical Path Tracing: An Alternative to Fault Simulation," *IEEE Design & Test of Computers*, Vol. 1, No. 1, pp. 83-93, February, 1984.

[Abramovici *et al.* 1986] M. Abramovici, J. J. Kulikowski, P. R. Menon, and D. T. Miller, "SMART and FAST: Test Generation for VLSI Scan-Design Circuits," *IEEE Design & Test of Computers*, Vol. 3, No. 4, pp. 43-54, August, 1986.

[Agrawal 1981] V. D. Agrawal, "Sampling Techniques for Determining Fault Coverage in LSI Circuits," *Journal of Digital Systems*, Vol. 5, No. 3, pp. 189-202, Fall, 1981.

[Agrawal *et al.* 1981] V. D. Agrawal, S. C. Seth, and P. Agrawal, "LSI Product Quality and Fault Coverage," *Proc. 18th Design Automation Conf.*, pp. 196-203, June, 1981.

[Antreich and Schulz 1987] K. J. Antreich and M. H. Schulz, "Accelerated Fault Simulation and Fault Grading in Combinational Circuits," *IEEE Trans. on Computer-Aided Design*, Vol. CAD-6, No. 9, pp. 704-712, September, 1987.

[Armstrong 1972] D. B. Armstrong, "A Deductive Method of Simulating Faults in Logic Circuits," *IEEE Trans. on Computers*, Vol. C-21, No. 5, pp. 464-471, May, 1972.

[Bose *et al.* 1982] A. K. Bose, P. Kozak, C-Y. Lo, H. N. Nham, E. Pacas-Skewes, and K. Wu, "A Fault Simulator for MOS LSI Circuits," *Proc. 19th Design Automation Conf.*, pp. 400-409, June, 1982.

[Bryant and Schuster 1983] R. E. Bryant and M. D. Schuster, "Fault Simulation of MOS Digital Circuits," *VLSI Design*, Vol. 4, pp. 24-30, October, 1983.

[Butler *et al.* 1974] T. T. Butler, T. G. Hallin, J. J. Kulzer, and K. W. Johnson, "LAMP: Application to Switching System Development," *Bell System Technical Journal*, Vol. 53, pp. 1535-1555, October, 1974.

[Chan and Law 1986] T. Chan and E. Law, "MegaFAULT: A Mixed-Mode, Hardware Accelerated Concurrent Fault Simulator," *Proc. Intn'l. Conf. on Computer-Aided Design*, pp. 394-397, November, 1986.

[Chang *et al.* 1974] H. Y. Chang, S. G. Chappell, C. H. Elmendorf, and L. D. Schmidt, "Comparison of Parallel and Deductive Simulation Methods," *IEEE Trans. on Computers*, Vol. C-23, No. 11, pp. 1132-1138, November, 1974.

[Chappell *et al.* 1974] S. G. Chappell, C. H. Elmendorf, and L. D. Schmidt, "LAMP: Logic-Circuit Simulators," *Bell System Technical Journal*, Vol. 53, pp. 1451-1476, October, 1974.

[Cheng and Yu 1989] W.-T. Cheng and M.-L. Yu, "Differential Fault Simulation — A Fast Method Using Minimal Memory," *Proc. 26th Design Automation Conf.*, pp. 424-428, June, 1989.

[Daniels and Bruce 1985] R. G. Daniels and W. C. Bruce, "Built-In Self-Test Trends in Motorola Microprocessors," *IEEE Design & Test of Computers*, Vol. 2, No. 2, pp. 64-71, April, 1985.

[Davidson and Lewandowski 1986] S. Davidson and J. L. Lewandowski, "ESIM/AFS — A Concurrent Architectural Level Fault Simulation," *Proc. Intn'l. Test Conf.*, pp. 375-383, September, 1986.

[Duba *et al.* 1988] P. A. Duba, R. K. Roy, J. A. Abraham, and W. A. Rogers, "Fault Simulation in a Distributed Environment," *Proc. 25th Design Automation Conf.*, pp. 686-691, June, 1988.

[Gai *et al.* 1986] S. Gai, F. Somenzi and E. Ulrich, "Advanced Techniques for Concurrent Multilevel Simulation," *Proc. Intn'l. Conf. on Computer-Aided Design*, pp. 334-337, November, 1986.

[Godoy and Vogelsberg 1971] H. C. Godoy and R. E. Vogelsberg, "Single Pass Error Effect Determination (SPEED)," *IBM Technical Disclosure Bulletin*, Vol. 13, pp. 3443-3344, April, 1971.

[Goel 1980] P. Goel, "Test Generation Costs Analysis and Projections," *Proc. 17th Design Automation Conf.*, pp. 77-84, June, 1980.

[Goel *et al.* 1986] P. Goel, C. Huang, and R. E. Blauth, "Application of Parallel Processing to Fault Simulation," *Proc. Intn'l. Conf. on Parallel Processing*, pp. 785-788, August, 1986.

[Harel and Krishnamurthy 1987] D. Harel and B. Krishnamurthy, "Is There Hope for Linear Time Fault Simulation?," *Digest of Papers 17th Intn'l. Symp. on Fault-Tolerant Computing*, pp. 28-33, July, 1987.

[Henckels *et al.* 1980] L. P. Henckels, K. M. Brown, and C-Y. Lo, "Functional Level, Concurrent Fault Simulation," *Digest of Papers 1980 Test Conf.*, pp. 479-485, November, 1980.

[Hong 1978] S. J. Hong, "Fault Simulation Strategy for Combinational Logic Networks," *Digest of Papers 8th Annual Intn'l Conf. on Fault-Tolerant Computing*, pp. 96-99, June, 1978.

[Huisman 1988] L. M. Huisman, "The Reliability of Approximate Testability Measures," *IEEE Design & Test of Computers*, Vol. 5, No. 6, pp. 57-67, December, 1988.

[Iyengar and Tang 1988] V. S. Iyengar and D. T. Tang, "On Simulating Faults in Parallel," *Digest of Papers 18th Intn'l. Symp. on Fault-Tolerant Computing*, pp. 110-115, June, 1988.

[Jain and Agrawal 1985] S. K. Jain and V. D. Agrawal, "Statistical Fault Analysis," *IEEE Design & Test of Computers*, Vol. 2, No. 1, pp. 38-44, February, 1985.

[Jain and Singer 1986] S. K. Jain and D. M. Singer, "Characteristics of Statistical Fault Analysis," *Proc. Intn'l. Conf. on Computer Design*, pp. 24-30, October, 1986.

[Kawai and Hayes 1984] M. Kawai and J. P. Hayes, "An Experimental MOS Fault Simulation Program CSASIM," *Proc. 21st Design Automation Conf.*, pp. 2-9, June, 1984.

[Ke *et al.* 1988] W. Ke, S. Seth, and B. B. Bhattacharya, "A Fast Fault Simulation Algorithm for Combinational Circuits," *Proc. Intn'l. Conf. on Computer-Aided Design*, pp. 166-169, November, 1988.

[Levendel 1980] Y. H. Levendel, private communication, 1980.

[Levendel and Menon 1980] Y. H. Levendel and P. R. Menon, "Comparison of Fault Simulation Methods — Treatment of Unknown Signal Values," *Journal of Digital Systems*, Vol. 4, pp. 443-459, Winter, 1980.

[Lo *et al.* 1987] C-Y. Lo, H. H. Nham, and A. K. Bose, "Algorithms for an Advanced Fault Simulation System in MOTIS," *IEEE Trans. on Computer-Aided Design*, Vol. CAD-6, No. 3, pp. 232-240, March, 1987.

[McCluskey and Buelow 1988] E. J. McCluskey and F. Buelow, "IC Quality and Test Transparency," *Proc. Intn'l. Test Conf.*, pp. 295-301, September, 1988.

[Menon and Chappell 1978] P. R. Menon and S. G. Chappell, "Deductive Fault Simulation with Functional Blocks," *IEEE Trans. on Computers*, Vol. C-27, No. 8, pp. 689-695, August, 1978.

[Menon *et al.* 1988] P. R. Menon, Y. H. Levendel, and M. Abramovici, "Critical Path Tracing in Sequential Circuits," *Proc. Intn'l. Conf. on Computer-Aided Design*, pp. 162-165, November, 1988.

[Moorby 1983] P. R. Moorby, "Fault Simulation Using Parallel Value Lists," *Proc. Intn'l. Conf. on Computer-Aided Design*, pp. 101-102, September, 1983.

[Narayanan and Pitchumani 1988] V. Narayanan and V. Pitchumani, "A Parallel Algorithm for Fault Simulation on the Connection Machine," *Proc. Intn'l. Test Conf.*, pp. 89-93, September, 1988.

[Ozguner *et al.* 1979] F. Ozguner, W. E. Donath, and C. W. Cha, "On Fault Simulation Techniques," *Journal of Design Automation & Fault-Tolerant Computing,*" Vol. 3, pp. 83-92, April, 1979.

[Rogers *et al.* 1987] W. A. Rogers, J. F. Guzolek, and J. Abraham, "Concurrent Hierarchical Fault Simulation," *IEEE Trans. on Computer-Aided Design,* Vol. CAD-6, No. 9, pp. 848-862, September, 1987.

[Roth *et al.* 1967] J. P. Roth, W. G. Bouricius, and P. R. Schneider, "Programmed Algorithms to Compute Tests to Detect and Distinguish Between Failures in Logic Circuits," *IEEE Trans. on Computers,* Vol. EC-16, No. 10, pp. 567-579, October, 1967.

[Saab and Hajj 1984] D. Saab and I. Hajj, "Parallel and Concurrent Fault Simulation of MOS Circuits," *Proc. Intn'l. Conf. on Computer Design,* pp. 752-756, October, 1984.

[Savir 1983] J. Savir, "Good Controllability and Observability Do Not Guarantee Good Testability," *IEEE Trans. on Computers,* Vol. C-32, No. 12, pp. 1198-1200, December, 1983.

[Schuler and Cleghorn 1977] D. M. Schuler and R. K. Cleghorn, "An Efficient Method of Fault Simulation for Digital Circuits Modeled from Boolean Gates and Memories," *Proc. 14th Design Automation Conf.,* pp. 230-238, June, 1977.

[Seshu 1965] S. Seshu, "On an Improved Diagnosis Program," *IEEE Trans. on Electronic Computers,* Vol. EC-12, No. 2, pp. 76-79, February, 1965.

[Seth and Agrawal 1984] S. C. Seth and V. D. Agrawal, "Characterizing the LSI Yield Equation from Wafer Test Data," *IEEE Trans. on Computer-Aided Design,* Vol. CAD-3, No. 2, pp. 123-126, April, 1984.

[Shen *et al.* 1985] J. P. Shen, W. Maly, and F. J. Ferguson, "Inductive Fault Analysis of MOS Integrated Circuits," *IEEE Design & Test of Computers,* Vol. 2, No. 6, pp. 13-26, December, 1985.

[Silberman and Spillinger 1986] G. M. Silberman and I. Spillinger, "The Difference Fault Model — Using Functional Fault Simulation to Obtain Implementation Fault Coverage," *Proc. Intn'l. Test Conf.,* pp. 332-339, September, 1986.

[Son 1985] K. Son, "Fault Simulation with the Parallel Value List Algorithm," *VLSI Systems Design,* Vol. 6, No. 12, pp. 36-43, December, 1985.

[Stein *et al.* 1986] A. J. Stein, D. G. Saab, and I. N. Hajj, "A Special-Purpose Architecture for Concurrent Fault Simulation," *Proc. Intn'l. Conf. on Computer Design*, pp. 243-246, October, 1986.

[Su and Cho 1972] S. Y. H. Su and Y-C. Cho, "A New Approach to the Fault Location of Combinational Circuits," *IEEE Trans. on Computers*, Vol. C-21, No. 1, pp. 21-30, January, 1972.

[Thompson and Szygenda 1975] E. W. Thompson and S. A. Szygenda, "Digital Logic Simulation in a Time-Based, Table-Driven Environment — Part 2. Parallel Fault Simulation," *Computer*, Vol. 8, No. 3, pp. 38-49, March, 1975.

[Ulrich and Baker 1974] E. G. Ulrich and T. G. Baker, "Concurrent Simulation of Nearly Identical Digital Networks," *Computer*, Vol. 7, No. 4, pp. 39-44, April, 1974.

[Wadsack 1984] R. L. Wadsack, "Design Verification and Testing of the WE32100 CPUs," *IEEE Design & Test of Computers*, Vol. 1, No. 3, pp. 66-75, August, 1984.

[Waicukauski *et al.* 1985] J. A. Waicukauski, E. B. Eichelberger, D. O. Forlenza, E. Lindbloom, and T. McCarthy, "Fault Simulation for Structured VLSI," *VLSI Systems Design*, Vol. 6, No. 12, pp. 20-32, December, 1985.

[Williams and Brown 1981] T. W. Williams and N. C. Brown, "Defect Level as a Function of Fault Coverage," *IEEE Trans. on Computers*, Vol. C-30, No. 12, pp. 987-988, December, 1981.

PROBLEMS

5.1

   a.   Show that a single-output combinational circuit has no independent faults.

   b.   Construct a circuit that has two independent faults.

   c.   Show that the number of primary outputs is a least upper bound on the number of independent faults.

5.2 Simulate the latch shown in Figure 5.36, assuming $\bar{y} = 0$, $y = 1$, and input sequences $\mathbf{S} = 10$ and $\mathbf{R} = 11$. Use a simple event-directed unit delay simulation process with 0,1 logic states. Assume a parallel simulator with the following faults.

Show the value of y and $\bar{y}$ for each time frame.

5.3 For the circuit and the fault set used in Example 5.1, determine the faults detected by the test 11010 by deductive simulation.

5.4 Flow chart the following procedures for processing a sequential table structure used to store fault lists:

fault	bit position
fault-free	0
S s-a-1	1
S s-a-0	2
$\bar{y}$ s-a-1	3
$\bar{y}$ s-a-0	4

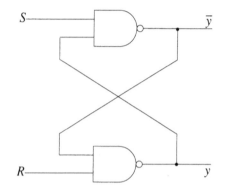

Figure 5.36

  a. set intersection assuming ordered lists;

  b. set intersection assuming unordered lists.

Compare the complexity of these two procedures.

5.5 For the latch shown in Figure 5.36, assume $\bar{y} = 0$, $y = 1$, and $S = R = 1$. Assume the initial fault lists associated with lines y, $\bar{y}$, R and S are L_y, $L_{\bar{y}}$, L_R, and L_S. Let S change to a 0. Determine the new output fault lists in terms of the given fault lists produced by this input event. Also, include all s-a-faults associated with this circuit.

5.6 Associate with each line α a list L_α^1, called the *one-list*, where fault $f \varepsilon L_\alpha^1$ if and only if line α in the circuit under fault f has the value 1. Note that $L_\alpha^1 = L_\alpha$ if line α in the fault-free circuit has the value 0, and $L_\alpha^1 = \bar{L}_\alpha$ if the line has the value 1. Show that for an AND (OR) gate with inputs a and b and output c, $L_c^1 = L_a^1 \cap L_b^1 \cup \{c \ s-a-1\}$ ($L_c^1 = L_a^1 \cup L_b^1 \cup \{c \ s-a-1\}$). What are the major advantages and disadvantages of carrying out fault analysis using L_α^1 rather than L_α?

5.7 Repeat the simulation carried out in Example 5.1 using the concurrent method. (Show the fault lists after simulating each vector.)

5.8 Consider the portion of a circuit shown in Figure 5.37, where $a = 0$.

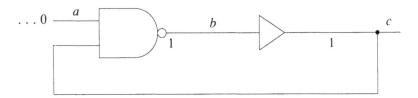

Figure 5.37 Simple circuit that oscillates for fault a s-a-1

Assume the initial conditions

$$L_a = L_A \cup \{a_1\}, \; L_b = \varnothing, \; L_c = \varnothing,$$

where L_A is an arbitrary fault list and $b_o \notin L_A$. Note that the fault a s-a-1 causes the circuit to oscillate.

    a.   Determine the oscillatory values for the fault set L_b.

    b.   Simulate this same case using concurrent simulation.

    c.   Compare the complexity of these two simulation procedures for this case.

5.9 For the circuit in Figure 5.38, determine the faults detected by the test 111 by

    a.   concurrent fault simulation (start with a collapsed set of faults)

    b.   critical path tracing.

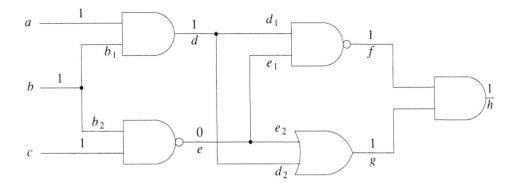

Figure 5.38

5.10   Let i denote a primary input of a fanout-free circuit and let p_i be the inversion parity of the path between i and the primary output. Let v_i be the value applied to i in a test t. Show that all primary inputs critical in t have the same sum $v_i \oplus p_i$.

5.11   Discuss the extensions needed to allow critical path tracing to handle partially specified vectors (i.e., any line may have value 0, 1 or x).

6. TESTING FOR SINGLE STUCK FAULTS

About This Chapter

In this chapter we examine test generation methods for SSFs. In Chapter 4 we discussed the wide applicability and the usefulness of the SSF model. Many TG methods dealing with other fault models extend the principles and techniques used for SSFs.

The TG process depends primarily on the type of testing experiment for which stimuli are generated. This chapter is devoted to off-line, edge-pin, stored-pattern testing with full comparison of the output results. On-line testing and compact testing are discussed in separate chapters.

6.1 Basic Issues

Test generation is a complex problem with many interacting aspects. The most important are

- the cost of TG;
- the quality of the generated test;
- the cost of applying the test.

The cost of TG depends on the complexity of the TG method. *Random TG* (RTG) is a simple process that involves only generation of random vectors. However, to achieve a high-quality test — measured, say, by the fault coverage of the generated test — we need a large set of random vectors. Even if TG itself is simple, determining the test quality — for example, by fault simulation — may be an expensive process. Moreover, a longer test costs more to apply because it increases the time of the testing experiment and the memory requirements of the tester. (Random vectors are often generated on-line by hardware and used with compact testing, but in this chapter we analyze their use only in the context of stored-pattern testing with full output comparison.)

RTG generally works without taking into account the function or the structure of the circuit to be tested. In contrast, *deterministic TG* produces tests by processing a model of the circuit. Compared to RTG, deterministic TG is more expensive, but it produces shorter and higher-quality tests. Deterministic TG can be manual or automatic. In this chapter we focus on *automatic TG* (ATG) methods.

Deterministic TG can be fault-oriented or fault-independent. In a *fault-oriented* process, tests are generated for specified faults of a fault universe (defined by an explicit fault model). *Fault-independent* TG works without targeting individual faults.

Of course, the TG cost also depends on the complexity of the circuit to be tested. Methods of reducing the complexity for testing purposes — referred to as *design for testability* techniques — form the subject of a separate chapter.

Figure 6.1 shows a general view of a deterministic TG system. Tests are generated based on a model of the circuit and a given fault model. The generated tests include both the stimuli to be applied and the expected response of the fault-free circuit. Some TG systems also produce diagnostic data to be used for fault location.

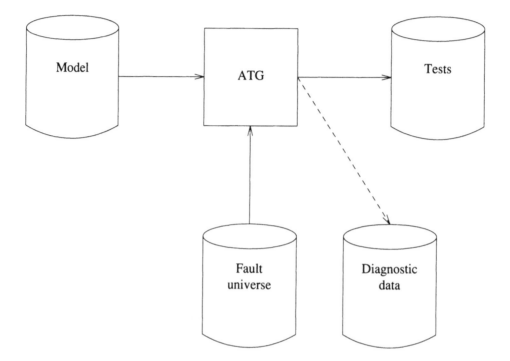

Figure 6.1   Deterministic test generation system

6.2 ATG for SSFs in Combinational Circuits

In this section we consider only gate-level combinational circuits composed of AND, NAND, OR, NOR, and NOT gates.

6.2.1 Fault-Oriented ATG

Fanout-Free Circuits

We use fanout-free circuits only as a vehicle to introduce the main concepts of ATG for general circuits. The two fundamental steps in generating a test for a fault l s-a-v are first, to *activate* (excite) *the fault*, and, second, to *propagate the resulting error* to a primary output (PO). Activating the fault means to set primary input (PI) values that cause line l to have value $\bar{v}$. This is an instance of the **line-justification** problem, which deals with finding an assignment of PI values that results in a desired value setting on a specified line in the circuit. To keep track of error propagation we must consider values in both the fault-free circuit N and the faulty circuit N_f defined by the target fault f. For this we define *composite logic values* of the form v/v_f, where v and v_f are values of the same signal in N and N_f. The composite logic values that represent errors — 1/0 and 0/1 — are denoted by the symbols D and $\bar{D}$ [Roth 1966].

The other two composite values — 0/0 and 1/1 — are denoted by 0 and 1. Any logic operation between two composite values can be done by separately processing the fault-free and faulty values, then composing the results. For example, $\bar{D} + 0 =$ $0/1 + 0/0 = 0{+}0/1{+}0 = 0/1 = \bar{D}$. To these four binary composite values we add a fifth value (x) to denote an unspecified composite value, that is, any value in the set $\{0,1,D,\bar{D}\}$. In practice, logic operations using composite values are defined by tables (see Figure 6.2). It is easy to verify that D behaves consistently with the rules of Boolean algebra, i.e., $D+\bar{D} = 1$, $D.\bar{D} = 0$, $D+D = D.D = D$, $\bar{D}.\bar{D} = \bar{D}+\bar{D} = \bar{D}$.

v/v_f			AND	0	1	D	$\bar{D}$	x		OR	0	1	D	$\bar{D}$	x
0/0	0		0	0	0	0	0	0		0	0	1	D	$\bar{D}$	x
1/1	1		1	0	1	D	$\bar{D}$	x		1	1	1	1	1	1
1/0	D		D	0	D	D	0	x		D	D	1	D	1	x
0/1	$\bar{D}$		$\bar{D}$	0	$\bar{D}$	0	$\bar{D}$	x		$\bar{D}$	$\bar{D}$	1	1	$\bar{D}$	x
			x	0	x	x	x	x		x	x	1	x	x	x

| (a) | (b) | (c) |

Figure 6.2   Composite logic values and 5-valued operations

Figure 6.3 shows the structure of an algorithm for generating a test for l s-a-v. It initializes all values to x and it performs the two basic steps, represented by the routines *Justify* and *Propagate*.

> **begin**
>   set all values to x
>   *Justify(l, $\bar{v}$)*
>   **if** $v = 0$ **then** *Propagate (l, D)*
>   **else** *Propagate (l, $\bar{D}$)*
> **end**

Figure 6.3   Test generation for the fault l s-a-v in a fanout-free circuit

Line justification (Figure 6.4) is a recursive process in which the value of a gate output is justified by values of the gate inputs, and so on, until PIs are reached. Let us consider a NAND gate with k inputs. There is only one way to justify a 0 output value, but to justify a 1 value we can select any one of the 2^k-1 input combinations that produce 1. The simplest way is to assign the value 0 to only one (arbitrarily selected) input and to leave the others unspecified. This corresponds to selecting one of the k primitive cubes of the gate in which the output is 1.

Justify (l, val)
begin
    set l to *val*
    if l is a PI **then return**
    /* l is a gate (output) */
    c = controlling value of l
    i = inversion of l
    inval = $val \oplus i$
    if (*inval* = $\bar{c}$)
        then for every input j of l
            Justify (j, inval)
        else
            begin
                select one input (j) of l
                Justify (j, inval)
            end
end

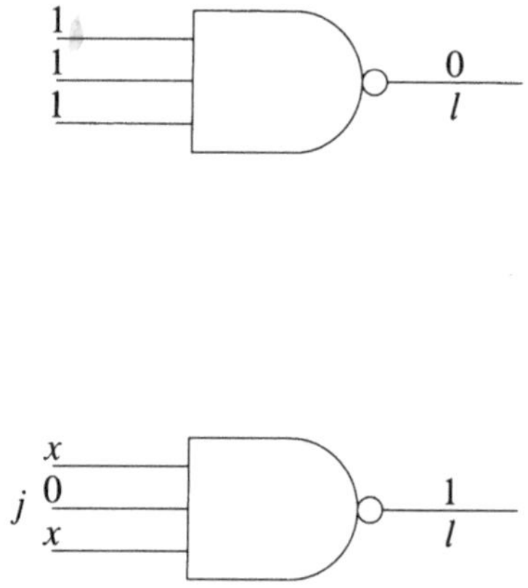

Figure 6.4 Line justification in a fanout-free circuit

To propagate the error to the PO of the circuit, we need to sensitize the unique path from l to the PO. Every gate on this path has exactly one input sensitized to the fault. According to Lemma 4.1, we should set all the other inputs of G to the noncontrolling value of the gate. Thus we *transform the error-propagation problem into a set of line-justification problems* (see Figure 6.5).

Example 6.1: Let us generate a test for the fault f s-a-0 in the circuit of Figure 6.6(a). The initial problems are *Justify(f,1)* and *Propagate(f,D)*. *Justify(f,1)* is solved by $a=b=1$. *Propagate(f,D)* requires *Justify(g,0)* and *Propagate(h,D)*. We solve *Justify(g,0)* by selecting one input of g — say, c — and setting it to 0. *Propagate(h,D)* leads to *Justify(i,1)*, which results in $e=0$. Now the error reaches the PO j. Figure 6.6(b) shows the resulting values. The generated test is $110x0$ (d can be arbitrarily assigned 0 or 1).   □

It is important to observe that *in a fanout-free circuit every line-justification problem can be solved independently of all the others*, because the sets of PIs that are eventually assigned to justify the required values are mutually disjoint.

Circuits with Fanout

Now we consider the general case of circuits with fanout and contrast it with the fanout-free case. We must achieve the same two basic goals — fault activation and error propagation. Again, fault activation translates into a line-justification problem. A first difference caused by fanout is that now we may have several ways to propagate an

Propagate (l, err)
/* *err* is D or $\overline{D}$ */
begin
    set *l* to *err*
    if *l* is PO **then return**
    k = the fanout of *l*
    c = controlling value of *k*
    i = inversion of *k*
    for every input *j* of *k* other than *l*
      Justify (j, $\overline{c}$)
    Propagate (k, err⊕i)
end

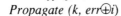

Figure 6.5   Error propagation in a fanout-free circuit

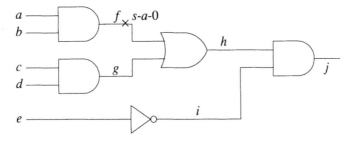

(a)

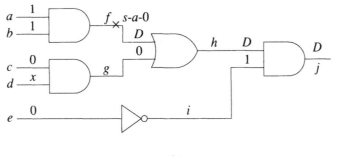

(b)

Figure 6.6

error to a PO. But once we select a path, we again reduce the error-propagation problem to a set of line-justification problems. *The fundamental difficulty caused by (reconvergent) fanout is that, in general, the resulting line-justification problems are no longer independent.*

Example 6.2: In the irredundant circuit of Figure 6.7 consider the fault G_1 *s-a*-1. To activate it we need to justify G_1=0. Now we have a choice of propagating the error via a path through G_5 or through G_6. Suppose we decide to select the former. Then we need to justify G_2 =1. The resulting set of problems — *Justify(G_1,0)* and *Justify(G_2,1)* — cannot be solved simultaneously, because their two unique solutions, $a=b=c=1$ and $a=d=0$, require contradictory values for *a*. This shows that the decision to propagate the error through G_5 was wrong. Hence we have to try an alternative decision, namely propagate the error through G_6. This requires $G_4 = 1$, which is eventually solved by c=1 and e=0. The resulting test is 111*x*0.   □

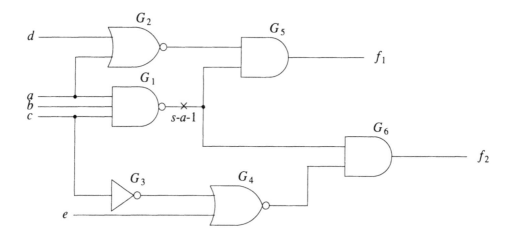

Figure 6.7

This example shows the need to explore alternatives for error propagation. Similarly, we may have to try different choices for line justification.

Example 6.3: Consider the fault *h s-a*-1 in the circuit of Figure 6.8. To activate this fault we must set *h*=0. There is a unique path to propagate the error, namely through *p* and *s*. For this we need *e*=*f*=1 and *q*=*r*=1. The value *q*=1 can be justified by *l*=1 or by *k*=1. First, let us try to set *l*=1. This leads to *c*=*d*=1. However, these two assignments, together with the previously specified *e*=1, would imply *r*=0, which leads to an inconsistency. Therefore the decision to justify *q*=1 by *l*=1 has been incorrect. Hence we must choose the alternative decision *k*=1, which implies *a*=*b*=1. Now the only remaining line-justification problem is *r*=1. Either *m*=1 or *n*=1 leads to consistent solutions.   □

Backtracking

We have seen that the search for a solution involves a *decision process*. Whenever there are several alternatives to justify a line or to propagate an error, we choose one of them to try. But in doing so we may select a decision that leads to an inconsistency

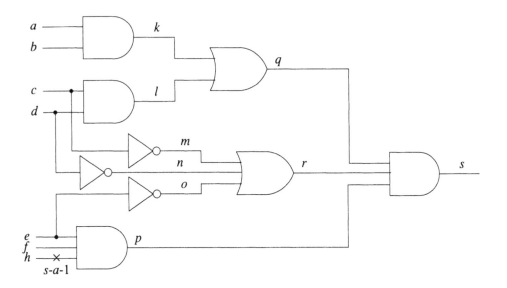

Figure 6.8

(also termed *contradiction* or *conflict*). Therefore in our search for a test we should use a **backtracking strategy** that allows a systematic exploration of the complete space of possible solutions and recovery from incorrect decisions. Recovery involves restoring the state of the computation to the state existing before the incorrect decision.

Usually the assignments performed as a result of a decision uniquely determine (imply) other values. The process of computing these values and checking for their consistency with the previously determined ones is referred to as **implication**. Figure 6.9 shows the progression of value computation for Example 6.3, distinguishing between values resulting from decisions and those generated by implication. The initial implications follow from the unique solutions to the fault-activation and error-propagation problems.

In most backtracking algorithms we must record all values assigned as a result of a decision, to be able to erase them should the decision lead to an inconsistency. In Example 6.3 all values resulting from the decision $l=1$ are erased when backtracking occurs.

Figure 6.10 outlines a recursive scheme of a backtracking TG algorithm for a fault. (This description is quite abstract, but more details will be provided later.) The original problems to be solved in generating a test for the fault l s-a-v are to justify a value $\bar{v}$ on l and to propagate the error from l to a PO. The basic idea is that if a problem cannot be directly solved, we recursively transform it into subproblems and try to solve these first. Solving a problem may result in SUCCESS or FAILURE.

First the algorithm deals with all the problems that have unique solutions and hence can be solved by implication. These are processed by the procedure *Imply_and_check*,

Decisions	Implications	
	$h=\overline{D}$	Initial implications
	$e=1$	
	$f=1$	
	$p=\overline{D}$	
	$r=1$	
	$q=1$	
	$o=0$	
	$s=\overline{D}$	
$l=1$		To justify $q=1$
	$c=1$	
	$d=1$	
	$m=0$	
	$n=0$	
	$r=0$	Contradiction
$k=1$	$a=1$	To justify $q=1$
	$b=1$	
$m=1$		To justify $r=1$
	$c=0$	
	$l=0$	

Figure 6.9   Computations for Example 6.3

which also checks for consistency. The procedure *Solve* reports FAILURE if the consistency check fails. SUCCESS is reported when the desired goal is achieved, namely when an error has been propagated to a PO and all the line-justification problems have been solved. Even in a consistent state, the algorithm may determine that no error can be further propagated to a PO; then there is no point in continuing the search, and FAILURE is reported.

If *Solve* cannot immediately determine SUCCESS or FAILURE, then it selects one currently unsolved problem. This can be either a line-justification or an error-propagation problem. At this point there are several alternative ways to solve the selected problem. The algorithm selects one of them and tries to solve the problem at the next level of recursion. This process continues until a solution is found or all possible choices have failed. If the initial activation of *Solve* fails, then the algorithm has failed to generate a test for the specified fault.

The selection of an unsolved problem to work on and the selection of an untried way to solve it, can be — in principle — arbitrary. That is, if the fault is detectable, we

Solve()
begin
    if *Imply_and_check()* = FAILURE **then return** FAILURE
    if (error at PO **and** all lines are justified)
      then return SUCCESS
    if (no error can be propagated to a PO)
      then return FAILURE
    select an unsolved problem
    repeat
      begin
        select one untried way to solve it
        if *Solve()* = SUCCESS **then return** SUCCESS
      end
    until all ways to solve it have been tried
    return FAILURE
end

Figure 6.10   General outline of a TG algorithm

will generate a test for it independent of the order in which problems and solutions are attempted. However, the selection process may greatly affect the efficiency of the algorithm as well as the test vector generated. Selection criteria are discussed in Section 6.2.1.3.

There are several fault-oriented TG algorithms whose structure is similar to that of *Solve*. In the following we first analyze concepts common to most of them, then we discuss specific algorithms in more detail.

6.2.1.1 Common Concepts

Decision Tree

The execution of the backtracking TG algorithm can be visualized with the aid of a *decision tree* (see Figure 6.11). A decision node (shown as a circle) denotes a problem that the algorithm is attempting to solve. A branch leaving a decision node corresponds to a decision, i.e., trying one of the available alternative ways to solve the problem. A FAILURE terminal node of the tree (shown as a square labeled F) indicates the detection of an inconsistency or encountering a state that precludes further error propagation. A SUCCESS terminal node (shown as a square labeled S) represents finding a test. The execution of the algorithm can be traced by a depth-first traversal of the associated decision tree. For example, in Figure 6.11(b), starting at the decision node $q=1$, we first follow the branch $l=1$ which reaches an F terminal node. Then we backtrack to the last decision node and take the other branch ($k=1$), which leads to a new decision node ($r=1$). Here the first decision ($m=1$) reaches an S terminal node.

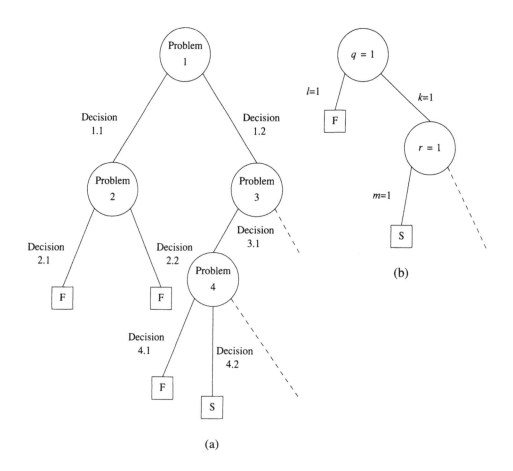

Figure 6.11   Decision trees (a) General structure (b) Decision tree for
                Example 6.3

Implicit Enumeration

An important property of the algorithm of Figure 6.10 is that it is *exhaustive*, that is, it
is guaranteed to find a solution (test) if one exists. Thus if the algorithm fails to
generate a test for a specified fault, then the fault is undetectable.

Example 6.4: Let us try to generate a test for the fault f s-a-0 in the circuit of
Figure 6.12(a). To justify f=1 we first try b=0 (see the decision tree in Figure 6.12(b)).
But this implies e=1, which precludes error propagation through gate h. Trying c=0
results in a similar failure. We can conclude that no test exists for f s-a-0.   □

The algorithm is guaranteed to find a test, if one exists, because it can *implicitly
enumerate all possible solutions*. The concept of implicit enumeration is best
understood by contrasting it to explicit enumeration, which (in this context) means to
repeatedly generate an input vector and to check whether it detects the target fault.
Using implicit enumeration we direct the search toward vectors that can satisfy the set

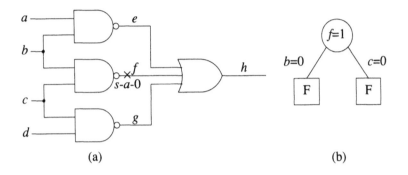

Figure 6.12   TG failure for an undetectable fault (a) Circuit (b) Decision tree

of constraints imposed by the set of lines whose values must be simultaneously justified. As the set of constraints grows during the execution of the algorithm, the set of vectors that can satisfy them becomes smaller and smaller. The advantage of implicit enumeration is that it bounds the search space and begins to do so early in the search process.

Let us compare implicit and explicit enumeration for Example 6.4. Using implicit enumeration, we start by limiting the search to the set of vectors that satisfy $f=1$. From this set, we first reject the subset of all vectors with $b=0$. Then we reject the subset of all vectors with $c=0$, and this concludes the search. Using explicit enumeration, we would generate and simulate all 2^4 input vectors.

Complexity Issues

Because of the exhaustive nature of the search process, the *worst-case complexity* of the algorithm in Figure 6.10 is *exponential*; i.e., the number of operations performed is an exponential function of the number of gates in the circuit. The worst-case behavior is characterized by many remade decisions; that is, much searching is done before a test is found or the target fault is recognized as undetectable. To minimize the total TG time, any practical TG algorithm is allowed to do only a limited amount of search; namely, the search is abandoned when the number of incorrect decisions (or the CPU time) reaches a specified limit. This may result in not generating tests for some detectable faults. The worst-case behavior has been observed mainly for undetectable faults [Cha *et al.* 1978].

The *best-case behavior* occurs when the result — generating a test or recognizing redundancy — is obtained without backtracking. This means either that the result is found only by implications (then the decision tree degenerates to one terminal node), or that only correct decisions are taken. Then the number of operations is a *linear* function of the number of gates. This is always the case for fanout-free circuits and for circuits without reconvergent fanout, because in these types of circuits no decision can produce a conflict. (Furthermore in such circuits all faults are detectable).

From analyzing the worst-case and the best-case behavior of the algorithm we can conclude that the key factor in controlling the complexity of our TG algorithm is to *minimize the number of incorrect decisions*. In Section 6.2.1.3 we shall discuss several heuristic techniques that help in achieving this goal. Here we introduce a simple principle that helps in minimizing the number of incorrect decisions by reducing the number of problems that require decisions (in other words, the algorithm will have fewer opportunities to make a wrong choice). This is the *maximum implications principle*, which requires one always to *perform as many implications as possible*. Clearly, by doing more implications, we either reduce the number of problems that otherwise would need decisions, or we reach an inconsistency sooner.

Example 6.5: Consider the circuit in Figure 6.13 and assume that at a certain stage in the execution of the TG algorithm we have to justify $f=0$ and $e=0$. Suppose that we justify $f=0$ by $c=0$. If we do not determine all the implications of this assignment, we are left with the problems of justifying $e=0$ and $c=0$. Both problems are solved, however, if we compute all implications. First $c=0$ implies $d=1$. This leaves only one way to justify $e=0$, namely by $b=0$, which in turn justifies $c=0$.   □

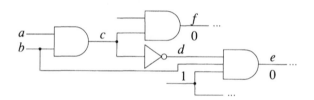

Figure 6.13

The D-Frontier

The *D-frontier* consists of all gates whose output value is currently x but have one or more error signals (either D's or $\overline{D}$'s) on their inputs. Error propagation consists of selecting one gate from the *D-frontier* and assigning values to the unspecified gate inputs so that the gate output becomes D or $\overline{D}$. This procedure is also referred to as the *D-drive* operation. If the *D-frontier* becomes empty during the execution of the algorithm, then no error can be propagated to a PO. Thus an empty *D-frontier* shows that backtracking should occur.

The J-Frontier

To keep track of the currently unsolved line-justification problems, we use a set called the *J-frontier*, which consists of all gates whose output value is known but is not implied by its input values. Let c be the controlling value and i be the inversion of a gate on the *J-frontier*. Then the output value is $c \oplus i$, at least two inputs must have value x, and no input can have value c.

The Implication Process

The tasks of the implication process (represented by the routine *Imply_and_check* in Figure 6.10) are

- Compute all values that can be uniquely determined by implication.

- Check for consistency and assign values.

- Maintain the *D-frontier* and the *J-frontier*.

We can view the implication process as a modified zero-delay simulation procedure. As in simulation, we start with some values to be assigned; these assignments may determine (imply) new values, and so on, until no more values are generated. All values to be assigned are processed via an *assignment queue* similar to the event queue used in simulation. Unlike simulation, where values only propagate forward (toward POs), here values may also propagate backward (toward PIs). An entry in the assignment queue has the form $(l, v', direction)$, where v' is the value to be assigned to line l and $direction \in \{backward, forward\}$. To generate a test for the fault l s-a-1, the initial two entries in the assignment queue are $(l, 0, backward)$ and $(l, \overline{D}, forward)$.

Imply_and_check retrieves in turn every entry in the assignment queue. The value v' to be assigned to l is first checked for consistency with the current value v of l (all values are initialized to x). An inconsistency is detected if $v \neq x$ and $v \neq v'$. (An exception is allowed for the faulty line, which gets a binary value to be propagated backward and an error value to be propagated forward.) A consistent value is assigned, then it is further processed according to its *direction*.

Backward propagation of values is illustrated in Figure 6.14. The right side of the figure shows the effects of the assignments made on the left side. An arrow next to a logic value shows the direction in which that value propagates. The assignment $a=0$ in Figure 6.14(c) causes a to be added to the *J-frontier*. Figure 6.14(d) shows how backward propagation on a fanout branch may induce forward propagation on other fanout branches of the same stem.

Similarly, Figures 6.15 and 6.16 illustrate forward propagation of values. Note in Figure 6.15(d) how forward propagation on a gate input may induce backward propagation on another input of the same gate.

If after all values have been propagated, the *D-frontier* contains only one entry — say, a — then the only way to propagate the error is through gate a. Figure 6.17 illustrates the implications resulting from this situation, referred to as *unique D-drive*.

Global Implications

Let us consider the example in Figure 6.18(a). The *D-frontier* is $\{d,e\}$, so we do not have unique D-drive. We can observe, however, that no matter how we will decide to propagate the error (i.e., through d or e or both), eventually we will reach a unique D-drive situation, because the error must propagate through g. Then we can make the implications shown in Figure 6.18(b) [Akers 1976, Fujiwara and Shimono 1983].

While the implications previously described can be characterized as *local*, as they consist in propagating values from one line to its immediate successors or

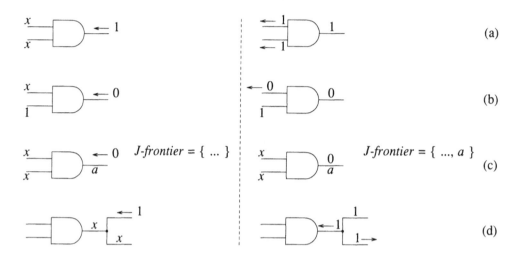

Figure 6.14   Backward implications

predecessors, the implication illustrated in Figure 6.18 can be characterized as *global*, since it involves a larger area of the circuit and reconvergent fanout.

Next we analyze other global implications used in the SOCRATES system [Schulz *et al.* 1988, Schulz and Auth 1989]. Consider the circuit in Figure 6.19. Assume that $F = 1$ has just been assigned by backward propagation. No other values can be determined by local implications. But we can observe that, no matter how we decide to justify $F = 1$ (by $D = 1$ or $E = 1$), in either case we will imply $B = 1$. Thus we can conclude that $F = 1$ implies $B = 1$. This implication is "learned" during the preprocessing phase of SOCRATES by the following analysis. Simulating $B = 0$, we determine that it implies $F = 0$. Then $\overline{(F=0)}$ implies $\overline{(B=0)}$, that is, $F = 1$ implies $B = 1$.

The type of learning illustrated above is called *static*, because the implications determined are valid independent of other values in the circuit. SOCRATES also performs *dynamic learning* to determine global implications enabled by previously assigned values. For example, in the circuit in Figure 6.20, $F = 0$ implies $B = 0$ when $A = 1$ (because $B = 1$ implies $F = 1$ when $A = 1$).

Reversing Incorrect Decisions

Consider the problem of justifying a 0 on the output of an AND gate with three inputs — a, b, and c — all currently with value x (see Figure 6.21). Let us assume that the first decision — $a=0$ — has been proven incorrect. This shows that, independent of the values of b and c, a cannot be 0. Therefore, we can conclude that a must be 1. Then before we try the next decision — $b=0$ — we should set $a=1$

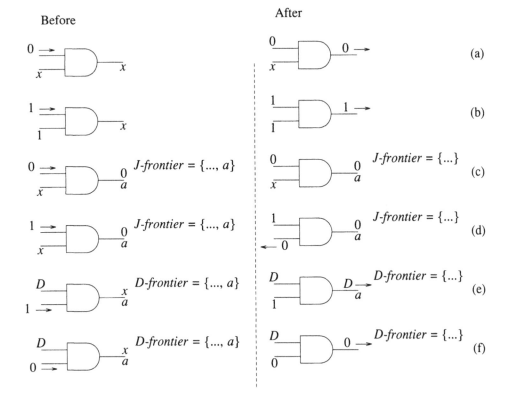

Figure 6.15   Forward implications (binary values)

(rather than leave $a=x$) and process $a=1$ as an implication. Similarly, if the decision $b=0$ fails as well, we should set both $a=1$ and $b=1$ before trying $c=0$. The benefit of this technique of *reversing incorrect decisions* [Cha *et al.* 1978] is an increase in the number of implications.

Error-Propagation Look-Ahead

Consider the circuit and the values shown in Figure 6.22. The *D-frontier* is $\{a,b\}$. We can observe that, independent of the way we may try to propagate the errors, eventually the *D-frontier* will become empty, as Ds cannot be driven through e or f. This future state can be identified by checking the following necessary condition for successful error propagation.

Let an *x-path* denote a path all of whose lines have value x. Let G be a gate on the *D-frontier*. The error(s) on the input(s) of G can propagate to a PO Z only if there exists at least one x-path between G and Z [Goel 1981].

Clearly, none of the gates on the *D-frontier* in Figure 6.22 satisfies this condition. The benefit of identifying this situation is that we can avoid all the decisions that are bound

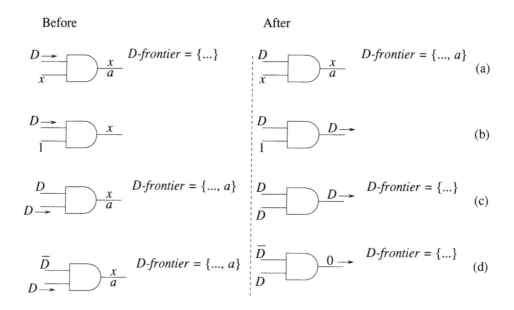

Figure 6.16   Forward implications (error values)

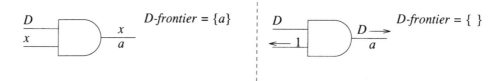

Figure 6.17   Unique *D*-drive

eventually to fail and their associated backtracking. Thus by using this look-ahead technique we may prune the decision tree by recognizing states from which any further decision will lead to a failure.

6.2.1.2 Algorithms

Many of the concepts presented in the previous section are common to a class of TG algorithms generally referred to as *path-sensitization algorithms*. In this section we discuss specific algorithms of this class.

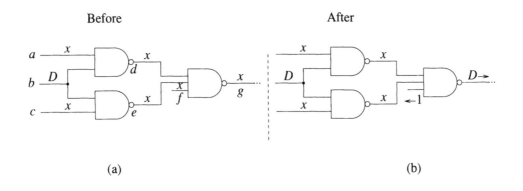

(a)   (b)

Figure 6.18   Future unique D-drive

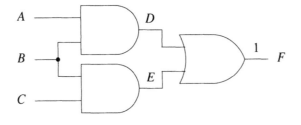

Figure 6.19   Global implication: $F = 1$ implies $B = 1$

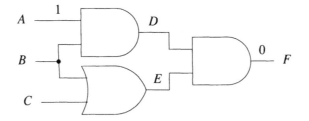

Figure 6.20   Global implication: $F = 0$ implies $B = 0$ when $A = 1$

The D-Algorithm

Figure 6.23 presents our version of the classical D-algorithm [Roth 1966, Roth *et al.* 1967]. It follows the general outline shown in Figure 6.10. For the sake of

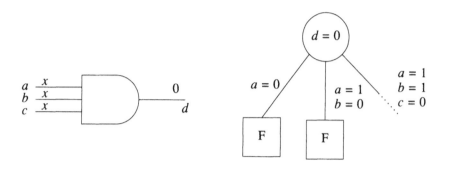

Figure 6.21   Reversing incorrect decisions

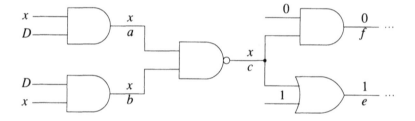

Figure 6.22   The need for look-ahead in error propagation

simplicity, we assume that error propagation is always given priority over justification problems; however, this assumption is not essential (see Section 6.2.1.3).

The term "assign" should be understood as "add the value to be assigned to the assignment queue" rather than an immediate assignment. Recall that all the assignments are made and further processed by *Imply_and_check*.

A characteristic feature of the D-algorithm is its ability to propagate errors on several reconvergent paths. This feature, referred to as *multiple-path sensitization*, is required to detect certain faults that otherwise (i.e., sensitizing only a single path) would not be detected [Schneider 1967].

Example 6.6: Let us apply the D-algorithm for the circuit and the fault shown in Figure 6.24(a). Figure 6.24(b) traces the value computation and Figure 6.24(c) depicts the decision tree. The content of a decision node corresponding to an error propagation problem shows the associated *D-frontier*. A branch emanating from such a decision node shows the decision taken, that is, the gate selected from the *D-frontier* for error propagation. (Remember that when backtracking occurs, the *D-frontier* should be restored to its state before the incorrect decision.) Note how the D-algorithm first tried to propagate the error solely through i, then through both i and

D-alg()
begin
    if *Imply_and_check()* = FAILURE **then return** FAILURE
    if (error not at PO) **then**
       begin
          if *D-frontier* = ∅ **then return** FAILURE
          repeat
            begin
               select an untried gate (*G*) from *D-frontier*
               c = controlling value of *G*
               assign $\bar{c}$ to every input of *G* with value *x*
               if *D-alg()* = SUCCESS **then return** SUCCESS
            end
          until all gates from *D-frontier* have been tried
          return FAILURE
       end
    /* error propagated to a PO */
    if *J-frontier* = ∅ **then return** SUCCESS
    select a gate (*G*) from the *J-frontier*
    c = controlling value of *G*
    repeat
       begin
          select an input (*j*) of *G* with value *x*
          assign *c* to *j*
          if *D-alg()* = SUCCESS **then return** SUCCESS
          assign $\bar{c}$ to *j*   /* reverse decision */
       end
    until all inputs of *G* are specified
    return FAILURE
end

Figure 6.23   The *D*-algorithm

k, and eventually succeeded when all three paths from *g* (through *i*, *k*, and *m*) were simultaneously sensitized.   □

The 9-V Algorithm

The 9-V algorithm [Cha *et al.* 1978] is similar to the *D*-algorithm. Its main distinguishing feature is the use of nine values [Muth 1976]. In addition to the five composite values of the *D*-algorithm, the 9-V algorithm employs four *partially specified composite values*. The *x* value of the *D*-algorithm is totally unspecified, that is, neither *v* nor v_f is known. For a partially specified composite value v/v_f, either *v* is binary and v_f is unknown (*u*) or vice versa. For example, $1/u$ represents both $1/0$ and $1/1$. This means that $1/u$ can be either *D* or 1. Figure 6.25 shows the partially

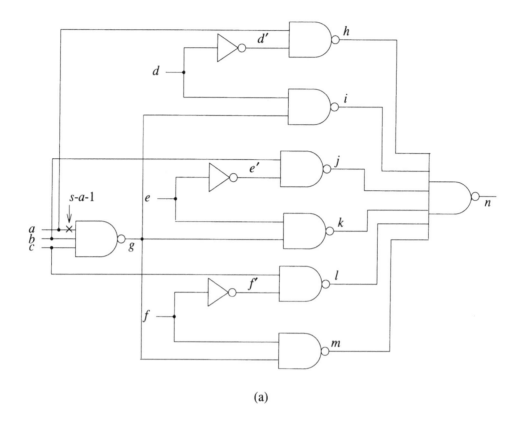

(a)

Figure 6.24

specified composite values and the sets of completely specified composite values they represent. The totally unspecified value x is u/u and represents the set $\{0,1,D,\overline{D}\}$.

A logic operation between two composite values can be carried out by separately processing the good and the faulty circuit values, and then composing the results. For example $D.x = 1/0 . u/u = (1.u)/(0.u) = u/0$. (In practice, logic operations using the nine composite values are defined by tables.) Note that using only the five values of the D-algorithm the result of $D.x$ is x. The 9-valued system provides more information as $D.x = u/0$ shows that the result is 0 or D.

When the 9-V algorithm tries to drive a D through a gate G with controlling value c, the value it assigns to the unspecified inputs of G corresponds to the set $\{\overline{c},D\}$. Similarly, the propagation of a $\overline{D}$ is enabled by values corresponding to the set $\{\overline{c},\overline{D}\}$. For example, to drive a $\overline{D}$ through an AND gate, the unspecified inputs are assigned a $u/1$ value (which is 1 or $\overline{D}$), and it is the task of the implication process to determine whether this value eventually becomes 1 or $\overline{D}$. A partially specified composite value u/b or b/u (where b is binary) assigned to a PI is immediately transformed to b/b, because the PI cannot propagate fault effects. The benefit of the flexibility provided by

Decisions	Implications	
	$a=0$	Activate the fault
	$h=1$	
	$b=1$	Unique D-drive through g
	$c=1$	
	$g=D$	
$d=1$		Propagate through i
	$i=\overline{D}$	
	$d'=0$	
$j=1$		Propagate through n
$k=1$		
$l=1$		
$m=1$		
	$n=D$	
	$e'=0$	
	$e=1$	
	$k=\overline{D}$	Contradiction
$e=1$		Propagate through k
	$k=\overline{D}$	
	$e'=0$	
	$j=1$	
$l=1$		Propagate through n
$m=1$		
	$n=D$	
	$f'=0$	
	$f=1$	
	$m=\overline{D}$	Contradiction
$f=1$		Propagate through m
	$m=\overline{D}$	
	$f'=0$	
	$l=1$	
	$n=D$	

(b)

Figure 6.24   (Continued)

the partially specified composite values is that it reduces the amount of search done for multiple path sensitization.

Example 6.7:   Let us redo the problem from Example 6.6 using the 9-V algorithm. Figure 6.26(a) traces the value computation and Figure 6.26(b) shows the corresponding decision tree. Now the same test is generated without backtracking.   □

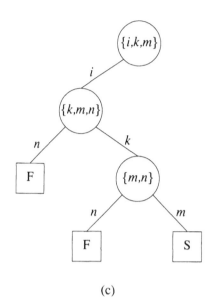

(c)

Figure 6.24   (Continued)

0/u	$\{0,\bar{D}\}$
1/u	$\{D,1\}$
u/0	$\{0,D\}$
u/1	$\{\bar{D},1\}$

Figure 6.25   Partially specified composite values

The main difference between the D-algorithm and 9-V algorithm can be summarized as follows. Whenever there are k possible paths for error propagation, the D-algorithm may eventually try all the 2^k-1 combinations of paths. The 9-V algorithm tries only one path at a time, but without precluding simultaneous error propagation on the other $k-1$ paths. This is made possible by the partially specified composite values that denote potential error propagation. Thus in a situation where the D-algorithm may enumerate up to 2^k-1 combinations of paths, the 9-V algorithm will enumerate at most k ways of error propagation.

Single-Path Sensitization

Experience has shown that faults whose detection is possible only with multiple-path sensitization are rare in practical circuits (see, for example, the results presented in

Decisions	Implications	
	$a=0$	Activate the fault
	$h=1$	
	$b=1$	Unique D-drive through g
	$c=1$	
	$g=D$	
	$i=u/1$	
	$k=u/1$	
	$m=u/1$	
$d=1$		Propagate through i
	$i=\bar{D}$	
	$d'=0$	
	$n=1/u$	
$l=u/1$		Propagate through n
$j=u/1$		
	$n=D$	
	$f'=u/0$	
	$f=1$	
	$f'=0$	
	$e'=u/0$	
	$e=1$	
	$e'=0$	
	$k=\bar{D}$	
	$m=\bar{D}$	

(a)

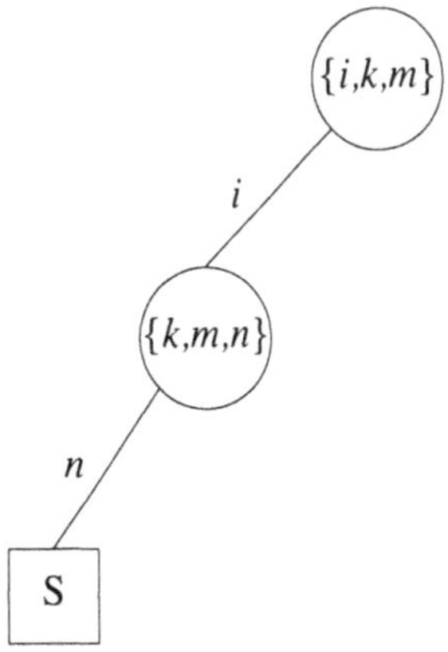

$\{i,k,m\}$

i

$\{k,m,n\}$

n

S

(b)

(a)

Figure 6.26   Execution trace of the 9-V algorithm on the problem of
Example 6.6

[Cha *et al.* 1978]). To reduce computation time, TG algorithms are often restricted to
single-path sensitization.

To restrict the D-algorithm (Figure 6.23) to single-path sensitization, it should be
modified as follows. After we select a gate from the *D-frontier* and propagate an
error to its output, we consider only that D or $\bar{D}$ for further propagation and ignore the
other gates from the *D-frontier*. In this way we force the algorithm to propagate errors
only on single paths.

PODEM

The goal of any (fault-oriented) TG algorithm is to find a test for a specified fault, that
is, an input vector that detects the fault. Although the test belongs to the space of all
input vectors, the search process of the algorithms discussed so far takes place in a
different space. A decision in this search process consists of selecting either a gate

from the *D-frontier* or a way of justifying the value of a gate from the *J-frontier*. These decisions are eventually mapped into PI values, but the search process is an indirect one.

PODEM (Path-Oriented Decision Making) [Goel 1981] is a TG algorithm characterized by a *direct search* process, in which decisions consist only of PI assignments. We have seen that the problems of fault activation and error propagation lead to sets of line-justification problems. PODEM treats a value v_k to be justified for line k as an *objective* (k,v_k) to be achieved via PI assignments [Snethen 1977]. A backtracing procedure (Figure 6.27) maps a desired objective into a PI assignment that is likely to contribute to achieving the objective. Let (j, v_j) be the PI assignment returned by *Backtrace* (k, v_k), and let p be the inversion parity of the path followed from k to j. All lines on this path have value x, and the value v_j to be assigned and the objective value v_k satisfy the relation $v_k = v_j \oplus p$. Note that no values are assigned during backtracing. Values are assigned only by simulating PI assignments.

Backtrace (k,v_k)
/* map objective into PI assignment */
begin
    $v = v_k$
    while k is a gate output
        begin
            i = inversion of k
            select an input (j) of k with value x
            $v = v \oplus i$
            $k = j$
        end
    /* k is a PI */
    return (k,v)
end

Figure 6.27   Backtracing of an objective

Example 6.8: Consider the circuit shown in Figure 6.28 and the objective $(f,1)$. Assume that *Backtrace(f,1)* follows the path (f,d,b) and returns $(b,1)$. Simulating the assignment $b=1$ does not achieve the objective $(f,1)$. Executing again *Backtrace(f,1)* results in following the path (f,d,c,a) and leads to $(a,0)$. Now simulating the assignment $a=0$ achieves $f=1$.   □

Objectives are selected (see Figure 6.29) so that first the target fault is activated; then the resulting error is propagated towards a PO.

Figure 6.30 outlines the overall structure of PODEM. It uses the same five values — 0, 1, x, D, and $\overline{D}$ — as the D-algorithm. Initially all values are x. Non-x values are generated only by simulating PI assignments. This 5-valued simulation is the task of the routine *Imply*, which also creates the initial D or $\overline{D}$ when the fault is activated, and maintains the *D-frontier*. At every level of recursion, PODEM starts by analyzing

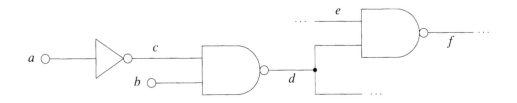

Figure 6.28

Objective()
begin
    /* the target fault is *l* s-a-v */
    if (the value of *l* is *x*) **then return** $(l, \bar{v})$
    select a gate (*G*) from the *D-frontier*
    select an input (*j*) of *G* with value *x*
    c = controlling value of *G*
    return $(j, \bar{c})$
end

Figure 6.29   Selecting an objective

values previously established, with the goal of identifying a SUCCESS or a FAILURE state. SUCCESS is returned if a PO has a *D* or $\bar{D}$ value, denoting that an error has been propagated to a PO. FAILURE is returned if the current values show that generating a test is no longer possible. This occurs when either of the following conditions applies:

- The target fault *l* s-a-v cannot be activated, since line *l* has value *v*.

- No error can be propagated to a PO, either because the *D-frontier* is empty or because the error propagation look-ahead shows that it will become empty.

If PODEM cannot immediately determine SUCCESS or FAILURE, it generates an objective (k, v_k) that is mapped by backtracing into a PI assignment. The assignment $j=v_j$ is then simulated by *Imply* and a new level of recursion is entered. If this fails, PODEM backtracks by reversing the decision $j=v_j$ to $j=\bar{v}_j$. If this also fails, then *j* is set to *x* and PODEM returns FAILURE.

The selection of a gate from the *D-frontier* (done in *Objective*) and the selection of an unspecified gate input (done in *Objective* and in *Backtrace*) can be, in principle, arbitrary. Selection criteria that tend to increase the efficiency of the algorithm are discussed in Section 6.2.1.3.

```
PODEM()
begin
    if (error at PO) then return SUCCESS
    if (test not possible) then return FAILURE
    (k,v_k) = Objective()
    (j,v_j) = Backtrace(k,v_k)  /* j is a PI */
    Imply (j,v_j)
    if PODEM() = SUCCESS then return SUCCESS
    /* reverse decision */
    Imply (j,v̄_j)
    if PODEM() = SUCCESS then return SUCCESS
    Imply (j,x)
    return FAILURE
end
```

Figure 6.30   PODEM

Example 6.9: Let us apply PODEM to the problem from Example 6.6. Figure 6.31(a) traces a possible execution of PODEM, showing the objectives, the PI assignments determined by backtracing objectives, the implications generated by simulating PI assignments, and the corresponding *D-frontier*. Note that the assignment $e=0$ causes the PO n to have a binary value, which makes the x-path check fail; this shows that generating a test for the target fault is no longer possible and leads to backtracking by reversing the incorrect decision. Also note that PODEM handles multiple-path sensitization without any special processing.   □

Since a decision in PODEM is choosing a value $v_j \in \{0,1\}$ to be assigned to the PI j, the decision tree of PODEM is a binary tree in which a node corresponds to a PI j to be assigned and a branch emanating from the node j is labeled with the selected value v_j. Figure 6.31(b) shows the decision tree for Example 6.9.

As illustrated by the structure of its decision tree, the PODEM search process is based on *direct implicit enumeration* of the possible input vectors. As in the *D*-algorithm, the search is exhaustive, such that FAILURE is eventually returned only if no test exists for the target fault (see Problem 6.12). (In practice, the amount of search is bounded by user-imposed limits.)

PODEM differs from the TG algorithms patterned after the schema given in Figure 6.10 in several aspects. In PODEM, values are computed only by forward implication of PI assignments. Consequently, the computed values are always self-consistent and all values are justified. Therefore, PODEM does not need

- consistency check, as conflicts can never occur;

- the *J-frontier*, since there are no values that require justification;

- backward implication, because values are propagated only forward.

Objective	PI Assignment	Implications	D-frontier	
a=0	a=0	h=1	g	
b=1	b=1		g	
c=1	c=1	g=D	i,k,m	
d=1	d=1	d'=0 i=$\overline{D}$	k,m,n	
k=1	e=0	e'=1 j=0 k=1 n=1	m	x-path check fails
	e=1	e'=0 j=1 k=$\overline{D}$ n=x	m,n	reversal
l=1	f=1	f'=0 l=1 m=$\overline{D}$ n=D		

(a)

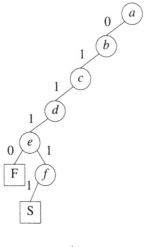

(b)

Figure 6.31   Execution trace of PODEM on the problem of Example 6.6

Another important consequence of the direct search process is that it allows PODEM to use a simplified backtracking mechanism. Recall that backtracking involves restoring the state of computation to that existing before an incorrect decision. In the TG algorithms previously described, state saving and restoring is an explicit (and time-consuming) process. In PODEM, since the state depends only on PI values and any decision deals only with the value of one PI, any change in state is easily computed by propagating a new PI value. Thus *backtracking is implicitly done by simulation* rather than by an explicit save/restore process.

Because of these advantages, PODEM is much simpler than other TG algorithms, and this simplicity is a key factor contributing to its success in practice. Experimental results presented in [Goel 1981] show that PODEM is generally faster than the *D*-algorithm. PODEM is the core of a TG system [Goel and Rosales 1981] successfully used to generate tests for large circuits.

FAN

The FAN (Fanout-Oriented TG) algorithm [Fujiwara and Shimono 1983] introduces two major extensions to the backtracing concept of PODEM:

- Rather than stopping at PIs, *backtracing in FAN may stop at internal lines.*

- Rather than trying to satisfy one objective, FAN uses a *multiple-backtrace* procedure that attempts to simultaneously satisfy a set of objectives.

The internal lines where FAN stops backtracing are defined as follows. A line that is reachable from (i.e., directly or indirectly fed by) at least one stem is said to be *bound*. A line that is not bound is said to be *free*. A *head line* is a free line that directly feeds a bound line. For example, in the circuit of Figure 6.32, *A, B, C, E, F, G, H,* and *J* are free lines, *K, L,* and *M* are bound lines, and *H* and *J* are head lines. Since the subcircuit feeding a head line *l* is fanout-free, a value of *l* can be justified without contradicting any other value previously assigned in the circuit. Thus backtracing can stop at *l,* and the problem of justifying the value of *l* can be postponed for the last stage of the TG algorithm. The following example illustrates how this technique may simplify the search process.

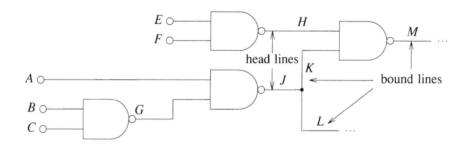

Figure 6.32   Example of head lines

Example 6.10: For the circuit of Figure 6.32, suppose that at some point in the execution of PODEM, we want to set *J*=0. Moreover, we assume that with the PI assignments previously made, setting *J*=0 causes the *D-frontier* to become empty and hence leads to a FAILURE state. Figure 6.33(a) shows the portion of the PODEM decision tree corresponding to this failing search. Since *J* is a head line, backtracing in

FAN stops at J and the assignment $J=0$ is tried *before* any attempt is made to justify $J=0$. This leads to the reduced decision tree in Figure 6.33(b).   □

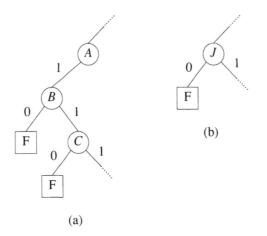

Figure 6.33   Decision trees (a) For PODEM (backtracing to PIs) (b) For FAN (backtracing to head lines)

Recall that the fault-activation and the error-propagation problems map into a set of line-justification problems. In PODEM, each one of these problems becomes in turn an objective that is individually backtraced. But a PI assignment satisfying one objective may preclude achieving another one, and this leads to backtracking. To minimize this search, the multiple backtrace procedure of FAN (*Mbacktrace* in Figure 6.34) starts with a set of objectives (*Current_objectives*) and it determines an assignment $k=v_k$ that is likely either to contribute to achieving a subset of the original objectives or to show that some subset of the original objectives cannot be simultaneously achieved.

The latter situation may occur only when different objectives are backtraced to the same stem with conflicting values at its fanout branches (see Figure 6.35). To detect this, *Mbacktrace* stops backtracing when a stem is reached and keeps track of the number of times a 0-value and a 1-value have been requested on the stem.

Mbacktrace processes in turn every current objective until the set *Current_objectives* is exhausted. Objectives generated for head lines reached during this process are stored in the set *Head_objectives*. Similarly, the set *Stem_objectives* stores the stems reached by backtracing. After all current objectives have been traced, the highest-level stem from *Stem_objectives* is analyzed. Selecting the highest-level stem guarantees that all the objectives that could depend on this stem have been backtraced. If the stem k has been reached with conflicting values (and if k cannot propagate the effect of the target fault), then *Mbacktrace* returns the objective (k, v_k), where v_k is the most requested value for k. Otherwise the backtracing is restarted from k. If no stems have been reached, then *Mbacktrace* returns an entry from the set *Head_objectives*.

```
Mbacktrace (Current_objectives)
begin
   repeat
      begin
         remove one entry (k,vₖ) from Current_objectives
         if k is a head line
            then add (k,vₖ) to Head_objectives
         else if k is a fanout branch then
               begin
                  j = stem(k)
                  increment number of requests at j for vₖ
                  add j to Stem_objectives
               end
            else /* continue tracing */
               begin
                  i = inversion of k
                  c = controlling value of k
                  if (vₖ⊕i = c) then
                     begin
                        select an input (j) of k with value x
                        add (j,c) to Current_objectives
                     end
                  else
                     for every input (j) of k with value x
                        add (j,c̄) to Current_objectives
               end
      end
   until Current_objectives = ∅
   if Stem_objectives ≠ ∅ then
      begin
         remove the highest-level stem (k) from Stem_objectives
         vₖ = most requested value of k
         if (k has contradictory requirements and
            k is not reachable from target fault)
            then return (k,vₖ)
         add (k,vₖ) to Current_objectives
         return Mbacktrace (Current_objectives)
      end
   remove one objective (k,vₖ) from Head_objectives
   return (k,vₖ)
end
```

Figure 6.34   Multiple backtrace

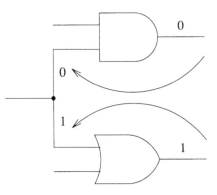

Figure 6.35   Multiple backtrace generating conflicting values on a stem

Example 6.11: Consider the circuit in Figure 6.36(a). Figure 6.36(b) illustrates the execution of *Mbacktrace*, starting with (*I*,1) and (*J*,0) as current objectives. After all the current objectives are exhausted for the first time, *Stems_objectives* = {*A*,*E*}. Since the highest-level stem (*E*) has no conflicting requirements, backtracing is restarted from *E*. Next time when *Current_objectives* = ∅, stem *A* is selected for analysis. Because *A* has been reached with 1-value from *A1* and with 0-value from *A2*, *Mbacktrace* returns (*A*,*v*), where *v* can be either 0 or 1 (both values have been requested only once).   □

Figure 6.37 outlines a recursive version of FAN. Its implication process (denoted by *Imply_and_check*) is the same as the one described in Section 6.2.1.1. Unlike PODEM, implication propagates values both forward and backward; hence unjustified values may exist. FAN recognizes SUCCESS when an error has been propagated to a PO and all the bound lines have been justified. Then it justifies any assigned head line; recall that this justification process cannot cause conflicts.

If FAN cannot immediately identify a SUCCESS or a FAILURE state, it marks all the unjustified values of bound lines as current objectives, together with the values needed for the *D*-drive operation through one gate from the *D-frontier*. From these objectives, *Mbacktrace* determines an assignment for a stem or a head line to be tried next. The decision process is similar to the one used in PODEM. Experimental results presented in [Fujiwara and Shimono 1983] show that FAN is more efficient than PODEM. Its increased speed is primarily caused by a significant reduction in backtracking.

Other Algorithms

Next we will discuss other TG algorithms which extend the concept of head lines by identifying higher-level lines whose values can be justified without conflicts and using them to stop the backtracing process.

A *total-reconvergence line* used in the TOPS (*Top*ological *S*earch) algorithm [Kirkland and Mercer 1987] is the output *l* of a subcircuit *C* such that all paths between any line in *C* and any PO go through *l* (see Figure 6.38). In other words, cutting *l* would

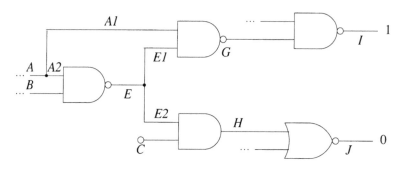

(a)

Current_objectives	Processed entry	Stem_objectives	Head_objectives
$(I,1),(J,0)$	$(I,1)$		
$(J,0),(G,0)$	$(J,0)$		
$(G,0),(H,1)$	$(G,0)$		
$(H,1),(A1,1),(E1,1)$	$(H,1)$		
$(A1,1),(E1,1),(E2,1),(C,1)$	$(A1,1)$	A	
$(E1,1),(E2,1),(C,1)$	$(E1,1)$	A,E	
$(E2,1),(C,1)$	$(E2,1)$	A,E	
$(C,1)$	$(C,1)$	A,E	C
$\varnothing$		A	C
$(E,1)$	$(E,1)$	A	C
$(A2,0)$	$(A2,0)$	A	C
$\varnothing$		A	C

(b)

Figure 6.36   Example of multiple backtrace

isolate C from the rest of the circuit. Clearly, a head line satisfies the definition of a total-reconvergence line; then C is a fanout-free subcircuit. However, the subcircuit C bounded by a total-reconvergence line may have fanout, but all such fanout must reconverge before I. Assuming that the function implemented by C is not a constant (0 or 1), any value assigned to I can be justified without conflicts with other values already assigned in the circuit. Thus the justification of a total-reconvergence line can be postponed in the same way as the justification of a head line. (Note that the justification of a total-reconvergence line may require backtracking).

Both the head lines and the total-reconvergence lines are found by a topological analysis of the circuit. A *backtrace-stop line* used in the FAST (*F*ault-oriented *A*lgorithm for *S*ensitized-path *T*esting) algorithm [Abramovici *et al.* 1986a] represents

FAN()
begin
    if *Imply_and_check()* = FAILURE **then return** FAILURE
    if (error at PO **and** all bound lines are justified) **then**
      begin
          justify all unjustified head lines
          return SUCCESS
      end
    if (error not at PO **and** *D-frontier* = ∅) **then return** FAILURE
    /* initialize objectives */
    add every unjustified bound line to *Current_objectives*
    select one gate (*G*) from the *D-frontier*
    c = controlling value of G
    for every input (*j*) of G with value x
      add $(j,\bar{c})$ to *Current_objectives*
    /* multiple backtrace */
    (i,v_i) = *Mbacktrace(Current_objectives)*
    Assign(i,v_i)
    if *FAN()* = SUCCESS **then return** SUCCESS
    Assign(i,\bar{v}_i) /* reverse decision */
    if *FAN()* = SUCCESS **then return** SUCCESS
    Assign(i,x)
    return FAILURE
end

Figure 6.37 FAN

another generalization of the head-line concept, based on an analysis that is both topological and functional. In FAST, a line *l* is a backtrace-stop for value *v*, if the assignment *l* = *v* can be justified without conflicts. For example, in Figure 6.39, *L* is a backtrace-stop for 0, because *L* = 0 can be justified without assigning any line with reconvergent fanout (by *F* = 0 and *A* = *B* = 1). Note that *L* is fed by reconvergent fanout, and it is not a total-reconvergence line. The identification of backtrace-stop lines will be explained in the next section.

6.2.1.3 Selection Criteria

The search process of any of the TG algorithms analyzed in this chapter involves decisions. A first type of decision is to select one of the several unsolved problems existing at a certain stage in the execution of the algorithm. A second type is to select one possible way to solve the selected problem. In this section we discuss *selection criteria* that are helpful in speeding up the search process. These selection criteria are based on the following principles:

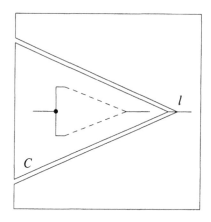

Figure 6.38   Total-reconvergence line

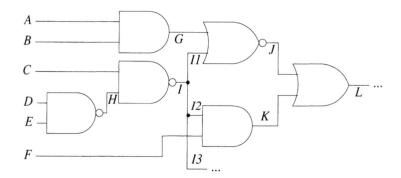

Figure 6.39

1. *Among different unsolved problems, first attack the most difficult one* (to avoid
    the useless time spent in solving the easier problems when a harder one cannot
    be solved).

2. *Among different solutions of a problem, first try the easiest one.*

Selection criteria differ mainly by the *cost functions* they use to measure "difficulty."
Typically, cost functions are of two types:

• *controllability measures,* which indicate the relative difficulty of setting a line to a
  value;

• *observability measures,* which indicate the relative difficulty of propagating an error
  from a line to a PO.

Controllability measures can be used both to select the most difficult line-justification problem (say, setting gate G to value v), and then to select among the unspecified inputs of G the one that is the easiest to set to the controlling value of the gate. Observability measures can be used to select the gate from the *D-frontier* whose input error is the easiest to observe. Note that the cost functions should provide only *relative measures*. (Controllability and observability measures have also been used to compute "testability" measures, with the goal of estimating the difficulty of generating tests for specific faults or for the entire circuit [Agrawal and Mercer 1982]. Our goal here is different; namely we are looking for measures to guide the decision process of a TG algorithm.)

Controllability measures can also be used to guide the backtracing process of PODEM. Consider the selection of the gate input in the procedure *Backtrace* (Figure 6.27). While this selection can be arbitrary, according to the two principles stated above it should be based on the value v needed at the gate input. If v is the controlling (noncontrolling) value of the gate, then we select the input that is the easiest (most difficult) to set to v.

Of course, the savings realized by using cost functions should be greater than the effort required to compute them. In general, cost functions are *static*, i.e., they are computed by a preprocessing step and are not modified during TG.

Distance-Based Cost Functions

Any cost function should show that PIs are the easiest to control and POs are the easiest to observe. Taking a simplistic view, we can consider that the difficulty of controlling a line increases with its distance from PIs, and the difficulty of observing a line increases with its distance from POs. Thus we can measure the controllability of a line by its level, and its observability by its minimum distance from a PO. Although these measures are crude, it is encouraging to observe that using them still gives better results than not using any cost functions (i.e., taking random choices).

Recursive Cost Functions

The main drawback of the distance-based cost functions is that they do not take the logic into account. In this section we present more complex cost functions that do not suffer from this shortcoming [Rutman 1972, Breuer 1978, Goldstein 1979].

First we discuss controllability measures. For every signal l we want to compute two cost functions, $C0(l)$ and $C1(l)$, to reflect, respectively, the relative difficulty of setting l to 0 and 1. Consider an AND gate (see Figure 6.40) and assume that we know the $C0$ and $C1$ costs for every input. What can we say about the costs of the setting the output X? To set the output X to 0 it is enough to set any input to 0, so we can select the easiest one. Thus:

$$C0(X) = min\{C0(A),C0(B),C0(C)\} \qquad (6.1)$$

To set X to 1, we must simultaneously set all its inputs to 1. If A, B, and C are independent (i.e., they do not depend on common PIs), then setting X to 1 is the union of three disjoint problems, so we can add the three separate costs:

$$C1(X) = C1(A)+C1(B)+C1(C) \qquad (6.2)$$

Figure 6.40

This formula may lead to less accurate results if applied when inputs of X are not independent because of reconvergent fanout. In Figure 6.41(a), B and C are identical signals for which the cost of controlling them simultaneously should be the same as the cost of controlling each one of them. In Figure 6.41(b), B and C are complementary signals that can never be simultaneously set to the same value, so the correct value of $C1(X)$ should show that setting $X=1$ is impossible.

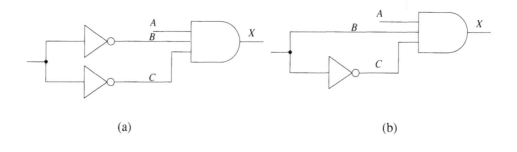

(a)   (b)

Figure 6.41

Nevertheless, to keep the computation of costs a simple process, we will use the simplifying assumption that costs of simultaneous line-setting problems are additive, and later we will try to compensate for potential inaccuracies by introducing *correction terms*.

Recursive formulas similar to (6.1) and (6.2) can be easily developed for other types of gates. The computation of controllability costs proceeds level by level. First $C0$ and $C1$ of every PI are set to 1. Then $C0$ and $C1$ are computed for every gate at level 1, then for every gate at level 2, and so on. In this way the costs of a line are computed only after the costs of its predecessors are known. Thus controllability costs are determined in one forward traversal of the circuit, and the algorithm is linear in the number of gates in the circuit.

Now let us discuss an observability measure $O(l)$ that reflects the relative difficulty of propagating an error from l to a PO. Consider again the AND gate in Figure 6.40 and assume that we know $O(X)$. What can we say about the cost of observing the input A? To propagate an error from A we must set both B and C to 1 (this propagates the error from A to X), and then we must propagate the error from X to a PO. If these three problems are independent, then

$$O(A) = C1(B)+C1(C)+O(X) \tag{6.3}$$

Applying formula (6.3) when the problems of setting B to 1, of setting C to 1, and of propagating the error from X to a PO are not independent leads to erroneous results (see Problem 6.14). Again, to keep the computation simple, we will use the simplifying assumption that problems are independent.

Now let us consider the problem of determining the observability of a stem X, knowing the observability of its fanout branches $X1$, $X2$, and $X3$. To propagate an error from X, we may choose any path starting at $X1$, $X2$, or $X3$. With the simplifying assumption that single-path propagation is possible, we can select the most observable path. Then

$$O(X) = min\{O(X1),O(X2),O(X3)\} \tag{6.4}$$

The computation of observability costs starts by setting the cost of every PO to 0. Then the circuit is traversed backward, applying formulas similar to (6.3) and (6.4) where appropriate. Now the cost of a line is computed only after the costs of all its successors are known. Note that computation of observability costs assumes that controllability costs are known. The algorithm is linear in the number of lines in the circuit.

Fanout-Based Cost Functions

We have seen that the existence of reconvergent fanout makes TG difficult. Consider the problem of justifying $X=0$ in the circuit in Figure 6.42(a). Clearly, we would like a selection criterion that would guide the TG algorithm to choose the solution $B=0$ rather than $A=0$, because A has fanout and the side-effects of setting A to 0 may cause a conflict. As Figure 6.42(b) shows, it is not enough to look only at the fanout count of the lines directly involved in the decision (A and B).

A fanout-based controllability measure $C(l)$ reflects the relative potential for conflicts resulting from assigning a value to line l. The cost $C(l)$ should depend both on the fanout count of l and on the fanout count of the predecessors of l. Such a measure was first defined in [Putzolu and Roth 1971] as

$$C(l) = \sum_i C(i)+f_l -1 \tag{6.5}$$

where the summation is for all inputs of l, and f_l is the fanout count of l. The cost of a PI l is f_l-1. Note that $C(l) = 0$ iff l is a free line (i.e., it is not reachable from a stem). Thus a 0 cost denotes a line that can be justified without conflicts. Applying (6.5) to the circuit in Figure 6.42(b), we obtain $C(A) = 0$, $C(B) = 2$, and $C(X) = 2$. Based on these values, a TG algorithm will select $A = 0$ to justify $X = 0$.

Formula (6.5) does not distinguish between setting a line to 0 and to 1. We obtain more accurate fanout-based measures if we use two controllability cost functions, $C0(l)$ and $C1(l)$, to reflect the relative potential for conflict resulting from setting l respectively to 0 and to 1 [Rutman 1972]. For an AND gate we have

$$C0(l) = min_i\{C0(i)\}+f_l-1 \tag{6.6}$$

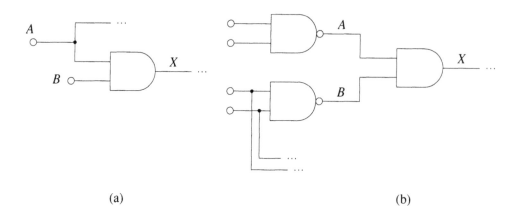

(a)   (b)

Figure 6.42

and

$$C1(l) = \sum_i C1(i) + f_l - 1 \tag{6.7}$$

For a PI l, its $C0$ and $C1$ costs are set to f_l-1. Formulas (6.6) and (6.7) also have the property that a 0 value indicates an assignment that can be done without conflicts. Thus these measures can identify the backtrace-stop lines used by FAST. Applying these measures to the circuit in Figure 6.42(a), we obtain

$$C0(A)=C1(A)=1, \; C0(B)=C1(B)=0, \; C0(X)=0, \; C1(X)=1.$$

Thus we correctly identify that X can be set to 0 without conflicts, even if X is fed by a stem.

Let us rewrite (6.1) and (6.2) in a more general form:

$$C0(l) = \min_i \{C0(i)\} \tag{6.1a}$$

$$C1(l) = \sum_i C1(i) \tag{6.2a}$$

and compare them with (6.6) and (6.7). We can observe that they are almost identical, except for the term f_l-1. So we can consider (6.6) and (6.7) as being extensions of (6.1a) and (6.2a), with a *correction term* added to reflect the influence of fanout.

The measure presented in [Abramovici *et al.* 1986a] takes into account that only reconvergent fanout can cause conflicts and introduces correction terms that reflect the extent to which fanout is reconvergent.

Let us consider the circuit in Figure 6.43. The correction term for both $C0(A)$ and $C1(A)$ used in formulas (6.6) and (6.7) has value 1, since A has a fanout count of 2. But if we analyze the effect of setting A to 0 and to 1, we can see that $A=0$ has a much greater potential for conflicts than $A=1$. This is because $A=0$ results in B, C, D, and E being set to binary values, while $A=1$ does not set any other gate output.

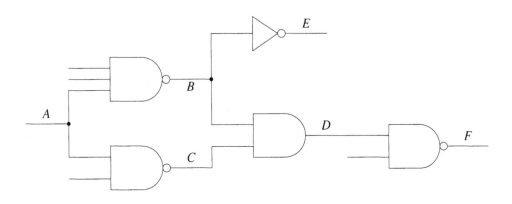

Figure 6.43

To account for this type of difference, Breuer[1978] has defined *side-effects cost functions*, $CS0(l)$ and $CS1(l)$, to reflect the relative potential for conflicts caused by setting l to 0 and 1 respectively. These functions are computed by simulating the assignment $l=v$ ($v\in\{0,1\}$) in a circuit initialized to an all-x state, then accounting for its effects as follows:

1. A gate whose output is set to a binary value increases the cost by 1.

2. A gate with n inputs, whose output remains at x but which has m inputs set to a binary value, increases the cost by m/n.

For the circuit in Figure 6.43, the side-effects costs of setting A are $CS0(A)=4\frac{1}{2}$ and $CS1(A)=\frac{1}{3}+\frac{1}{2}=\frac{5}{6}$.

Using the side-effects cost functions as correction terms we obtain (for an AND gate)

$$C0(l) = \min_{i}\{C0(i)\}+CS0(l) \qquad (6.8)$$

and

$$C1(l) = \sum_{i}C1(i)+CS1(l) \qquad (6.9)$$

Concluding Remarks on Cost Functions

The "best" controllability and observability measures could be obtained by actually solving the corresponding line-justification and error-propagation problems and measuring their computational effort. But such measures are useless, because their computation would be as expensive as the TG process they are meant to guide. Keeping the cost of computing the measures significantly lower than the TG cost requires that we accept less than totally accurate measures.

Suppose we compare two cost functions, A and B, and we find that A performs better than B for some circuits. For other circuits, however, using B leads to better results.

This type of "inconsistency" is inherent in the heuristic nature of cost functions. To compensate for this effect, Chandra and Patel [1989] suggest switching among cost functions during TG.

An approach that combines the computation of cost functions with a partial test generation is presented in [Ratiu *et al.* 1982]. Ivanov and Agarwal [1988] use *dynamic* cost functions, which are updated during test generation based on the already assigned values.

6.2.2 Fault-Independent ATG

The goal of the fault-oriented algorithms discussed in the previous section is to generate a test for a specified target fault. To generate a set of tests for a circuit, such an algorithm must be used with procedures for determining the initial set of faults of interest, for selecting a target fault, and for maintaining the set of remaining undetected faults. In this section we describe a fault-independent algorithm whose goal is to *derive a set of tests that detect a large set of SSFs without targeting individual faults.*

Recall that half of the SSFs along a path critical in a test t are detected by t. Therefore it seems desirable to generate tests that produce long critical paths. Such a method of *critical-path TG* was first used in the LASAR (*L*ogic *A*utomated *S*timulus *A*nd *R*esponse) system [Thomas 1971]. A different critical-path TG algorithm is described in [Wang 1975].

The basic steps of a critical-path TG algorithm are

1. Select a PO and assign it a critical 0-value or 1-value (the value of a PO is always critical).

2. Recursively justify the PO value, trying to justify any critical value on a gate output by critical values on the gate inputs.

Figure 6.44 illustrates the difference between justifying a critical value and a noncritical value for an AND gate with three inputs. Critical values are shown in bold type (**0** and **1**). Unlike the primitive cubes used to justify a noncritical value, the input combinations used to justify a critical value — referred to as *critical cubes* — are always completely specified. This is because changing the value of a critical gate input should change the value of the gate output.

Example 6.12: Consider the fanout-free circuit in Figure 6.45(a). We start by assigning $G=0$. To justify $G=0$, we can choose $(E,F)=$**01** or $(E,F)=$**10**. Suppose we select the first alternative. Then $E=0$ is uniquely justified by $(A,B)=$**11**, and $F=1$ is uniquely justified by $(C,D)=$**00**. Now we have generated the test $(A,B,C,D)=$1100; Figure 6.45(b) shows the critical paths in this test.

Recall that our objective is to generate a set of tests for the circuit. To derive a new test with new critical lines, we return to the last gate where we had a choice of critical input combinations and select the next untried alternative. This means that now we justify $G=0$ by $(E,F)=$**10**. To justify $E=1$ we can arbitrarily select either $A=0$ or $B=0$. Exploring both alternatives to justify $F=$**0** results in generating two new tests, shown in Figures 6.45(c) and (d).

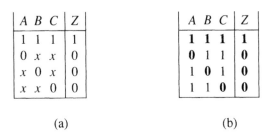

A	B	C	Z
1	1	1	1
0	x	x	0
x	0	x	0
x	x	0	0

A	B	C	Z
1	**1**	**1**	1
0	1	1	0
1	**0**	1	0
1	1	**0**	**0**

(a)   (b)

Figure 6.44   Line justification for an AND gate (a) By primitive cubes (b) By critical cubes

Now we have exhausted all the alternative ways to justify $G=0$. Two additional tests can be obtained by starting with $G=1$ (see Figures 6.45(e) and (f)). The five generated tests form a complete test set for SSFs.   □

Figure 6.46 outlines a recursive procedure *CPTGFF* for critical-path TG for fanout-free circuits. The set *Critical* contains the critical lines to be justified and their values. To generate a complete set of tests for a fanout-free circuit whose PO is Z, we use

> add (Z,0) to *Critical*
> *CPTGFF()*
> add (Z,1) to *Critical*
> *CPTGFF()*

(see Problem 6.18).

CPTGFF justifies in turn every entry in the set *Critical*. New levels of recursion are entered when there are several alternative ways to justify a critical value. Noncritical values are justified (without decisions) using the procedure *Justify* presented in Figure 6.4. A new test is recorded when all lines have been justified.

The decision process and its associated backtracking mechanism allow the systematic generation of all possible critical paths for a given PO value. As we did for the fault-oriented TG algorithms, we use a decision tree to trace the execution of the algorithm. A decision node (shown as a circle) corresponds to a critical line-justification problem for which several alternatives are available. Each alternative decision is denoted by a branch leaving the decision node. A terminal node (shown as a square) is reached when all lines have been justified. The number of terminal nodes is the number of tests being generated. Figure 6.47 gives the decision trees for Example 6.12.

The algorithm can be easily extended to multiple-output circuits that do not have reconvergent fanout (that is, every fanout branch of a stem leads to a different PO). Here, every cone of a PO is a fanout-free circuit that can be processed in turn as before. However, one problem appears because of logic common to two (or more) PO cones. Consider a line *l* that feeds two POs, *X* and *Y*. Suppose that we first generate all critical paths with $X=0$ and with $X=1$. When we repeat the procedure for *Y*, at some point we will have to justify $l=0$ or $l=1$. But all the possible ways to do this

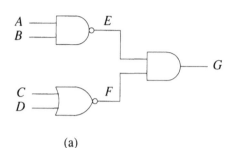

(a)

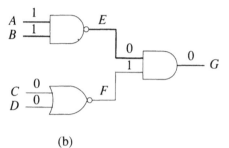

(b)

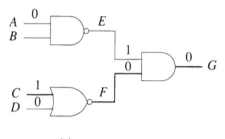

(c)

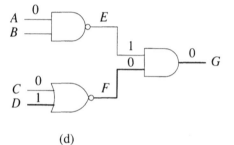

(d)

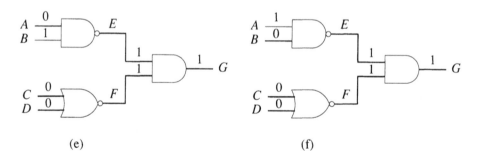

(e)   (f)

Figure 6.45   Example of critical-path TG

have been explored in the cone of X, and repeating them may generate many unnecessary tests that do not detect any faults not detected by previously generated tests. This can easily be avoided by not trying to make critical a line l that has already been marked as critical in a previous test (see Problem 6.19).

In attempting to extend critical-path TG to circuits with reconvergent fanout, we encounter problems caused by

- conflicts,
- self-masking,

```
CPTGFF()
begin
    while (Critical ≠ ∅)
        begin
            remove one entry (l,val) from Critical
            set l to val
            mark l as critical
            if l is a gate output then
                begin
                    c = controlling value of l
                    i = inversion of l
                    inval = val ⊕ i
                    if (inval = c̄)
                        then for every input j of l
                            add (j,c̄) to Critical
                        else
                            begin
                                for every input j of l
                                    begin
                                        add (j,c) to Critical
                                        for every input k of l other than j
                                            Justify (k,c̄)
                                        CPTGFF()
                                    end
                                return
                            end
                end
        end
    /* Critical = ∅ */
    record new test
    return
end
```

Figure 6.46   Critical-path TG for fanout-free circuits

- multiple-path sensitization,

- overlap among PO cones.

In the remainder of this section we discuss these problems, but because of the level of detail involved, we will not present a complete critical-path TG algorithm.

Conflicts appear from attempts to solve a set of simultaneous line-justification problems. As in fault-oriented TG algorithms, conflicts are dealt with by backtracking. A new aspect of this problem is that some conflicts may be caused by trying to justify a critical value by critical cubes. Compared with primitive cubes, critical cubes specify

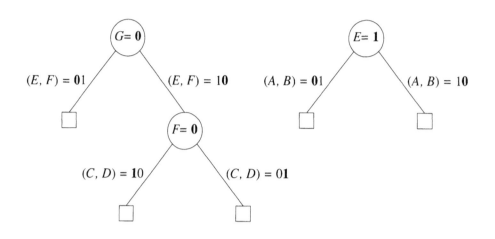

Figure 6.47   Decision trees for Example 6.12

more values, which may lead to more conflicts. Therefore when all attempts to justify a critical line by critical cubes have failed, we should try to justify it by primitive cubes. Here, in addition to recovery from incorrect decisions, backtracking also achieves a systematic exploration of critical paths.

We have seen that reconvergent fanout may lead to self-masking, which occurs when a stem is not critical in a test t, although one or more of its fanout branches are critical in t. Therefore, when a fanout branch is made critical, this does not imply that its stem is also critical. One solution is to determine the criticality of the stem by an analysis similar to that done by the critical-path tracing method described in Chapter 5. Another solution, relying on the seldomness of self-masking, is to continue the TG process by assuming that the stem is critical. Because this strategy ignores the possibility of self-masking, it must simulate every generated test to determine the faults it detects.

Tests generated by critical-path methods may fail to detect faults that can be detected only by multiple-path sensitization. For example, for the circuit in Figure 6.48, a critical-path TG algorithm will not produce any test in which both P and Q have value 1. But this is the only way to detect the fault B s-a-0. In practice, however, this situation seldom occurs, and even when it does, the resulting loss of fault coverage is usually insignificant.

Because of overlap among the PO cones, a critical-path TG algorithm may repeatedly encounter the same problem of justifying a critical value v for a line l. To avoid generating unnecessary tests, we can adopt the strategy of not trying to make line l have a critical value v if this has been done in a previously generated test. Using this strategy in circuits with reconvergent fanout may preclude the detection of some faults, but this problem is insignificant in practical circuits.

Compared to fault-oriented TG, critical-path TG offers the following advantages:

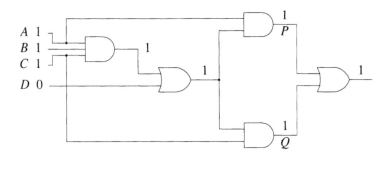

Figure 6.48

- The SSFs detected by the generated tests are immediately known without using a separate fault simulation step, as is required with a fault-oriented algorithm.

- A new test is generated by modifying the critical paths obtained in the previous test. This may avoid much duplicated effort inherent in a fault-oriented algorithm. For example, suppose that a critical-path algorithm has created the critical path shown in Figure 6.49, in which *A*=**0**. The next test that makes *B*=**0** is immediately obtained by switching (*A,B*)=**01** to (*A,B*)=**10**. By contrast, a fault-oriented TG algorithm treats the generation of tests for the targets *A s-a*-1 and *B s-a*-1 as unrelated problems, and hence it would duplicate the effort needed to propagate the error from *C* to *Z*.

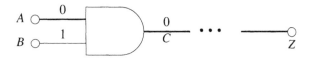

Figure 6.49

Avoiding the duplication of effort caused by repeated error propagation along the same paths is also one of the goals of the *subscripted D-algorithm* [Airapetian and McDonald 1979, Benmehrez and McDonald 1983], which tries to *generate simultaneously a family of tests* to detect SSFs associated with the same gate by using *flexible signals*. The output of an *n*-input target gate is assigned the flexible signal D_0, while its inputs are assigned the flexible signals D_i, $1 \leq i \leq n$ (see Figure 6.50(a)). The D_0 is propagated to a PO in the same way as a D in the D-algorithm, and the D_i signals are propagated towards PIs. A D_i on the output of a gate G is propagated by setting all the currently unassigned inputs of G to value D_i or $\bar{D}_i$, depending on the inversion of G. In general, conflicts among the D_i values preclude the simultaneous propagation of all of them. But usually a subset of the D_is can be propagated. For example, in the circuit of Figure 6.50(b), the algorithm starts by trying to propagate D_1, D_2, and D_3 from the inputs of gate W. (Because W is a PO, the D_0 value has already reached a PO.) But the reconvergent fanout of Q and R causes conflicts that are resolved by replacing D_2 by a 0. (More details on the propagation of D_is and on the conflict resolution can be found in [Benmehrez and McDonald 1983].) Then only D_1 and D_3 are propagated. The vector $\overline{D}_1 11 \overline{D}_3$ represents a family of tests where D_1 and D_3 can be independently replaced by 0 or 1 values. Setting P, S in turn to 01, 10, and 00, we obtain three (out of the four) critical cubes of gate W: **100**, **001**, and **000**.

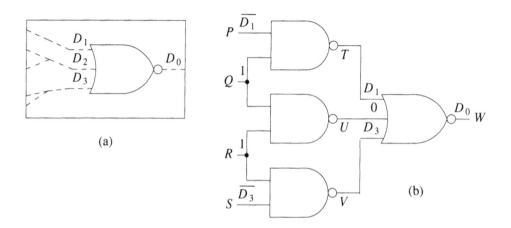

(a)

(b)

Figure 6.50   The subscripted D-algorithm (a) Flexible signals (b) Generating a family of tests

Unlike a fault-oriented algorithm, a fault-independent algorithm cannot identify undetectable faults.

6.2.3 Random Test Generation

RTG is not a truly random process of selecting input vectors, but a *pseudorandom* one. This means that vectors are generated by a deterministic algorithm such that their statistical properties are similar to those of a randomly selected set.

The main advantage of RTG is the ease of generating vectors. Its main disadvantage is that a randomly generated test set that detects a set of faults is much larger (usually 10 times or more) than a deterministically generated test set for the same set of faults. This raises the problem of how to determine the quality of a test set obtained by RTG, because conventional methods based on fault simulation may become too costly. In this section we analyze statistical methods that estimate the quality of a test set based on probabilities of detecting faults with random vectors. A related problem is to determine the number of randomly generated vectors needed to achieve a given test quality.

Initially we assume that the *input vectors are uniformly distributed*, that is, each one of the 2^n possible input vectors of a circuit with n PIs is equally likely to be generated.

This means that every PI has equal probability of having a 0-value or a 1-value. We also assume that the *input vectors are independently generated*. Then the same vector may appear more than once in the generated sequence. However, most pseudorandom vector generators work such that an already generated vector is not repeated. This mechanism leads to test sets that are smaller than those generated under the assumption of independent vectors [Wagner *et al.* 1987].

6.2.3.1 The Quality of a Random Test

The quality of a random test set should express the *level of confidence* we have in interpreting the results of its application. Clearly, if at least one applied test fails, then the circuit under test is faulty. But if all the tests pass, how confident can we be that the circuit is indeed fault-free? (By fault-free we mean that none of the detectable SSFs is present.) This level of confidence can be measured by the probability that the applied tests detect every detectable SSF. Thus for a test sequence of length N, we define its *testing quality* t_N as the probability that all detectable SSFs are detected by applying N random vectors. In other words, t_N is the probability that N random vectors will contain a complete test set for SSFs.

A different way of measuring the level of confidence in the results of random testing is to consider the faults individually. Namely, if all the N applied tests pass, how confident are we that the circuit does not contain a fault f? This can be measured by the probability d_N^f that f is detected (at least once) by applying N random vectors; d_N^f is called the *N-step detection probability of f*. The *detection quality* d_N of a test sequence of length N is the lowest N-step detection probability among the SSFs in the circuit:

$$d_N = \min_{f} d_N^f \qquad (6.10)$$

The difference between the testing quality t_N and the detection quality d_N of a test sequence of length N is that t_N is the probability of detecting every fault, while d_N is the probability of detecting the fault that is most difficult to detect [David and Blanchet 1976]. Note that $t_N < d_N$.

6.2.3.2 The Length of a Random Test

Now we discuss the problem of determining the length N required to achieve a level of confidence of at least c. Usually c relates to the detection quality, so N is chosen to satisfy

$$d_N \geq c \qquad (6.11)$$

This is justified since a test sequence long enough to detect the most difficult fault with probability c will detect any other fault f with a probability $d_N^f \geq c$.

The *N-step escape probability* e_N^f of a fault f is the probability that f remains undetected after applying N random vectors. Clearly:

$$d_N^f + e_N^f = 1 \qquad (6.12)$$

Let T_f be the set of all tests that detect f in a circuit with n PIs. Then the probability d_f that a random vector detects f is

$$d_f = |T_f| / 2^n \qquad (6.13)$$

This is d_1^f, the one-step detection probability of f; we will refer to it as the *detection probability* of f.

The probability of f remaining undetected after one vector is $e_1^f = 1 - d_f$. Because the input vectors are independent, the N-step escape probability of f is

$$e_N^f = (1 - d_f)^N \qquad (6.14)$$

Let d_{min} be the lowest detection probability among the SSFs in the circuit. Then the required N to achieve a detection quality of at least c is given by

$$1 - (1 - d_{min})^N \geq c \qquad (6.15)$$

The smallest value of N that satisfies this inequality is

$$N_d = \left\lceil \frac{ln(1-c)}{ln(1-d_{min})} \right\rceil \qquad (6.16)$$

(For values of $d_{min} \ll 1$, $ln(1 - d_{min})$ can be approximated by $-d_{min}$.)

This computation of N assumes that the detection probability d_{min} of the most difficult fault is known. Methods of determining detection probability of faults are discussed in the next section.

If the desired level of confidence c relates to the testing quality (rather than to the detection quality), then N is chosen to satisfy

$$t_N \geq c \qquad (6.17)$$

Savir and Bardell [1984] have derived the following formula to estimate an upper bound on the smallest value of N that achieves a testing quality of at least c:

$$N_t \approx \left\lceil \frac{ln(1-c) - ln(k)}{ln(1-d_{min})} \right\rceil \qquad (6.18)$$

where k is the number of faults whose detection probability is in the range $[d_{min}, 2d_{min}]$. Faults whose detection probability is $2d_{min}$ or greater do not significantly affect the required test length.

Example 6.13: Figure 6.51 tabulates several values of N_d and N_t for different values of c and k for a circuit in which $d_{min} = 0.01^*$. For example, to detect any fault with a probability of at least 0.95 we need to apply at least $N_d = 300$ random vectors. If the circuit has only $k=2$ faults that are difficult to detect (i.e., their detection probability is in the range [0.01, 0.02]), we need to apply at least $N_t = 369$ vectors to have a probability of at least 0.95 of detecting every fault.

For $d_{min} = 0.001$, all the N_d and N_t values in Figure 6.51 are increased by a factor of 10.   □

---

*   This value should be understood as an approximate value rather than an exact one (see Problem 6.20).

c	N_d	k	N_t
0.95	300	2	369
		10	530
0.98	392	2	461
		10	622

Figure 6.51

6.2.3.3 Determining Detection Probabilities

We have seen that to estimate the length of a random test sequence needed to achieve a specified detection quality, we need to know d_{min}, the detection probability of the most difficult fault. In addition, to estimate the length needed to achieve a specified testing quality, we need to know the number k of faults whose detection probability is close to d_{min}. In this section we will discuss the problem of determining the detection probability of faults.

A Lower Bound on d_{min}

The following result [David and Blanchet 1976, Schedletsky 1977] provides an easy-to-derive lower bound on d_{min}.

Lemma 6.1: In a multiple-output combinational circuit, let n_{max} be the largest number of PIs feeding a PO. Then

$$d_{min} \geq 1/2^{n_{max}}.$$

Proof: Let f be a detectable SSF in a circuit with n PIs. There exists at least one test that detects f such that some PO, say Z_k, is in error. Let n_k be the number of PIs feeding Z_k. Because the values of the other $n-n_k$ PIs are irrelevant for the detection of f at Z_k, there exist at least 2^{n-n_k} vectors that detect f. Then the detection probability of f is at least $2^{n-n_k}/2^n = 1/2^{n_k}$. The lower bound is obtained for the PO cone with the most PIs.   □

This lower bound, however, is usually too conservative, since by using it for the value of d_{min} in formula (6.16), we may often obtain $N_d > 2^n$, which means that we would need more vectors for random testing than for exhaustive testing [Schedletsky 1977].

Detection Probabilities Based on Signal Probabilities

The *signal probability* of a line l, p_l, is defined as the probability that l is set to 1 by a random vector:

$$p_l = Pr(l=1) \tag{6.19}$$

An obvious relation exists between the signal probability of a PO and the detection probabilities of its stuck faults. Namely, the detection probability of a s-a-c fault on a PO Z is p_Z for $c=0$ and $1-p_Z$ for $c=1$. We will illustrate the general relation between signal probabilities and detection probabilities by reference to Figure 6.52 [Savir *et al.*

1984]. Suppose that in a single-output circuit there exists only one path between line l and the PO. Let f be the fault l $s\text{-}a\text{-}0$. A vector that detects f must set $l=1$ and also set all the values needed to sensitize the path ($A=1$, $B=0$, etc). To account for this set of conditions we introduce an auxiliary AND gate G, such that $G=1$ iff all the conditions required for detecting f are true. If $x=v$ is a required condition, then x is directly connected to G if $v=1$, or via an inverter if $v=0$. Thus the detection probability of f is given by the signal probability of G in the modified circuit:

$$d_f = p_G \qquad\qquad (6.20)$$

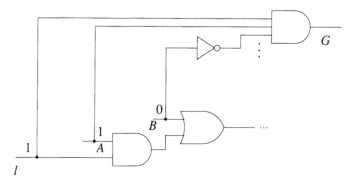

Figure 6.52

In general, there may be several paths along which an error caused by a fault f can propagate to a PO. Then the probability of detecting f propagating an error along only one of these paths is clearly a lower bound on the probability of detecting f. Denoting by G_k the auxiliary gate associated with the k-th propagation path, we have

$$d_f \geq p_{G_k} \qquad\qquad (6.21)$$

Thus signal probabilities can be used to compute detection probabilities or lower bounds on the detection probabilities. Now we turn to the problem of determining signal probabilities. Since signal probabilities of PIs are known (until now we have assumed that for every PI i, $p_i = 1/2$), let us consider the problem of computing p_Z for a gate output Z, knowing p_X of every input X of Z [Parker and McCluskey 1975].

Lemma 6.2: For an inverter with output Z and input X

$$p_Z = 1 - p_X \qquad\qquad (6.22)$$

Proof

$$p_Z = Pr(Z=1) = Pr(X=0) = 1 - p_X \qquad\qquad \square$$

Lemma 6.3: For an AND gate with output Z and inputs X and Y, if X and Y do not depend on common PIs then:

$$p_Z = p_X p_Y \qquad\qquad (6.23)$$

Proof

$$p_Z = Pr(Z=1) = Pr(X=1 \cap Y=1)$$

Because X and Y do not depend on common PIs, the events $X=1$ and $Y=1$ are independent. Thus

$$p_Z = Pr(X=1)Pr(Y=1) = p_X p_Y \qquad \qquad \square$$

Lemma 6.4: For an OR gate with output Z and inputs X and Y, if X and Y do not depend on common PIs, then

$$p_Z = p_X + p_Y - p_X p_Y \qquad \qquad (6.24)$$

Proof

$$p_Z = 1 - Pr(Z=0) = 1 - Pr(X=0 \cap Y=0)$$

Since the events $X=0$ and $Y=0$ are independent

$$p_Z = 1 - Pr(X=0)Pr(Y=0) = 1 - (1-p_X)(1-p_Y) = p_X + p_Y - p_X p_Y \qquad \square$$

Formulas (6.23) and (6.24) can be generalized for gates with more than two inputs. A NAND (NOR) gate can be treated as an AND (OR) gate followed by an inverter. Then we can compute signal probabilities in any circuit in which inputs of the same gate do not depend on common PIs; these are fanout-free circuits and circuits without reconvergent fanout. The time needed to compute all signal probabilities in such a circuit grows linearly with the number of gates.

Now let us consider a circuit that has only one stem, say A, with reconvergent fanout. Let Z be a signal that depends on A. Using conditional probabilities, we can express p_Z as

$$Pr(Z=1) = Pr(Z=1 \mid A=0)Pr(A=0) + Pr(Z=1 \mid A=1)Pr(A=1)$$

or

$$p_Z = Pr(Z=1 \mid A=0)(1-p_A) + Pr(Z=1 \mid A=1)p_A \qquad (6.25)$$

We can interpret this formula as follows. Let us create two circuits, N^0 and N^1, obtained from the original circuit N by permanently setting $A=0$ and $A=1$ respectively. Then $Pr(Z=1 \mid A=0)$ is simply p_Z computed in N^0, and $Pr(Z=1 \mid A=1)$ is p_Z computed in N^1. The result of this conceptual splitting is that both N^0 and N^1 are free of reconvergent fanout so we can apply the formulas (6.23) and (6.24).

Example 6.14: Let us compute signal probabilities for the circuit in Figure 6.53(a). The only gates whose inputs depend on common PIs are Z_1 and Z_2; the computation for all the other gates is straightforward and their values are shown in Figure 6.53(a). Figures 6.53(b) and (c) show the values obtained for $A=0$ and $A=1$ respectively (the computation is done only for the lines affected by A). Finally we compute p_{Z_1} and p_{Z_2} using formula (6.25)

$$p_{Z_1} = \frac{1}{2} \cdot \frac{1}{4} + \frac{5}{8} \cdot \frac{3}{4} = \frac{19}{32}$$

$$p_{Z_2} = \frac{1}{2} \cdot \frac{1}{4} + \frac{11}{16} \cdot \frac{3}{4} = \frac{41}{64}$$

□

Note that for every line affected by the stem A we compute two signal probabilities, one for $A=0$ and one for $A=1$. If we generalize this approach for a circuit that has k stems with reconvergent fanout, for every line we may have to compute and store up to 2^k signal probabilities. This exponential growth renders this type of approach impractical for large circuits. Different methods for computing exact signal probabilities [Parker and McCluskey 1975, Seth *et al.* 1985] suffer from the same problem. An approximate method is described in [Krishnamurthy and Tollis 1989].

The *cutting algorithm* [Savir *et al.* 1984] reduces the complexity of computation by computing a range $[p_l^L, p_l^U]$ rather than an exact value for the signal probability p_l, such that $p_l \in [p_l^L, p_l^U]$. Its basic mechanism, illustrated in Figure 6.54, consists of cutting $k-1$ fanout branches of a stem having k fanout branches; the cut fanout branches become PIs with unknown signal probabilities and are assigned the range $[0,1]$. Only reconvergent fanout branches should be cut. The resulting circuit is free of reconvergent fanout, so signal probabilities are easy to compute. Whenever ranges are involved, the computations are done separately for the two bounds.

Example 6.15: Figure 6.55 shows the signal probability ranges computed by the cutting algorithm for the circuit in Figure 6.53(a). Note that the fanout branch that remains connected to the stem inherits its signal probability. We can check that the exact values of p_{Z_1} and p_{Z_2} determined in Example 6.14 do fall within the ranges computed for p_{Z_1} and p_{Z_2} by the cutting algorithm.

□

Various extensions of the cutting algorithm that compute tighter bounds on signal probabilities are described in [Savir *et al.* 1984].

Finding the Difficult Faults

Usually the maximal length of a test sequence N_{max} is limited by the testing environment factors such as the capabilities of the tester or the maximum time allocated for a testing experiment. Given N_{max} and a desired detection quality c, from (6.15) we can derive the required lower bound d_L on the detection probability of every fault in the circuit. In other words, if $d_f \geq d_L$ for every fault f, then testing the circuit with N_{max} vectors will achieve a detection quality of at least c. Then, given a circuit, we want to determine whether it contains "difficult" faults whose detection probability is lower than d_L.

The following result shows that the difficult faults, if any, are among the checkpoint faults of the circuit.

Lemma 6.5: The fault with the smallest detection probability in a circuit is one of its checkpoint faults.

Proof: For any fault g on a line that is not a checkpoint (fanout branch or PI), we can find at least one checkpoint fault f such that g dominates f. Thus the number of tests that detect f is smaller than or equal to the number of tests that detect g. Hence $p_f \leq p_g$.

□

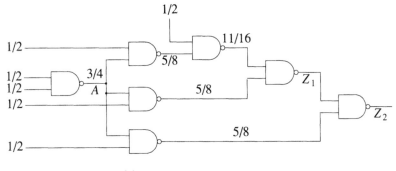

(a)

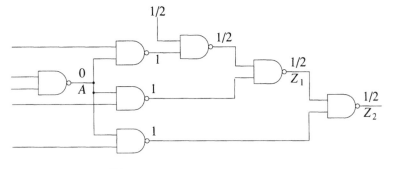

(b)

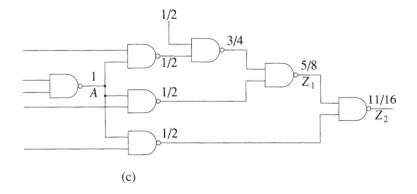

(c)

Figure 6.53

Figure 6.56 presents a procedure for determining the difficult faults [Savir *et al.* 1984]. Note that only single propagation paths are analyzed. A fault that can be detected only by multiple-path sensitization will be marked as difficult (see Problem 6.27). Redundant faults will also be among the difficult ones (see Problem 6.28).

Once the difficult faults of a circuit are identified, the circuit can be modified so that their detection probabilities become acceptable [Eichelberger and Lindbloom 1983].

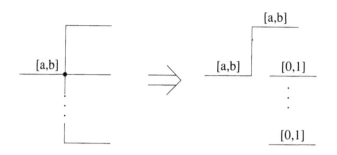

Figure 6.54 Basic transformation in the cutting algorithm

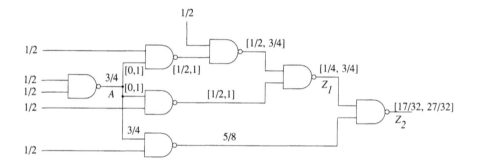

Figure 6.55 Signal probabilities computed by the cutting algorithm

6.2.3.4 RTG with Nonuniform Distributions

Until now we have assumed that input vectors are uniformly distributed; hence the signal probability of every PI i is $p_i=0.5$. This uniform distribution, however, is not necessarily optimal; that is, different p_i values may lead to shorter test sequences.

For example, Agrawal and Agrawal [1976] showed that for fanout-free circuits composed of NAND gates with the same number (n) of inputs, the optimal value p_{opt} of p_i is given by

$$p_i = 1-p_i^n \tag{6.26}$$

For $n=2$, $p_{opt}=0.617$. For such a circuit with 10 levels, one would need about 8000 random vectors to achieve a detection quality of 0.99 using $p_i=0.5$, while using p_{opt} the same detection quality can be achieved with about 700 vectors.

Similar experimental results, showing that nonuniform distributions lead to shorter random test sequences, have also been obtained for general circuits. An even better approach is to allow different PIs to have different p_i values. Such random vectors are

```
for every checkpoint fault f
    begin
        repeat
            begin
                select an untried propagation path P for f
                introduce an auxiliary AND gate G such that
                    pG is the detection probability of f along P
                compute pGᴸ, the lower bound of pG
            end
        until pGᴸ≥dL or all propagation paths have been analyzed
        if pGᴸ<dL then mark f as difficult
    end
```

Figure 6.56 Procedure for finding the difficult faults

said to be *weighted* or *biased*. Methods for computing the p_i values are presented in [Wunderlich 1987, Waicukauski *et al.* 1989].

Adaptive RTG methods dynamically modify the p_i values trying to reach "optimal" values. This requires an adaptation mechanism based on monitoring the TG process to determine "successful" random vectors. The method used in [Parker 1976] grades the random vectors by the number of faults they detect. Then the frequency of 1-values on PIs is measured in the set of tests that detected the most faults, and the p_i values are set according to these frequencies. A different method, based on monitoring the rate of variation of the activity count generated by changing PI values, is presented in [Schnurmann *et al.* 1975]. Lisanke *et al.* [1987] describe an adaptive method in which the p_i values are changed with the goal of minimizing a testability cost function.

6.2.4 Combined Deterministic/Random TG

Because RTG is oblivious to the internal model of the circuit, faults that have small detection probabilities require long sequences for their detection. But some of these faults may be easy to detect using a deterministic algorithm. As a simple example, consider an AND gate with 10 inputs. The detection probability of any input fault is 1/1024. However, generating tests for these faults is a simple task for any deterministic TG algorithm. In this section we analyze TG methods that combine features of deterministic critical-path TG algorithms with those of RTG.

RAPS

Random Path Sensitization (RAPS) [Goel 1978] attempts to create random critical paths between PIs and POs (see Figure 6.57). Initially all values are x. RAPS starts by randomly selecting a PO Z and a binary value v. Then the objective (Z,v) is mapped into a PI assignment (i,v_i) by a *random backtrace* procedure *Rbacktrace*. *Rbacktrace* is similar to the *Backtrace* procedure used by PODEM (Figure 6.27), except that the selection of a gate input is random. (In PODEM this selection is guided by controllability costs). The assignment $i=v_i$ is then simulated (using

3-valued simulation), and the process is repeated until the value of Z becomes binary. After all the POs have binary values, a second phase of RAPS assigns those PIs with x values (if any), such that the likelihood of creating critical paths increases (see Problem 6.29).

> **repeat**
>   **begin**
>   randomly select a PO (Z) with x value
>   randomly select a value v
>   **repeat**
>   **begin**
> $(i,v_i) = Rbacktrace\ (Z,v)$
> $Simulate\ (i,v_i)$
>   **end**
>   **until** Z has binary value
>   **end**
> **until** all POs have binary values
> **while** some PIs have x values
>   **begin**
> G = highest level gate with x input values
> c = controlling value of G
>   randomly select an input (j) of G with x value
>   **repeat**
>   **begin**
> $(i,v_i) = Rbacktrace\ (j,\bar{c})$
> $Simulate\ (i,v_i)$
>   **end**
>   **until** j has binary value
>   **end**

Figure 6.57   RAPS

The entire procedure is repeated to generate as many tests as needed. A different set of critical paths is likely to be generated in every test. Figure 6.58(b) traces a possible execution of RAPS for the circuit in Figure 6.58(a). In general, test sets generated by RAPS are smaller and achieve a higher fault coverage than those obtained by RTG.

Figure 6.59 illustrates one problem that may reduce the efficiency of RAPS. First $A = 0$ has been selected to justify $PO1 = 0$, then $B = 0$ to justify $PO2 = 0$. But the latter selection precludes propagation of any fault effects to $PO1$.

Another problem is shown in Figure 6.60. Because of the random selection process of RAPS, half of the tests in which $Z = 1$ will have $B = 0$. But only one test with $B = 0$ (and $C = 1$) is useful, because none of the subsequent tests with $B = 0$ will detect any new faults on the PO Z.

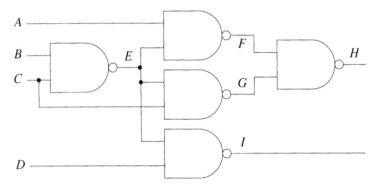

(a)

	Objective	Path backtraced	PI assignment	Simulation
t_1	$H = 0$	H, F, A H, G, C	$A = 0$ $C = 0$	$F = 1$ $E = 1, G = 1, H = 0$
	$I = 1$	I, D	$D = 0$	$I = 1$
	$B = 1$		$B = 1$	
t_2	$H = 0$	H, G, E, B H, F, E, C	$B = 1$ $C = 1$	$E = 0, F = 1, G = 1, H = 0, I = 1$
	$D = 1$		$D = 1$	
	$A = 1$		$A = 1$	
t_3	$I = 0$	I, E, C I, D	$C = 0$ $D = 1$	$E = 1, G = 1$ $I = 1$
	$H = 1$	H, F, A	$A = 1$	$F = 0, H = 1$
	$B = 1$		$B = 1$	

(b)

Figure 6.58 Example of random path sensitization

A similar problem occurs when all faults that can be detected at a PO Z when $Z = v$ have been detected by the already generated tests. In half of the subsequent tests, RAPS will continue trying to set $Z = v$, which, in addition to being useless, may also prevent detection of faults at other POs that depend on PIs assigned to satisfy $Z = v$.

Figure 6.61 illustrates a problem in the second stage of RAPS. Because the value of d is justified by $a = 0$, while $b = c = x$, RAPS will try to justify 1's for b and c. But

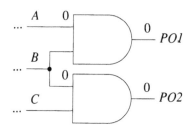

Figure 6.59

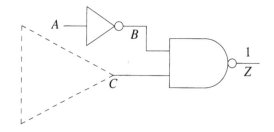

Figure 6.60

setting $b = c = 1$ is useless, because no additional faults can be detected through d; in addition, $c = 1$ requires $e = 0$, which will preclude detecting any other faults through h.

SMART

SMART (*S*ensitizing *M*ethod for *A*lgorithmic *R*andom *T*esting) is another combined deterministic/random TG method [Abramovici *et al.* 1986a], which corrects the problems encountered by RAPS. SMART generates a vector incrementally and relies on a close interaction with fault simulation. The partial vector generated at every step is simulated using critical-path tracing (CRIPT), whose results guide the test generation process. This interaction is based on two byproducts of CRIPT: *stop lines* and *restart gates*. Stop lines delimit areas of the circuit where additional fault coverage cannot be obtained, while restart gates point to areas where new faults are likely to be detected with little effort.

A line l is a 0(1)-*stop line* for a test set T, if T detects all the faults that can cause l to take value 1(0). For example, consider the circuit and the test shown in Figure 6.62. This test detects all the faults that can set $E = 1$, $E2 = 1$ and $G = 0$; hence E and $E2$ are 0-stops and G is a 1-stop. A line l that is not a 0(1)-stop is said to be 0(1)-*useful*, because there are some new faults that could be detected when l has value 0(1). In

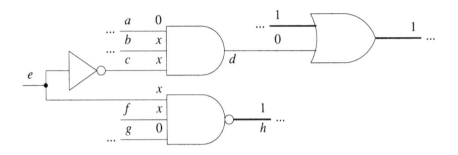

Figure 6.61

Figure 6.62, *E1* is 0-useful, because the still undetected fault *E1* s-a-1 requires *E1* = 0 for detection.

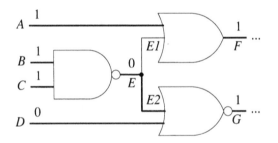

Figure 6.62   Example of stop lines

A line *l* becomes a *v*-stop for a test set *T* if *l* has had a critical value *v* and

1.   If *l* is a fanout branch, its stem is a *v*-stop.

2.   If *l* is the output of a gate with inversion *i*, all the gate inputs are ($v \oplus i$)-stops.

A PI becomes a *v*-stop when it has had a critical value *v*. While TG progresses and the set *T* grows, more and more lines become stops and they advance from PIs toward POs. CRIPT uses stop lines to avoid tracing paths in any area bounded by a *v*-stop line *l* in any test in which *l* has value *v*.

A *restart gate* (in a test *t*) must satisfy the following conditions:

1.   Its output is critical (in *t*) but none of its inputs is critical.

2.   Exactly one input has the controlling value *c*, and this input is *c*-useful.

For the example in Figure 6.63, if we assume that *C* is 0-useful, then *G* is a restart gate. To make the same test detect additional faults we should set *E* = *D* = 1; then *C* becomes critical and the faults that make $C = 1$ are likely to be detected along the existing critical path from G. No special effort is required to identify restart gates, since they are a subset of the gates where CRIPT terminates its backtracing. (The CRIPT algorithm described in Chapter 5 handles only binary values; for this application it needs to be extended to process partially specified vectors).

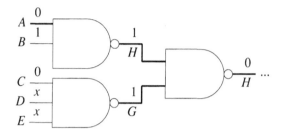

Figure 6.63   Example of restart gate

SMART (see Figure 6.64) starts by randomly selecting a PO objective (Z,v) such that Z is v-useful. Then the objective (Z,v) is mapped into a PI assignment (i, v_i) by the procedure *Useful_rbacktrace*. When a value v_j is needed for a gate input, *Useful_rbacktrace* gives preference to those inputs that are v_j-useful and have value x. The partial vector generated so far is fault simulated by CRIPT, which backtraces critical paths from the POs that have been set to binary values. CRIPT also determines stop lines and restart gates. Then SMART repeatedly picks a restart gate, tries to justify noncontrolling values on its inputs with value x, and reruns CRIPT. Now CRIPT restarts its backtracing from the new POs that have binary values (if any) and from the restart gates.

Unlike RAPS, SMART tries to create critical paths leading to the chosen PO Z before processing other POs. Thus in the example shown in Figure 6.59, after justifying $PO1 = 0$ by $A = 0$, the gate $PO1$ is a restart gate; hence SMART will try to justify $B = 1$ before processing $PO2$. Then $PO2 = 0$ will be justified by $C = 0$.

Using stop lines avoids the problem illustrated in Figure 6.60. After the first test that sets $B = 0$ and $C = 1$, B becomes a 0-stop and will no longer be used to justify $Z = 1$.

In the example in Figure 6.61, keeping track of restart gates allows SMART to select h (and ignore d, which is not a restart gate), and continue by trying to set $a = f = 1$.

Experimental results [Abramovici *et al.* 1986a] show that, compared with RAPS, SMART achieves higher fault coverage with smaller test sets and requires less CPU time.

6.2.5 ATG Systems

Requirements

In general, a TG algorithm is only a component of an *ATG system*, whose overall goal is to generate a test set for the SSFs in a circuit. Ideally, we would like to have

```
while (useful POs have x values)
    begin
        randomly select a useful PO (Z) with x value
        if (Z is both 0- and 1-useful)
            then randomly select a value v
        else v = the useful value of Z
        repeat
            begin
                (i, vᵢ) = Useful_rbacktrace (Z, v)
                Simulate (i, vᵢ)
            end
        until Z has binary value
        CRIPT()
        while (restart_gates ≠ ∅)
            begin
                remove a gate (G) from restart_gates
                c = controlling value of G
                for every input (j) of G with value x
                    repeat
                        begin
                            (i, vᵢ) = Rbacktrace (j, c̄)
                            Simulate (i, vᵢ)
                        end
                    until j has binary value
                CRIPT()
            end
    end
```

Figure 6.64   SMART

- the fault coverage of the generated tests as high as possible,

- the cost of generating tests (i.e., the CPU time) as low as possible,

- the number of generated tests as small as possible.

These requirements follow from economic considerations. The fault coverage directly affects the quality of the shipped products. The number of generated tests influences the cost of applying the tests, because a larger test set results in a long test-application time. Unlike the cost of generating tests, which is a one-time expense, the test time is an expense incurred every time the circuit is tested. Of course, the above ideal requirements are conflicting, and in practice the desired minimal fault coverage, the maximal TG cost allowed, and the maximal number of vectors allowed are controlled by user-specified limits.

The desired fault coverage should be specified with respect to the detectable faults, because otherwise it may be impossible to achieve it. For example, in a circuit where

4 percent of the faults are redundant, a goal of 97 percent absolute fault coverage would be unreachable. The fault coverage for detectable faults is computed by

$$\frac{number\ of\ detected\ faults}{total\ number\ of\ faults\ -\ number\ of\ undetectable\ faults}$$

that is, the faults proven redundant are excluded from the fault universe. Thus it is important that the ATG algorithm can identify redundant faults.

Another question to consider is whether the ATG system should collapse faults and, if so, what fault-collapsing technique it should use. Fault collapsing is needed for systems relying on a fault simulator whose run-time grows with the number of simulated faults. Fault collapsing can be based on equivalence and/or dominance relations. The latter can be used only if the ATG system does not produce fault-location data. Heap and Rogers [1989] have shown that computing the fault coverage based on a collapsed set of faults is less accurate than computing based on the original (uncollapsed) set; they presented a method to compute the correct fault coverage from the results obtained from simulating the collapsed set. In Chapter 4 we have shown that detecting all checkpoint faults results in detecting all SSFs in the circuit. Some ATG systems start only with the checkpoint faults (which can be further collapsed using equivalence and dominance relations). Abramovici *et al.* [1986b] have shown that this set may not be sufficient in redundant circuits, where detecting all the detectable checkpoint faults does not guarantee the detection of all the detectable faults. They presented a procedure that determines additional faults to be consideredso that none of the potentially detectable faults is overlooked. Note that fault collapsing is not needed in a system using critical-path tracing, which deals with faults only implicitly and does not require fault enumeration.

Structure

Most ATG systems have a 2-phase structure (Figure 6.65), which combines two different methods. The first phase uses a low-cost, fault-independent procedure, providing an initial test set which detects a large percentage of faults (typically, between 50 and 80 percent). In the second phase, a fault-oriented algorithm tries to generate tests for the faults left undetected after the first phase [Breuer 1971, Agrawal and Agrawal 1972].

Figure 6.66 outlines the structure of the first phase of an ATG system. *Generate_test* can be a pseudorandom generation routine or a combined deterministic/random procedure (RAPS or SMART). The generated test *t* is fault simulated and the function *value* determines a figure of merit usually based on the number of new (i.e., previously undetected) faults detected by *t*. Depending on its value, which is analyzed by the function *acceptable*, *t* is added to the test set or rejected. A test that does not detect any new fault is always rejected. A characteristic feature of a sequence of vectors generated in the first phase is that the number of new faults detected per test is initially large but decreases rapidly. Hence the number of rejected vectors will increase accordingly. The role of the function *endphase1* is to decide when the first phase becomes inefficient and the switch to the use of a deterministic fault-oriented TG algorithm should occur.

Although RTG produces a vector faster than a combined deterministic/random method, a vector generated by RAPS or SMART detects more new faults than a pseudorandom

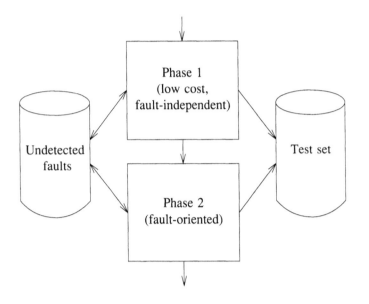

Figure 6.65   Two-phase ATG system

vector. To reach the same fault coverage as RAPS or SMART, RTG needs more vectors, many of which will be rejected. If every generated vector is individually fault simulated, the total computational effort (test generation and test evaluation) is greater when using RTG. Groups of vectors generated by RTG or RAPS can be evaluated in parallel using a fault simulator based on parallel-pattern evaluation, thus significantly reducing the CPU time spent in fault simulation. SMART cannot use parallel-pattern evaluation, because the results obtained in simulating one vector are used in guiding the generation of the next one. However, SMART achieves higher fault coverage with fewer vectors than RTG or RAPS.

Many techniques can be used to implement the functions *value, acceptable* and *endphase1*. The simplest techniques define the value of a vector as the number of new faults it detects, accept any test that detects more than k new faults, and stop the first phase when more than b consecutive vectors have been rejected (k and b are user-specified limits). More sophisticated techniques are described in [Goel and Rosales 1981, Abramovici *et al.* 1986a].

Figure 6.67 outlines the structure of the second phase of an ATG system. The initial set of target faults is the set of faults undetected at the end of the first phase. Since some faults may cause the TG algorithm to do much search, the computational effort allocated to a fault is controlled by a user-specified limit; this can be a direct limit on the CPU time allowed to generate a test or an indirect one on the amount of backtracking performed.

ATG algorithms spend the most time when dealing with undetectable faults, because to prove that a fault is undetectable the algorithm must fail to generate a test by

```
repeat
   begin
      Generate_test(t)
      fault simulate t
      v = value(t)
      if acceptable(v) then add t to the test set
   end
until endphase1()
```

Figure 6.66   First phase of an ATG system

```
repeat
   begin
      select a new target fault f
      try to generate a test (t) for f
      if successful then
         begin
            add t to the test set
            fault simulate t
            discard the faults detected by t
         end
   end
until endphase2()
```

Figure 6.67   Second phase of an ATG system

exhaustive search. Thus identifying redundant faults without exhaustive search can significantly speed up TG. Two types of redundancy identification techniques have been developed. *Static techniques* [Menon and Harihara 1989] work as a preprocessing step to TG, while *dynamic techniques* [Abramovici *et al.* 1989] work during TG.

A good way of selecting a new target fault f (i.e., a still undetected fault that has not been a target before) is to choose one that is located at the lowest possible level in the circuit. In this way, if the test generation for f is successful, the algorithm creates a long path sensitized to f, so the chances of detecting more still undetected faults along the same path are increased. Detecting more new faults per test results in shorter test sequences.

A different method is to select the new target fault from a precomputed set of *critical faults* [Krishnamurthy and Sheng 1985]. The critical faults are a maximal set of faults with the property that there is no test that can detect two faults from this set simultaneously.

Of course, the second phase stops when the set of target faults is exhausted. In addition, both the second and the first phase may also be stopped when the generated sequence becomes too long, or when a desired fault coverage is achieved, or when too much CPU time has been spent; these criteria are controlled by user-imposed limits.

Test Set Compaction

Test vectors produced by a fault-oriented TG algorithm are, in general, partially specified. Two tests are *compatible* if they do not specify opposite values for any PI. Two compatible tests t_i and t_j can be combined into one test $t_{ij}=t_i \cap t_j$ using the intersection operator defined in Figure 2.3. (The output vectors corresponding to t_i and t_j are compatible as well — see Problem 6.30.) Clearly, the set of faults detected by t_{ij} is the union of the sets of faults detected by t_i and t_j. Thus we can replace t_i and t_j by t_{ij} without any loss of fault coverage. This technique can be iterated to reduce the size of a test set generated in the second phase of an ATG system; this process is referred to as *static compaction*.

In general, the test set obtained by static compaction depends on the order in which vectors are processed. For example, let us consider the following test set:

$$
\begin{aligned}
t_1 &= 01x \\
t_2 &= 0x1 \\
t_3 &= 0x0 \\
t_4 &= x01
\end{aligned}
$$

If we first combine t_1 and t_2, we obtain the set

$$
\begin{aligned}
t_{12} &= 011 \\
t_3 &= 0x0 \\
t_4 &= x01
\end{aligned}
$$

which cannot be further compacted. But if we start by combining t_1 and t_3, then t_2 and t_4 can also be combined, producing the smaller test set

$$
\begin{aligned}
t_{13} &= 010 \\
t_{24} &= 001.
\end{aligned}
$$

However, the use of an optimal static compaction algorithm (that always yields a minimal test set) is too expensive. The static compaction algorithms used in practice are based on heuristic techniques.

Static compaction is a postprocessing operation done after all vectors have been generated. By contrast, in *dynamic compaction* every partially specified vector is processed immediately after its generation, by trying to assign PIs with x values so that it will detect additional new faults [Goel and Rosales 1979]. Figure 6.68 outlines such a dynamic compaction technique that is activated after a test t has been successfully generated for the selected target fault (see Figure 6.67). The function *promising* decides whether t is a good candidate for dynamic compaction; for example, if the percentage of PIs with x values is less than a user-specified threshold, t is not processed for compaction. While t is "promising," we select a secondary target fault g and try to extend t (by changing some of the PIs with x values) such that it will detect *g*. This is done by attempting to generate a test for *g* without changing the already assigned PI values.

> **while** *promising(t)*
>   **begin**
>   select a secondary target fault *g*
>   try to extend *t* to detect *g*
>   **end**

Figure 6.68   Dynamic compaction

Experimental results have shown that dynamic compaction produces smaller test sets than static compaction with less computational effort. Note that dynamic compaction cannot be used with a fault-independent TG algorithm.

The most important problem in dynamic compaction is the selection of secondary target faults. Goel and Rosales [1980] compare an arbitrary selection with a technique based on analyzing the values determined by the partially specified test *t* with the goal of selecting a fault for which TG is more likely to succeed. Their results show that the effort required by this additional analysis pays off both in terms of test length and the overall CPU time.

A better dynamic compaction method is presented in [Abramovici *et al.* 1986a]. It relies on a close interaction with fault simulation by critical-path tracing, whose results are used to select secondary target faults already activated by the partially generated vector and/or whose effects can be easily propagated to a PO. When simulating a partially specified vector, critical-path tracing also identifies restart gates, which point to areas where additional faults are likely to be easily detected. In the example in Figure 6.61, if *g* is 0-useful, *h* is a restart gate. A good secondary target fault is likely to be found in the area bounded by *g*.

6.2.6 Other TG Methods

Algebraic Methods

The TG methods discussed in this chapter can be characterized as *topological*, as they are based on a structural model of a circuit. A different type of methods, referred to as *algebraic*, use Boolean equations to represent the circuit. (A review of algebraic methods appears in [Breuer and Friedman 1976]). Algebraic methods are impractical for large circuits.

Extensions for Tristate Logic

Extensions of the conventional test generation algorithms to handle circuits with tristate busses are presented in [Breuer 1983, Itazaki and Kinoshita 1986]. An additional problem for test generation in this type of circuits is to avoid conflicts caused by the simultaneous enabling of two or more bus drivers.

TG for Module-Level Circuits

Until now we have considered only gate-level circuits. The previously discussed TG algorithms can be extended to process more complex components, referred to as *modules* (for example, exclusive-ORs, multiplexers, etc.). The internal gate-level structure of a module is ignored. While for gates we had simple and uniform procedures for line justification and error propagation, for every type of module we need to precompute and store all the possible solutions to the problems of justifying an output value and of propagating errors from inputs to outputs. For some modules, a set of solutions may also be incrementally generated by special procedures, that, on demand, provide the next untried solution from the set.

Example 6.16: Consider a module that implements the function $Z=a\overline{b}+b\overline{c}$, whose Karnaugh map is shown in Figure 6.69(a). The solutions to the problems of justifying $Z=1$ and $Z=0$ are given by the primitive cubes of Z, shown in Figure 6.69(b). The primitive cubes with $Z=1(0)$ correspond to the prime implicants of $Z(\overline{Z})$.

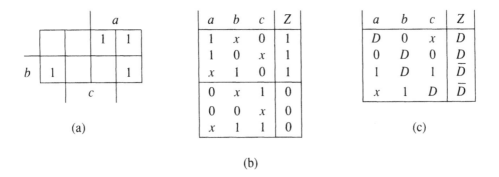

a	b	c	Z
1	x	0	1
1	0	x	1
x	1	0	1
0	x	1	0
0	0	x	0
x	1	1	0

a	b	c	Z
D	0	x	D
0	D	0	D
1	D	1	$\overline{D}$
x	1	D	$\overline{D}$

(a)   (b)   (c)

Figure 6.69   (a) Karnaugh map (b) Primitive cubes (c) Propagation D-cubes

Now let us assume that $a=D$ and we want to determine the minimal input conditions that will propagate this error to Z. In other words, we look for values of b and c such that Z changes value when a changes from 1 to 0. This means finding pairs of cubes of Z of the form $1v_bv_c \mid v_z$ and $0v_b'v_c' \mid \overline{v}_z$, where the values of b and c are compatible. From such a pair we derive a cube of the form $D(v_b \cap v_b')(v_c \cap v_c') \mid D'$, where $D'=D(\overline{D})$ if $v_z=1(0)$. For example, from the primitive cubes $10x \mid 1$ and $00x \mid 0$ we construct the cube $D0x \mid D$; this tells us that setting $b=0$ propagates a D from a to Z. Such a cube is called a *propagation D-cube* [Roth 1980]. Figure 6.69(c) lists the propagation D-cubes dealing with the propagation of one error. (The propagation D-cube $D00 \mid D$, obtained by combining $1x0 \mid 1$ and $00x \mid 0$, is covered by $D0x \mid D$ and does not have to be retained). Propagation D-cubes for multiple errors occurring on the inputs of Z can be similarly derived.   □

Clearly, for every propagation D-cube there exists a corresponding one in which all the D components are complemented. For example, if $x1D \mid \overline{D}$ is a propagation D-cube, then so is the cube $x1\overline{D} \mid D$.

A TG algorithm for gate-level circuits uses built-in procedures for line justification and error propagation. In a TG algorithm for module-level circuits, whenever an output line of a module has to be justified (or errors have to be propagated through a module), the table of primitive cubes (or propagation D-cubes) associated with the type of the module is searched to find a cube that is compatible with the current line values. The intersection operator defined in Figure 2.3 is now extended to composite logic values as shown in Figure 6.70.

$\cap$	0	1	D	$\overline{D}$	x
0	0	$\varnothing$	$\varnothing$	$\varnothing$	0
1	$\varnothing$	1	$\varnothing$	$\varnothing$	1
D	$\varnothing$	$\varnothing$	D	$\varnothing$	D
$\overline{D}$	$\varnothing$	$\varnothing$	$\varnothing$	$\overline{D}$	$\overline{D}$
x	0	1	D	$\overline{D}$	x

Figure 6.70   5-valued intersection operator

The following example illustrates module-level TG.

Example 6.17: The circuit shown in Figure 6.71 uses two modules of the type described in Figure 6.69. Assume that the target fault is P s-a-1, which is activated by P=0. To propagate the error $P = \overline{D}$ through $M1$, we intersect the current values of $M1$ ($QPRS = x\overline{D}x|x$) with the propagation D-cubes given in Figure 6.69(c), and with those obtained by complementing all Ds and $\overline{D}$s. The first consistent propagation D-cube is $0\overline{D}0|\overline{D}$. Thus we set $Q = R = 0$ to obtain $S = \overline{D}$. To check the effect of these values on $M2$, we intersect its current values ($RQOT = 00x|x$) with the primitive cubes given in Figure 6.69(b). A consistent intersection is obtained with the second cube with $Z = 0$. Hence $T = 0$, which precludes error propagation through gate V. Then we backtrack, looking for another propagation D-cube for $M1$. The cube $1\overline{D}1|D$ is consistent with the current values of $M1$. The values $R=Q=1$ do not determine a value for T. To propagate $S=D$, we need $T=1$. Then we intersect the values of $M2$ ($RQOT = 11x|1$) with the primitive cubes with $Z=1$. The first cube is consistent and implies $O = 0$. The generated vector is $PQRO = 0110$.   □

Compared with its gate-level model, a module-level model of a circuit has a simpler structure. This results from having fewer components and lines, and also from eliminating the internal fanout structure of modules. (A circuit using exclusive-OR modules offers a good example of this simplification.) Another advantage of a module-level model occurs in circuits containing many instances of the same module type, because the preprocessing required to determine the primitive cubes and the propagation D-cubes is done only once for all the modules of the same type. While a simpler circuit structure helps the TG algorithm, the components are now more complex and their individual processing is more complicated. Because of this trade-off, the question of whether TG at the module level is more efficient than TG at the gate level does not have a general answer [Chandra and Patel 1987].

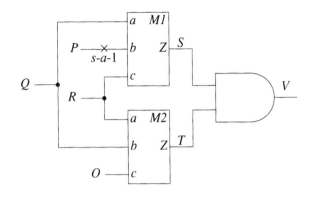

Figure 6.71   Module-level circuit

In general, modules are assumed to be free of internal faults, and TG at the module level is limited to the pin-fault model. This limitation can be overcome in two ways. The first approach is to use a hierarchical mixed-level model, in which one module at a time is replaced by its gate-level model. This allows generating tests for faults internal to the expanded module. The second approach is to preprocess every type of module by generating tests for its internal faults. To apply a precomputed test to a module M embedded in a larger circuit, it is necessary to justify the values required at the inputs of M using the modules feeding M, and at the same time, to propagate the error(s) from the output(s) of M to POs using the modules fed by M. A design for testability technique that allows precomputed tests to be easily applied to embedded modules is discussed in [Batni and Kime 1976]. Other approaches to module-level TG are described in [Somenzi *et al.* 1985, Murray and Hayes 1988, Renous *et al.* 1989].

6.3 ATG for SSFs in Sequential Circuits

6.3.1 TG Using Iterative Array Models

In this section we show how the TG methods for combinational circuits can be extended to sequential circuits. First we restrict ourselves to synchronous sequential circuits whose components are gates and clocked flip-flops. The extension is based on the modeling technique introduced in Section 2.1.2, which transforms a synchronous sequential circuit into an iterative combinational array (see Figure 2.9). One cell of this array is called a *time frame*. In this transformation a F/F is modeled as a combinational element having an additional input q to represent its current state and an additional output q^+ to represent its next state, which becomes the current state in the next time frame. An input vector of the iterative combinational array represents an input sequence for the sequential circuit. Initially we assume that all F/Fs are directly driven by the same clock line which is considered to be fault-free. With these assumptions we can treat the clock line as an implicit line in our model.

Figure 6.72(b) shows the structure of one time frame of an iterative array model corresponding to the circuit in Figure 6.72(a) that uses *JK* F/Fs. The gate-level model for the *JK* F/F implements the equations

$$q^+ = J\bar{q} + \bar{K}q$$
$$y = q$$
$$\bar{y} = \bar{q}$$

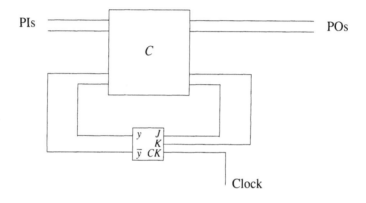

(a)

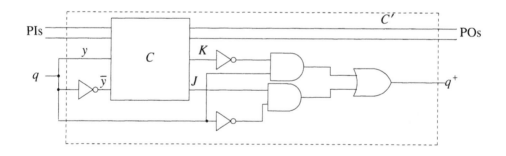

Figure 6.72   Sequential circuit with *JK* F/F (a) General model (b) Model for
one time frame

If we are not interested in faults internal to F/Fs, we can model the circuit that realizes q^+ as a module with three inputs (*J*, *K*, and *q*).

Because all time frames have an identical structure, there is no need to actually construct the complete model of the iterative array. Thus the TG algorithms can use the same structural model for every time frame; however, signal values in different time frames should be separately maintained.

Since the circuit C', obtained by adding the F/F models to the original combinational circuit C, is also a combinational circuit, we can apply any TG algorithm discussed in Section 6.2. First let us consider a fault-oriented algorithm. A generated test vector t for C' may specify values both for PIs and for q variables; the latter must be justified in the previous time frame. Similarly, t may not propagate an error to a PO but to a q^+ variable. Then the error must be propagated into the next time frame. Thus the search process may span several time frames, going both backward and forward in time. A major difficulty arises from the lack of any *a priori* knowledge about the number of time frames needed in either direction.

When dealing with multiple time frames, the target fault is present in every time frame. In other words, the original single fault is equivalent to a multiple fault in the iterative array model. Consequently, an error value (D or $\overline{D}$) may propagate onto the faulty line itself (this cannot happen in a combinational circuit with a single fault). If the faulty line is s-a-c, and the value propagating onto it is v/v_f, the resulting composite value is v/c (see Figure 6.73). Note that when $v=c$, the faulty line stops the propagation of fault effects generated in a previous time frame.

Value propagated onto line l	Fault of line l	Resulting value of line l
D	s-a-0	D
D	s-a-1	1
$\overline{D}$	s-a-0	0
$\overline{D}$	s-a-1	$\overline{D}$

Figure 6.73   Result of a fault effect propagating to a faulty line

When generating a test sequence for a fault it is never necessary to enter the same state twice. Moreover, allowing a previously encountered state to be entered again will cause the program to go into an infinite loop. To avoid this, any TG algorithm for sequential circuits includes some mechanism to monitor the sequence of states and to initiate backtracking when states are repeated.

TG from a Known Initial State

The most general form of a TG algorithm for sequential circuits should assume that the initial state in the fault-free and in the faulty circuit is unknown, because F/F states are arbitrary when power is first applied to a circuit. First we outline a less general algorithm that assumes that the initial state vector $q(1)$ is known both in the fault-free and the faulty circuit (see Figure 6.74). Here the search process goes only forward in time. Since we desire to generate a test sequence as short as possible, we will try a sequence of length r only if we fail to generate a sequence of length $r-1$. Thus when we consider an iterative array of r time frames, we ignore the POs of the first $r-1$ frames. We also ignore the q^+ outputs of the last frame. Once the iterative array model is built, we can use any of the fault-oriented TG algorithms presented in

Section 6.2.1 (some minor modification is required to account for the existence of the faulty line in every frame).

$r=1$
repeat
    begin
            build model with r time frames
            ignore the POs in the first $r-1$ frames
            ignore the q^+ outputs in the last frame
            $q(1)$ = given initial state
            if (test generation is successful) **then return** SUCCESS
            /* no solution with r frames */
            $r = r + 1$
    end
    until $r = f_{max}$
    return FAILURE

(a)

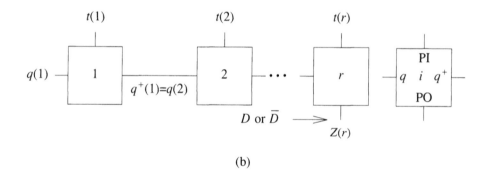

(b)

Figure 6.74   TG from a known initial state (a) General procedure (b) Iterative array model

An important question is when to stop increasing the number of time frames. In other words, we would like to know the maximal length of a possible test sequence for a fault. For a circuit with n F/Fs, for every one of the 2^n states of the fault-free circuit, the faulty circuit may be in one of its 2^n states. Hence no test sequence without repeated states can be longer then 4^n. This bound, however, is too large to be useful in practice, so the number of time frames is bounded by a (much smaller) user-specified limit f_{max}.

While conceptually correct, the algorithm outlined in Figure 6.74 is inefficient, because when the array with r frames is created, all the computations previously done in the first $r-1$ frames are ignored. To avoid this waste, we save the partial sequences that can be extended in future frames. Namely, while searching for a sequence S_r of length r, we save every such sequence that propagates errors to the q^+ variables of the

r-th frame (and the resulting state vector $q^+(r)$) in a set SOL_r of partial solutions. If no sequence S_r propagates an error to a PO, the search for a sequence S_{r+1} starts from one of the saved states $q^+(r)$.

Example 6.18: Consider the sequential circuit in Figure 6.75(a) and the fault a s-a-1. We will derive a test sequence to detect this fault, assuming the initial state is $q_1=q_2=0$. Figure 6.75(b) shows the model of one time frame.

Time frame 1: We apply $q_1=q_2=0$, which creates a $\overline{D}$ at the location of the fault. The vector $I(1)=1$ propagates the error to q_2^+. Because the fault effect does not reach Z, we save the sequence $S_1 = (1)$ and the state $q^+(1) = (0,\overline{D})$ in the set SOL_1.

Time frame 2: We apply $(0,\overline{D})$ to the q lines of frame 2. The D-frontier is $\{G_1,G_3,G_4\}$. Selecting G_1 or G_4 for error propagation results in $I(2)=1$ and the next state $q^+(2)=(\overline{D},\overline{D})$. Trying to propagate the error through G_3 results in $I(2)=0$ and the next state $q^+(2)=(0,D)$.

Time Frame 3: Now the set SOL_2 contains two partial solutions: (1) the sequence $(1,1)$ leading to the state $(\overline{D},\overline{D})$ and (2) the sequence $(1,0)$ leading to the state $(0,D)$. Trying the first solution, we apply $(\overline{D},\overline{D})$ to the q lines of frame 3, which results in the D-frontier $\{Z,G_1,G_2,G_3,G_4\}$. Selecting Z for error propagation we obtain $I(3)=1$ and $Z=D$. The resulting test sequence is $I=(1,1,1)$.   □

With the given initial state, it may not always be possible to activate the fault in the first time frame; then a sequence is needed for activation.

Generation of Self-Initializing Test Sequences

The algorithm outlined in Figure 6.76(a) handles the general case when the initial state is unknown. The fault is activated in one time frame (temporarily labeled 1), and the resulting error is propagated to a PO going forward in time using $r \geq 1$ frames (see Figure 6.76(b)). If some q values of frame 1 are binary, these are justified going backward in time using p time frames temporarily labeled $0, -1, ..., -(p-1)$. This process succeeds when the q values in the first time frame are all x. Such a test sequence is said to be *self-initializing*. (After p is known, the time frames can be correctly labeled $1,...,p,p+1,...,p+r$.)

We will try a sequence of length $p + r + 1$ only when we fail to generate a sequence of length $p + r$. To reuse previous computations, the algorithm saves all partial solutions of the form $(q(-(p-1)); S_{p+r}; q^+(r))$, where S_{p+r} is a sequence obtained with $p + r$ frames that propagates errors to $q^+(r)$.

Example 6.19: Let us derive a self-initializing test sequence for the fault Z s-a-0 in the circuit of Figure 6.77(a). Figure 6.77(b) shows the structure of one time frame, using a module to realize the function $q^+=J\overline{q}+q\overline{K}$. The primitive cubes of this module are given in Figure 6.69(b) (substitute J,q,K,q^+ for a,b,c,Z).

Time frame 1: The only test that detects Z s-a-0 is $(I,q_1)=10$. Since Z has value D, an error is propagated to the PO using $r=1$ frames. Since $q_1(1)\neq x$, we need $p\geq1$.

Time frame 0: To justify $q_1^+(0)=0$, we first try $J_1=0$ and $K_1=1$ (see the decision tree in Figure 6.78(a)). Both solutions justifying $J_1=0$ involve assignments to q_2. The other two solutions justifying q_1^+ involve assignments to q_1. Hence there is no solution with

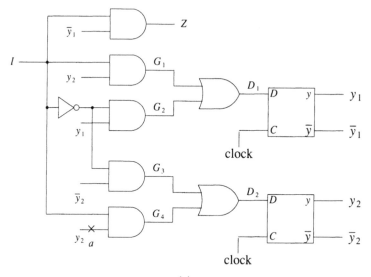

(a)

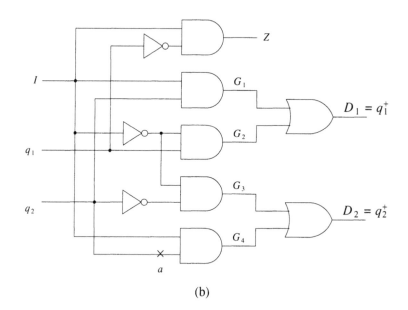

(b)

Figure 6.75

$p=1$. Next we try $p=2$ and we return to the first solution for $J_1=0$, namely $I=0$ and $q_2=1$.

Time frame -1: To justify $q_2^+=1$, we first try $J_2=1$ and $K_2=0$, which are both satisfied by setting $I=0$. Since all lines are justified while q_1 and q_2 of this time frame are both

$r = 1$
$p = 0$
repeat
    begin
        build model with $p + r$ time frames
        ignore the POs in the first $p + r - 1$ frames
        ignore the q^+ outputs in the last frame
        if (test generation is successful **and** every q input in the first frame has
            value x) **then return** SUCCESS
        increment r or p
    end
until $(r+p=f_{max})$
return FAILURE

(a)

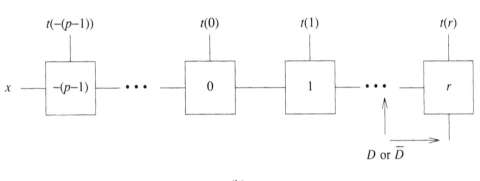

(b)

Figure 6.76   Deriving a self-initializing test sequence (a) General procedure
                   (b) Iterative array model

x, we have obtained the self-initializing test sequence $I=(0,0,1)$ with $p=2$ and $r=1$. Figure 6.78(b) shows the corresponding iterative array.   □

Other Algorithms

A major complication in the algorithm outlined in Figure 6.76 results from the need of going both forward and backward in time. One approach that avoids this problem is the EBT (Extended Backtrace) method [Marlett 1978], whose search process goes only backward in time. EBT first selects a path from the fault site to a PO (this path may involve several time frames), then sensitizes the path starting from the PO. After the error propagation succeeds, the fault is activated by justifying the needed value at the fault site. All line-justification problems that cannot be solved in the current time frame are continued in the previous frame. Thus all vectors are generated in the order

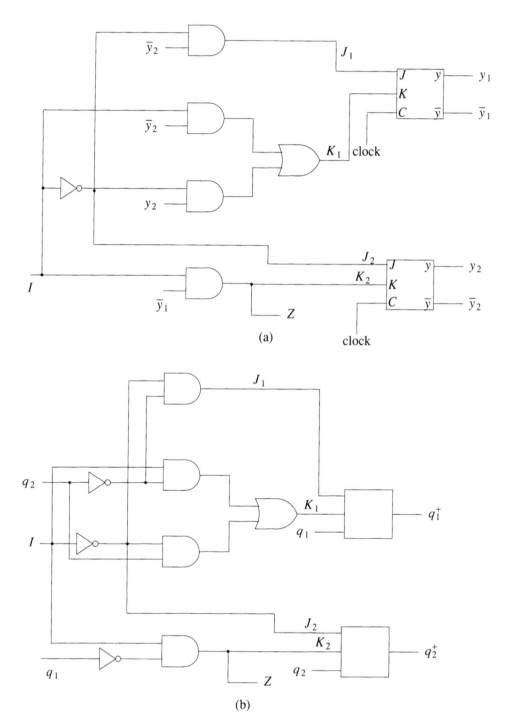

(a)

(b)

Figure 6.77

opposite to that of application. The same principle is used in the algorithms described in [Mallela and Wu 1985] and [Cheng and Chakraborty 1989].

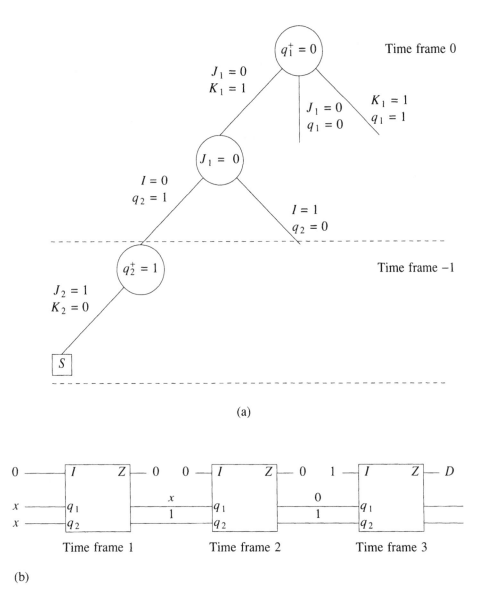

Figure 6.78   Example 6.19 (a) Decision tree (b) Iterative array

Critical-path TG algorithms for combinational circuits have also been extended to synchronous sequential circuits using the iterative array model [Thomas 1971, Breuer and Friedman 1976].

The method described in [Ma *et al.* 1988] relies on the existence of a *reset state* and assumes that every test sequence starts from the reset state. Like the algorithm given in Figure 6.76(a), it begins by activating the fault in time frame 1 and propagating the error forward to a PO, using several time frames if necessary. The state required in frame 1, $q(1)$, is justified by a sequence that brings the circuit from the reset state to $q(1)$. To help find this sequence, a preprocessing step determines a *partial state-transition graph* by systematically visiting states reachable from the reset state. Thus transfer sequences from the reset state to many states are precomputed. If the state $q(1)$ is not in the partial state diagram, then a state-justification algorithm searches for a transfer sequence from the reset state to $q(1)$. This search process systematically visits the states from which $q(1)$ can be reached, and stops when any state included in the partial state-transition graph is encountered.

Logic Values

Muth [1976] has shown that an extension of the D-algorithm restricted to the values $\{0,1,D,\bar{D},x\}$ may fail to generate a test sequence for a detectable fault, while an algorithm using the complete set of nine composite values would be successful.

With the nine composite values, a line in both the fault-free and the faulty circuit may assume one of the three values $\{0,1,u\}$. Because of the limitations of the 3-valued logic (discussed in Chapter 3), it is possible that a TG algorithm using the 9-valued logic may not be able to derive a self-initializing test, even when one does exist. (A procedure guaranteed to produce an initializing sequence, when one exists, requires the use of multiple unknown values.)

We have seen that for a self-initializing test sequence, all the q variables of the first time frame have value x. Some systems use a different value U to denote an initial unknown state. The difference between U and x is that $x \cap a = a$, while $U \cap a = \emptyset$; in other words, an x can be changed during computation, but an U cannot.

One way to avoid the complex problem of initialization is to design sequential circuits that can be easily initialized. The simplest technique is to provide each F/F with a common reset (or preset) line. Then one vector can initialize the fault-free circuit and most of the faulty circuits. Such design techniques to simplify testability will be considered in greater detail in a separate chapter.

Reusing Previously Solved Problems

An additional difficulty in TG for sequential circuits is the existence of "impossible" states, that is, states never used in the fault-free operation. For example, a 5-state counter implemented with three F/Fs may have three states that cannot be entered. Trying to justify one of these invalid states during TG is bound eventually to fail and may consume a significant amount of time. To help alleviate this problem, a table of invalid and unjustifiable states can be constructed. Initially, the user may make entries in this table. Then the system itself will enter the states whose justification failed or could not be completed within a user-specified time limit. Before attempting to justify a state $\mathbf{q}$, the TG system will compare it with the entries in the invalid-states table; backtracking is initiated if $\mathbf{q}$ is covered by an invalid state. For example, if $0xx1$ could not be justified (within the allowed limits), then it is useless to try to justify $01x1$, which is a more stringent restriction on the state space.

This concept of keeping track of previous state-justification problems can be extended to successfully solved problems [Rutman 1972]. For example, suppose that we want to justify the state $\mathbf{q}=0x1x$, and that the state $011x$ has been previously justified. Clearly, we can again apply the sequence used to justify a state covered by **q**. Hence we construct a table of solved state-justification problems and their corresponding solutions. Whenever we wish to justify a state **q**, we search this table to see if **q** (or a state covered by **q**) has been justified before. If so, the stored justification sequence is reused.

Cost Functions

The controllability and observability cost functions discussed in Section 6.2.1.3 have been extended to sequential circuits [Rutman 1972, Breuer 1978, Goldstein 1979]. The computation of controllability costs starts by assigning $C0(l)=C1(l)=\infty$ for every line l (in practice, ∞ is represented by a large integer). Next the $C0(i)$ and $C1(i)$ values of every PI i are changed to f_i-1, where f_i is the fanout count of i (using fanout-based cost functions). Then *cost changes* are further propagated in the same way logic events are propagated in simulation. To compute the new cost of the output of a gate in response to cost changes on gate inputs, we use formulas similar to (6.6) and (6.7). Note that for an AND gate l, $C1(l)$ will keep the value ∞ until every input of the gate l has a finite $C1$ value. F/Fs are processed based on their equations. The controllability costs of the state variables and of the lines affected by them may change several times during this process, until all costs reach their stable values. The convergence of the cost computation is guaranteed because when a cost changes, its value always decreases. Usually the costs reach their stable values in only a few iterations.

After the controllability costs are determined, observability costs are computed by a similar iterative process. We start by setting $O(l)=\infty$ for every line l and changing the observability cost of every PO to 0. Now changes in the observability costs are propagated going backward through the circuit, applying formulas similar to (6.3) and (6.4).

Selection of Target Faults

The structure of a TG system for sequential circuits using a fault-oriented TG algorithm is similar to that outlined in Figure 6.67, except that a test sequence (rather than one vector) is generated to detect the target fault. A self-initializing sequence is necessary only for the first target fault. For any other target fault, TG starts from the state existing at the end of the previously generated sequence.

Every generated sequence is simulated to determine all the faults it detects. An important byproduct of this fault simulation is that it also determines the undetected faults whose effects have been propagated to state variables. This information is used by selecting one such fault as the next target; this "opportunistic" selection saves the effort involved in activating the fault and propagating an error to a state variable.

Clock Logic and Multiple Clocks

Until now we have assumed that all F/Fs are directly driven by the same fault-free clock line. In practice, however, synchronous sequential circuits may have multiple clock lines and clocks may propagate through logic on their way to F/Fs. Figure 6.79(a) shows the general structure of such a circuit and Figure 6.79(b) gives an example of clock logic. (We assume that clock PIs do not feed data inputs of F/Fs.)

To handle circuits with this structure, we use F/F models where the clock line appears explicitly. Figure 6.80 illustrates the model for a D F/F that implements the equation

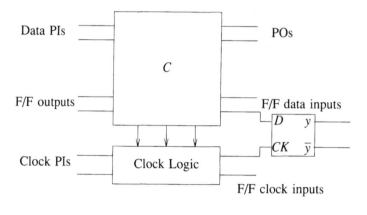

(a)

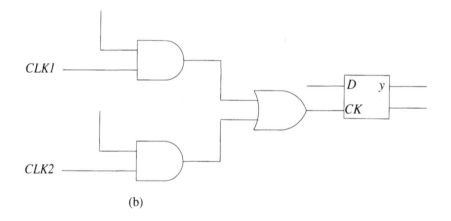

(b)

Figure 6.79   (a) Synchronous circuit with clock logic (a) General structure
(b) Example of clock logic

$q^+=CD+\overline{C}q$. Note that in this model the F/F maintains its current state if its clock is not active.

This modeling technique allows the clock logic to be treated as a part of the combinational circuit. A 1(0)-value specified for a clock PI i in a vector of the generated test sequence means that a clock pulse should (should not) be applied to i in that vector. Although the values of the data PIs and of the clock PIs are specified by the same vector, the tester must delay the clock pulses to ensure that the F/F clock inputs change only after the F/F data inputs have settled to their new values.

For a F/F for which clocking is the only means to change its state, certain faults in the clock logic preclude clock propagation and make the F/F keep its initial state. A conventional ATG procedure cannot detect such a fault, because its fault effects are of the form $1/u$ or $0/u$ (never D or $\overline{D}$), which, when propagated to a PO, indicate only a

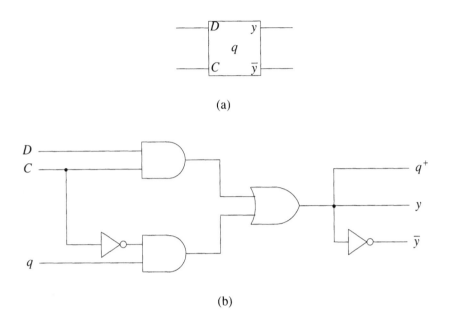

(a)

(b)

Figure 6.80   (a) D flip-flop (b) Model with explicit clock line

potential detection. The method described in [Ogihara *et al.* 1988] takes advantage of the F/F maintaining its initial state, say q_i, and tries to propagate both a $1/q_i$ and a $0/q_i$ fault effect to a PO (they may propagate at different times). Then one of these two fault effects is guaranteed to indicate a definite error.

Complexity Issues

TG for sequential circuits is much more complex than for combinational circuits, mainly because both line justification and error propagation involve multiple time frames. In the worst case the number of time frames needed is an exponential function of the number of F/Fs.

An important structural property of a sequential circuit is the presence of *cycles*, where a cycle is a loop involving F/Fs [Miczo 1983]. TG for a sequential circuit without cycles is not much more difficult than for a comparable combinational circuit. The complexity of TG for a sequential circuit with cycles is directly related to its cyclic structure, which can be characterized by

- the number of cycles;

- the number of F/Fs per cycle;

- the number of cycles a F/F is involved in (reflecting the degree of interaction between cycles) [Lioy *et al.* 1989].

We cannot expect TG for large sequential circuits with complex cyclic structures to be practically feasible. Practical solutions rely on design for testability techniques.

Several such techniques (to be discussed in Chapter 9) transform a sequential circuit into a combinational one during testing. Cheng and Agrawal [1989] describe a technique whose goal is to simplify the cyclic structure of a sequential circuit.

Asynchronous Circuits

TG for asynchronous circuits is considerably more difficult than for synchronous circuits. First, asynchronous circuits often contain races and are susceptible to improper operation caused by hazards. Second, to obtain an iterative array model of the circuit, all feedback lines must be identified; this is a complex computational task. Finally, correct circuit operation often depends on delays intentionally placed in the circuit, but none of the TG algorithms previously discussed takes delays into account.

Note that an asynchronous circuit may go through a sequence of states in response to an input change. We assume that the PO values are measured only after a stable state is reached and that PIs are not allowed to change until the circuit has stabilized. Based on these assumptions, we can use the modeling technique illustrated in Figure 6.81 [Breuer 1974]. Here the time scale is divided into *frames* and *phases*. A time frame i corresponds to the application of one input vector $x(i)$. (Unlike synchronous circuits, two consecutive vectors must differ in at least one variable). Every frame i is composed of one or more phases to reflect the sequence of changes of the state variables. All the phases of the same frame receive the same vector $x(i)$. PO values are observed only after stability has been reached ($y=Y$) in the last phase of a frame. One problem is that the number of phases in a frame is not known *a priori*.

The TG procedures for synchronous circuits can be extended to handle the model in Figure 6.81 and can be further modified so that the generated test will be free of hazards [Breuer and Harrison 1974]. In practice, however, ATG for large asynchronous circuits is not feasible.

6.3.2 Simulation-Based TG

The principle of simulation-based TG [Seshu 1965] is to explore the space of input vectors as follows:

1. Generate and (fault) simulate trial vectors.

2. Based on the simulation results, evaluate trial vectors according to some cost function.

3. Select the "best" trial vector and add it to the test sequence.

Initially we can start with an arbitrary (or user-specified) vector. Trial vectors are usually selected among the "neighbors" of the current input vector t, i.e., vectors different from t in one bit. The selection process is random, so this method is in part similar to RTG. The cost function depends on the objective of the current phase of the TG process [Agrawal *et al.* 1988]:

1. *Initialization:* Here the objective is to set all F/Fs to binary values. The cost function is the number of F/Fs in an unknown state. (This phase does not require fault simulation.)

2. *Test Generation for Groups of Faults:* Here the objective is to detect as many faults as possible, hence all undetected faults are simulated. For every activated

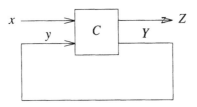

(a)

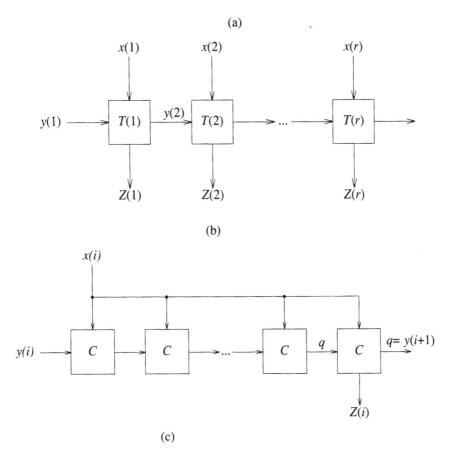

(b)

(c)

Figure 6.81   (a) Asynchronous circuit (b) Iterative array model (c) Structure
of time frame $T(i)$

fault i we compute its cost c_i as the shortest distance between the gates on its
D-frontier and POs. The distance is the weighted number of elements on a path,
with a F/F having a heavier weight than a gate. The cost c_i represents a measure
of how close is fault i to being detected. The cost function is $\sum c_i$, where the
summation is taken over a subset of faults containing the faults having the
lowest costs.

3. *Test Generation for Individual Faults:*   Here we target one of the remaining
undetected faults and the objective is to generate a test sequence to detect it.
The cost function is a *dynamic testability measure*, designed to measure the
additional effort required to detect the fault, starting from the current state.

In all three phases, a "good" trial vector should reduce the cost. During the
initialization phase, reducing the cost means initializing more F/Fs. When the cost
becomes 0, all F/Fs have binary values. In the other two phases, reducing the cost
means bringing a group of faults, or an individual fault, closer to detection. In
principle, we should evaluate all possible trial vectors and select the one leading to the
minimal cost. (In a circuit with n PIs, n trial vectors can be obtained by 1-bit changes
to the current vector.) In practice, we can apply a "greedy" strategy, which accepts the
first vector that reduces the cost. Rejecting a trial vector that does not reduce the cost
means restoring the state of the problem (values, fault lists, etc.) to the one created by
the last accepted vector; this process is similar to backtracking. At some point, it may
happen that no trial vector can further reduce the cost, which shows that the search
process has reached a local minimum. In this case a change of strategy is needed,
such as (1) accept one vector that increases the cost, (2) generate a new vector that
differs from the current one in more than one bit, (3) go to the next phase.

An important advantage of simulation-based TG is that it can use the same delay
model as its underlying fault simulator. It can also handle asynchronous circuits.
Because its main component is a conventional fault simulator, simulation-based TG is
easier to implement than methods using the iterative array model.

SOFTG (*Simulator-Oriented Fault Test Generator*) is another simulation-based TG
method [Snethen 1977]. SOFTG also generates a test sequence in which each vector
differs from its predecessor in one bit. Rather than randomly selecting the bit to
change, SOFTG determines it by a backtrace procedure similar to the one used in
PODEM.

6.3.3 TG Using RTL Models

The TG methods described in the previous section work with a circuit model
composed of gates and F/Fs. By contrast, a designer or a test engineer usually views
the circuit as an interconnection of higher-level components, such as counters,
registers, and multiplexers. Knowing how these components operate allows simple and
compact solutions for line-justification and error-propagation problems. For example,
justifying a 1 in the second bit of a counter currently in the all-0 state can be done by
executing two increment operations. But a program working with a low-level
structural model (in which the counter is an interconnection of gates and F/Fs) cannot
use this type of functional knowledge.

In this section we present *TG methods that use RTL models* for components (or for the
entire circuit). They generate tests for SSFs and can be extended to handle stuck
RTL variables. (TG for functional faults is the topic of Chapter 8.)

The main motivation for RTL models is to speed up TG for sequential circuits and to
increase the size of the circuits that can be efficiently processed. A second motivation
is the use of "off-the-shelf" commercial components for which gate-level models are
not available, but whose RTL descriptions are known or can be derived from their
functional specifications.

6.3.3.1 Extensions of the *D*-Algorithm

To extend the *D*-algorithm to circuits containing RTL components, we keep its overall structure, but we extend its basic operations — line-justification, error-propagation, and implication. We will consider devices of moderate complexity, such as shift registers and counters [Breuer and Friedman 1980]. The function of such a device is described using the following primitive RTL operators:

- *Clear* (reset),
- parallel *Load*,
- *Hold* (do nothing),
- shift *Left*,
- shift *Right*,
- increment (count *Up*),
- decrement (count *Down*).

Figure 6.82(b) describes the operation of the bidirectional shift register shown in Figure 6.82(a), where **Y** and **y** denote the next and the current state of the register, *Clock* = ↑ represents a 0 to 1 transition of *Clock*, and *Clock* ≠ ↑ means that the consecutive values of *Clock* are 00 or 11 or 10. The outputs **Q** always follow the state variables **y**.

The primitive RTL operators are implemented by built-in *generic algorithms*, which are applicable to any n-bit register. Different types of components are characterized by different sets of conditions that activate the operators; for example negative versus positive clock, or synchronous versus asynchronous clear. For every type of device we use a *translation table* similar to the one in Figure 6.82(b) to provide the mapping between operators and the values of the control lines.

Implication

A first type of implication is to compute **Y** given **y**, **A** and the values of the control lines. For this we first determine what operation will be executed. When some control lines have x values, the device may execute one of a set of *potential operations*. For example, if $(Reset, Clock, S_0, S_1) = 1 \uparrow 1\ x$, the shift register may execute either *Right* or *Load*. Even if we do not know the operation executed, some of the resulting values can be implied by the following *union algorithm*, obtained by taking the common results of *Right* and *Load*:

$$
\begin{aligned}
&\textbf{if } A_i = y_{i+1} \textbf{ then } Y_i = A_i \\
&\textbf{else } Y_i = x \qquad\qquad\qquad i = 1, \ldots, n-1 \\
&\textbf{if } A_n = A_{n+1} \textbf{ then } Y_n = A_n \\
&\textbf{else } Y_n = x
\end{aligned}
$$

There are 20 union algorithms for the shift register of Figure 6.82 [Breuer and Friedman 1980].

A second type of implication is to determine what operation could have been executed, given the values of **Y**. For example, if the values A_i are compatible with the values Y_i,

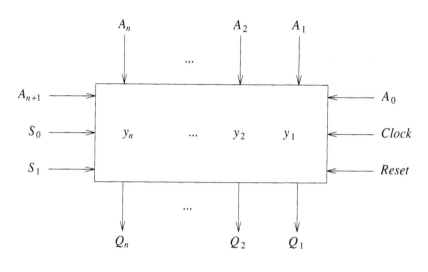

(a)

Control lines				Operation	Algorithm	
Reset	Clock	S_0	S_1			
0	x	x	x	Clear	$Y_i = 0$	$i = 1, ..., n$
1	$\neq\uparrow$	x	x	Hold	$Y_i = y_i$	$i = 1, ..., n$
1	$\uparrow$	0	0	Hold		
1	$\uparrow$	0	1	Left	$Y_i = y_{i-1}$ $i = 2, ..., n$ $Y_1 = A_0$	
1	$\uparrow$	1	0	Right	$Y_i = y_{i+1}$ $i = 1, ..., n-1$ $Y_n = A_{n+1}$	
1	$\uparrow$	1	1	Load	$Y_i = A_i$	$i = 1, ..., n$

(b)

Figure 6.82 Shift register (a) Module-level view (b) Generic algorithms

i.e., $Y_i \cap A_i \neq \emptyset$ for $i = 1, ..., n$, then *Load* is a potential operation. Figure 6.83 summarizes these implications for a shift register.

A third type of implication is to determine the values of **A** or **y**, knowing **Y** and the operation executed. This is done using the concept of *inverse operations*. That is, if $\mathbf{Y} = g(\mathbf{y})$, then $\mathbf{y} = g^{-1}(\mathbf{Y})$. Figure 6.84 presents the inverse operations for a shift register.

Note that $Clear^{-1}$ is undefined, because the previous values y_i cannot be determined once the register has been cleared.

Conditions		Potential operation
$Y_i \cap y_i \neq \varnothing$	$i = 1, ..., n$	Hold
$Y_i \cap A_i \neq \varnothing$	$i = 1, ..., n$	Load
$Y_i \cap y_{i-1} \neq \varnothing$ $\quad$ $Y_1 \cap A_0 \neq \varnothing$	$i = 2, ..., n$	Left
$Y_i \cap y_{i+1} \neq \varnothing$ $\quad$ $Y_n \cap A_{n+1} \neq \varnothing$	$i = 1, ..., n-1$	Right
$Y_i \neq 1$	$i = 1, ..., n$	Clear

Figure 6.83   Potential operations for a shift register

Inverse Operation	Algorithm	
$Hold^{-1}$	$y_i = Y_i$	$i = 1, ..., n$
$Left^{-1}$.	$y_i = Y_{i+1}$ $\quad$ $A_0 = Y_1$	$i = 1, ..., n-1$
$Right^{-1}$	$y_i = Y_{i-1}$ $\quad$ $A_{n+1} = Y_n$	$i = 2, ..., n$
$Load^{-1}$	$A_i = Y_i$	$i = 1, ..., n$

Figure 6.84   Inverse operations for a shift register

Example 6.20:   For the shift register of Figure 6.82, assume that $n = 4$, (*Reset*, *Clock*, S_0, S_1) = $1 \uparrow x\,0$, $\mathbf{Y} = (Y_4, Y_3, Y_2, Y_1) = x\,x\,0\,x$, $\mathbf{y} = (y_4, y_3, y_2, y_1) = x\,1\,x\,0$ and $\mathbf{A} = (A_5, A_4, A_3, A_2, A_1, A_0) = x\,x\,x\,x\,x\,x$. From the values of the control lines and Figure 6.82(b), we determine that the potential operations are *Right* and *Hold*. Checking their conditions given in Figure 6.83, we find that *Hold* is possible but *Right* is not (since $Y_2 \cap y_3 = \varnothing$). Hence we conclude that the operation is *Hold*. This implies $S_0 = 0$, $Y_3 = y_3 = 1$ and $Y_1 = y_1 = 0$. Using the inverse operator $Hold^{-1}$, we determine that $y_2 = Y_2 = 0$.   □

Line Justification

For line justification, we first determine the set of potential operations the device can execute; among these we then select one that can produce the desired result.

Example 6.21:   Assume we wish to justify $\mathbf{Y} = 0110$ and that currently $\mathbf{y} = x\,0\,x\,x$, $\mathbf{A} = x\,x\,x\,x\,x\,x$, and (*Reset*, *Clock*, S_0, S_1) = $1 \uparrow x\,x$. From the values of the control lines we determine that the potential operations are {*Hold*, *Left*, *Right*, *Load*}. *Hold* is not possible because $Y_3 \cap y_3 = \varnothing$. *Right* is not possible since $Y_2 \cap y_3 = \varnothing$. Thus we have two solutions:

1. Execute *Left*. For this we set $(S_0, S_1) = 01$. By applying $Left^{-1}$ to $\mathbf{Y} = 0110$, we determine that $(y_2, y_1) = 11$ and $A_0 = 0$.

2. Execute *Load*. For this we set $(S_0, S_1) = 11$. By applying $Load^{-1}$ to $\mathbf{Y} = 0110$, we determine that $(A_4, A_3, A_2, A_1) = 0110$.

Before selecting one of these two solutions, we can imply $S_1 = 1$, which is common to both of them.   □

Error Propagation

We consider only single error-propagation problems, i.e., we have a D or $\overline{D}$ affecting one input of the shift register. The simple case occurs when the error affects a data line A_i. Then to propagate the error through the shift register we execute

1. *Left*   if $i = 0$;

2. *Load*   if $1 \leq i \leq n$;

3. *Right*   if $i = n + 1$.

To propagate an error that affects a control line, we have to make the shift register in the faulty circuit f execute an operation O_f different from the operation O done by the shift register in the good circuit, and to ensure that the results of O and O_f are different. Figure 6.85 gives the conditions needed to propagate a D on a control line and the resulting *composite operation* O/O_f. The conditions needed to propagate a $\overline{D}$ are the same, but the resulting composite operation is reversed (that is, if D produced O_1/O_2, $\overline{D}$ will produce O_2/O_1). The notation $Clock = \uparrow/\neq\uparrow$ means that $Clock$ has a 0-to-1 transition in the good circuit but not in the faulty circuit (i.e., the consecutive composite values of $Clock$ can be $0D$ or $\overline{D}1$ or $\overline{D}D$).

Reset	*Clock*	S_0	S_1	O/O_f
D	x	x	x	{*Hold,Left,Right,Load*}/*Clear*
1	$\uparrow$	D	0	*Right*/*Hold*
1	$\uparrow$	D	1	*Load*/*Left*
1	$\uparrow$	0	D	*Left*/*Hold*
1	$\uparrow$	1	D	*Load*/*Right*
1	$\uparrow/\neq\uparrow$	1	x	{*Load,Right*}/*Hold*
1	$\uparrow/\neq\uparrow$	x	1	{*Load,Left*}/*Hold*

Figure 6.85   Propagation of errors on control lines of a shift register

Example 6.22:   Assume we want to propagate a D from S_1 and that currently $\mathbf{y} = 0000$ and $\mathbf{A} = 010000$. From Figure 6.85 we have two potential solutions:

1) (*Reset*, *Clock*, S_0) = $1 \uparrow 0$ with the composite operation *Left*/*Hold*. This is unacceptable, since the result is $\mathbf{Y} = 0000$.

2) *(Reset, Clock, S_0)* = 1 ↑ 1 with the composite operation *Load/Right*. This produces **Y** = *D*000.   □

An ATG system that successfully implements the concepts introduced in this section is described in [Shteingart *et al.* 1985]. Methods for error propagation through language constructs used in RTLs are presented in [Levendel and Menon 1982, Norrod 1989].

6.3.3.2 Heuristic State-Space Search

SCIRTSS (*Sequential Circuit Test Search* System) [Hill and Huey 1977] is an ATG system that generates tests for synchronous circuits for which both a low-level model (gate and F/F) and an RTL-level model are available. Based on the low-level model, SCIRTSS employs a TG algorithm for combinational circuits to generate a test for a target SSF in one time frame. Then the search for finding sequences to propagate an error to a PO and to bring the circuit into the needed state is done using the RTL model.

SCIRTSS assumes an RTL model that provides a clear separation between the data and the control parts of the circuit, as shown in Figure 6.86. This separation allows the control inputs to be processed differently from the data inputs.

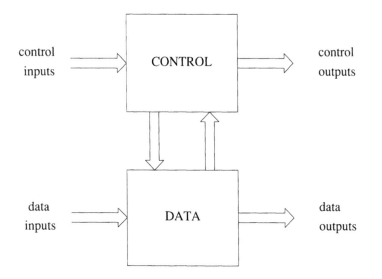

Figure 6.86   Model with separation between data and control

We discuss the mechanism employed by SCIRTSS in searching for an input sequence to propagate an error stored in a state variable to a PO. From the current state (of both the good and the faulty circuit), SCIRTSS simulates the RTL description to compute the next states for a set of input vectors. All possible combinations of control inputs are applied, while data inputs are either user-specified or randomly generated. This strategy assumes that, compared to the data inputs, the control inputs have a stronger influence in determining the next state of the circuit, and their number is much smaller.

A state for which the set of next states has been computed is said to be *expanded*. Based on the results of RTL simulation, SCIRTSS determines whether the desired goal (propagating an error to a PO) has been achieved in one of the newly generated states. If not, one unexpanded state is selected and the expansion process repeats.

Figure 6.87 illustrates the first two steps of such a search process. The expanded states are 0 and 2. The selection of an unexpanded state is based on *heuristic functions* whose purpose is to guide the search on the path most likely to succeed. This type of search is patterned after techniques commonly used in artificial intelligence. The heuristic value H_n associated with a state n is given by the expression

$$H_n = G_n + w\, F_n$$

where

- G_n is the length of the sequence leading to n (in Figure 6.87, $G_1 = G_2 = \cdots = G_k = 1$, $G_{k+1} = G_{k+2} = 2$).

- F_n is the main heuristic function that will be further discussed.

- w is a weight that determines the extent to which the search is to be directed by F_n.

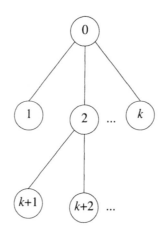

Figure 6.87   SCIRTSS search process

The unexpanded state with the smallest H_n is selected for expansion. Note that for $w = 0$ the search degenerates to breadth-first search. In general, F_n is a linear combination of different heuristics

$$F_n = \sum_i w_i\, F_{ni}$$

where w_i is the weight of the heuristic F_{ni}.

A useful heuristic in error-propagation problems is the *fault proliferation function*, expressed as

$$1 - \frac{the\ number\ of\ F/Fs\ with\ error\ values\ in\ state\ n}{total\ number\ of\ F/Fs}$$

Other heuristic functions [Huey 1979] are based on

- distance to a goal node;

- probabilities of reaching different states.

After generating a test sequence for a fault, SCIRTSS fault simulates the generated sequence on the low-level model. Then it discards all the detected faults. The results of the fault simulation are also used to select opportunistically the next target fault among the faults whose effects are propagated to state variables.

SCIRTSS has been successfully applied to generate tests for moderately complex sequential circuits. Its distinctive feature is the use of an RTL model as a basis for heuristic search procedures. Its performance is strongly influenced by the sophistication of the user, who has to provide heuristic weights and some data vectors.

6.3.4 Random Test Generation

RTG for sequential circuits is more complicated than for combinational circuits because of the following problems:

- Random sequences may fail to properly initialize a circuit that requires a specific initialization sequence.

- Some control inputs, such as clock and reset lines, have much more influence on the behavior of the circuit than other inputs. Allowing these inputs to change randomly as the other ones do, may preclude generating any useful sequences.

- The evaluation of a vector cannot be based only on the number of new faults it detects. A vector that does not detect any new faults, but brings the circuit into a state from which new faults are likely to be detected, should be considered useful. Also a vector that may cause races or oscillations should be rejected.

Because of the first two problems, a *semirandom* process is often used, in which pseudorandom stimuli are combined with user-specified stimuli [Schuler *et al.* 1975]. For example, the user may provide initialization sequences and deterministic stimuli for certain inputs (e.g., clock lines). The user may also specify inputs (such as reset lines) that should change infrequently and define their relative rate of change. In addition, the user is allowed to define the signal probabilities of certain inputs.

Having nonuniform signal probabilities is especially useful for control inputs. An adaptive procedure that dynamically changes the signal probabilities of control inputs is described in [Timoc *et al.* 1983].

To determine the next vector to be appended to the random sequence, the RTG method described in [Breuer 1971] generates and evaluates Q candidate random vectors. Each vector t is evaluated according to the function

$$v(t) = a\ v_1 + b(v_2 - v_3)$$

where

- v_1 is the number of new faults detected by t.

- v_2 is the number of new faults whose effects are propagated to state variables by t (but are not detected by t).

- v_3 is the number of faults whose effects were propagated to state variables before the application of t, but were "lost" by t (and have not yet been detected).

- a and b are weight factors.

Thus the value $v(t)$ of a vector t is also influenced by its potential for future detections. At every step the vector with the largest value is selected (this represents a "local" optimization technique).

The more candidate vectors are evaluated, the greater is the likelihood of selecting a better one. But then more time is spent in evaluating more candidates. This trade-off can be controlled by an adaptive method that monitors the "profit" obtained by increasing Q and increments Q as long as the profit is increasing [Breuer 1971].

6.4 Concluding Remarks

Efficient TG for large circuits is an important practical problem. Some of the new research directions in this area involve hardware support and Artificial Intelligence (AI) techniques.

Hardware support for TG can be achieved by a special-purpose architecture using pipelining [Abramovici and Menon 1983] or by a general-purpose multiprocessing architecture [Motohara *et al.* 1986, Chandra and Patel 1988, Patil and Banerjee 1989]. The latter can take advantage of the available parallelism in several ways:

- *concurrent fault processing*, where each processor executes the same test generation job for a subset of faults;

- *concurrent decision processing*, where different processors simultaneously explore different branches from the same node in a decision tree;

- *concurrent guidance processing*, where different processors target the same fault using different cost functions for guidance.

AI techniques are applied to TG in an attempt to emulate features of the reasoning process of expert test engineers when generating tests. In contrast with an ATG algorithm, which is oblivious to the functionality of the circuit and looks only at a narrow portion of its structure at a time, a test engineer relies heavily on high-level knowledge about the intended behavior of the circuit and of its components, and uses a global view of its hierarchical structure. Another difference is the ability of the test engineer to recognize a family of similar problems and to take advantage of their similarity by providing a common solution mechanism; for example, the same control sequence can be repeated with different data patterns to set a register to different values. In contrast, an ATG algorithm would repeatedly regenerate the control sequence for every required data pattern.

HITEST [Robinson 1983, Bending 1984] is a knowledge-based system that combines algorithmic procedures with user-provided knowledge. A combinational test generator (based on PODEM) is used to generate a test in one time frame. For state-justification and error-propagation problems, HITEST relies on user-provided solutions. Such solutions are often obvious for engineers but may be difficult to derive by a conventional test generator. For example, the knowledge specifying how to bring an up-down counter into a desired state would (1) indicate values needed for passive inputs (such as ENABLE, RESET), (2) determine, by comparing the desired and the current states, the number of clock pulses to be applied and the necessary setting for the UP/DOWN control input. The knowledge is stored and maintained using AI techniques.

Unlike an ATG algorithm, which explores the space of possible operations of a circuit, a test engineer searches the more restricted space of intended operations, i.e., the operations the circuit was designed to perform. Similarly, the system described by Shirley *et al.* [1987] uses knowledge about the intended behavior of the circuit to reduce the search domain. Rather than being provided by the user, the knowledge is acquired by analyzing traces obtained during a symbolic simulation of the circuit. In addition to the constant values propagated by a conventional simulator, a symbolic simulator also propagates variables, and its results are expressions describing the behavior of the circuit for all the possible data represented by variables. For example, if X and Y are variables assigned to the A and B inputs of an adder, its output will be set to $X+Y$, and if $X=0$, the output will be Y. During test generation, the desired objectives are matched with the simulation traces. If the objective for the adder is to propagate an error from the B input, the solution $A=0$ will be found by matching with the simulation traces, which also recorded the events that had set $A=0$; these events are retrieved and reused as part of the generated test.

Other applications of AI techniques to TG problems are described in [Krishnamurthy 1987, Singh 1987].

TG techniques have other applications beyond TG. Sometimes one has two different combinational circuits supposed to implement the same function. For example, $C1$ is a manual design and $C2$ is a circuit automatically synthesized from the same specifications. Or $C1$ is an existing circuit and $C2$ is a redesign of $C1$ using a different technology. Of course, we can simulate $C1$ and $C2$ with the same stimuli and compare the results, but a simulation-based process usually cannot provide a complete check. *Logic verification* can automatically compare the two circuits and either prove that they implement the same function or generate vectors that cause different responses. The basic mechanism [Roth 1977] is to combine $C1$ and $C2$ to create a composite circuit M (see Figure 6.88), then to use a line-justification algorithm to try to set each PO of M in turn to 1. If the line justification succeeds, then the generated vector sets two corresponding POs in $C1$ and $C2$ to different values; otherwise no such vector exists and the two POs have the same function.

Abadir *et al.* [1988] have shown that a complete test set for SSFs in a combinational circuit is also useful in *detecting* many typical *design errors*, such as missing or extra inverters, incorrect gate types, interchanged wires, and missing or extra wires.

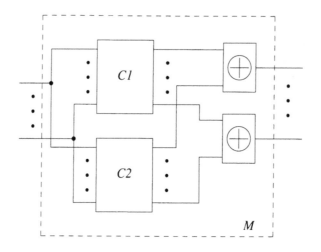

Figure 6.88   Composite circuit for proving equivalence between *C1* and *C2*

REFERENCES

[Abadir *et al.* 1988] M. S. Abadir, J. Ferguson, and T. E. Kirkland, "Logic Design
       Verification via Test Generation," *IEEE Trans. on Computer-Aided Design*,
       Vol. 7, No. 1, pp. 138-148, January, 1988.

[Abramovici and Menon 1983] M. Abramovici and P. R. Menon, "A Machine for
       Design Verification and Testing Problems," *Proc. Intn'l. Conf. on
       Computer-Aided Design*, pp. 27-29, September, 1983.

[Abramovici *et al.* 1986a] M. Abramovici, J. J. Kulikowski, P. R. Menon, and D. T.
       Miller, "SMART and FAST: Test Generation for VLSI Scan-Design Circuits,"
       IEEE Design & Test of Computers, Vol. 3, No. 4, pp. 43-54, August, 1986.

[Abramovici *et al.* 1986b] M. Abramovici, P. R. Menon, and D. T. Miller, "Checkpoint
       Faults Are Not Sufficient Target Faults for Test Generation," *IEEE Trans. on
       Computers*, Vol. C-35, No. 8, pp. 769-771, August, 1986.

[Abramovici and Miller 1989] M. Abramovici and D. T. Miller, "Are Random Vectors
       Useful in Test Generation?," *Proc. 1st European Test Conf.*, pp. 22-25, April,
       1989.

[Abramovici *et al.* 1989] M. Abramovici, D. T. Miller, and R. K. Roy, "Dynamic
       Redundancy Identification in Automatic Test Generation," *Proc. Intn'l. Conf.
       on Computer-Aided Design*, November, 1989 (to appear).

[Agrawal and Agrawal 1972] V. D. Agrawal and P. Agrawal, "An Automatic Test
       Generation System for ILLIAC IV Logic Boards," *IEEE Trans. on Computers*,
       Vol. C-21, No. 9, pp. 1015-1017, September, 1972.

[Agrawal and Agrawal 1976] P. Agrawal and V. D. Agrawal, "On Monte Carlo Testing of Logic Tree Networks," *IEEE Trans. on Computers*, Vol. C-25, No. 6, pp. 664-667, June, 1976.

[Agrawal and Mercer 1982] V. D. Agrawal and M. R. Mercer, "Testability Measures — What Do They Tell Us?," *Digest of Papers 1982 Intn'l. Test Conf.*, pp. 391-396, November, 1982.

[Agrawal *et al.* 1988] V. D. Agrawal, H. Farhat, and S. Seth, "Test Generation by Fault Sampling," *Proc. Intn'l. Conf. on Computer Design*, pp. 58-61, October, 1988.

[Agrawal *et al.* 1989] V. D. Agrawal, K. Cheng, and P. Agrawal, "A Directed Search Method for Test Generation Using a Concurrent Simulator," *IEEE Trans. on Computer-Aided Design*, Vol. 8, No. 2, pp. 131-138, February, 1989.

[Airapetian and McDonald 1979] A. N. Airapetian and J. F. McDonald, "Improved Test Set Generation Algorithm for Combinational Logic Control," *Digest of Papers 9th Annual Intn'l. Symp. on Fault-Tolerant Computing*, pp. 133-136, June, 1979.

[Akers 1976] S. B. Akers, "A Logic System for Fault Test Generation," *IEEE Trans. on Computers*, Vol. C-25, No. 6, pp. 620-630, June, 1976.

[Batni and Kime 1976] R. P. Batni and C. R. Kime, "A Module-Level Testing Approach for Combinational Networks," *IEEE Trans. on Computers*, Vol. C-25, No. 6, pp. 594-604, June, 1976.

[Bellon *et al.* 1983] C. Bellon, C. Robach, and G. Saucier, "An Intelligent Assistant for Test Program Generation: The SUPERCAT System," *Proc. Intn'l. Conf. on Computer-Aided Design*, pp. 32-33, September, 1983.

[Bending 1984] M. T. Bending, "Hitest: A Knowledge-Based Test Generation System," *IEEE Design & Test of Computers*, Vol. 1, No. 2, pp. 83-92, May, 1984.

[Benmehrez and McDonald 1983] C. Benmehrez and J. F. McDonald, "Measured Performance of a Programmed Implementation of the Subscripted D-Algorithm," *Proc. 20th Design Automation Conf.*, pp. 308-315, June, 1983.

[Bhattacharya and Hayes 1985] D. Bhattacharya and J. P. Hayes, "High-Level Test Generation Using Bus Faults," *Digest of Papers 15th Annual Intn'l. Symp. on Fault-Tolerant Computing*, pp. 65-70, June, 1985.

[Breuer 1983] M. A. Breuer, "Test Generation Models for Busses and Tri-State Drivers," *Proc. IEEE ATPG Workshop*, pp. 53-58, March, 1983.

[Breuer 1971] M. A. Breuer, "A Random and an Algorithmic Technique for Fault Detection Test Generation for Sequential Circuits," *IEEE Trans. on Computers*, Vol. C-20, No. 11, pp. 1364-1370, November, 1971.

[Breuer 1974] M. A. Breuer, "The Effects of Races, Delays, and Delay Faults on Test Generation," *IEEE Trans. on Computers*, Vol. C-23, No. 10, pp. 1078-1092, October, 1974.

[Breuer 1978] M. A. Breuer, "New Concepts in Automated Testing of Digital Circuits," *Proc. EEC Symp. on CAD of Digital Electronic Circuits and Systems*, pp. 69-92, North-Holland Publishing Co., November, 1978.

[Breuer and Friedman 1976] M. A. Breuer and A. D. Friedman, *Diagnosis & Reliable Design of Digital Systems*, Computer Science Press, Rockville, Maryland, 1976.

[Breuer and Friedman 1980] M. A. Breuer and A. D. Friedman, "Functional Level Primitives in Test Generation," *IEEE Trans. on Computers*, Vol. C-29, No. 3, pp. 223-235, March, 1980.

[Breuer and Harrison 1974] M. A. Breuer and L. M. Harrison, "Procedures for Eliminating Static and Dynamic Hazards in Test Generation," *IEEE Trans. on Computers*, Vol. C-23, No. 10, pp. 1069-1078, October, 1974.

[Cha *et al.* 1978] C. W. Cha, W. E. Donath, and F. Ozguner, "9-V Algorithm for Test Pattern Generation of Combinational Digital Circuits," *IEEE Trans. on Computers*, Vol. C-27, No. 3, pp. 193-200, March, 1978.

[Chandra and Patel 1987] S. J. Chandra and J. H. Patel, "A Hierarchical Approach to Test Vector Generation," *Proc. 24th Design Automation Conf.*, pp. 495-501, June, 1987.

[Chandra and Patel 1988] S. J. Chandra and J. H. Patel, "Test Generation in a Parallel Processing Environment," *Proc. Intn'l. Conf. on Computer Design*, pp. 11-14, October, 1988.

[Chandra and Patel 1989] S. J. Chandra and J. H. Patel, "Experimental Evaluation of Testability Measures for Test Generation," *IEEE Trans. on Computer-Aided Design*, Vol. 8, No. 1, pp. 93-98, January, 1989.

[Cheng and Agrawal 1989] K.-T. Cheng and V. D. Agrawal, "An Economical Scan Design for Sequential Logic Test Generation," *Digest of Papers 19th Intn'l. Symp. on Fault-Tolerant Computing*, pp. 28-35, June, 1989.

[Cheng and Chakraborty 1989] W.-T. Cheng and T. Chakraborty, "Gentest — An Automatic Test-Generation System for Sequential Circuits," *Computer*, Vol. 22, No. 4, pp. 43-49, April, 1989.

[David and Blanchet 1976] R. David and G. Blanchet, "About Random Fault Detection of Combinational Networks," *IEEE Trans. on Computers*, Vol. C-25, No. 6, pp. 659-664, June, 1976.

[Eichelberger and Lindbloom 1983] E. E. Eichelberger and E. Lindbloom, "Random-Pattern Coverage Enhancement and Diagnosis for LSSD Logic Self-Test," *IBM Journal of Research and Development*, Vol. 27, No. 3, pp. 265-272, March, 1983

[Fujiwara and Shimono 1983] H. Fujiwara and T. Shimono, "On the Acceleration of Test Generation Algorithms," *IEEE Trans. on Computers*, Vol. C-32, No. 12, pp. 1137-1144, December, 1983.

[Goel 1978] P. Goel, "RAPS Test Pattern Generator," *IBM Technical Disclosure Bulletin*, Vol. 21, No. 7, pp. 2787-2791, December, 1978.

[Goel 1981] P. Goel, "An Implicit Enumeration Algorithm to Generate Tests for Combinational Logic Circuits," *IEEE Trans. on Computers*, Vol C-30, No. 3, pp. 215-222, March, 1981.

[Goel and Rosales 1979] P. Goel and B. C. Rosales, "Test Generation & Dynamic Compaction of Tests," *Digest of Papers 1979 Test Conf.*, pp. 189-192, October, 1979.

[Goel and Rosales 1980] P. Goel and B. C. Rosales, "Dynamic Test Compaction with Fault Selection Using Sensitizable Path Tracing," *IBM Technical Disclosure Bulletin*, Vol. 23, No. 5, pp. 1954-1957, October, 1980.

[Goel and Rosales 1981] P. Goel and B. C. Rosales, "PODEM-X: An Automatic Test Generation System for VLSI Logic Structures," *Proc. 18th Design Automation Conf.*, pp. 260-268, June, 1981.

[Goldstein 1979] L. H. Goldstein, "Controllability/Observability Analysis of Digital Circuits," *IEEE Trans. on Circuits and Systems*, Vol. CAS-26, No. 9, pp. 685-693, September, 1979.

[Heap and Rogers 1989] M. A. Heap and W. A. Rogers, "Generating Single-Stuck-Fault Coverage from a Collapsed-Fault Set," *Computer*, Vol. 22, No. 4, pp. 51-57, April, 1989

[Hill and Huey 1977] F. J. Hill and B. Huey, "SCIRTSS: A Search System for Sequential Circuit Test Sequences," *IEEE Trans. on Computers*, Vol. C-26, No. 5, pp. 490-502, May, 1977.

[Huey 1979] B. M. Huey, "Heuristic Weighting Functions for Guiding Test Generation Searches," *Journal of Design Automation & Fault-Tolerant Computing*, Vol. 3, pp. 21-39, January, 1979.

[Itazaki and Kinoshita 1986] N. Itazaki and K. Kinoshita, "Test Pattern Generation for Circuits with Three-state Modules by Improved Z-algorithm," *Proc. Intn'l. Test Conf.*, pp. 105-108, September, 1986.

[Ivanov and Agarwal 1988] A. Ivanov and V. K. Agarwal, "Dynamic Testability Measures for ATPG," *IEEE Trans. on Computer-Aided Design*, Vol. 7, No. 5, pp. 598-608, May, 1988.

[Kirkland and Mercer 1987] T. Kirkland and M. R. Mercer, "A Topological Search Algorithm for ATPG," *Proc. 24th Design Automation Conf.*, pp. 502-508, June, 1987.

[Krishnamurthy and Sheng 1985] B. Krishnamurthy and R. L. Sheng, "A New Approach to the Use of Testability Analysis in Test Generation," *Proc. Intn'l. Test Conf.*, pp. 769-778, November, 1985.

[Krishnamurthy 1987] B. Krishnamurthy, "Hierarchical Test Generation: Can AI Help?," *Proc. Intn'l. Test Conf.*, pp. 694-700, September, 1987.

[Krishnamurthy and Tollis 1989] B. Krishnamurthy and I. G. Tollis, "Improved Techniques for Estimating Signal Probabilities," *IEEE Trans. on Computers*, Vol. 38, No. 7, pp. 1041-1045, July, 1989.

[Levendel and Menon 1982] Y. H. Levendel and P. R. Menon, "Test Generation Algorithms for Computer Hardware Description Languages," *IEEE Trans. on Computers*, Vol. C-31, No. 7, pp. 577-588, July, 1982.

[Lioy 1988] A. Lioy, "Adaptive Backtrace and Dynamic Partitioning Enhance a ATPG," *Proc. Intn'l. Conf. on Computer Design*, pp. 62-65, October, 1988.

[Lioy *et al.* 1989] A. Lioy, P. L. Montessoro, and S. Gai, "A Complexity Analysis of Sequential ATPG," *Proc. Intn'l. Symp. on Circuits and Systems*, pp. 1946-1949, May, 1989.

[Lisanke *et al.* 1987] R. Lisanke, F. Brglez, A. J. de Geus, and D. Gregory, "Testability-Driven Random Test-Pattern Generation," *IEEE Trans. on Computer-Aided Design*, Vol. CAD-6, No. 6, November, 1987.

[Ma *et al.* 1988] H.-K. T. Ma, S. Devadas, A. R. Newton, and A. Sangiovanni-Vincentelli, "Test Generation for Sequential Circuits," *IEEE Trans. on Computer-Aided Design*, Vol. 7, No. 10, pp. 1081-1093, October, 1988.

[Mallela and Wu 1985] S. Mallela and S. Wu, "A Sequential Circuit Test Generation System," *Proc. Intn'l. Test Conf.*, pp. 57-61, November, 1985.

[Marlett 1978] R. A. Marlett, "EBT: A Comprehensive Test Generation Technique for Highly Sequential Circuits," *Proc. 15th Design Automation Conf.*, pp. 335-339, June, 1978.

[Menon and Harihara 1989] P. R. Menon and M. R. Harihara, "Identification of Undetectable Faults in Combination Circuits," *Proc. Intn'l. Conf. on Computer Design*, October, 1989 (to appear).

[Miczo 1983] A. Miczo, "The Sequential ATPG: A Theoretical Limit," *Proc. Intn'l. Test Conf.*, pp. 143-147, October, 1983.

[Motohara *et al.* 1986] A. Motohara, K. Nishimura, H. Fujiwara, and I. Shirakawa, "A Parallel Scheme for Test-Pattern Generation," *Proc. Intn'l. Conf. on Computer-Aided Design*, pp. 156-159, November, 1986.

[Murray and Hayes 1988] B. T. Murray and J. P. Hayes, "Hierarchical Test Generation Using Precomputed Tests for Modules," *Proc. Intn'l. Test Conf.*, pp. 221-229, September, 1988.

[Muth 1976] P. Muth, "A Nine-Valued Circuit Model for Test Generation," *IEEE Trans. on Computers*, Vol. C-25, No. 6, pp. 630-636, June, 1976.

[Norrod 1989] F. E. Norrod, "An Automatic Test Generation Algorithm for Hardware Description Languages," *Proc. 26th Design Automation Conf.*, pp. 429-434, June, 1989.

[Ogihara *et al.* 1988] T. Ogihara, S. Saruyama, and S. Murai, "Test Generation for Sequential Circuits Using Individual Initial Value Propagation," *Proc. Intn'l. Conf. on Computer-Aided Design*, pp. 424-427, November, 1988.

[Parker 1976] K. P. Parker, "Adaptive Random Test Generation," *Journal of Design Automation and Fault-Tolerant Computing*, Vol. 1, No. 1, pp. 62-83, October, 1976.

[Parker and McCluskey 1975] K. P. Parker and E. J. McCluskey, "Probabilistic Treatment of General Combinational Networks," *IEEE Trans. on Computers*, Vol. C-24, No. 6, pp. 668-670, June, 1975.

[Patil and Banerjee 1989] S. Patil and P. Banerjee, "A Parallel Branch and Bound Algorithm for Test Generation," *Proc. 26th Design Automation Conf.*, pp. 339-344, June, 1989.

[Putzolu and Roth 1971] G. R. Putzolu and T. P. Roth, "A Heuristic Algorithm for the Testing of Asynchronous Circuits," *IEEE Trans. on Computers*, Vol. C-20, No. 6, pp. 639-647, June, 1971.

[Ratiu *et al.* 1982] I. M. Ratiu, A. Sangiovanni-Vincentelli, and D. O. Pederson "VICTOR: A Fast VLSI Testability Analysis Program," *Digest of Papers 1982 Intn'l. Test Conf.*, pp. 397-401, November, 1982.

[Renous *et al.*] R. Renous, G. M. Silberman, and I. Spillinger, "Whistle — A Workbench for Test Development of Library-Based Designs," *Computer*, Vol. 22, No. 4, pp. 27-41, April, 1989.

[Robinson 1983] G. D. Robinson, "HITEST-Intelligent Test Generation," *Proc. Intn'l. Test Conf.*, pp. 311-323, October, 1983.

[Roth 1966] J. P. Roth, "Diagnosis of Automata Failures: A Calculus and a Method," *IBM Journal of Research and Development*, Vol. 10, No. 4, pp. 278-291, July, 1966.

[Roth *et al.* 1967] J. P. Roth, W. G. Bouricius, and P. R. Schneider, "Programmed Algorithms to Compute Tests to Detect and Distinguish Between Failures in Logic Circuits," *IEEE Trans. on Electronic Computers*, Vol. EC-16, No. 10, pp. 567-579, October, 1967.

[Roth 1977] J. P. Roth, "Hardware Verification," *IEEE Trans. on Computers*, Vol. C-26, No. 12, pp. 1292-1294, December, 1977.

[Roth 1980] J. P. Roth, *Computer Logic, Testing and Verification*, Computer Science Press, Rockville, Maryland, 1980.

[Rutman 1972] R. A. Rutman, "Fault Detection Test Generation for Sequential Logic by Heuristic Tree Search," *IEEE Computer Group Repository*, Paper No. R-72-187, 1972.

[Savir and Bardell 1984] J. Savir and P. H. Bardell, "On Random Pattern Test Length," *IEEE Trans. on Computers*, Vol. C-33, No. 6, pp. 467-474, June, 1984.

[Savir *et al.* 1984] J. Savir, G. S. Ditlow, and P. H. Bardell, "Random Pattern Testability," *IEEE Trans. on Computers*, Vol. C-33, No. 1, pp. 79-90, January, 1984.

[Schneider 1967] R. R. Schneider, "On the Necessity to Examine D-Chains in Diagnostic Test Generation," *IBM Journal of Research and Development*, Vol. 11, No. 1, p. 114, January, 1967.

[Schnurmann *et al.* 1975] H. D. Schnurmann, E. Lindbloom, and R. G. Carpenter, "The Weighted Random Test-Pattern Generator," *IEEE Trans. on Computers*, Vol. C-24, No. 7, pp. 695-700, July, 1975.

[Schuler *et al.* 1975] D. M. Schuler, E. G. Ulrich, T. E. Baker, and S. P. Bryant, "Random Test Generation Using Concurrent Fault Simulation," *Proc. 12th Design Automation Conf.*, pp. 261-267, June, 1975.

[Schulz *et al.* 1988] M. H. Schulz, E. Trischler, and T. M. Sarfert, "SOCRATES: A Highly Efficient Automatic Test Pattern Generation System," *IEEE Trans. on Computer-Aided Design*, Vol. 7, No. 1, pp. 126-137, January, 1988.

[Schulz and Auth 1989] M. Schulz and E. Auth, "Improved Deterministic Test Pattern Generation with Applications to Redundancy Identification," *IEEE Trans. on Computer-Aided Design*, Vol. 8, No. 7, pp. 811-816, July, 1989.

[Seshu 1965] S. Seshu, "On an Improved Diagnosis Program," *IEEE Trans. on Electronic Computers*, Vol. EC-12, No. 2, pp. 76-79, February, 1965.

[Seth *et al.* 1985] S. C. Seth, L. Pan, and V. D. Agrawal, "PREDICT — Probabilistic Estimation of Digital Circuit Testability," *Digest of Papers 15th Annual Intn'l. Symp. on Fault-Tolerant Computing*, pp. 220-225, June, 1985.

[Shedletsky 1977] J. J. Shedletsky, "Random Testing: Practicality vs. Verified Effectiveness," *Proc. 7th Annual Intn'l. Conf. on Fault-Tolerant Computing*, pp. 175-179, June, 1977.

[Shirley *et al.* 1987] M. Shirley, P. Wu, R. Davis, and G. Robinson, "A Synergistic Combination of Test Generation and Design for Testability," *Proc. Intn'l. Test Conf.*, pp. 701-711, September, 1987.

[Shteingart *et al.* 1985] S. Shteingart, A. W. Nagle, and J. Grason, "RTG: Automatic Register Level Test Generator," *Proc. 22nd Design Automation Conf.*, pp. 803-807, June, 1985.

[Silberman and Spillinger 1988] G. M. Silberman and I. Spillinger, "G-RIDDLE: A Formal Analysis of Logic Designs Conducive to the Acceleration of Backtracking," *Proc. Intn'l. Test Conf.*, pp. 764-772, September, 1988.

[Singh 1987] N. Singh, *An Artificial Intelligence Approach to Test Generation*, Kluwer Academic Publishers, Norwell, Massachusetts, 1987.

[Snethen 1977] T. J. Snethen, "Simulator-Oriented Fault Test Generator," *Proc. 14th Design Automation Conf.*, pp. 88-93, June, 1977.

[Somenzi *et al.* 1985] F. Somenzi, S. Gai, M. Mezzalama, and P. Prinetto, "Testing Strategy and Technique for Macro-Based Circuits," *IEEE Trans. on Computers*, Vol. C-34, No. 1, pp. 85-90, January, 1985.

[Thomas 1971] J. J. Thomas, "Automated Diagnostic Test Programs for Digital Networks," *Computer Design*, pp. 63-67, August, 1971.

[Timoc *et al.* 1983] C. Timoc, F. Stott, K. Wickman and L. Hess, "Adaptive Probabilistic Testing of a Microprocessor," *Proc. Intn'l. Conf. on Computer-Aided Design*, pp. 71-72, September, 1983.

[Wagner *et al.* 1987] K. D. Wagner, C. K. Chin, and E. J. McCluskey, "Pseudorandom Testing," *IEEE Trans. on Computers*, Vol. C-36, No. 3, pp. 332-343, March, 1987.

[Waicukauski *et al.* 1989] J. A. Waicukauski, E. Lindbloom, E. B. Eichelberger, and O. P. Forlenza, "WRP: A Method for Generating Weighted Random Test Patterns," *IBM Journal of Research and Development*, Vol. 33, No. 2, pp. 149-161, March, 1989.

[Wang 1975] D. T. Wang, "An Algorithm for the Generation of Test Sets for Combinational Logic Networks," *IEEE Trans. on Computers*, Vol. C-24, No. 7, pp. 742-746, July, 1975.

[Wang and Wei 1986] J.-C. Wang and D.-Z. Wei, "A New Testability Measure for Digital Circuits," *Proc. Intn'l. Test Conf.*, pp. 506-512, September, 1986.

[Wunderlich 1987] H.-J. Wunderlich, "On Computing Optimized Input Probabilities for Random Tests," *Proc. 24th Design Automation Conf.*, pp. 392-398, June, 1987.

PROBLEMS

6.1 For the circuit of Figure 6.89, generate a test for the fault g s-a-1. Determine all the other faults detected by this test.

6.2 Modify the TG algorithm for fanout-free circuits (given in Figures 6.3, 6.4, and 6.5) so that it tries to generate a test that will detect as many faults as possible in addition to the target fault.

6.3 Use only implications to show that the fault f s-a-0 in the circuit of Figure 6.12(a) is undetectable.

6.4 Construct the truth table of an XOR function of two inputs using the five logic values 0, 1, x, D, and $\overline{D}$.

6.5 Can a gate on the *D-frontier* have both a D and a $\overline{D}$ among its input values?

6.6 For the circuit of Figure 6.90, perform all possible implications starting from the given values.

6.7 For the circuit of Figure 6.91, perform all possible implications starting from the given values.

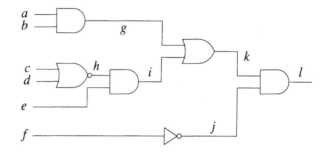

Figure 6.89

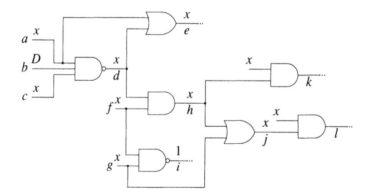

Figure 6.90

6.8 Consider the circuit and the values shown in Figure 6.92. Let us assume that trying to justify $d=0$ via $a=0$ has lead to an inconsistency. Perform all the implications resulting from reversing the incorrect decision.

6.9 Define the *D-frontier* for the 9-V algorithm.

6.10 Using the five logic values of the *D*-algorithm, the consistency check works as follows. Let v' be the value to be assigned to a line and v be its current value. Then v' and v are consistent if $v=x$ or $v=v'$. Formulate a consistency check for the 9-V algorithm.

6.11 Outline a TG algorithm based on single-path sensitization. Apply the algorithm to the problem from Example 6.6.

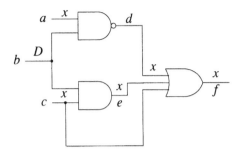

Figure 6.91

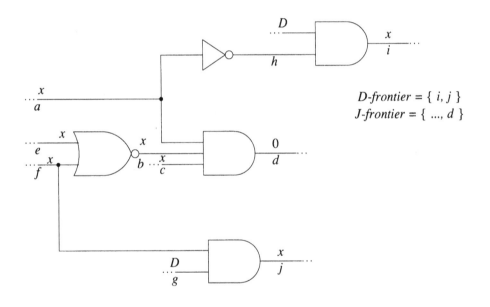

D-frontier = { i, j }
J-frontier = { ..., d }

Figure 6.92

6.12 Apply PODEM to the problem from Example 6.4. The D-algorithm can determine that a fault is undetectable without backtracking (i.e., only by implications). Is the same true for PODEM?

6.13 Apply FAN to the problem from Example 6.4.

6.14 Discuss the applicability of formula (6.3) to the circuit in Figure 6.93.

6.15 Extend formulas (6.1) ... (6.7) for different gate types.

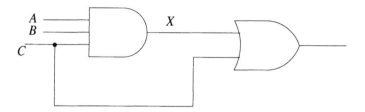

Figure 6.93

6.16 Show that in a fanout-free circuit, if we start by setting $C0$ and $C1$ of every PI to 1, and then recursively compute controllability costs, $C0(l)$ and $C1(l)$ represent the minimum number of PIs we have to assign binary values to set l to 0 and 1, respectively.

6.17 Compute controllability and observability costs for every line in the circuit in Figure 6.7.

    a. Use the (generalized) formulas (6.1) to (6.4).

    b. Repeat, by introducing fanout-based correction terms in controllability formulas.

    c. Use the costs computed in a. to guide PODEM in generating a test for the fault $G_1 s\text{-}a\text{-}1$.

    d. Repeat, using the costs computed in b.

6.18 Prove that the test set generated by using the procedure *CPTGFF* (Figure 6.46) for a fanout-free circuit is complete for SSFs.

6.19

    a. Modify the procedure *CPTGFF* (Figure 6.46) for use in multiple-output circuits without reconvergent fanout.

    b. Apply critical-path TG for the circuit in Figure 6.94.

6.20 Can the detection probability of a fault in a combinational circuit be 1/100?

6.21 Show that the expected fault coverage of a random test sequence of length N is greater than

    a. its testing quality t_N;

    b. its detection quality d_N.

6.22

    a. Compute signal probabilities in the circuit in Figure 6.94.

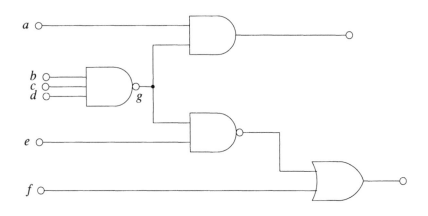

Figure 6.94

    b.   Compute the detection probability for the fault g s-a-0.

6.23 Consider a fanout-free circuit with L levels composed of NAND gates. Each gate has n inputs. Let p_0 be the signal probability of every PI. Because of symmetry, all signals at level l have the same signal probability p_l.

    a.   Show that

$$p_l = 1 - (p_{l-1})^n$$

    b.   Show that the smallest detection probability of a fault is

$$d_{\min} = r \prod_{k=0}^{L-1} (p_k)^{n-1}$$

          where $r = \min\{p_0, 1-p_0\}$

    c.   For $L=2$ and $n=2$, determine the value p_0 that maximizes $d_{\min}$.

6.24 Use formula (6.25) to compute p_z for

    a. $Z = A \cdot A$
    b. $Z = A \cdot \overline{A}$
    c. $Z = A + A$
    d. $Z = A + \overline{A}$

6.25 Let Z be the output of an OR gate whose inputs, X and Y, are such that $X.Y = 0$ (i.e., are never simultaneously 1). Show that

$$p_Z = p_A + p_B$$

6.26 Let Z be the output of a combinational circuit realizing the function f. Let us express f as a sum of minterms, i.e., $f = \displaystyle\sum_{i=1}^{k} m_i$.

a. If $p_i = P_r(m_i = 1)$, show that

$$p_Z = \sum_{i=1}^{k} p_i$$

b. For the circuit in Figure 6.95, compute p_Z as a function of p_A, p_B, and p_C.

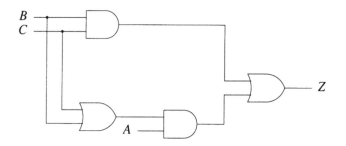

Figure 6.95

c. Let $p_A = p_B = p_C = q$. Plot p_Z as a function of q.

6.27 Determine lower bounds on the detection probability of every checkpoint fault for the circuit in Figure 6.48.

6.28 Determine the lower bound on the detection probability of the fault Y s-a-0 in the circuit in Figure 6.96.

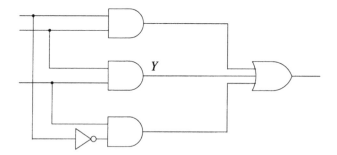

Figure 6.96

6.29

a. Show that the gate G selected in the second phase of the procedure RAPS (Figure 6.57) has at least one input with controlling value.

  b. When G has more than one input with controlling value, do we gain anything by trying to justify noncontrolling values on the other inputs?

6.30 Show that if two input vectors (of the same circuit) are compatible, then their corresponding output vectors are also compatible.

6.31 Determine the primitive cubes and the propagation D-cubes for an exclusive-OR module with two inputs.

6.32 Derive a test sequence for the fault Z s-a-0 in the circuit of Figure 6.75.

  a. Assume an initial state $q_1 = q_2 = 1$.

  b. Assume an unknown initial state.

6.33 Derive a self-initializing test sequence for the fault J_1 s-a-1 in the circuit of Figure 6.77.

6.34 Try to use the TG procedure given in Figure 6.76 to generate a self-initializing test sequence for the fault R s-a-1 in the circuit of Figure 6.97. Explain why the procedure fails.

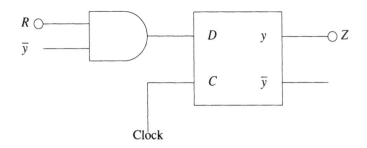

Figure 6.97

6.35 For the shift register of Figure 6.82, determine the union algorithm executed for $(Reset, Clock, S_0, S_1) = 1 \uparrow x\ 0$.

6.36 For the shift register of Figure 6.82, assume that $n = 4$, $(Reset, Clock, S_0, S_1) = 1 \uparrow x\ 1$, $\mathbf{y} = 0\ 1\ x\ x$ and $\mathbf{A} = x\ x\ x\ x\ x\ x$. Determine how to justify $\mathbf{Y} = 1\ 1\ x\ 0$.

7. TESTING FOR BRIDGING FAULTS

About This Chapter

Shorts between normally unconnected signals are manufacturing defects that occur often in chips and on boards. Shorts are modeled as bridging faults. In this chapter we present fault simulation and test generation methods for bridging faults. These methods rely on an analysis of relations between single stuck faults and bridging faults.

7.1 The Bridging-Fault Model

Bridging faults (BFs) are caused by shorts between two (or more) normally unconnected signal lines. Since the lines involved in a short become equipotential, all of them have the same logic value. For a shorted line i, we have to distinguish between the value one could actually observe on i and the value of i as determined by its source element; the latter is called *driven value*. Figure 7.1 shows a general model of a BF between two lines x and y. We denote such a BF by $(x.y)$ and the function introduced by the BF by $Z(x,y)$. The fanout of Z is the union of the fanouts of the shorted signals. Note that the values of x and y in this model are their driven values, but these are not observable in the circuit.

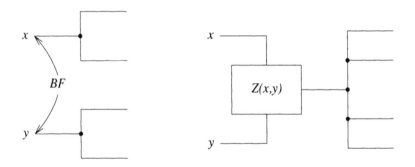

Figure 7.1   BF Model

The function Z has the property that $Z(a,a)=a$. What happens when x and y have opposite values depends on the technology. For example, in MOS the value of $Z(a,\bar{a})$ is, in general, indeterminate (i.e., the resulting voltage does not correspond to any logic value). In this section we will consider only BFs for which the values of the shorted lines are determinate. In many technologies (such as TTL or ECL), when two shorted lines have opposite driven values, one value (the *strong* one) overrides the other. If $c \in \{0,1\}$ is the strong value, then $Z(0,1)=Z(1,0)=c$, and the function introduced by the BF is AND if $c=0$ and OR if $c=1$.

Unlike in the SSF model, in the BF model there is no need to distinguish between a stem and its fanout branches, since they always have the same values. Thus the BFs we shall consider are defined only between gate outputs and/or primary inputs.

If there exists (at least) one path between x and y, then a BF $(x.y)$ creates one or more feedback loops. Such a fault is referred to as a *feedback bridging fault* (FBF). A BF that does not create feedback is referred to as a *nonfeedback bridging fault* (NFBF). *An FBF transforms a combinational circuit into a sequential one* (see Figure 7.2). Moreover, if a feedback loop involves an odd number of inversions, the circuit may oscillate (see Figure 7.3). If the delay along the loop is small, the resulting oscillations have very high frequency and may cause the affected signals to assume indeterminate logic values. Such values may confuse the testing equipment.

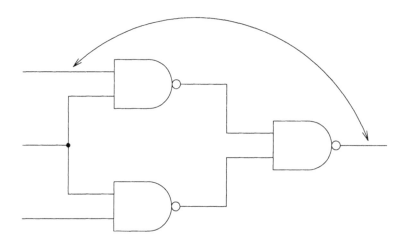

Figure 7.2   Example of an FBF

The *multiple bridging-fault (MBF) model* represents shorts involving more than two signals. An MBF with $p>2$ shorted lines can be thought of as composed of $p-1$ BFs between two lines. For example, an MBF among lines i, j, and k can be represented by the BFs $(i.j)$ and $(j.k)$. This model assumes that only one group of lines are shorted. For example, we cannot have both $(a.b)$ and $(c.d)$ present in the circuit. Although masking relations may occur among the components of an MBF [Mei 1974], most MBFs are detected by the tests designed to detect their component BFs. Also MBFs are less likely to occur. Thus in the following we shall consider only BFs between two lines.

The number of theoretically possible BFs in a circuit with G gates and I primary inputs is $b = \begin{bmatrix} G+I \\ 2 \end{bmatrix}$. For $G \gg I$, $b = G(G-1)/2$, so the dominant factor is G^2. However, this figure assumes that a short may occur between any two lines. In most circumstances a

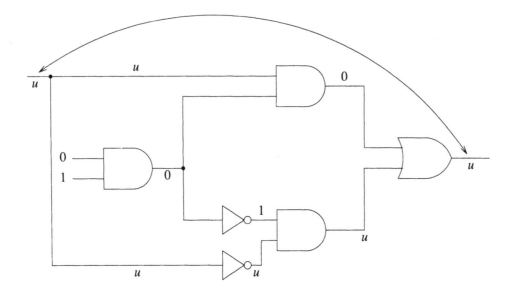

Figure 7.3   Oscillation induced by an FBF

short only involves physically adjacent lines. Let N_i be the "neighborhood" of i, i.e., the set of lines physically adjacent to a line i. Then the number of *feasible* BFs is

$$b = \frac{1}{2} \sum_i |N_i|$$

Denoting by k the average size of N_i, we have

$$b = \frac{1}{2}(G+I)k$$

In general, k increases with the average fanout count f. Let us assume that every stem and fanout branch has, on the average, r distinct neighbors, where r depends on the layout. Then

$$b = \frac{1}{2}(G+I)r(1+f)$$

Comparing this result with the number of SSFs derived in Chapter 4, we can conclude that *the number of feasible BFs is of the same order of magnitude as the number of SSFs in a circuit.* But since r is usually greater than 2, the number of feasible BFs is usually greater. Moreover, the number of SSFs to be analyzed can be reduced by using structural equivalence and dominance relations, but similar collapsing techniques have not been developed for BFs.

However, the BF model is often used *in addition to* the SSF model. To take advantage of this, we will first analyze relations between the detection of SSFs and the detection of BFs. Then we will present fault simulation and test generation methods for BFs that exploit these relations, so that *the processing of the two fault models can be combined.* In this way BF testing can be done with a small additional effort beyond that required for testing SSFs.

The following analysis considers only combinational circuits. Although we discuss only AND BFs, the results can be extended to OR BFs by interchanging 0s and 1s.

7.2 Detection of Nonfeedback Bridging Faults

Theorem 7.1: [Williams and Angell 1973] A test t detects the AND NFBF $(x.y)$ iff either t detects x s-a-0 and sets $y = 0$, or t detects y s-a-0 and sets $x = 0$.

Proof

a. Assume the first condition holds, namely t detects x s-a-0 and sets $y = 0$. Hence t activates x s-a-0 (by setting $x = 1$) and the resulting error propagates to a primary output. The same effect occurs in the presence of $(x.y)$, while the value of y does not change. Therefore t detects $(x.y)$. The proof for the second condition is symmetric.

b. If neither of the above conditions holds, then we shall prove that t does not detect $(x.y)$. Two cases are possible:

   Case 1: x and y have the same value. Then $(x.y)$ is not activated.

   Case 2: x and y have different values. Assume (without loss of generality) that $y = 0$. Then an error is generated at x. But since t does not detect x s-a-0, this error does not propagate to any primary output.

   Therefore, in either case t does not detect $(x.y)$ □

We emphasize that lines x and y involved in a BF are gate outputs and/or primary inputs, *not* fanout branches. As illustrated in Figure 7.4, Theorem 7.1 does not hold if applied to fanout branches. Although both h s-a-0 and m s-a-0 are undetectable, the AND BF $(h.m)$ is detected by the test 110, because the fault effect on h propagates from the stem b (Theorem 7.1 applies for b and c).

In general, there is no guarantee that a complete test set for SSFs will satisfy the conditions of Theorem 7.1 for every NFBF. However, detection can be guaranteed for certain types of BFs involving inputs of the same gate [Friedman 1974]. This is important because shorts between inputs of the same gate are likely to occur.

Corollary 7.1: Let x and y be signals without fanout. If x and y are inputs to the same OR or NOR gate, then the AND BF $(x.y)$ dominates both x s-a-0 and y s-a-0.

Proof: From Figure 7.5 we can observe that the values required to detect x s-a-0 (and similarly, those required to detect y s-a-0) satisfy the conditions of Theorem 7.1. Thus the AND BF $(x.y)$ dominates both x s-a-0 and y s-a-0.   □

Note that, in general, this dominance relation does not extend to sequential circuits. For example, in the circuit of Figure 7.6 (starting with $y=1$), consider the OR BF $(x_1.x_2)$, which would dominate x_2 s-a-1 in a combinational circuit. Although the test sequence shown detects x_2 s-a-1, it does not detect $(x_1.x_2)$.

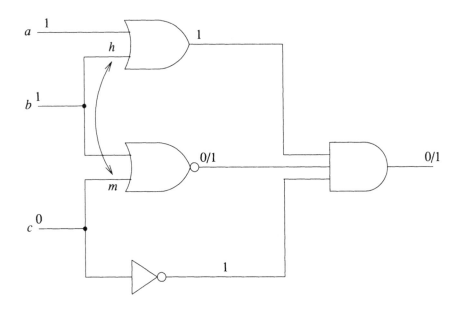

Figure 7.4

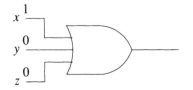

Figure 7.5

Now consider an AND BF between inputs of an OR gate, with only one of the shorted inputs having fanout. An argument similar to that used to prove Corollary 7.1 (refer to Figure 7.7) leads to the following result.

Corollary 7.2: Let x be a line with fanout and y a line without. If x and y are inputs to the same OR or NOR gate, the AND BF $(x.y)$ dominates y s-a-0. □

BFs between inputs of the same gate that do not satisfy the condition of Corollaries 7.1 or 7.2 are not guaranteed to be detected by complete test sets for SSFs.

[Friedman 1974] has conjectured that in an irredundant combinational circuit, all BFs between inputs of the same gate, where both inputs have fanout, are detectable. To our knowledge, no counterexample to this conjecture has yet been found.

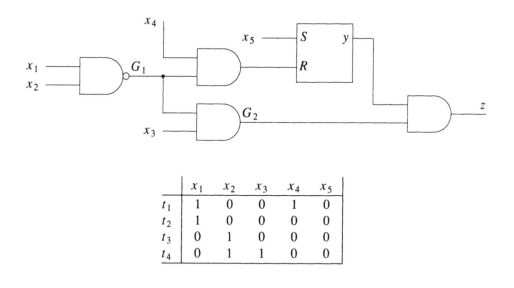

	x_1	x_2	x_3	x_4	x_5
t_1	1	0	0	1	0
t_2	1	0	0	0	0
t_3	0	1	0	0	0
t_4	0	1	1	0	0

Figure 7.6

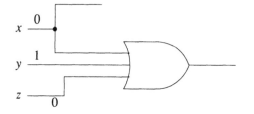

Figure 7.7

7.3 Detection of Feedback Bridging Faults

An FBF is created when there exists at least one path between the two shorted lines. We refer to the two lines involved in an FBF as the *back line b* and the *front line f*, where b is the line with the lower level (i.e., is closer to the primary inputs). Since a combinational circuit becomes a sequential one in the presence of an FBF, in general we need a test sequence to detect an FBF. However, we shall show that *in many cases an FBF can be detected by only one test* [Abramovici and Menon 1985].

Theorem 7.2: A test t that detects f s-a-0 and sets $b = 0$ detects the AND FBF (b,f).

Proof: Independent of the state of the circuit before test t is applied, b is driven to 0, which is the strong value for the AND BF (see Figure 7.8). The driven value of f is 1 and hence an error is generated at f. Since the value of b is the same as in the

fault-free circuit, the error propagates on the same path(s) as the error generated by
f s-a-0. Therefore t detects (b,f).   □

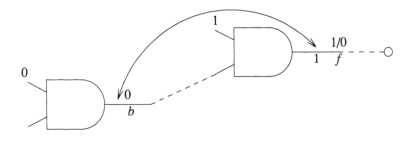

Figure 7.8

In the next theorem the roles of b and f are reversed.

Theorem 7.3: A test t that detects b s-a-0, and sets f=0 without sensitizing f to
b s-a-0, detects the AND FBF (b,f).

Proof: Since f is not sensitized to b s-a-0 by t, the value of f does not depend on the
value of b (see Figure 7.9). Independent of the state of the circuit (with the FBF
present) before t is applied, the driven values of b and f are, respectively, 1 and 0.
Then t activates (b,f) and the error propagates from b along the path(s) sensitized by t
to detect b s-a-0. Therefore t detects (b,f).   □

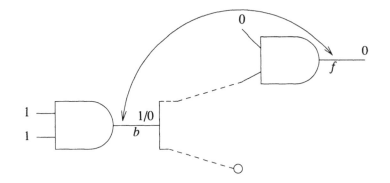

Figure 7.9

Note that a test t that propagates b s-a-0 through f and sets f=0 induces an oscillation
along the loop created by (b,f) (see Figure 7.3); we will refer to such a BF as
potentially oscillating.

Theorems 7.2 and 7.3 are valid independent of the number of paths between b and f and of their inversion parity. Now we shall prove that when all the paths between b and f have even parity, the conditions for the detection of $(b.f)$ can be relaxed.

Corollary 7.3: Let $(b.f)$ be an AND FBF such that all paths between b and f have even inversion parity. A test t that either detects f s-a-0 and sets $b = 0$, or detects b s-a-0 and sets $f = 0$, also detects $(b.f)$.

Proof: The first condition is covered by Theorem 7.2. Let us analyze the second condition, namely t detects b s-a-0 and sets $f = 0$. We shall prove that line f is not sensitized to the fault b s-a-0 by t. Let us assume the contrary. Then, according to Corollary 4.1, the value of f in t should be $1 \oplus 0 = 1$; but this contradicts the assumption that t sets $f = 0$. Hence f does not lie on any path sensitized to b s-a-0 by t. Then this case is covered by Theorem 7.3, and therefore t detects $(b.f)$.   □

All the preceding results in this section only give sufficient conditions for the detection of FBFs by one test. But an FBF can also be detected by a sequence of tests, none of which individually satisfies these conditions. The next example shows such a case.

Example 7.1: Consider the AND FBF $(a.z)$ in the circuit in Figure 7.10, and the tests $t_1 = 1010$ and $t_2 = 1110$. The only path between a and z has even inversion parity. Neither t_1 nor t_2 satisfies the conditions of Corollary 7.3. Nevertheless the sequence (t_1, t_2) detects $(a.z)$, because the 0 value of z in t_1 overrides the 1 value applied to a in t_2, and the resulting error propagates to z.   □

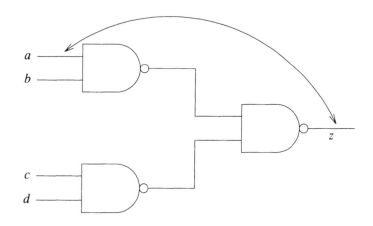

Figure 7.10

For certain cases we can derive conditions that are both necessary and sufficient for the detection of an FBF by one test.

Corollary 7.4: Let $(b.f)$ be an AND FBF such that any path between b and a primary output goes through f. A test t detects $(b.f)$ iff t detects f s-a-0 and sets $b=0$.

Proof: Sufficiency follows from Theorem 7.2. To prove necessity, we show that if t does not detect f s-a-0 or does not set $b=0$, then it does not detect (b,f).

Case 1: Suppose that t does not detect f s-a-0. Since any path between b and a primary output goes through f, an error generated by (b,f) can propagate only from f. Then t must detect f s-a-1. Hence, t sets $f=0$. Since 0 is the strong value for an AND BF, in the presence of (b,f), f can either maintain value 0 or it can oscillate. If $f = 0$, no error is generated. If f oscillates, no definite error is generated.

Case 2: Suppose that t detects f s-a-0 but sets $b=1$. Again no error is generated at f.

Therefore in either case t does not detect (b,f).   □

Corollary 7.5: Let (b,f) be an AND FBF where b and f are such that $f=1$ whenever $b=0$ (in the fault-free circuit). A test t detects (b,f) iff t detects f s-a-0 and sets $b=0$.

Proof: Sufficiency follows from Theorem 7.2. To prove necessity, two cases must be considered.

Case 1: Suppose that t sets $b=1$. If t sets $f=1$, then no error is generated. If t sets $f=0$, this would make $b=0$ because of the BF; but in the fault-free circuit $b=0$ implies $f=1$, and this is also true in the presence of (b,f). Hence the loop created by (b,f) oscillates and no definite error is generated.

Case 2: Suppose that t sets $b=0$ but does not detect f s-a-0. By assumption $b=0$ implies $f=1$. This means that t activates f s-a-0 but does not propagate its effect to a primary output. Similarly, the error generated by (b,f) appears only at f but is not propagated to a primary output.

Therefore, in either case t does not detect (b,f).   □

An example of an FBF satisfying Corollary 7.5 is an FBF between an input and the output of a NAND gate.

The following result is a consequence of Corollary 7.4 and allows us to identify a type of FBF that is *not* detectable by one test.

Corollary 7.6: No single test can detect an AND FBF (b,f) such that every path between b and a primary output goes through f, and $f=0$ whenever $b=0$.

For example, if b is an input of an AND gate whose output is f, and b does not have other fanout, then (b,f) is not detectable by one test.

7.4 Bridging Faults Simulation

In this section we discuss simulation of BFs. It is possible to explicitly simulate BFs by a process similar to the simulation of SSFs, based on determining when BFs are activated and on propagating their fault effects. This approach, however, cannot efficiently handle large circuits, because

- A BF is both structurally and functionally more complex than an SSF.

- The number of all feasible BFs is larger than that of SSFs.

Thus explicit fault simulation of BFs would be much more expensive than that of SSFs.

In the following we present an *implicit simulation method for BFs* in combinational circuits [Abramovici and Menon 1985]. This method uses the relations between the detection of SSFs and that of BFs to *determine the BFs detected by a test set without explicitly simulating BFs*. This is done by monitoring the occurrence of these relations during the simulation of SSFs.

Since we analyze the detection of BFs only by single tests, the method is approximate in the sense that it will not recognize as detected those BFs detected only by sequences of tests. Hence the computed fault coverage may be pessimistic; that is, the actual fault coverage may be better than the computed one.

We assume that layout information is available such that for every signal x (primary input or gate output) we know its neighborhood N_x consisting of the set of all lines that can be shorted to x. We consider only AND BFs. Note that for every BF $(x.y)$ we have both $x \in N_y$ and $y \in N_x$.

All the results in the previous sections relate the detection of AND BFs to the detection of s-a-0 faults. If a test that detects x s-a-0 also detects the BF $(x.y)$, we say that $(x.y)$ is *detected based on x*. To simplify the simulation of BFs, we construct a *reduced neighborhood N_x'* obtained by deleting from N_x all lines y such that $(x.y)$ is undetectable or undetectable based on x. If x and y are inputs to the same AND or NAND gate and they do not have other fanouts, then $(x.y)$ is undetectable and we remove y from N_x (and x from N_y). The same is done if x and y satisfy the conditions of Corollary 7.6. If $(x.y)$ is an FBF with y being the front line, and the conditions of Corollary 7.4 or Corollary 7.5 apply, then $(x.y)$ can be detected only based on y, and we remove y from N_x.

Next we partition the remaining lines in N_x' into two sets, M_x and M_x^*. The set M_x^* contains all lines y that are successors of x, such that at least one path between x and y has odd inversion parity, and M_x contains all other lines in N_x'. The reason for this partitioning is that if $y \in M_x$, detecting x s-a-0 when $y=0$ is sufficient for the detection of $(x.y)$. But if $y \in M_x^*$, then $(x.y)$ is a potentially oscillating FBF that is detected based on x only if the effect of x s-a-0 does not propagate to y.

The processing required during SSF simulation to determine the detected BFs can be summarized as follows. After simulating each test t, analyze every fault x s-a-0 detected by t. For every line $y \in M_x$ with value 0, mark $(x.y)$ as detected. For every line $y \in M_x^*$ with value 0, mark $(x.y)$ as detected if the effect of x s-a-0 does not propagate to y.

Determining whether the effect of x s-a-0 propagates to y depends on the method used by SSF simulation. In deductive simulation we can simply check whether x s-a-0 appears in the fault list of y. In concurrent simulation we must also check whether the values of y in the good circuit and in the circuit with x s-a-0 are different. In critical path tracing we have to determine whether x is reached by backtracing critical paths from y.

A BF $(x.y)$ may be discarded as soon as it is detected. This means removing y from M_x or M_x^* *and* removing x from M_y or M_y^*. A fault x s-a-0, however, should be retained until all BFs that are detectable based on x are detected, i.e., when M_x and M_x^* become empty. If equivalent SSFs have been collapsed, whenever a SSF fault is detected, all the s-a-0 faults equivalent to it should be checked. A SSF fault should be discarded only after all the BFs whose detection may be based on its equivalent s-a-0 faults have been detected.

Delaying fault discarding beyond first detection increases the cost of explicit fault simulation algorithms, such as deductive and concurrent. Therefore, the most suitable SSF simulation technique for determining the detected BFs is critical path tracing, which is an implicit fault simulation method that does not discard or collapse faults. Although the additional checks for fault-effect propagation are more expensive in critical path tracing than in the explicit fault simulation methods, the experimental results presented in [Abramovici and Menon 1985] show that in practice these checks are not needed, because most of the potentially oscillating FBFs are detected based on their front lines.

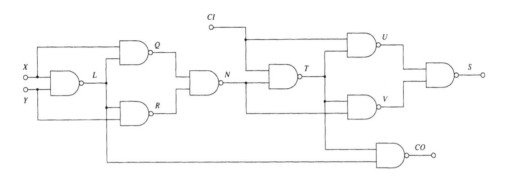

Figure 7.11

Example 7.2: Consider the adder in Figure 7.11. The rows of the array in Figure 7.12(a) show the sets M and M^* of every line under the assumption that any two lines are neighbors. This is an unrealistic worst case, but it has the merit that it does not bias the example toward any particular layout. Neighbors in M are denoted by "•" and those in M^* by "*". For example, $M_R = \{X,Y,CI,L,T\}$ and $M_R^* = \{U,V,S,CO\}$. Note that N does not appear in M_R^* because all paths between R and a PO pass through N. However, $R \in M_N$. Similarly, Q is not included in M_R because $(Q.R)$ is obviously undetectable, and Q, R, T, V, and CO do not appear in M_L^* because they are set to 1 whenever $L=0$. Figure 7.12(b) shows the applied tests (which detect all SSFs) and the values in the fault-free circuit.

The detection of BFs is summarized in Figure 7.12(c). A number i in row x and column y shows that the SSF x s-a-0 and the BF $(x.y)$ are detected by the test t_i. An "×" indicates that $(x.y)$ is detected based on y (this test is found in row y and column x). For example, t_1 detects L s-a-0. Since X,Y,CI, and N have value 0 in t_1, we write 1 in the corresponding entries. Since S has value 0 and the effect of L s-a-0 does not reach S, we also write 1 in the entry for $(L.S)$. We also enter an "×" in the entries for $(CI.L)$, $(N.L)$, and $(S.L)$.

	X	Y	CI	L	Q	R	N	T	U	V	S	CO
X		•	•			•	*	*	*	*	*	*
Y	•		•		•		*	*	*	*	*	*
CI	•	•		•	•	•	•			•	*	•
L	•	•	•				•		•		*	
Q	•	•	•	•				•	*	*	*	*
R	•	•	•	•				•	*	*	*	*
N	•	•	•	•	•	•			•		*	•
T	•	•	•	•	•	•	•				•	
U	•	•	•	•	•	•	•	•				•
V	•	•	•	•	•	•	•	•				•
S	•	•	•	•	•	•	•	•	•	•		•
CO	•	•	•	•	•	•	•	•	•	•	•	

(a)

	X	Y	CI	L	Q	R	N	T	U	V	S	CO
t_1	0	0	0	1	1	1	0	1	1	1	0	0
t_2	1	0	0	1	0	1	1	1	1	0	1	0
t_3	0	1	1	1	1	0	1	0	1	1	0	1
t_4	1	1	0	0	1	1	0	1	1	1	0	1
t_5	1	1	1	0	1	1	0	1	0	1	1	1

(b)

	X	Y	CI	L	Q	R	N	T	U	V	S	CO
X	—	2	2	—	—	×	×	×	×	×		2
Y	×	—	4	—	×	—	×	×	×	×	×	
CI	×	×	—	×	×	×	×	—	—	×	×	×
L	1	1	1	—	—	—	1	—	×	—	1	—
Q	1	1	1	4	—	—	—	×			×	1
R	1	1	1	4	—	—	—		×	×		2
N	3	2	2	×	2	3	—	—	×	—	×	2
T	1	1	1	5	2		1	—	—	—	1	—
U	1	1	1	4		3	1	3	—	—	—	1
V	1	1	1	4		3	1	3	—	—	—	1
S		2	2	×	2		5	×	5	2	—	2
CO	×		4	4	×	×	×	3	×	×	×	—

(c)

Figure 7.12

In this example 91 percent (58 out of 64) of the considered BFs are detected by single tests. The fault coverage is actually higher, because two of the six BFs declared as undetected — $(X.S)$ and $(Y.CO)$ — are in fact detected, respectively, by the sequences (t_1, t_2) and (t_2, t_3). $\hspace{1cm}$ □

The experimental results presented in [Abramovici and Menon 1985], involving simulation of up to 0.5 million BFs, show that

1. Tests with high coverage for SSFs (about 95 percent) detected, on the average, 83 percent of all possible BFs. The actual BF coverage may be higher, because BFs detected by sequences and potentially oscillating BFs detected based on back lines were not accounted for.

2. Although for most circuits, tests with high SSF coverage also achieved good BF coverage, this was not always the case. For example, a test set with 98 percent SSF coverage obtained only 51 percent BF coverage. Thus the BF coverage does not always follow the SSF coverage, and it must be separately determined.

3. On the average, 75 percent of the potentially oscillating BFs were detected based on their front lines. Thus, the pessimistic approximation introduced in the BF coverage by ignoring their detection based on back lines is negligible.

4. In large circuits, the additional time required for the implicit simulation of the feasible BFs is a small fraction of the time needed for SSF simulation (by critical path tracing).

7.5 Test Generation for Bridging Faults

The relations established in Sections 7.2 and 7.3 can also be used to extend a TG algorithm for SSFs so that it will generate tests for BFs as well.

Often when generating a test for a SSF, there are lines whose values are not constrained. We can take advantage of this freedom by setting values that contribute to the detection of BFs. Namely, for every fault x s-a-0 detected by the partially specified test, we can try to find lines $y \in M_x$ such that $(x.y)$ is still undetected and the value of y is unspecified. Then justifying $y=0$ creates the conditions for detecting $(x.y)$.

After the test generation for SSFs is complete, we may still have undetected BFs. Let $(x.y)$ be one of these BFs. If $y \in M_x$, we try to derive a test for x s-a-0, with the added constraint $y=0$. (This, of course, should be attempted only if x s-a-0 has been detected during SSF test generation.) Similarly, if $x \in M_y$, we try to derive a test that detects y s-a-0 and sets $x=0$. These types of operations require only minor changes to a test generation program for SSFs.

When $y \in M_x^*$, the potentially oscillating FBF $(x.y)$ can be detected by trying to generate a test for x s-a-0 while setting $y=0$, with the additional restriction that the effect of x s-a-0 should not propagate to y. However, because most potentially oscillating FBFs can be detected based on their front lines, we can overlook the case $y \in M_x^*$ without any significant loss in the fault coverage.

Example 7.3: Let us continue Example 7.2 by generating a test for $(Q.U)$, which is one of the BFs not detected by the analyzed tests. Since $Q \in M_U$, we try to generate a test for U s-a-0 while setting $Q=0$. The unique solution to this problem is the test $(X,Y,CI) = 101$, which also detects the previously undetected BFs $(Q.V)$ and $(Y.CO)$. □

Although most FBFs can be detected by one vector, some FBFs require a sequence for their detection. Detection of FBFs by sequences has been studied by [Xu and Su 1985].

7.6 Concluding Remarks

The BF model is a "nonclassical" fault model. BFs cannot be "represented" by SSFs, because, in general, BFs are not equivalent to SSFs. Nevertheless, the methods presented in this chapter show that by analyzing the dominance (rather than equivalence) relations between these two types of faults, the processing of the two fault models can be combined, and BFs can be dealt with at a small additional cost over that required for testing SSFs.

REFERENCES

[Abramovici and Menon 1985] M. Abramovici and P. R. Menon, "A Practical Approach to Fault Simulation and Test Generation for Bridging Faults," *IEEE Trans. on Computers*, Vol. C-34, No. 7, pp. 658-663, July, 1985.

[Friedman 1974] A. D. Friedman, "Diagnosis of Short-Circuit Faults in Combinational Logic Circuits," *IEEE Trans. on Computers*, Vol. C-23, No. 7, pp. 746-752, July, 1974.

[Malaiya 1986] Y. K. Malaiya, "A Detailed Examination of Bridging Faults," *Proc. Intn'l. Conf. on Computer Design*, pp. 78-81, 1986.

[Mei 1974] K. C. Y. Mei, "Bridging and Stuck-At Faults," *IEEE Trans. on Computers*, Vol. C-23, No. 7, pp. 720-727 July, 1974.

[Williams and Angell 1973] M. J. Y. Williams and J. B. Angel, "Enhancing Testability of Large-Scale Integrated Circuits via Test Points and Additional Logic," *IEEE Trans. on Computers*, Vol. C-22, No. 1, pp. 46-60, January, 1973.

[Xu and Su 1985] S. Xu and S. Y. H. Su, "Detecting I/O and Internal Feedback Bridging Faults," *IEEE Trans. on Computers*, Vol. C-34, No. 6, pp. 553-557, June, 1985.

PROBLEMS

7.1 For the circuit in Figure 7.13 determine the BFs detected by each of the tests 00, 01, 10, and 11.

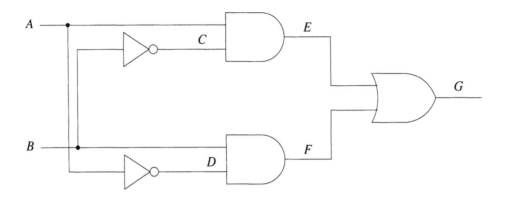

Figure 7.13

7.2   Find a single test that detects the FBF ($a.z$) of Example 7.1.

7.3   For Example 7.2, determine the potentially oscillating BFs that are detected based on their front lines.

7.4   For the circuit in Figure 7.14 show that the AND FBF ($A.Z$) is not detectable by one test. Find a two-test sequence that detects ($A.Z$).

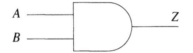

Figure 7.14

8. FUNCTIONAL TESTING

About This Chapter

The test generation methods studied in the previous chapters are based on a structural model of a system under test and their objective is to produce tests for structural faults such as stuck faults or bridging faults. But detailed structural models of complex devices are usually not provided by their manufacturers. And even if such models were available, the structure-based test generation methods are not able to cope with the complexity of VLSI devices. In this chapter we examine *functional testing methods* that are *based on a functional model* of the system. (Functional test generation methods for SSFs using RTL models for components are described in Chapter 6.) We discuss only external testing methods.

8.1 Basic Issues

A functional model reflects the functional specifications of the system and, to a great extent, is independent of its implementation. Therefore functional tests derived from a functional model can be used not only to check whether physical faults are present in the manufactured system, but also as design verification tests for checking that the implementation is free of design errors. Note that tests derived from a structural model reflecting the implementation cannot ascertain whether the desired operation has been correctly implemented.

The objective of functional testing is to *validate the correct operation of a system with respect to its functional specifications*. This can be approached in two different ways. One approach assumes specific functional fault models and tries to generate tests that detect the faults defined by these models. By contrast, the other approach is not concerned with the possible types of faulty behavior and tries to derive tests based only on the specified fault-free behavior. Between these two there is a third approach that defines an implicit fault model which assumes that almost any fault can occur. Functional tests detecting almost any fault are said to be *exhaustive*, as they must completely exercise the fault-free behavior. Because of the length of the resulting tests, exhaustive testing can be applied in practice only to small circuits. By using some knowledge about the structure of the circuit and by slightly narrowing the universe of faults guaranteed to be detected, we can obtain *pseudoexhaustive tests* that can be significantly shorter than the exhaustive ones. The following sections discuss these three approaches to functional testing: (1) functional testing without fault models, (2) exhaustive and pseudoexhaustive testing, and, (3) functional testing using specific fault models.

8.2 Functional Testing Without Fault Models

8.2.1 Heuristic Methods

Heuristic, or ad hoc, functional testing methods simply attempt to exercise the functions of the system. For example, a functional test for a flip-flop may include the following:

1. Validate that the flip-flop can be set (0 to 1 transition) and reset (1 to 0 transition).

2. Validate that the flip-flop can hold its state.

The following example [Chiang and McCaskill 1976] illustrates a heuristic testing procedure for a microprocessor.

Example 8.1: The operation of a microprocessor is usually defined by its architectural block diagram together with its instruction set. Figure 8.1 shows the block diagram of the INTEL 8080. We assume that an external tester is connected to the data, address, and control busses of the 8080. The tester supplies the 8080 with instructions to be executed and checks the results.

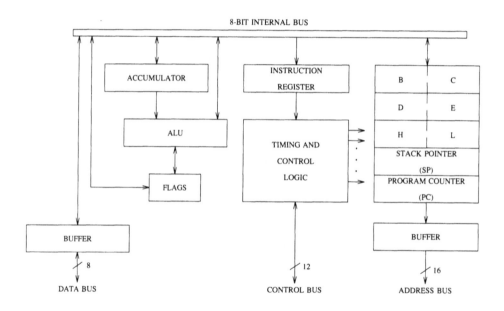

Figure 8.1   Architecture of the INTEL 8080 microprocessor

A typical functional test consists of the following steps:

1. test of the program counter (PC):

   a. The tester resets the 8080; this also clears the PC.

   b. The tester places a NOP (no operation) instruction on the data bus and causes the 8080 to execute it repeatedly. The repeated execution of NOP causes the PC to be incremented through all its 2^{16} states. The content of the PC is available for checking on the address bus.

2. test of registers H and L:

   a. The tester writes 8-bit patterns into H and L using MVI (move-immediate) instructions.

      b.   Executing an PCHL instruction transfers the contents of H and L into the PC. Here we rely on having checked the PC in step 1.

This sequence is repeated for all the possible 256 8-bit patterns.

   3.   test of registers B,C,D, and E:

In a similar fashion, the tester writes an 8-bit pattern into a register $R \in \{B,C,D,E\}$. Then R is transferred into the PC via H or L (R cannot be directly transferred into the PC). Here we take advantage of the testing of the PC and HL done in steps 1 and 2. These tests are executed for all the 256 8-bit patterns.

   4.   test of the stack pointer (SP):

The SP is incremented and decremented through all its states and accessed via the PC.

   5.   test of the accumulator:

All possible patterns are written into the accumulator and read out. These can be done directly or via previously tested registers.

   6.   test of the ALU and flags:

This test exercises all arithmetic and logical instructions. The operands are supplied directly or via already tested registers. The results are similarly accessed. The flags are checked by conditional jump instructions whose effect (the address of the next instruction) is observed via the PC.

   7.   test of all previously untested instructions and control lines.   ☐

An important issue in the functional testing of a microprocessor is whether its instruction set is *orthogonal*. An orthogonal instruction set allows every operation that can be used in different addressing modes to be executed in every possible addressing mode. This feature implies that the mechanisms of op-code decoding and of address computation are independent. If the instruction set is not orthogonal, then every operation must be tested for all its addressing modes. This testing is not necessary for an orthogonal instruction set, and then the test sequences can be significantly shorter.

The Start-Small Approach

Example 8.1 illustrates the *start-small* (or *bootstrap*) approach to functional testing of complex systems, in which the testing done at a certain step uses components and/or instructions tested in previous steps. In this way the tested part of the system is gradually extended. The applicability of the start-small approach depends on the degree of overlap among the components affected by different instructions. Its objective is to simplify the fault location process. Ideally, if the first error occurs at step i in the test sequence, this should indicate a fault in the new component(s) and/or operation(s) tested in step i.

A technique for ordering the instructions of a microprocessor according to the start-small principle is presented in [Annaratone and Sami 1982]. The *cardinality* of an instruction is defined as the number of registers accessed during the execute phase of an instruction (i.e., after the fetch-and-decode phase). Thus NOP, which has only the fetch-and-decode phase, has cardinality 0. Instructions are also graded according to their *observability*; the observability of an instruction shows the extent to which the results of the register operations performed by the instruction are directly observable at the primary outputs of the microprocessor. Instructions are tested in increasing order of their cardinality. In this way the instructions affecting fewer registers are tested first (a classical "greedy" algorithm approach). Among instructions of the same cardinality, priority is given to those with higher observability.

The Coverage Problem

The major problem with heuristic functional testing is that the quality of the obtained functional tests is unknown. Indeed, without a fault model it is extremely difficult to develop a rigorous quality measure.

Consequently an important question is whether a heuristically derived functional test does a good job of detecting physical faults. For cases in which a low-level structural model reflecting the implementation of the system is available, experience shows that the fault coverage of a typical heuristic functional test is likely to be in the 50 to 70 percent range. Hence, in general, such a test does not achieve a satisfactory fault coverage, but it may provide a good basis that can be enhanced to obtain a higher-quality test. However, low-level structural models are usually not available for complex systems built with off-the-shelf components.

Heuristic measures can be used to estimate the "completeness" of a test with respect to the control flow of the system. These measures are based on monitoring the activation of operations in an RTL model [Noon 1977, Monachino 1982]. For example, if the model contains a statement such as

$$\textbf{if } x \textbf{ then } operation_1 \textbf{ else } operation_2$$

the technique determines if the applied test cases make the condition x both true and false during simulation. A "complete" test is required to exercise all possible exits from decision blocks. One measure is the ratio between the number of exits taken during simulation and the number of all possible exits. A second, more complicated measure traces decision paths, i.e., combinations of consecutive decisions. For example, if the above statement is followed by

$$\textbf{if } y \textbf{ then } operation_3 \textbf{ else } operation_4$$

then there are four possible decision paths of length 2 (see Figure 8.2), corresponding to the four combinations of values of x and y. The second measure is the ratio between the number of decision paths of length k traversed during simulation and the total number of such paths. Note that data operations are not addressed by these measures.

An important aspect of functional testing, often overlooked by heuristic methods, is that in addition to verifying the specified operations of a system, it is also necessary to check that unintended operations do not occur. For example, in addition to a correct transfer of data into register R1, the presence of a fault may cause the same data to be written into register R2. In such a case, checking only the desired operation — as is usually done by heuristic methods — is clearly not enough.

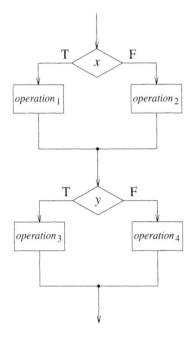

Figure 8.2   Illustration for decision paths tracing

8.2.2 Functional Testing with Binary Decision Diagrams

In Chapter 2 we have introduced *binary decision diagrams* as a functional modeling tool. In this section we describe a functional testing technique based on binary decision diagrams [Akers 1978].

First we recall the procedure used to determine the value of a function f, given its binary decision diagram and values of its inputs. We enter the diagram on the branch labeled f. At an internal node whose label is i we take the left or the right branch depending on whether the value of the variable i is 0 or 1. The *exit value* of the path followed during the traversal is the value, or the value of the variable, encountered at the end of the path. The *inversion parity* of the path is the number of inverting dots encountered along the path, taken modulo 2. For a traversal along a path with exit value v and inversion parity p, the value of the function f is $v \oplus p$. For example, consider the *JK* F/F and its diagram shown in Figure 8.3 (q represents the state of the F/F). The value of y along the path determined by $S=0$, $R=0$, $C=1$, and $q=1$ is $K \oplus 1 = \bar{K}$.

A traversal of a binary decision diagram implies a certain setting of the variables encountered along the path. Such a traversal is said to define a *mode of operation* of the device. A path whose exit value is x denotes an illegal mode of operation, in which the output value cannot be predicted. For the *JK* F/F of Figure 8.3, setting $S=1$ and $R=1$ is illegal. Figure 8.4 lists the five legal modes of operation of the *JK* F/F corresponding to the five paths with non-x exit values.

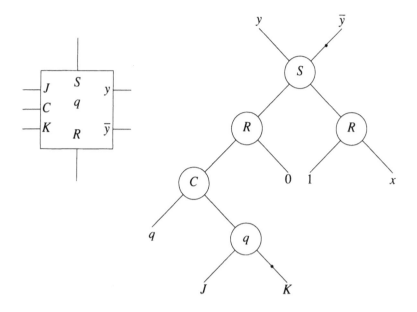

Figure 8.3   Binary decision diagram for a *JK* F/F

S	R	C	q	J	K	y
0	1	*x*	*x*	*x*	*x*	0
1	0	*x*	*x*	*x*	*x*	1
0	0	0	*q*	*x*	*x*	*q*
0	0	1	0	*J*	*x*	*J*
0	0	1	1	*x*	*K*	$\overline{K}$

Figure 8.4   Testing experiments for a *JK* F/F

Every (legal) mode of operation can be viewed as defining a testing *experiment* that partitions the variables of a function into three disjoint sets:

- *fixed variables*, whose binary values determine the path associated with the mode of operation;

- *sensitive variables*, which directly determine the output value;

- *unspecified variables*, which do not affect the output (their values are denoted by *x*).

An experiment provides a *partial specification* of the function corresponding to a particular mode of operation. In Figure 8.3, *y* is a function *y(S,R,C,q,J,K)*. The partial specification of *y* given by the third experiment in Figure 8.4 is *y(0,0,0,q,x,x)=q*.

One can show that the set of experiments derived by traversing all the paths corresponding to an output function provides a *complete specification* of the function. That is, every possible combination of variables is covered by one (and only one) experiment. In addition, the experiments are *disjoint*; i.e., every pair of experiments differ in the value of at least one fixed variable.

Some binary decision diagrams may contain adjacent nodes, such as A and B in Figure 8.5, where both branches from A reconverge at B. Consider two traversals involving node A, one with $A=0$ and the other with $A=1$. Clearly the results of these two traversals are complementary. Then we can combine the two traversals into one and treat A as a sensitive variable of the resulting experiment. Let v be the result of the traversal with $A=0$. The result of the combined traversal is $v \oplus A$.

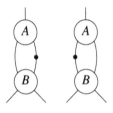

Figure 8.5

Example 8.2: For the diagram of Figure 8.6, let us derive an experiment by following the path determined by $A=0$, $D=1$, and $C=1$. The exit variable of the path is G, and its other sensitive variables are B and F. The result of the traversal with $B=0$ and $F=0$ is $G \oplus 1 = \overline{G}$. The result of the combined traversal is $\overline{G} \oplus B \oplus F$. The partial specification of the function $f(A,B,C,D,E,F,G)$ provided by this experiment is $f(0,B,1,1,x,F,G)=\overline{G} \oplus B \oplus F$.   □

It is possible that different nodes in a binary decision diagram have the same variable. Hence when we traverse such a diagram we may encounter the same variable more than once. For example, in the diagram of Figure 8.6, when following the path determined by $A=1$, $C=0$, and $D=1$, we reach again a node whose variable is C. Because C has already been set to 0, here we must continue on the left branch. Otherwise, if we continue on the right branch, the resulting experiment would not be *feasible*, because it would specify contradictory values for C. Therefore the traversal process should ensure that only feasible experiments are generated.

An experiment derived by traversing a binary decision diagram is not in itself a test, but only a partial functional specification for a certain mode of operation. Experiments can be used in different ways to generate tests. Since the output value will change with any change of a sensitive variable of an experiment, a useful procedure is to generate all the combinations of sensitive variables for every mode of operation. If the sensitive variables are inputs of the function, this strategy tends to create input-output paths along which many internal faults are also likely to be detected.

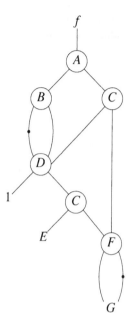

Figure 8.6

Some variables of a function may not appear as sensitive in a set of experiments. This is the case for S and R in Figure 8.4. Now we will illustrate a procedure that generates a test in which an output is made sensitive to a variable s. The principle is to combine two experiments, e_1 and e_2, in which s is the only fixed variable with opposite values, and the output values are (or can be made) complementary. In other words, we need two paths that diverge at a node whose variable is s and lead to complementary exit values. The combined experiment has all the fixed values of e_1 and e_2 (except s). Assume that, using the diagram of Figure 8.3, we want to make y sensitive to the value of S. When $S=1$, the only legal experiment requires $R=0$ and produces $y=1$. To make y sensitive to S, we look for a path with $S=0$ leading to $y=0$. (Note that we cannot set $R=1$, since R has been set to 0.) One such path requires $C=0$ and $q=0$. This shows that to make the output sensitive to S we should first reset the F/F.

As an example of functional testing using binary decision diagrams, Akers [1978] has derived the complete set of experiments for a commercial 4-bit ALU with 14 PIs and 8 POs.

Testing of circuits composed of modules described by binary decision diagrams is discussed in [Abadir and Reghbati 1984]. The examples include a commercial 1-bit microprocessor slice. Chang *et al.* [1986] assume a functional fault model in which a fault can alter the path defined by an experiment. Their procedure generates tests with high fault coverage for SSFs. They also show that the test set can be reduced by taking into account a limited amount of structural information.

The main advantage of a binary decision diagram is that it provides a complete and succinct functional model of a device, from which a complete set of experiments — corresponding to every mode of operation of the device — can be easily derived.

8.3 Exhaustive and Pseudoexhaustive Testing

The Universal Fault Model

Exhaustive tests detect all the faults defined by the *universal fault model*. This implicit fault model assumes that any (permanent) fault is possible, except those that increase the number of states in a circuit. For a combinational circuit N realizing the function $Z(x)$, the universal fault model accounts for any fault f that changes the function to $Z_f(x)$. The only faults not included in this model are those that transform N into a sequential circuit; bridging faults introducing feedback and stuck-open faults in CMOS circuits belong to this category. For a sequential circuit, the universal fault model accounts for any fault that changes the state table without creating new states. Of course, the fault universe defined by this model is not enumerable in practice.

8.3.1 Combinational Circuits

To test all the faults defined by the universal fault model in a combinational circuit with n PIs, we need to apply all 2^n possible input vectors. The exponential growth of the required number of vectors limits the practical applicability of this *exhaustive testing* method only to circuits with less than 20 PIs. In this section we present *pseudoexhaustive testing* methods that can test almost all the faults defined by the universal fault model with significantly less than 2^n vectors.

8.3.1.1 Partial-Dependence Circuits

Let $O_1, O_2, ..., O_m$ be the POs of a circuit with n PIs, and let n_i be the number of PIs feeding O_i. A circuit in which no PO depends on all the PIs (i.e., $n_i < n$ for all i), is said to be a *partial-dependence circuit*. For such a circuit, pseudoexhaustive testing consists in applying all 2^{n_i} combinations to the n_i inputs feeding every PO O_i [McCluskey 1984].

Example 8.3: The circuit of Figure 8.7(a) has three PIs, but every PO depends only on two PIs. While exhaustive testing requires eight vectors, pseudoexhaustive testing can be done with the four vectors shown in Figure 8.7(b).   □

Since every PO is exhaustively tested, the only faults that a pseudoexhaustive test *may* miss are those that make a PO dependent on additional PIs. For the circuit of Figure 8.7(a), the bridging fault $(a.c)$ is not detected by the test set shown. With the possible exception of faults of this type, all the other faults defined by the universal fault model are detected.

Because of the large number of test vectors required in practical circuits, pseudoexhaustive testing is not used in the context of stored-pattern testing; its main use is in circuits employing built-in self-test, where the vectors are generated by hardware embedded in the circuit under test. Pseudoexhaustive testing is discussed in detail in Chapter 11.

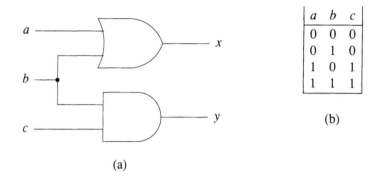

(a)

Figure 8.7   (a) Partial dependence circuit (b) Pseudoexhaustive test set

8.3.1.2 Partitioning Techniques

The pseudoexhaustive testing techniques described in the previous section are not applicable to *total-dependence circuits*, in which at least one PO depends on all PIs. Even for a partial-dependence circuit, the size of a pseudoexhaustive test set may still be too large to be acceptable in practice. In such cases, pseudoexhaustive testing can be achieved by *partitioning techniques* [McCluskey and Bozorgui-Nesbat 1981]. The principle is to partition the circuit into *segments* such that the number of inputs of every segment is significantly smaller than the number of PIs of the circuit. Then the segments are exhaustively tested.

The main problem with this technique is that, in general, the inputs of a segment are not PIs and its outputs are not POs. Then we need a means to control the segment inputs from the PIs and to observe its outputs at the POs. One way to achieve this, referred to as *sensitized partitioning*, is based on sensitizing paths from PIs to the segment inputs and from the segment outputs to POs, as illustrated in the following example.

Example 8.4: Consider the circuit in Figure 8.8(a). We partition it into four segments. The first segment consists of the subcircuit whose output is h. The other three segments consist, respectively, of the gates g, x, and y. Figure 8.8(b) shows the eight vectors required to test exhaustively the segment h and to observe h at the PO y. Since $h=1$ is the condition needed to observe g at the PO x, we can take advantage of the vectors 5 through 8 in which $h=1$ to test exhaustively the segment g (see Figure 8.8(c)). We also add vectors 9 and 10 to complete the exhaustive test of the segment y. Analyzing the tests applied so far to the segment x, we can observe that the missing combinations are those in which $h=0$; these can be applied by using vectors 4 and 9.

Figure 8.8(d) shows the resulting test set of 10 vectors. This compares favorably with the $2^6=64$ vectors required for exhaustive testing or with the $2^5=32$ vectors needed for pseudoexhaustive testing without partitioning.   □

In an example presented in [McCluskey and Bozorgui-Nesbat 1981], sensitized partitioning is applied to a commercial 4-bit ALU. This is a total-dependence circuit

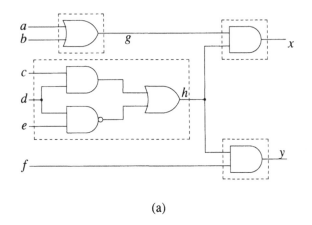

(a)

Figure 8.8

with 14 PIs, that would require 2^{14} vectors for exhaustive testing. Its pseudoexhaustive test set based on sensitized partitioning has only 356 vectors.

A pseudoexhaustive test set based on sensitized partitioning detects any fault that changes the truth table of a segment. Since a circuit may have several possible partitions, this fault model depends, to a certain extent, on the chosen set of segments. Partitioning a circuit so that the size of its pseudoexhaustive test set is minimal is an *NP*-complete problem. A partitioning algorithm based on simulated annealing is described in [Shperling and McCluskey 1987].

Note that partitioning techniques assume knowledge of the internal structural model.

8.3.2 Sequential Circuits

For a sequential circuit, the universal fault model accounts for any fault that modifies the state table of the circuit without increasing its number of states. An input sequence that detects every fault defined by this model must distinguish a given *n*-state sequential machine from all other machines with the same inputs and outputs and at most *n* states. The existence of such a *checking sequence* is guaranteed by the following theorem [Moore 1956].

Theorem 8.1: For any reduced, strongly connected*, *n*-state sequential machine *M*, there exists an input-output sequence pair that can be generated by *M*, but cannot be generated by any other sequential machine *M'* with *n* or fewer states.   □

---

* A reduced machine does not have any equivalent states. In a strongly connected machine any state can be reached from any other state.

	a	b	c	d	e	f	g	h	x	y
1			0	0	0	1		1		1
2			0	0	1	1		1		1
3			0	1	0	1		1		1
4			0	1	1	1		0		0
5			1	0	0	1		1		1
6			1	0	1	1		1		1
7			1	1	0	1		1		1
8			1	1	1	1		1		1

(b)

	a	b	c	d	e	f	g	h	x	y
1			0	0	0	1		1		1
2			0	0	1	1		1		1
3			0	1	0	1		1		1
4			0	1	1	1		0		0
5	0	0	1	0	0	1	0	1	0	1
6	0	1	1	0	1	1	1	1	1	1
7	1	0	1	1	0	1	1	1	1	1
8	1	1	1	1	1	1	1	1	1	1
9			0	1	1	0		0		0
10			0	0	0	0		1		0

(c)

	a	b	c	d	e	f	g	h	x	y
1			0	0	0	1		1		1
2			0	0	1	1		1		1
3			0	1	0	1		1		1
4	0	0	0	1	1	1	0	0	0	0
5	0	0	1	0	0	1	0	1	0	1
6	0	1	1	0	1	1	1	1	1	1
7	1	0	1	1	0	1	1	1	1	1
8	1	1	1	1	1	1	1	1	1	1
9	0	1	0	1	1	0	1	0	0	0
10			0	0	0	0		1		0

(d)

Figure 8.8   (Continued)

Since the generation of a checking sequence is based on the state table of the circuit, this exhaustive testing approach is applicable only to small circuits. We will consider checking sequences only briefly; a comprehensive treatment of this topic can be found in [Friedman and Menon 1971].

A checking sequence for a sequential machine M consists of the following three phases:

1.   initialization, i.e., bringing M to a known starting state;

2.   verifying that M has n states;

3.   verifying every entry in the state table of M.

Initialization is usually done by a *synchronizing sequence*, which, independent of the initial state of M, brings M in a unique state.

The derivation of a checking sequence is greatly simplified if M has a *distinguishing sequence*. Let Z_i be the output sequence generated by M, starting in state q_i, in response to an input sequence X_D. If Z_i is unique for every $i=1,2...,n$, then X_D is a distinguishing sequence. (A machine M that does not have a distinguishing sequence can be easily modified to have one by adding one output.) The importance of X_D is that by observing the response of M to X_D we can determine the state M was in when X_D was applied.

To verify that M has n distinct states, we need an input/output sequence that contains n subsequences of the form X_D/Z_i for $i=1,2,...,n$. We can then use X_D to check various entries in the state table. A transition $N(q_i,x)=q_j$, $Z(q_i,x)=z$, is verified by having two input/output subsequences of the following form:

$$X_D X' X_D / Z_p Z' Z_i \quad \text{and} \quad X_D X' x X_D / Z_p Z' z Z_j$$

The first one shows that $X_D X'$ takes M from state q_p to state q_i. Based on this, we can conclude that when input x is applied in the second subsequence, M is in state q_i. The last X_D verifies that x brings M to state q_j.

The sequences that verify that M has n distinct states and check every entry in the state table can often be overlapped to reduce the length of the checking sequence.

8.3.3 Iterative Logic Arrays

Pseudoexhaustive testing techniques based on partitioning are perfectly suited for circuits structured as *iterative logic arrays* (ILAs), composed from identical *cells* interconnected in a regular pattern as shown in Figure 8.9. The partitioning problem is naturally solved by using every cell as a segment that is exhaustively tested. We consider *one-dimensional* and *unilateral* ILAs, that is, ILAs where the connections between cells go in only one direction. First we assume that cells are combinational. A ripple-carry adder is a typical example of such an ILA.

Some ILAs have the useful property that they can be pseudoexhaustively tested with a number of tests that does not depend on the number of cells in the ILA. These ILAs are said to be *C-testable* [Friedman 1973]. The following example shows that the ripple-carry adder is C-testable.

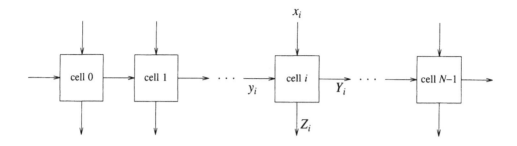

Figure 8.9 Iterative Logic Array

Example 8.5: Consider the truth table of a cell of the ripple-carry adder (Figure 8.10). In this truth table, we separate between the values of the cell inputs that are PIs (X and Y) and those that are provided by the previous cell (CI). An entry in the table gives the values of CO and S. We can observe that $CO=CI$ in six entries. Thus if we apply the X,Y values corresponding to one of these entries to every cell, then every cell will also receive the same CI value. Hence these six tests can be applied concurrently to all cells. In the remaining two entries, $CO=\overline{CI}$. The tests corresponding to these two entries can be applied to alternating cells, as shown in Figure 8.11. Hence a ripple-carry adder of arbitrary size can be pseudoexhaustively tested with only eight tests; therefore, it is a C-testable ILA.   □

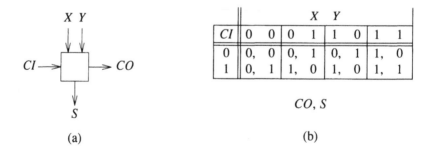

Figure 8.10 A cell of a ripple-carry adder

Let us denote an input vector of an ILA with N cells by $(y_0\ x_0\ x_1\ \cdots\ x_{N-1})$, where y_0 is the y input applied to the first cell, and x_i ($i=0,1,...,N-1$) is the x input applied to cell i. Let us denote the corresponding output vector by $(Z_0\ Z_1\ \cdots\ Z_{N-1})$, where Z_i is the response at the primary outputs of cell i. Consider a sequential circuit constructed from one cell of the array by connecting the Y outputs back to the y inputs (Figure 8.12). If this circuit starts in state y_0, its response to the input sequence $(x_0,x_1,...,x_{N-1})$ is the sequence $(Z_0,Z_1,...,Z_{N-1})$. Hence we can use this sequential circuit to represent the ILA. (This is the inverse of the technique described in

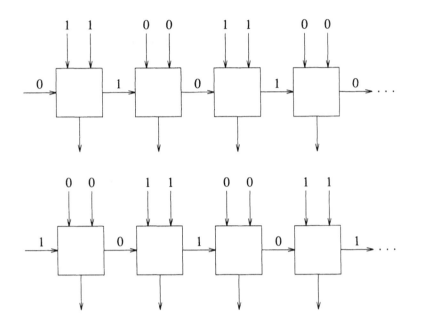

Figure 8.11   Tests with $CO=\overline{CI}$ for a ripple-carry adder

Chapter 6, which uses an ILA to model a sequential circuit.) Verifying the truth table of a cell in the ILA corresponds to verifying every entry in the state table of the equivalent sequential circuit. The truth table of the cell of the ripple-carry adder, shown in Figure 8.10(b), can be interpreted as the state table of the sequential circuit modeling the ripple-carry adder.

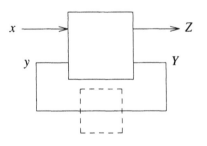

Figure 8.12

Let us consider the checking of an entry (y,x) of the truth table of an ILA cell. To have the total number of tests independent of the number of cells in the ILA, the same input combination (y,x) must be simultaneously applied to cells spaced at regular intervals — say, k — along the array. (Example 8.5 has $k=1$ and $k=2$). For the sequential circuit modeling the ILA, this means that the k-step sequence starting with x, which is applied in state y, must bring the circuit back to state y. Then one test applies the input combination (y,x) to every k-th cell, and k tests are sufficient to apply it to every cell in the ILA. The following result [Friedman 1973, Sridhar and Hayes 1979] characterizes a C-testable ILA.

Theorem 8.2: An ILA is C-testable iff its equivalent sequential circuit has a reduced state table, and for every transition from any state y there exists a sequence that brings the circuit back to state y. ☐

The state diagram of a sequential circuit modeling a C-testable ILA is either a strongly connected graph, or it is composed of disjoint strongly connected graphs.

Now we discuss the generation of tests to check an entry (y,x) in the state table representing a C-testable ILA. To verify the next state $N(y,x)=p$, we use a *set of pairwise distinguishing sequences* [Sridhar and Hayes 1979]. A *pairwise distinguishing sequence* for a pair of states p and q, $DS(p,q)$, is an input sequence for which the output sequences produced by the circuit starting in states p and q are different. (Because the state table is assumed to be reduced, $DS(p,q)$ exists for every pair of states p and q). For example, for the state table of Figure 8.13, $DS(y_0,y_1)=1$ and $DS(y_1,y_2)=01$. A set of pairwise distinguishing sequences for a state p, $SDS(p)$, is a set of sequences such that for every state $q \neq p$, there exists a $DS(p,q) \in SDS(p)$. In other words, $SDS(p)$ contains sequences that distinguish between p and any other state. For the state table of Figure 8.13, $SDS(y_1)=\{1,01\}$, where 1 serves both as $DS(y_1,y_0)$ and as $DS(y_1,y_3)$.

	x			
	0		1	
y_0	y_1,	0	y_3,	1
y_1	y_1,	0	y_2,	0
y_2	y_0,	0	y_2,	0
y_3	y_0,	0	y_3,	1

Figure 8.13

The following procedure checks the entry (y,x). (Assume that the sequential circuit starts in state y.)

1. Apply x. This brings the circuit to state p.

2. Apply D_i, one of the sequences in $SDS(p)$. Assume D_i brings the circuit to state r_i.

3. Apply $T(r_i,y)$, a transfer sequence that brings the circuit from state r_i to state y (such a sequence always exists for a C-testable ILA).

4. Reapply the sequence $I_i=(x,D_i,T(r_i,y))$ as many times as necessary to have an input applied to every cell in the ILA. This step completes one test vector t_i.

Let k_i be the length of I_i. The test t_i applies (y,x) to every k_i-th cell of the ILA.

5. Apply k_i-1 shifted versions of t_i to the array.

6. Repeat steps 1 through 5 for every other sequence D_i in $SDS(p)$.

Example 8.6:   We illustrate the above procedure by deriving tests to check the entry $(y_0,0)$ of the state table of Figure 8.13 modeling a C-testable ILA. Let $p=N(y_0,0)=y_1$ and $SDS(y_1)=\{1,01\}$. The first group of tests use $D_1=1$, $r_1=y_2$, and $T(y_2,y_0)=0$. Hence $I_1=(010)$ and the first test is $t_1=y_0$ $(010)^*$, where the notation $(S)^*$ denotes $SSS...$ (y_0 is applied to the first cell). Since $k_1=3$, we apply two shifted versions of t_1 to the ILA; we should also make sure that y_0 is applied to the first cell tested in each test. The two new tests are $y_20(010)^*$ and $y_210(010)^*$.

The second group of tests use $D_2=01$. Similarly we obtain $t_2=y_0(0010)^*$. Since $k_2=4$, we need three shifted versions of t_2 to complete this step. The seven obtained tests check the entry $(y_0,0)$ in all cells of the ILA, independent of the number of cells in the ILA.   □

The test set for the entire ILA merges the tests for all the entries in the truth table of a cell.

Extensions and Applications

The ILAs we have considered so far are unilateral, one-dimensional, and composed of combinational cells with directly observable outputs. The fault model we have used assumes that the faulty cell remains a combinational circuit. In the following we will briefly mention some of the work based on more general assumptions.

The problem of testing *ILAs whose cells do not have directly observable outputs* (i.e., only the outputs of the last cell are POs) is studied in [Friedman 1973, Parthasarathy and Reddy 1981].

The testing of *two-dimensional ILAs*, whose structure is shown in Figure 8.14, is discussed in [Kautz 1967, Menon and Friedman 1971, Elhuni *et al.* 1986]. Shen and Ferguson [1984] analyze the design of VLSI multipliers constructed as two-dimensional arrays and show how these designs can be modified to become C-testable.

Cheng and Patel [1985] develop a pseudoexhaustive test set for a ripple-carry adder, assuming an *extended fault model* that also includes faults that transform a faulty cell into a sequential circuit with two states.

Testing of *bilateral ILAs*, where interconnections between cells go in both directions, is examined in [Sridhar and Hayes 1981b].

Sridhar and Hayes [1981a] have shown how the methods for testing combinational ILAs can be extended to *ILAs composed of sequential cells* having the structure given in Figure 8.15.

Sridhar and Hayes [1979, 1981a, 1981b] have studied the testing of *bit-sliced systems*, whose general structure is shown in Figure 8.16. Such a system performs operations on n-bit words by concatenating operations on k-bit words done by $N=n/k$ identical cells called *slices*. The operations are determined by s control lines distributed to

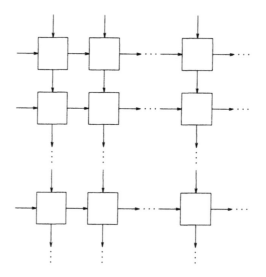

Figure 8.14   A two-dimensional ILA

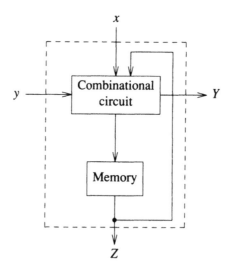

Figure 8.15

every slice. In general, a slice is a sequential device and the connections between adjacent slices are bilateral. This structure is typical for bit-sliced (micro)processors, where k is usually 2, 4, or 8. To model a bit-sliced system as an ILA, we can view it as composed of 2^s ILAs, each having a different basic cell implementing one of the 2^s possible operations.

control

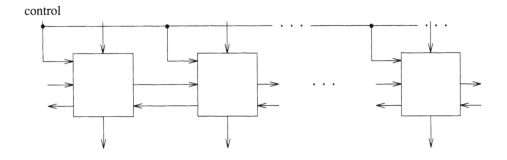

Figure 8.16   Bit-sliced processor

In practical bit-sliced systems, a slice is usually defined as an interconnection of small combinational and sequential modules. Figure 8.17 shows the 1-bit processor slice considered in [Sridhar and Hayes 1979, 1981a], which is similar to commercially available components. For this type of structure we apply a *hierarchical pseudoexhaustive testing approach*. First, we derive a pseudoexhaustive test T for a slice, such that T exhaustively tests every module. Because modules within a slice are small, it is feasible to test them exhaustively. The test T obtained for the slice of Figure 8.17 has 114 vectors. Second, we derive a pseudoexhaustive test T_A for the entire array, such that T_A applies T to every slice. A bit-sliced processor based on the slice of Figure 8.17 is C-testable and can be pseudoexhaustively tested with 114 vectors independent of its number of slices.

8.4 Functional Testing with Specific Fault Models
8.4.1 Functional Fault Models

Functional faults attempt to represent the effect of physical faults on the operation of a functionally modeled system. The set of functional faults should be *realistic*, in the sense that the faulty behavior induced by them should generally match the faulty behavior induced by physical faults. A functional fault model can be considered *good* if tests generated to detect the faults it defines also provide a high coverage for the SSFs in the detailed structural model of the system. (Because we do not know the comprehensiveness of a functional fault model, we cannot use the functional fault coverage of a test as a meaningful test quality measure.)

A functional fault model can be explicit or implicit. An *explicit model* identifies each fault individually, and every fault may become a target for test generation. To be useful, an explicit functional fault model should *define a reasonably small fault universe*, so that the test generation process will be computationally feasible. In contrast, an *implicit model* identifies classes of faults with "similar" properties, so that all faults in the same class can be detected by similar procedures. The advantage of an implicit fault model is that it does not require explicit enumeration of faults within a class.

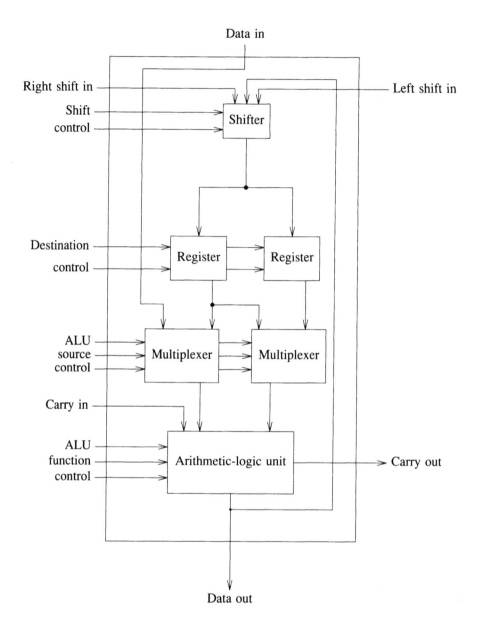

Figure 8.17   1-bit processor slice

Addressing Faults

Many operations in a digital system are based on decoding the address of a desired item. Typical examples include

- addressing a word in a memory;

- selecting a register according to a field in the instruction word of a processor;

- decoding an op-code to determine the instruction to be executed.

The common characteristic of these schemes is the use of an n-bit address to select one of 2^n possible items. *Functional addressing faults* represent the effects of physical faults in the hardware implementing the selection mechanism on the operation of the system. Whenever item i is to be selected, the presence of an addressing fault may lead to

- selecting no item,

- selecting item j instead of i,

- selecting item j in addition to i.

More generally, a set of items $\{j_1, j_2, ..., j_k\}$ may be selected instead of, or in addition to, i.

An important feature of this fault model is that it forces the test generation process to check that the intended function is performed and also that no extraneous operations occur. This fundamental aspect of functional testing is often overlooked by heuristic methods.

Addressing faults used in developing implicit fault models for microprocessors are examined in the following section.

8.4.2 Fault Models for Microprocessors

In this section we introduce functional fault models for microprocessors. Test generation procedures using these fault models are presented in the next section.

Graph Model for Microprocessors

For functional test generation, a microprocessor can be modeled by a graph based on its architecture and instruction set [Thatte and Abraham 1980, Brahme and Abraham 1984]. Every user-accessible register is represented by a node in the graph. Two additional nodes, labeled IN and OUT, denote the connections between the microprocessor and the external world; typically, these are the data, address, and control busses connecting the microprocessor to memory and I/O devices. When the microprocessor is under test, the tester drives the IN node and observes the OUT node. A directed edge from node A to node B shows that there exists an instruction whose execution involves a transfer of information from node A to node B.

Example 8.7: Let us consider a hypothetical microprocessor with the following registers:

- A — accumulator,

- PC — program counter,

- SP — stack pointer holding the address of the top of a last-in first-out data stack,

- R1 — general-purpose register,

- R2 — scratch-pad register,

- SR — subroutine register to save return addresses (assume that no subroutine nesting is allowed),

- IX — index register.

Figure 8.18 shows the graph model for this microprocessor. The table of Figure 8.19 illustrates the mapping between some of the instructions of the microprocessor and the edges of the graph. (The notation (R) denotes the contents of the memory location addressed by register R.) Note that an edge may correspond to several instructions and an instruction may create several edges.   □

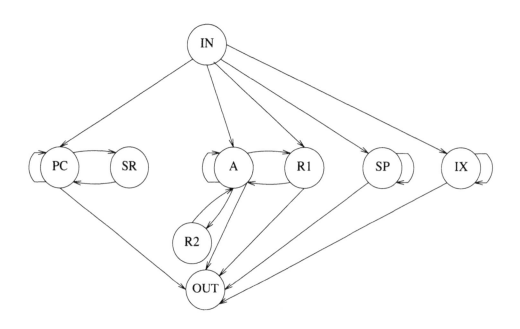

Figure 8.18   Graph model for a microprocessor

Fault Classes

Faults affecting the operation of a microprocessor can be divided into the following classes [Thatte and Abraham 1980]:

1. addressing faults affecting the register-decoding function;

2. addressing faults affecting the instruction-decoding and instruction-sequencing function;

3. faults in the data-storage function;

4. faults in the data-transfer function;

5. faults in the data-manipulation function.

Instruction	Operation	Edge(s)
MVI R, *a* where R∈{A,R1,SP,IX}	R←*a*	IN→R
MOV Ra,Rb where Ra,Rb∈{A,R1,R2}	Ra←Rb	Ra→Rb
ADD A,R1	A←A+R1	A→A R1→A
JMP *a*	PC←*a*	IN→PC PC→OUT
ADD A,(IX)	A←A+(IX)	IX→OUT IN→A A→A
CALL *a*	SR←PC PC←*a*	PC→SR IN→PC PC→OUT
RET	PC←SR	SR→PC PC→OUT
PUSH R where R∈{A,R1}	SP←R SP←SP+1	SP→OUT R→OUT SP→SP
POP R where R∈{A,R1}	SP←SP-1 R←(SP)	SP→SP SP→OUT IN→R
INCR R where R∈{A,SP,IX}	R←R+1	R→R
MOV (IX), R where R∈{A,R1}	(IX)←R	IX→OUT R→OUT

Figure 8.19 Instruction set

The overall fault model for the microprocessor allows for any number of faults in only one of these classes.

8.4.2.1 Fault Model for the Register-Decoding Function

Let us denote the decoding (selection) of a register R by a function $f_D(R)$ whose fault-free value is R for any register R of the microprocessor. Whenever an instruction accesses a register R, an addressing fault affecting the register-decoding function leads to one of the following [Thatte and Abraham 1980]:

1. No register is accessed.

2. A set of registers (that may or may not include R) is accessed.

The first case is represented by $f_D(R)=\emptyset$, where $\emptyset$ denotes a nonexistent register. Here an instruction trying to write R will not change its contents, and an instruction trying to read R will retrieve a ONE or ZERO vector (depending on technology), independent of the contents of R; a ONE (ZERO) vector has all its bits set to 1(0).

In the second case, $f_D(R)$ represents the set of registers erroneously accessed. Here an instruction trying to write data d in R will write d in every register in $f_D(R)$, and an instruction trying to read R will retrieve the bit-wise AND or OR (depending on technology) of the contents of all registers in $f_D(R)$.

Example 8.8: For the microprocessor of Example 8.7, consider the fault $f_D(SR)=\emptyset$. The CALL a instruction will correctly jump to a, but it will not save the return address in SR. This fault will be detected by executing RET, since at that time a ZERO or ONE vector will be loaded into the PC.

With the fault $f_D(R2)=R1$, the instruction MOV R2,R1 will behave like NOP, and the instruction MOV A,R2 will result in transferring R1 into A. In the presence of the fault $f_D(R2) = \{R1,R2\}$, the instruction MOV R2,R1 executes correctly, but MOV A,R2 will transfer R1*R2 into A, where "*" denotes the bit-wise AND or OR operator.

$\square$

Note that a fault that causes $f_D(R_i)=R_j$ and $f_D(R_j)=R_i$ does not affect the correct operation (since it only relabels R_i and R_j); such a fault is undetectable.

8.4.2.2 Fault Model for the Instruction-Decoding and Instruction-Sequencing Function

Microprogramming Model for Instruction Execution

An instruction can be viewed as a *sequence of microinstructions*, where *every microinstruction consists of a set of microorders* which are executed in parallel. Microorders represent the elementary data-transfer and data-manipulation operations; i.e., they constitute the basic building blocks of the instruction set. This *microprogramming model* [Parthasarathy *et al.* 1982, Annaratone and Sami 1982, Brahme and Abraham 1984] is an abstract model, applicable regardless of whether the microprocessor is actually microprogrammed.

For example, the instruction ADD A,R1 can be viewed as a sequence of the following three microinstructions: (1) two (parallel) microorders to bring the contents of A and R1 to the ALU inputs, (2) an ADD microorder, and (3) a microorder to load the ALU result into A. The same instruction can also be considered as one microorder [Brahme and Abraham 1984]. We emphasize that the process of test generation (to be described in the next section) does not require the knowledge of the structure of instructions in terms of microinstructions and microorders.

Fault Model

The microprogramming model for the instruction execution allows us to define a comprehensive fault model for the instruction-decoding and instruction-sequencing function. Namely, addressing faults affecting the execution of an instruction I may cause one or more of the following fault effects [Brahme and Abraham 1984]:

1. One or more microorders are not activated by the microinstructions of I.

2. Microorders are erroneously activated by the microinstructions of I.

3. A different set of microinstructions is activated instead of, or in addition to, the microinstructions of I.

This fault model is general, as it allows for partial execution of instructions and for execution of "new" instructions, not present in the instruction set of the microprocessor.

A fault affecting an instruction I is *simple* if at most one microorder is erroneously activated during the execution of I (any number of microorders may be inactive). Thus there exists a one-to-one correspondence between the set of microorders and the set of simple faults.

Two microorders are said to be *independent* if neither of them modifies the source registers used by the other. Two simple faults are independent if the microorders they activate are independent. Our fault model allows for any number of simple faults, provided that they are (pairwise) independent. An example of a fault not included in this fault model is a fault that erroneously activates the sequence of microorders (ADD A,R1 ; MOV R2,A); such a fault is said to be *linked*.

8.4.2.3 Fault Model for the Data-Storage Function

The fault model for the data-storage function [Thatte and Abraham 1980] is a straightforward extension of the stuck-fault model. It allows any register in the microprocessor to have any number of bits s-a-0 or s-a-1.

8.4.2.4 Fault Model for the Data-Transfer Function

The data-transfer function implements all the data transfers among the nodes of the graph model of a microprocessor ("transfer" means that data are moved without being modified). Recall that an edge from node A to node B may correspond to several instructions that cause a data transfer from A to B. Although the same hardware may implement (some of) these data transfers, to make our model independent of implementation, we assume that every instruction causing a transfer $A \rightarrow B$ defines a separate *logical transfer path* $A \rightarrow B$, and each such transfer path may be independently faulty. For the microprocessor of Example 8.7, this model implies that the IN$\rightarrow$PC transfer caused by a JMP a instruction may be faulty, while the same transfer caused by CALL a can be fault-free.

The fault model for the data-transfer function [Thatte and Abraham 1980] assumes that any line in a transfer path can be s-a-0 or s-a-1, and any two lines in a transfer path may be shorted.

8.4.2.5 Fault Model for the Data-Manipulation Function

The data-manipulation function involves instructions that change data, such as arithmetic and logic operations, register incrementing or decrementing, etc. It is practically impossible to establish a meaningful functional fault model for the data-manipulation function without any knowledge of the structure of the ALU or of the other functional units involved (shifter, incrementer, etc.). The usual approach in functional test generation is to assume that tests for the data-manipulation function are developed by some other techniques, and to provide means to apply these tests and observe their results. For example, if we know that the ALU is implemented as an

ILA, then we can derive a pseudoexhaustive test set for it using the methods described in Section 8.3.3. If the data-transfer function is checked first, this test set can be applied by loading the required operands in registers, executing the corresponding arithmetic or logic operations, and reading out the results. The data paths providing operands to the ALU and transferring the result from the ALU are checked together with the ALU.

8.4.3 Test Generation Procedures

In this section we examine test generation procedures for the functional fault models introduced in the previous section. These fault models are implicit and the test generation procedures deal with classes of faults without identifying the individual members of a class.

8.4.3.1 Testing the Register-Decoding Function

Because faults causing $f_D(R_i)=R_j$ and $f_D(R_j)=R_i$ are undetectable, our goal in testing the register-decoding function is to check that for every register R_i of the microprocessor, the size of the set $f_D(R_i)$ is 1 [Thatte and Abraham 1980]. This guarantees that none of the detectable addressing faults affecting the register-decoding function is present.

The testing procedure involves writing and reading of registers. For every register R_i, we predetermine a sequence of instructions WRITE(R_i) that transfers data from the IN node to R_i, and a sequence READ(R_i) that transfers the contents of R_i to the OUT node. Whenever there exist several ways to write or read a register, we select the shortest sequence.

Example 8.9: For the microprocessor of Example 8.7, we have the following sequences:

WRITE(A) = (MVI A,a)
READ(A) = (MOV (IX), A)

WRITE(R2) = (MVI A,a; MOV R2, A)
READ(R2) = (MOV A, R2; MOV (IX), A)

WRITE(SR) = (JMP a; CALL b)
READ(SR) = (RET)

                                                                                        □

With every register R_i we associate a *label* $l(R_i)$, which is the length of the sequence READ(R_i); here $l(R_i)$ represents the shortest "distance" from R_i to the OUT node. Using the READ sequences of Example 8.9, $l(A)=l(SR)=1$ and $l(R2)=2$.

The strategy of the testing procedure is to construct gradually a set A of registers such that

1. $f_D(R_i) \neq \varnothing$, for every $R_i \in A$

2. $f_D(R_i) \cap f_D(R_j) = \varnothing$, for every $R_i, R_j \in A$.

Eventually, A will contain all the registers of the microprocessor, and then these conditions will imply that $|f_D(R_i)|=1$ for every R_i.

Figure 8.20 presents the testing procedure (*Decode_regs*) to detect addressing faults affecting the register-decoding function. This procedure is executed by an external tester that supplies the instructions for WRITE and READ sequences and checks the data retrieved by READ sequences. According to the start-small principle, registers are added to A in increasing order of their labels. The same ordering is used for reading out the registers in A.

```
Decode_regs()
begin
    A=∅
    add a register with label 1 to A
    for every register R_i not in A
        begin
            for Data=ZERO, ONE
                begin
                    for every register R_j∈A  WRITE(R_j) with Data
                    WRITE(R_i) with Data
                    for every register R_j∈A  READ(R_j)
                    READ(R_i)
                end
            add R_i to A
        end
end
```

Figure 8.20   Testing the register-decoding function

Theorem 8.3:   If procedure *Decode_regs* executes without detecting any errors (in the data retrieved by READ sequences), then the register-decoding function is free of (detectable) addressing faults.

Proof: We will prove that successful completion of *Decode_regs* shows that $|f_D(R_i)|=1$ for every register R_i. The proof is by induction. Let R_i be a register with the lowest label among the registers currently not in A. Let us assume that all the registers currently in A have nonempty and disjoint sets $f_D(R_j)$. We want to show that this remains true after we add R_i to A.

If $f(R_i)=\varnothing$, READ(R_i) returns ZERO or ONE. Since READ(R_i) is executed twice, once expecting ZERO and once expecting ONE, one of them will detect the error. Hence if both READ(R_i) sequences execute without detecting errors, then $f(R_i)\neq\varnothing$.

If for some $R_j\in A$, $f_D(R_j)\cap f_D(R_i)\neq\varnothing$, then WRITE($R_j$) writes *Data* into a register whose contents are later changed to $\overline{Data}$ by WRITE(R_i). This error will be detected by one of the two READ(R_j) sequences. Hence if both READ(R_j) sequences execute without detecting errors, then $f_D(R_j)\cap f_D(R_i)=\varnothing$. Note that since $l(R_i)\geq l(R_j)$ for every $R_j\in A$, reading of R_j does not require routing of R_j through R_i.

The basis of induction, for the initial situation when A contains only one register, can be proved by similar arguments. Eventually, A will contain all the registers of the microprocessor. Then the relations $f_D(R_i)\neq\varnothing$ and $f_D(R_i)\cap f_D(R_j)\neq\varnothing$ imply that

$| f_D(R_i) | = 1$ for every R_i. Therefore successful completion of procedure *Decode_regs* shows that the register-decoding function is free of (detectable) addressing faults. □

Let n_R be the number of registers. The number of WRITE and READ sequences generated by procedure *Decode_regs* is proportional to n_R^2. Thus if all registers can be directly written and read out, i.e., every WRITE and READ sequence consists of only one instruction, then the number of instructions in the generated test sequence is also proportional to n_R^2. For an architecture with "deeply buried" registers, the worst-case length of the test sequence approaches n_R^3.

8.4.3.2 Testing the Instruction-Decoding and Instruction-Sequencing Function

Our goal is to detect all simple faults affecting the execution of any instruction. For this we have to ensure that any simple fault affecting an instruction I causes errors either in the data transferred to the OUT node or in a register that can be read after I is executed. This should be true if microorders of I are not activated and/or if additional (independent) microorders are erroneously activated. Missing microorders are generally easy to detect, as any instruction that does not activate all its microorders can be easily made to produce an incorrect result. To detect the execution of additional microorders, we associate different data patterns, called *codewords*, with different registers of the microprocessor [Abraham and Parker 1981, Brahme and Abraham 1984]. Let cw_i denote the codeword associated with register R_i. The set of codewords should satisfy the property that *any single microorder operating on codewords should either produce a noncodeword, or load a register R_i with a codeword cw_j of a different register.*

For n_R registers each having n bits, a set of codewords satisfying the above property can be obtained using a p-out-of-n code, where each codeword has exactly p bits set to 1, and $\binom{n}{p} \geq n_R$ (see Problem 8.10). Figure 8.21 illustrates a 5-out-of-8 codeword set.

cw_1	0 1 1 0 1 1 1 0
cw_2	1 0 0 1 1 1 1 0
cw_3	0 1 1 0 1 1 0 1
cw_4	1 0 0 1 1 1 0 1
cw_5	0 1 1 0 1 0 1 1
cw_6	1 0 0 1 1 0 1 1
cw_7	0 1 1 0 0 1 1 1
cw_8	1 0 0 1 0 1 1 1

Figure 8.21   A 5-out-of-8 codeword set

Example 8.10:   Consider eight registers R1...R8, loaded with the codewords given in Figure 8.21. We will illustrate how several fault-activated microorders produce noncodewords:

1.   ADD R1,R3 results in R1 having the noncodeword 11011011.

2. EXCHANGE R5,R7 results in both R5 and R7 having incorrect codewords.

3. OR R7,R8 produces the noncodeword 11110111 in R7.

Note that operations performed on noncodewords may produce a codeword. For example, if R4 and R5 have the noncodewords 00010101 and 10011001, then OR R4,R5 results in R4 having its correct codeword.   □

If all registers are loaded with the proper codewords, simple faults affecting the execution of an instruction I will either cause an incorrect result of I or cause a register to have a noncodeword or the codeword of a different register. Hence to detect these faults, all registers must be read out after executing I. For a register that cannot be read directly, its READ sequence should be *nondestructive*, i.e., it should not change the contents of any other register (with the possible exception of the program counter).

Example 8.11:   For the microprocessor of Example 8.7, the READ(R2) sequence used in the previous section — (MOV A,R2 ;   MOV (IX),A) — destroys the contents of A. A nondestructive READ(R2), which saves and restores A, is

$$READ(R2)=(PUSH \ A; \ MOV \ A,R2; \ MOV \ (IX),A; \ POP \ A)$$
    □

According to the start-small principle, we will check the READ sequences before checking the other instructions. We classify the faults affecting READ sequences according to the type of microorder they activate:

- type 1:   microorders operating on one register, for example, increment, negate, or rotate;

- type 2:   microorders causing a data transfer between two registers, for example, move or exchange;

- type 3:   microorders executing arithmetic or logic operations on two source registers, for example, add.

Let S_1 be the set of registers modified by microorders of type 1. Let S_2 be the set of register pairs (R_j, R_k) involved in microorders of type 2, where R_k and R_j are, respectively, the source and the destination registers. Let S_3 be the set of register triples (R_j, R_k, R_l) involved in microorders of type 3, where R_k and R_l are the two source registers, and R_j is the destination register.

Figures 8.22, 8.23, and 8.24 present the testing procedures (*Read1*, *Read2*, and *Read3*) to detect, respectively, faults of type 1, type 2, and type 3 affecting the execution of READ sequences. Each procedure starts by loading codewords in registers. Since at this point we cannot assume that WRITE sequences are fault-free, the testing procedures take into account that some registers may not contain their codewords.

Theorem 8.4:   If procedures *Read1*, *Read2*, and *Read3* execute without detecting any errors, then all READ sequences are fault-free.

Proof: We will prove only that procedure *Read3* detects all faults of type 3 affecting READ sequences. The proofs for procedures *Read1* and *Read2* are similar.

Consider a fault of type 3 that, during the execution of READ(R_i), erroneously activates a microorder that uses R_k and R_l as source registers and modifies R_j. If R_k and R_l now have their proper codewords, the first READ(R_j) detects a noncodeword

Read1()
begin
   for every R_i WRITE(R_i) with cw_i
   for every R_i
     for every $R_j \in S_1$
      begin
        READ (R_i)
        READ (R_j)
        READ (R_i)
        READ (R_j)
      end
end

Figure 8.22   Testing for type 1 faults

Read2()
begin
   for every R_i WRITE(R_i) with cw_i
   for every R_i
     for every $(R_j, R_k) \in S_2$
      begin
        READ (R_i)
        READ (R_j)
        READ (R_k)
        READ (R_i)
        READ (R_j)
      end
end

Figure 8.23   Testing for type 2 faults

(or an incorrect codeword) in R_j. However, if the microorder activated by READ(R_i) operates on incorrect data in R_k and/or R_l, then it may produce the correct codeword in R_j, so the first READ(R_j) does not detect an error. Next, READ(R_k) checks the contents of R_k; but even if R_k was not loaded with its codeword, it is possible that the first READ(R_j) changed R_k to its correct value. Similarly, READ(R_l) either detects an error or shows that R_l has its correct value. But READ(R_l) may change the contents of R_k. If the next READ(R_k) and READ(R_l) do not detect errors, we can be sure that both R_k and R_l have now their correct codewords. Thus when the second READ(R_i) is executed, the fault-activated microorder produces incorrect data in R_j, and the second READ(R_j) detects this error. Therefore, if no errors are detected, we can conclude that

```
Read3()
begin
    for every Rᵢ  WRITE(Rᵢ) with cwᵢ
    for every Rᵢ
        for every (Rⱼ,Rₖ,Rₗ) ∈ S₃
            begin
                READ (Rᵢ)
                READ (Rⱼ)
                READ (Rₖ)
                READ (Rₗ)
                READ (Rₖ)
                READ (Rₗ)
                READ (Rᵢ)
                READ (Rⱼ)
            end
end
```

Figure 8.24   Testing for type 3 faults

READ(R_i) is free of faults of type 3. As *Read3* repeats this test for every R_i, eventually all faults of type 3 affecting READ sequences are detected.   □

We have assumed that all simple faults are independent. Brahme and Abraham [1984] present extensions of the *Read* procedures to detect linked faults as well.

The *Read* procedures also detect some faults affecting WRITE sequences. Figure 8.25 outlines a testing procedure (*Load*) to detect all faults affecting WRITE sequences. *Load* assumes that READ sequences are fault-free.

Procedure *Instr*, given in Figure 8.26, checks for faults affecting the execution of every instruction in the instruction set of the microprocessor. *Instr* assumes that both WRITE and READ sequences are fault-free. The question of whether every instruction should be tested for every addressing mode is answered depending on the orthogonality of the instruction set.

The complete test sequence for the instruction-decoding and instruction-sequencing function of the microprocessor is obtained by executing the sequence of procedures *Read1*, *Read2*, *Read3*, *Load*, and *Instr*.

The worst-case length of the test sequence checking the READ sequences is proportional to n_R^4, and the length of the test sequence checking the execution of every instruction is proportional to $n_R n_I$, where n_I is the number of instructions in the instruction set.

```
Load()
begin
    for every R_i WRITE(R_i) with cw_i
    for every R_i
        for every R_j
            begin
                READ(R_j)
                WRITE(R_i) with cw_i
                READ(R_j)
            end
end
```

Figure 8.25   Testing WRITE sequences

```
Instr()
begin
    for every instruction I
        begin
            for every R_i  WRITE(R_i) with cw_i
            execute I
            for every R_i  READ(R_i)
        end
end
```

Figure 8.26   Testing all instructions

8.4.3.3 Testing the Data-Storage and Data-Transfer Functions

The data-storage and data-transfer functions are tested together, because a test that detects stuck faults on lines of a transfer path $A \rightarrow B$, also detects stuck faults in the registers corresponding to the nodes A and B.

Consider a sequence of instructions that activates a sequence of data transfers starting at the IN node and ending at the OUT node. We refer to such a sequence as an *IN/OUT transfer*. For the microprocessor of Example 8.7, the sequence (MVI A,a; MOV R1,A; MOV(IX),R1) is an IN/OUT transfer that moves data a along the path IN$\rightarrow$A$\rightarrow$R1$\rightarrow$OUT.

A test for the transfer paths and the registers involved in an IN/OUT transfer consists of repeating the IN/OUT transfer for different data patterns, so that

1. Every bit in a transfer path is set to both 0 and 1.

2. Every pair of bits is set to complementary values [Thatte and Abraham 1980].

Figure 8.27 shows a set of 8-bit data patterns satisfying these requirements.

```
1 1 1 1 1 1 1 1
1 1 1 1 0 0 0 0
1 1 0 0 1 1 0 0
1 0 1 0 1 0 1 0
0 0 0 0 0 0 0 0
0 0 0 0 1 1 1 1
0 0 1 1 0 0 1 1
0 1 0 1 0 1 0 1
```

Figure 8.27   Data patterns for checking an 8-bit transfer path

Clearly, for every transfer path involved in an IN/OUT transfer, a test sequence constructed in this way detects all the stuck faults on the lines of the transfer path and all the shorts between any two of its lines. The complete test for the data-storage and data-transfer functions consists of a set of IN/OUT transfers, such that every transfer path of the microprocessor is involved in at least one IN/OUT transfer.

8.4.4 A Case Study

Thatte and Abraham [1980] derived test sequences for an 8-bit microprocessor, based on functional fault models and testing procedures of the type presented in this chapter. These test sequences, consisting of about 9000 instructions, were evaluated by fault simulation using a gate and flip-flop model of the microprocessor. A sample of about 2200 SSFs were simulated. The obtained fault coverage was 96 percent, which is an encouraging result.

8.5 Concluding Remarks

Functional testing methods attempt to reduce the complexity of the test generation problem by approaching it at higher levels of abstraction. However, functional testing has not yet achieved the level of maturity and the success of structural testing methods. In this section we review some of the limitations and difficulties encountered in functional testing.

Although binary decision diagrams provide functional models that are easily used for test generation, their applicability in modeling complex systems has not been proven.

Pseudoexhaustive testing is best suited for circuits having a regular ILA-type structure. For arbitrary combinational circuits where at least one PO depends on many PIs, the number of tests required for pseudoexhaustive testing becomes prohibitive. Partitioning techniques can reduce the number of tests, but they rely on knowledge of the internal structure of the circuit, and their applicability to large circuits is hindered by the lack of good partitioning algorithms.

Explicit functional fault models are likely to produce a prohibitively large set of target faults. Implicit functional fault models have been successfully used in testing RAMs (see [Abadir and Reghbati 1983] for a survey) and in testing programmable devices such as microprocessors, for which test patterns can be developed as sequences of instructions. The test generation process is based on the architecture and the instruction set of a microprocessor and produces test procedures that detect classes of faults without requiring explicit fault enumeration. This process has not been automated and cannot generate tests for the data manipulation function.

Functional testing is an actively researched area. Some other approaches not examined in this chapter are presented in [Lai and Siewiorek 1983], [Robach and Saucier 1980], [Shen and Su 1984], [Lin and Su 1985], [Su and Hsieh 1981], and [Renous *et al.* 1989].

In general, functional testing methods are tightly coupled with specific functional modeling techniques. Thus the applicability of a functional testing method is limited to systems described via a particular modeling technique. Because there exist many widely different functional modeling techniques, it is unlikely that a generally applicable functional testing method can be developed. Moreover, deriving the functional model used in test generation is often a manual, time-consuming and error-prone process [Bottorff 1981].

Another major problem in functional testing is the lack of means for evaluating the effectiveness of test sequences at the functional level.

REFERENCES

[Abadir and Reghbati 1983] M. S. Abadir and H. K. Reghbati, "Functional Testing of Semiconductor Random Access Memories," *Computing Surveys*, Vol. 15, No. 3, pp. 175-198, September, 1983.

[Abadir and Reghbati 1984] M. S. Abadir and H. K. Reghbati, "Test Generation for LSI: A Case Study," *Proc. 21st Design Automation Conf.*, pp. 180-195, June, 1984.

[Abraham and Parker 1981] J. A. Abraham and K. P. Parker, "Practical Microprocessor Testing: Open and Closed Loop Approaches," *Proc. COMPCON Spring 1981*, pp. 308-311, February, 1981.

[Akers 1978] S. B. Akers, "Functional Testing With Binary Decision Diagrams," *Journal of Design Automation & Fault-Tolerant Computing*, Vol. 2, pp. 311-331, October, 1978.

[Akers 1985] S. B. Akers, "On the Use of Linear Sums in Exhaustive Testing," *Digest of Papers 15th Annual Intn'l. Symp. on Fault-Tolerant Computing*, pp. 148-153, June, 1985.

[Annaratone and Sami 1982] M. A. Annaratone and M. G. Sami, "An Approach to Functional Testing of Microprocessors," *Digest of Papers 12th Annual Intn'l. Symp. on Fault-Tolerant Computing*, pp. 158-164, June, 1982.

[Brahme and Abraham 1984] D. Brahme and J. A. Abraham, "Functional Testing of Microprocessors," *IEEE Trans. on Computers*, Vol. C-33, No. 6, pp. 475-485, June, 1984.

[Barzilai *et al.* 1981] Z. Barzilai, J. Savir, G. Markowsky, and M. G. Smith, "The Weighted Syndrome Sums Approach to VLSI Testing," *IEEE Trans. on Computers*, Vol. C-30, No. 12, pp. 996-1000, December, 1981.

[Bottorff 1981] P. S. Bottorff, "Functional Testing Folklore and Fact," *Digest of Papers 1981 Intn'l. Test Conf.*, pp. 463-464, October, 1981.

[Chang *et al.* 1986] H. P. Chang, W. A. Rogers, and J. A. Abraham, "Structured Functional Level Test Generation Using Binary Decision Diagrams," *Proc. Intn'l. Test Conf.*, pp. 97-104, September, 1986.

[Chen 1988] C. L. Chen, "Exhaustive Test Pattern Generation Using Cyclic Codes," *IEEE Trans. on Computers*, Vol. C-37, No. 2., pp. 225-228, February, 1988.

[Cheng and Patel 1985] W. T. Cheng and J. H. Patel, "A Shortest Length Test Sequence for Sequential-Fault Detection in Ripple Carry Adders," *Proc. Intn'l. Conf. on Computer-Aided Design*, pp. 71-73, November, 1985.

[Chiang and McCaskill 1976] A. C. L. Chiang and R. McCaskill, "Two New Approaches Simplify Testing of Microprocessors," *Electronics*, Vol. 49, No. 2, pp. 100-105, January, 1976.

[Elhuni *et al.* 1986] H. Elhuni, A. Vergis, and L. Kinney, "C-Testability of Two-Dimensional Iterative Arrays," *IEEE Trans. on Computer-Aided Design*, Vol. CAD-5, No. 4, pp. 573-581, October, 1986.

[Friedman 1973] A. D. Friedman, "Easily Testable Iterative Systems," *IEEE Trans. on Computers*, Vol. C-22, No. 12, pp. 1061-1064, December, 1973.

[Friedman and Menon 1971] A. D. Friedman and P. R. Menon, *Fault Detection in Digital Circuits*, Prentice Hall, Englewood Cliffs, New Jersey, 1971.

[Kautz 1967] W. H. Kautz, "Testing for Faults in Cellular Logic Arrays," *Proc. 8th Symp. Switching and Automata Theory*, pp. 161-174, 1967.

[Lai and Siewiorek 1983] K. W. Lai and D. P. Siewiorek, "Functional Testing of Digital Systems," *Proc. 20th Design Automation Conf.*, pp. 207-213, June, 1983.

[Lin and Su 1985] T. Lin and S. Y. H. Su, "The *S*-Algorithm: A Promising Solution for Systematic Functional Test Generation," *IEEE Trans. on Computer-Aided Design*, Vol. CAD-4, No. 3, pp. 250-263, July, 1985.

[McCluskey 1984] E. J. McCluskey, "Verification Testing — A Pseudoexhaustive Test Technique," *IEEE Trans. on Computers*, Vol. C-33, No. 6, pp. 541-546, June, 1984.

[McCluskey and Bozorgui-Nesbat 1981] E. J. McCluskey and S. Bozorgui-Nesbat, "Design for Autonomous Test," *IEEE Trans. on Computers*, Vol. C-30, No. 11, pp. 866-875, November, 1981.

[Menon and Friedman 1971] P. R. Menon and A. D. Friedman, "Fault Detection in Iterative Logic Arrays," *IEEE Trans. on Computers*, Vol. C-20, No. 5, pp. 524-535, May, 1971.

[Monachino 1982] M. Monachino, "Design Verification System for Large-Scale LSI Designs," *Proc. 19th Design Automation Conf.*, pp. 83-90, June, 1982.

[Moore 1956] E. F. Moore, "Gedanken Experiments on Sequential Machines," in *Automata Studies*, pp. 129-153, Princeton University Press, Princeton, New Jersey, 1956.

[Noon 1977] W. A. Noon, "A Design Verification and Logic Validation System," *Proc. 14th Design Automation Conf.*, pp. 362-368, June, 1977.

[Parthasarathy and Reddy 1981] R. Parthasarathy and S. M. Reddy, "A Testable Design of Iterative Logic Arrays," *IEEE Trans. on Computers*, Vol. C-30, No. 11, pp. 833-841, November, 1981.

[Parthasarathy *et al.* 1982] R. Parthasarathy, S. M. Reddy, and J. G. Kuhl, "A Testable Design of General Purpose Microprocessors," *Digest of Papers 12th Annual Intn'l. Symp. on Fault-Tolerant Computing*, pp. 117-124, June, 1982.

[Renous *et al.* 1989] R. Renous, G. M. Silberman, and I. Spillinger, "Whistle: A Workbench for Test Development of Library-Based Designs," *Computer*, Vol. 22, No. 4, pp. 27-41, April, 1989.

[Robach and Saucier 1980] C. Robach and G. Saucier, "Microprocessor Functional Testing," *Digest of Papers 1980 Test Conf.*, pp. 433-443, November, 1980.

[Shen and Ferguson 1984] J. P. Shen and F. J. Ferguson, "The Design of Easily Testable VLSI Array Multipliers," *IEEE Trans. on Computers*, Vol. C-33, No. 6, pp. 554-560, June, 1984.

[Shen and Su 1984] L. Shen and S. Y. H. Su, "A Functional Testing Method for Microprocessors," *Proc. 14th Intn'l. Conf. on Fault-Tolerant Computing*, pp. 212-218, June, 1984.

[Shperling and McCluskey 1987] I. Shperling and E. J. McCluskey, "Circuit Segmentation for Pseudo-Exhaustive Testing via Simulated Annealing," *Proc. Intn'l. Test Conf.*, pp. 58-66, September, 1987.

[Sridhar and Hayes 1979] T. Sridhar and J. P. Hayes, "Testing Bit-Sliced Microprocessors," *Digest of Papers 9th Annual Intn'l. Symp. on Fault-Tolerant Computing*, pp. 211-218, June, 1979.

[Sridhar and Hayes 1981a] T. Sridhar and J. P. Hayes, "A Functional Approach to Testing Bit-Sliced Microprocessors," *IEEE Trans. on Computers*, Vol. C-30, No. 8, pp. 563-571, August, 1981.

[Sridhar and Hayes 1981b] T. Sridhar and J. P. Hayes, "Design of Easily Testable Bit-Sliced Systems," *IEEE Trans. on Computers*, Vol. C-30, No. 11, pp. 842-854, November, 1981.

[Su and Hsieh 1981] S. Y. H. Su and Y. Hsieh, "Testing Functional Faults in Digital Systems Described by Register Transfer Language," *Digest of Papers 1981 Intn'l. Test Conf.*, pp. 447-457, October, 1981.

[Tang and Woo 1983] D. T. Tang and L. S. Woo, "Exhaustive Test Pattern Generation with Constant Weight Vectors," *IEEE Trans. on Computers*, Vol. C-32, No. 12, pp. 1145-1150, December, 1983.

[Thatte and Abraham 1980] S. M. Thatte and J. A. Abraham, "Test Generation for Microprocessors," *IEEE Trans. on Computers*, Vol. C-29, No. 6, pp. 429-441, June, 1980.

PROBLEMS

8.1 Show that a heuristic functional test that does not detect a detectable stuck fault on a PI of the circuit does not completely exercise its operation.

8.2 Generate all the feasible experiments for the binary decision diagram of Figure 8.6.

8.3 Using the binary decision diagram of Figure 8.6, generate a test in which f is sensitive to C.

8.4 Derive a pseudoexhaustive test set for the circuit of Figure 8.8(a) by partitioning it into the following three segments: (1) the subcircuit whose output is h, (2) gate y, and (3) gates g and x. Compare your results with those obtained in Example 8.4.

8.5 Show that any pseudoexhaustive test set based on a sensitized partitioning of a combinational circuit N detects all detectable SSFs in N.

8.6 Analyze the state tables shown in Figure 8.28 to determine whether the ILAs they represent are C-testable.

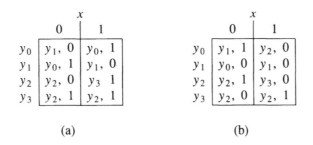

	x			x	
	0	1		0	1
y_0	y_1, 0	y_0, 1	y_0	y_1, 1	y_2, 0
y_1	y_0, 1	y_1, 0	y_1	y_0, 0	y_1, 0
y_2	y_2, 0	y_3 1	y_2	y_2, 1	y_3, 0
y_3	y_2, 1	y_2, 1	y_3	y_2, 0	y_2, 1

(a)   (b)

Figure 8.28

8.7 Derive tests that verify the entry $(y_1,1)$ in all the cells of the ILA represented by the state table of Figure 8.13.

8.8 The ILA of Figure 8.29 is a realization of an n-bit parity function. Its basic cell implements an exclusive-OR function. Show that the ILA is C-testable with four tests.

(Because the cells do not have direct outputs, analyze fault propagation through a cell using propagation D-cubes.)

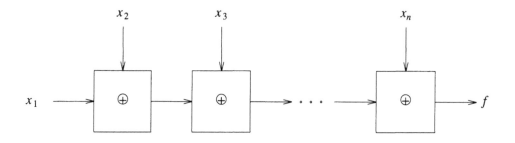

Figure 8.29

8.9 Consider a gate-level model of a 2-to-4 decoder. Show that any SSF leads to the following faulty behavior: for any input vector, instead of, or in addition to the expected output line, some other output is activated, or no output is activated.

8.10 Show that a p-out-of-n codeword set satisfies the required property of codewords with respect to any microorder of the form R1←R1*R2, where "*" denotes an ADD, AND, OR, or XOR operation.

9. DESIGN FOR TESTABILITY

About This Chapter

Previous chapters have dealt with the complexity of test generation and algorithms for constructing a test for a complex circuit. Test complexity can be converted into costs associated with the testing process. There are several facets to this cost, such as the cost of test pattern generation, the cost of fault simulation and generation of fault location information, the cost of test equipment, and the cost related to the testing process itself, namely the time required to detect and/or isolate a fault. Because these costs can be high and may even exceed design costs, it is important that they be kept within reasonable bounds. One way to accomplish this goal is by the process of *design for testability* (DFT), which is the subject of this chapter. The model of test being considered here is that of external testing.

The testability of a circuit is an abstract concept that deals with a variety of the costs associated with testing. By increasing the testability of a circuit, it is implied that some function of these costs is being reduced, though not necessarily each individual cost. For example, scan designs may lower the cost of test generation but increase the number of I/O pins, area, and test time.

Controllability, *observability*, and *predictability* are the three most important factors that determine the complexity of deriving a test for a circuit. The first two concepts have been discussed in Chapter 6. *Predictability* is the ability to obtain known output values in response to given input stimuli. Some factors affecting predictability are the initial state of a circuit, races, hazards, and free-running oscillators. Most DFT techniques deal with ways for improving controllability, observability, and predictability.

Section 9.1 deals briefly with some measures associated with testing, as well as methods for computing numeric values for controllability and observability. Section 9.2 discusses ad hoc DFT techniques; Section 9.3 deals with ad hoc scan-based DFT techniques. Sections 9.4 through 9.7 present structured scan-based designs, and Section 9.8 deals with board-level and system-level DFT approaches. Section 9.9 deals with advanced scan concepts, and Section 9.10 deals with the JTAG/IEEE 1149.1 proposed standards relating to boundary scan.

9.1 Testability

Testability is a design characteristic that influences various costs associated with testing. Usually it allows for (1) the status (normal, inoperable, degraded) of a device to be determined and the isolation of faults within the device to be performed quickly, to reduce both the test time and cost, and (2) the cost-effective development of the tests to determine this status. *Design for testability techniques* are design efforts specifically employed to ensure that a device is testable.

Two important attributes related to testability are controllability and observability. *Controllability* is the ability to establish a specific signal value at each node in a circuit by setting values on the circuit's inputs. *Observability* is the ability to determine the signal value at any node in a circuit by controlling the circuit's inputs and observing its outputs.

The degree of a circuit's controllability and observability is often measured with respect to whether tests are generated randomly or deterministically using some ATG algorithm. For example, it may be difficult (improbable) for a random test pattern generation to drive the output of a 10-input AND gate to a 1, but the D-algorithm could solve this problem with one table lookup. Thus, the term "random controllability" refers to the concept of controllability when random tests are being used. In general, a circuit node usually has poor random controllability if it requires a unique input pattern to establish the state of that node. A node usually has poor controllability if a lengthy sequence of inputs is required to establish its state.

Circuits typically difficult to control are decoders, circuits with feedback, oscillators, and clock generators. Another example is a 16-bit counter whose parallel inputs are grounded. Starting from the reset state, 2^{15} clock pulses are required to force the most significant bit to the value of 1.

A circuit often has poor random observability if it requires a unique input pattern or a lengthy complex sequence of input patterns to propagate the state of one or more nodes to the outputs of the circuit. Less observable circuits include sequential circuits; circuits with global feedback; embedded RAMs, ROMs, or PLAs; concurrent error-checking circuits; and circuits with redundant nodes.

The impact of accessibility on testing leads to the following general observations:

- Sequential logic is much more difficult to test than combinational logic.

- Control logic is more difficult to test than data-path logic.

- Random logic is more difficult to test than structured, bus-oriented designs.

- Asynchronous designs or those with unconstrained timing signals are much more difficult to test than synchronous designs that have easily accessible clock generation and distribution circuits.

9.1.1 Trade-Offs

Most DFT techniques deal with either the resynthesis of an existing design or the addition of extra hardware to the design. Most approaches require circuit modifications and affect such factors as area, I/O pins, and circuit delay. The values of these attributes usually increase when DFT techniques are employed. Hence, a critical balance exists between the amount of DFT to use and the gain achieved. Test engineers and design engineers usually disagree about the amount of DFT hardware to include in a design.

Increasing area and/or logic complexity in a VLSI chip results in increased power consumption and decreased yield. Since testing deals with identifying faulty chips, and decreasing yield leads to an increase in the number of faulty chips produced, a careful balance must be reached between adding logic for DFT and yield. The relationship among fault coverage, yield, and defect level is illustrated in Figure 5.1. Normally yield decreases linearly as chip area increases. If the additional hardware required to support DFT does not lead to an appreciable increase in fault coverage, then the defect level will increase. In general, DFT is used to reduce test generation costs, enhance the quality (fault coverage) of tests, and hence reduce defect levels. It can also affect test length, tester memory, and test application time.

Normally all these attributes are directly related to one another. Unfortunately, when either deterministic or random test patterns are being used, there is no reliable model for accurately predicting, for a specific circuit, the number of test vectors required to achieve a certain level of fault coverage. For many applications, test development time is a critical factor. By employing structured DFT techniques described later in this chapter, such time can often be substantially reduced, sometimes from many months to a few weeks. This reduction can significantly affect the time to market of a product and thus its financial success. Without DFT, tests may have to be generated manually; with DFT they can be generated automatically. Moreover, the effectiveness of manually generated tests is often poor.

The cost of test development is a one-time expense, and it can be prorated over the number of units tested. Hence, on the one hand, for a product made in great volume, the per unit cost of test development may not be excessively high. On the other hand, test equipment is expensive to purchase, costs a fixed amount to operate, and becomes obsolete after several years. Like test development, ATE-associated costs must be added to the cost of the product being tested. Hence, test time can be a significant factor in establishing the cost of a unit. Again, a conflict exists in that to reduce test application costs, shorter tests should be used, which leads to reduced fault coverage and higher defect levels. This is particularly true when testing microprocessors and large RAMs.

If a faulty chip is put into a printed circuit board, then the board is faulty. Identifying a faulty chip using board-level tests is usually 10 to 20 times more costly than testing a chip. This does not imply that chip-level testing should be done on all chips. Clearly this decision is a function of the defect level associated with each type of chip.

When a faulty board is put into a system, system-level tests are required to detect and isolate the faulty board. This form of testing again is about 10 to 20 times more expensive than board-level testing in identifying a faulty board. Normally, board yield is low enough to warrant that all boards be tested at the board level.

9.1.2 Controllability and Observability

In the early 70s attempts were made to quantify the concepts of controllability and observability. A precise definition for these abstract concepts and a formal method for computing their value would be desirable. The idea was to modify the design of a circuit to enhance its controllability and observability values. This modification would lead to a reduction in deterministic test generation costs. This analysis process must be relatively inexpensive, since if the computation were too involved, then the cost savings in ATG would be offset by the cost of analysis and DFT. Unfortunately, no such easily computable formal definition for these concepts has as yet been proposed. What is commonly done is that a procedure for computing controllability and observability is proposed, and then these concepts are defined by means of the procedure.

The pioneering work in this area was done by Rutman [1972] and independently by Stephenson and Grason [1976] and Grason [1979]. This work relates primarily to deterministic ATG. Rutman's work was refined and extended by Breuer [1978].

These results were then popularized in the papers describing the Sandia Controllability/Observability Analysis Program (SCOAP) [Goldstein 1979, Goldstein and Thigen 1980]. This work, in turn, formed the basis of several other systems that compute deterministic controllability and observability values, such as TESTSCREEN [Kovijanic 1979, 1981], CAMELOT (Computer-Aided Measure for Logic Testability) [Bennetts *et al.* 1980], and VICTOR (VLSI Identifier of Controllability, Testability, Observability, and Redundancy) [Ratiu *et al.* 1982].

The basic concepts used in these systems have previously been presented in Section 6.2.1.3. These programs compute a set of values for each line in a circuit. These values are intended to represent the relative degree of difficulty for computing an input vector or sequence for each of the following problems:

1. setting line x to a 1 (1-controllability);

2. setting line x to a 0 (0-controllability);

3. driving an error from line x to a primary output (observability).

Normally large values imply worse testability than smaller ones. Given these values, it is up to the designer to decide if they are acceptable or not. To reduce large values, the circuit must be modified; the simplest modifications deal with adding test points and control circuitry. These techniques will be illustrated later in this chapter. Normally the controllability and observability values of the various nodes in a circuit are combined to produce one or more testability values for the circuit.

The problem here is two-fold. First, the correlation between testability values and test generation costs has not been well established [Agrawal and Mercer 1982]. Second, it is not clear how to modify a circuit to reduce the value of these testability measures. Naive rule-of-thumb procedures, such as add test points to lines having the highest observability values and control circuitry to lines having the highest controllability values, are usually not effective. A method for automatically modifying a design to reduce several functions of these testability values, such as the maximum value and the sum of the values, has been developed by Chen and Breuer [1985], but its computational complexity is too high to be used in practice. Their paper introduces the concept of *sensitivity*, which is a measure of how the controllability and observability values of the entire circuit change as one modifies the controllability and/or observability of a given line.

Testability measures can also be derived for testing with random vectors. Here these measures deal with the probability of a random vector setting a specific node to the value of 0 or 1, or propagating an error from this node to a primary output. For this case too there is not a strong correlation between testability values and test generation costs [Savir 1983]. In summary, the testability measures that exist to date have not been useful in guiding the design process.

Recall that controllability and observability figures are used to guide decision making in deterministic ATG algorithms (see Chapter 6). These concepts have proven very useful in this context [Lioy and Messalama 1987, Chandra and Patel 1989].

9.2 Ad Hoc Design for Testability Techniques

In this section we will present several well known ad hoc designs for testability techniques. Many of these techniques were developed for printed circuit boards; some are applicable to IC design. They are considered to be ad hoc (rather than algorithmic) because they do not deal with a total design methodology that ensures ease of test generation, and they can be used at the designer's option where applicable. Their goal is to increase controllability, observability, and/or predictability.

The DFT techniques to be discussed in this section deal with the following concepts:

- test points,

- initialization,

- monostable multivibrators (one-shots),

- oscillators and clocks,

- counters/shift registers,

- partitioning large circuits,

- logical redundancy,

- breaking global feedback paths.

9.2.1 Test Points

Rule: Employ test points to enhance controllability and observability.

There are two types of test points, referred to as control points (CP) and observation points (OP). *Control points* are primary inputs used to enhance controllability; *observation points* are primary outputs used to enhance observability.

Figure 9.1(a) shows a NOR gate G buried within a large circuit. If this circuit is implemented on a printed circuit board, then the signal G can be routed to two pins, A and A', and a removable wire, called a *jumper*, can be used to connect A to A' externally. By removing the jumper, the external test equipment can monitor the signal G while applying arbitrary signals to A' (see Figure 9.1(b)). Thus A acts as an observation point and A' as a control point. Several other ways of adding test points, applicable to both boards and chips, are shown in Figure 9.1.

In Figure 9.1(c), a new control input CP has been added to gate G, forming the gate G^*. If $CP = 0$, then $G^* = G$ and the circuit operates in its normal way. For $CP = 1$, $G^* = 0$; i.e., we have forced the output to 0. This modified gate is called a *0-injection circuit* (*0-I*). The new observation test point OP makes the signal G^* directly observable.

Figure 9.1(d) illustrates the design of a 0/1 injection circuit, denoted by *0/1-I*. Here G has been modified by adding the input $CP1$, and a new gate G' has been added to the circuit. If $CP1 = CP2 = 0$, then $G' = G$, which indicates normal operation. $CP1 = 1$ inhibits the normal signals entering G^*, sets $G^* = 0$, hence $G' = CP2$. Thus G' can be easily controlled to either a 0 or a 1. Figure 9.1(e) shows the use of a multiplexer (MUX) as a 0/1 injection circuit, where $CP2$ is the select line. In general, if G is an arbitrary signal line, then inserting either an AND gate or a NOR gate in this line

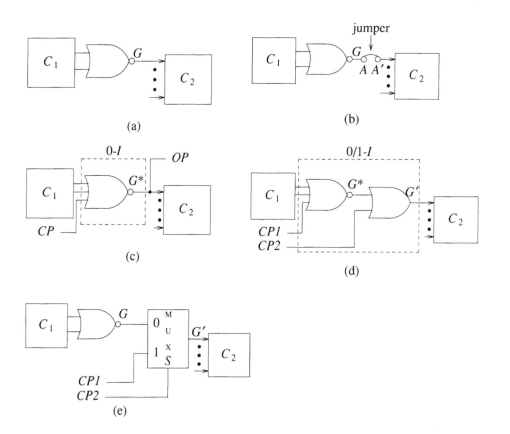

Figure 9.1   Employing test points (a) Original circuit (b) Using edge pins to
achieve controllability and observability (c) Using a CP for
0-injection and an OP for observability (d) Using a 0/1 injection
circuit (e) Using a MUX for 0/1 injection

creates a 0-injection circuit; inserting an OR or NAND gate creates a 1-injection
circuit. Inserting a series of two gates, such as NOR-NOR, produces a *0/1-injection
circuit.*

The major constraint associated with using test points is the large demand on I/O pins.
This problem can be alleviated in several ways. To reduce output pins, a multiplexer
can be used, as shown in Figure 9.2. Here the $N = 2^n$ observation points are replaced
by a single output Z and n inputs required to address a selected observation point. The
main disadvantage of this technique is that only one observation point can be observed
at a time; hence test time increases. If many outputs must be monitored for each input
vector applied, then the UUT clock must be stopped while the outputs are sampled.
For dynamic logic this can lead to problems. To reduce the pin count even further, a
counter can be used to drive the address lines of the multiplexer. Now, for each input
test vector, each observation point will be sampled in turn. The counter must be
clocked separately from the rest of the circuit. This DFT technique clearly suggests
one trade-off between test time and I/O pins.

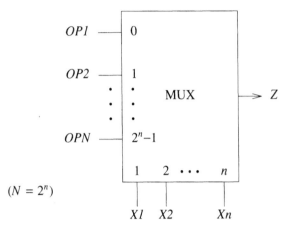

Figure 9.2 Multiplexing monitor points

A similar concept can be used to reduce input pin requirements for control inputs (see Figure 9.3). The values of the $N = 2^n$ control points are serially applied to the input Z, while their address is applied to $X1$, $X2$, ..., Xn. Using a demultiplexer, these N values are stored in the N latches that make up register R. Again, a counter can be used to drive the address lines of the demultiplexer. Also, N clock times are required between test vectors to set up the proper control values.

Another method that can be used, together with a multiplexer and demultiplexer, to reduce I/O overhead is to employ a shift register. This approach will be discussed in Section 9.3 dealing with scan techniques.

It is also possible to time-share the normal I/O pins to minimize the number of additional I/O pins required to be test points. This is shown in Figure 9.4. In Figure 9.4(a) a multiplexer has been inserted before the primary output pins of the circuit. For one logic value of the select line, the normal functional output signals are connected to the output pins; for the other value, the observation test points are connected to the output pins.

In Figure 9.4(b), a demultiplexer (DEMUX) is connected to the n primary inputs of a circuit. Again, by switching the value on the select lines, the data on the inputs can be sent to either the normal functional inputs or the control test points. This register R is used to hold the data at control test points while data are applied to the normal functional inputs.

The selection of signals to be made easily controllable or observable is based on empirical experience. Examples of good candidates for control points are as follows:

1. control, address, and data bus lines on bus-structured designs;

2. enable/hold inputs to microprocessors;

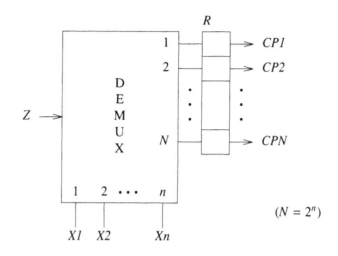

Figure 9.3   Using a demultiplexer and latch register to implement control points

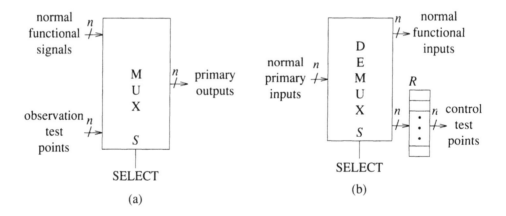

Figure 9.4   Time-sharing I/O ports

3. enable and read/write inputs to memory devices;

4. clock and preset/clear inputs to memory devices such as flip-flops, counters, and shift registers;

5. data select inputs to multiplexers and demultiplexers;

6. control lines on tristate devices.

Examples of good candidates for observation points are as follows:

1. stem lines associated with signals having high fanout;

2. global feedback paths;

3. redundant signal lines;

4. outputs of logic devices having many inputs, such as multiplexers and parity generators;

5. outputs from state devices, such as flip-flops, counters, and shift registers;

6. address, control, and data busses.

A heuristic procedure for the placement of test points is presented in [Hayes and Friedman 1974].

9.2.2 Initialization

Rule: Design circuits to be easily initializable.

Initialization is the process of bringing a sequential circuit into a known state at some known time, such as when it is powered on or after an initialization sequence is applied. The need for ease of initialization is dictated in part by how tests are generated. In this section we assume that tests are derived by means of an automatic test pattern generator. Hence, circuits requiring some clever input-initialization sequence devised by a designer should be avoided, since such sequences are seldom derived by ATG software. Later the concept of scan designs will be introduced. Such designs are easy to initialize.

For many designs, especially board designs using SSI and MSI components, initialization can most easily be accomplished by using the asynchronous preset (*PR*) or clear (*CLR*) inputs to a flip-flop. Figure 9.5 illustrates several common ways to connect flip-flop preset and clear lines for circuit initialization. In Figure 9.5(a), the preset and clear lines are driven from input pins; Figure 9.5(b) shows the same concept, but here the lines are tied via a pull-up resistor to V_{cc}, which provides for some immunity to noise. In Figure 9.5(c), the preset is controlled from an input pin while the clear line is deactivated. The configuration shown in Figure 9.5(d), which is used by many designers, is potentially dangerous because when the circuit is powered on, a race occurs and the flip-flop may go to either the 0 or 1 state.

When the preset or clear line is driven by logic, a gate can be added to achieve initialization, as shown in Figure 9.6.

To avoid the use of edge pins, preset and/or clear circuitry can be built in the circuit itself, as shown in Figure 9.7. When power is applied to the circuit, the capacitor charges from 0 to V_{cc}, and the initial low voltage on Z is used either to preset or clear flip-flops.

If the circuitry associated with the initialization of a circuit fails, test results are usually unpredictable. Such failures are often difficult to diagnose.

9.2.3 Monostable Multivibrators

Rule: Disable internal one-shots during test.

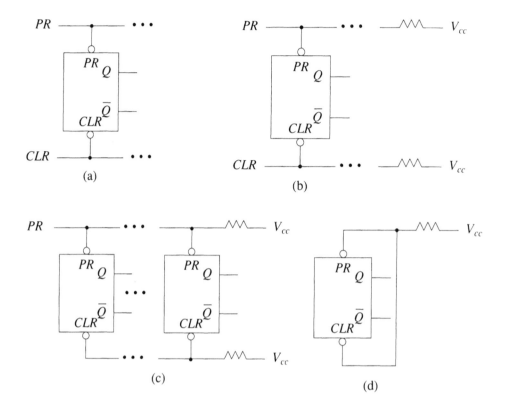

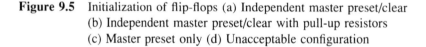

Figure 9.5   Initialization of flip-flops (a) Independent master preset/clear
(b) Independent master preset/clear with pull-up resistors
(c) Master preset only (d) Unacceptable configuration

Monostable multivibrators (one-shots) produce pulses internal to a circuit and make it difficult for external test equipment to remain in synchronization with the circuit being tested. There are several ways to solve this problem. On a printed circuit board, a jumper can be used to gain access to the I/O of a one-shot to disable it, control it, and observe its outputs, as well as to control the circuit normally driven by the one-shot. (See Figure 9.8(a)). Removing jumper 1 on the edge connector of the PCB allows the one-shot to be activated by the ATE via pin $A2$. The output of C_1 is also observable via pin $A1$. Pin $B2$ can be driven by the ATE, and the output of the one-shot can be observed via pin $B1$.

One can also gain controllability and observability of a one-shot and its surrounding logic by adding injection circuitry, as shown in Figure 9.8(b). In this circuit, E is used to observe the output of the one-shot; A is used to deactivate the normal input to the one-shot; B is used to trigger the one-shot during testing; C is used to apply externally generated pulses to C_2 to replace those generated by the one-shot; and D is used to select the input appropriate to the multiplexer.

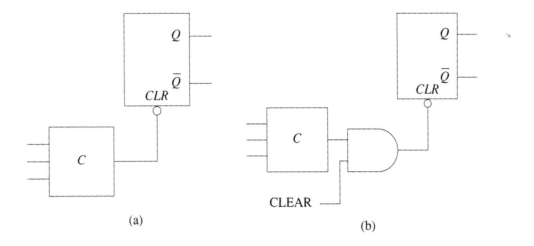

Figure 9.6   (a) Flip-flop without explicit clear (b) Flip-flop with explicit clear

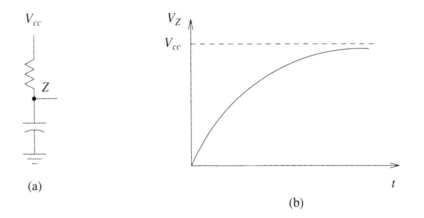

Figure 9.7   Built-in initialization signal generator

9.2.4 Oscillators and Clocks

Rule: Disable internal oscillators and clocks during test.

The use of free-running oscillators or clocks leads to synchronization problems similar to those caused by one-shots. The techniques to gain controllability and observability are also similar. Figure 9.9 shows one such DFT circuit configuration.

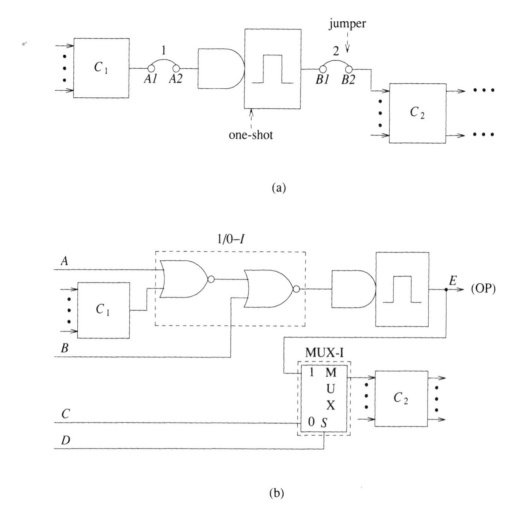

(a)

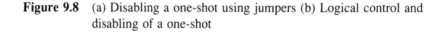

(b)

Figure 9.8 (a) Disabling a one-shot using jumpers (b) Logical control and
disabling of a one-shot

9.2.5 Partitioning Counters and Shift Registers

Rule: Partition large counters and shift registers into smaller units.

Counters, and to a lesser extent shift registers, are difficult to test because their test
sequences usually require many clock cycles. To increase their testability, such
devices should be partitioned so that their serial input and clock are easily controllable,
and output data are observable. Figure 9.10(a) shows a design that does not include
testability features and where the register R has been decomposed into two parts, and
Figure 9.10(b) shows a more testable version of this design. For example, $R1$ and $R2$
may be 16-bit registers. Here the gated clock from C can be inhibited and replaced by
an external clock. The serial inputs to $R1$ and $R2$ are easily controlled and the serial
output of $R2$ is easily observable. As a result, $R1$ and $R2$ can be independently tested.

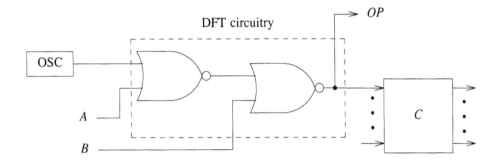

Figure 9.9 Testability logic for an oscillator

A 16-bit counter could take up to $2^{16} = 65536$ clock cycles to test. Partitioned into two 8-bit counters with enhanced controllability and observability, the new design can be tested with just $2 \times 2^8 = 512$ clock cycles. If the two partitions are tested simultaneously, even less time is required.

9.2.6 Partitioning of Large Combinational Circuits

Rule: Partition large circuits into small subcircuits to reduce test generation cost.

Since the time complexity of test generation and fault simulation grows faster than a linear function of circuit size, it is cost-effective to partition large circuits to reduce these costs. One general partitioning scheme is shown in Figure 9.11. In Figure 9.11(a) we show a large block of combinational logic that has been partitioned into two blocks, C_1 and C_2; many such partitions exist for multiple-output circuits. The bit widths of some of the interconnects are shown, and we assume, without loss of generality, that $p < m$ and $q \le n$. Figure 9.11(b) shows a modified version of this circuit, where multiplexers are inserted between the two blocks, and $A'(C')$ represents a subset of the signals in $A(C)$. For $T_1 T_2 = 00$ (normal operation), the circuit operates the same way as the original circuit, except for the delay in the multiplexers. For $T_1 T_2 = 01$, C_1 is driven by the primary inputs A and C'; the outputs F and D are observable at F' and G', respectively. Hence C_1 can be tested independently of C_2. Similarly C_2 can be tested independently of C_1. Part of the multiplexers are also tested when C_1 and C_2 are tested. However, not all paths through the multiplexers are tested, and some global tests are needed to ensure 100 percent fault coverage. For example, the path from D to C_2 is not tested when C_1 and C_2 are tested separately.

This partitioning scheme enhances the controllability and observability of the inputs and outputs, respectively, associated with C_1 and C_2, and hence reduces the complexity of test generation. Assume test generation requires n^2 steps for a circuit having n gates, and that C has 10000 gates and can be partitioned into two circuits C_1 and C_2 of size 5000 gates each. Then the test generation for the unpartitioned version of C requires $(10^4)^2 = 10^8$ steps, while the test generation for C_1 and C_2 requires only $2 \times 25 \times 10^6 = 5 \times 10^7$ or half the time.

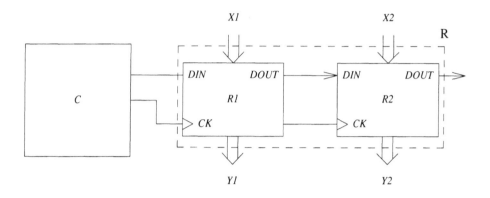

(a)

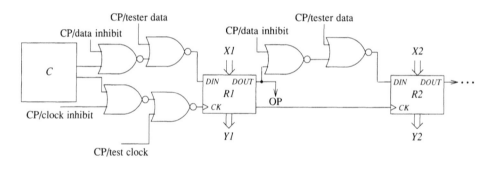

(b)

Figure 9.10   Partitioning a register

Assume it is desired to test this circuit exhaustively, i.e., by applying all possible inputs to the circuit. Let $m = n = s = 8$, and $p = q = 4$. Then to test C_1 and C_2 as one unit requires $2^{8+8+8} = 2^{24}$ test vectors. To test C_1 and C_2 individually requires $2^{8+8+4} = 2^{20}$ test vectors. The basic concept can be iterated so as to partition a circuit into more than two blocks and thus reduce test complexity by a greater factor.

9.2.7 Logical Redundancy

Rule: Avoid the use of redundant logic.

Recall that redundant logic introduces faults which in combinational logic, and in most instances in sequential logic, are not detectable using static tests. There are several reasons such logic should be avoided whenever possible. First, if a redundant fault occurs, it may invalidate some test for nonredundant faults. Second, such faults cause difficulty in calculating fault coverage. Third, much test generation time can be spent

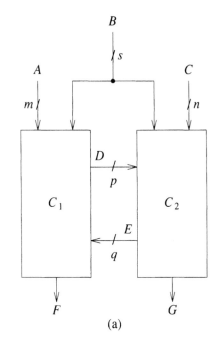

(a)

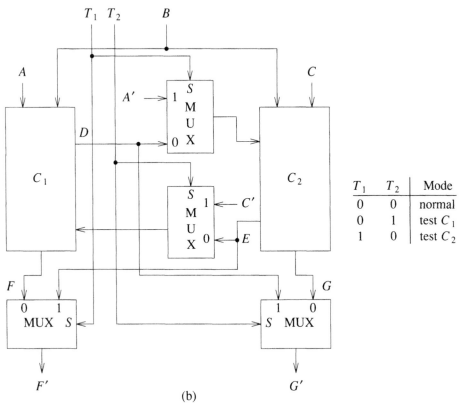

(b)

Figure 9.11 Partitioning to reduce test generation cost

in trying to generate a test for a redundant fault, assuming it is not known *a priori* that it is redundant. Unfortunately, here a paradox occurs, since the process of identifying redundancy in a circuit is NP-hard. Fourth, most designers are not aware of redundancy introduced inadvertently and hence cannot easily remove it. Finally, note that redundancy is sometimes intentionally added to a circuit. For example, redundant logic is used to eliminate some types of hazards in combinational circuits. It is also used to achieve high reliability, such as in Triple Modular Redundant (TMR) systems. For some of these cases, however, test points can be added to remove the redundancy during testing without inhibiting the function for which the redundancy is provided.

9.2.8 Global Feedback Paths

Rule: Provide logic to break global feedback paths.

Consider a global feedback path in a circuit. This path may be part of either a clocked or an unclocked loop. Observability can be easily obtained by adding an observation test point to some signal along this path. Controllability can be achieved by using injection circuits such as those shown in Figure 9.1.

The simplest form of asynchronous logic is a latch, which is an essential part of most flip-flop designs. In this section, we are concerned with larger blocks of logic that have global feedback lines. Problems associated with hazards are more critical in asynchronous circuits than in synchronous circuits. Also, asynchronous circuits exhibit unique problems in terms of races. The test generation problem for such circuits is more difficult than for synchronous circuits, mainly because signals can propagate through the circuitry one or more times before the circuit stabilizes. For these reasons, asynchronous circuits other than latches should be avoided when possible. If it is not feasible to eliminate such circuits, then the global feedback lines should be made controllable and observable as described previously.

Further information on ad hoc design-for-test techniques can be found in [Davidson 1979], [Grason and Nagel 1981], [Lippman and Donn 1979], and [Writer 1975].

9.3 Controllability and Observability by Means of Scan Registers

By using test points one can easily enhance the observability and controllability of a circuit. But enhancement can be costly in terms of I/O pins. Another way to enhance observability and/or controllability is by using a *scan register* (SR). A scan register is a register with both shift and parallel-load capability. The storage cells in the register are used as observation points and/or control points. The use of scan registers to replace I/O pins deals with a trade-off between test time, area overhead, and I/O pins. Figure 9.12 shows a generic form of a scan storage cell (SSC) and corresponding scan register. Here, when $\overline{N}/T = 0$ (normal mode), data are loaded into the scan storage cell from the data input line (D); when $\overline{N}/T = 1$ (test mode), data are loaded from S_i. A scan register R shifts when $\overline{N}/T = 1$, and loads data in parallel when $\overline{N}/T = 0$. Loading data into R from line S_{in} when $\overline{N}/T = 1$ is referred to as a *scan-in* operation; reading data out of R from line S_{out} is referred to as a *scan-out* operation.

We will next look a little closer at the number of variations used to inject and/or observe test data using scan registers.

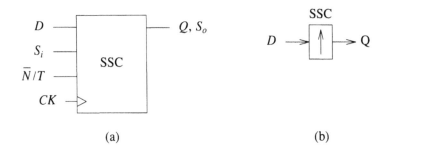

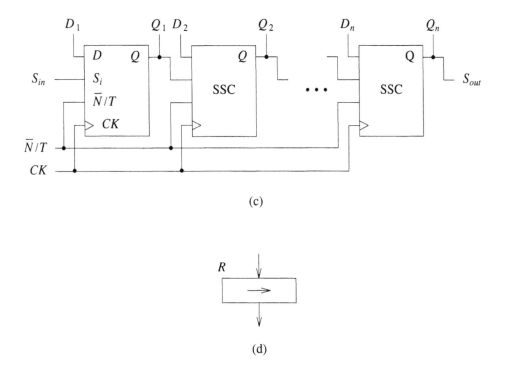

Figure 9.12   (a) A scan storage cell (SSC) (b) Symbol for a SSC (c) A scan register (SR) or shift register chain (d) Symbol for a scan register

Simultaneous Controllability and Observability

Figure 9.13(a) shows two complex circuits C_1 and C_2. They can be either combinational or sequential. Only one signal (Z) between C_1 and C_2 is shown.

Figure 9.13(b) depicts how line Z can be made both observable and controllable using a scan storage cell. Data at Z can be loaded into the SSC and observed by means of a scan-out operation. Data can be loaded into the SSC via a scan-in operation and then injected onto line Z'. Simultaneous controllability and observability can be achieved. That is, the scan register can be preloaded with data to be injected into the circuit.

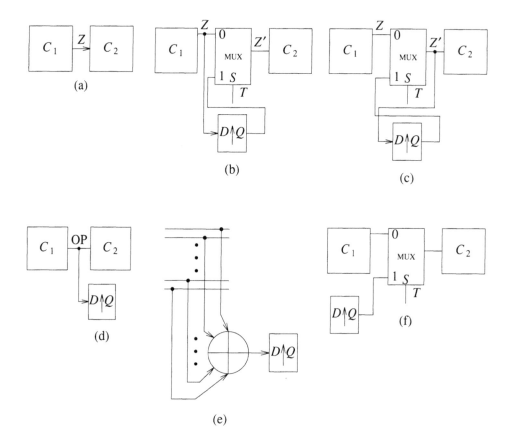

Figure 9.13   (a) Normal interconnect (b) Simultaneous C/O (c) Separate C/O
(d) Observability circuit (e) Compacting data (f) Controllability
circuit

The circuit can be run up to some time t. At time $t + 1$, if $T = 1$, then the data in the
SSC will be injected onto Z'; if $\overline{N}/T = 0$, the data at Z will be loaded into the SSC.

Nonsimultaneous Controllability and Observability

In Figure 9.13(c) we see a variant on the design just presented. Here we can either
observe the value Z' by means of the scan register or control the value of line Z'.
Both cannot be done simultaneously.

Figure 9.14 shows a more complex scan storage cell for use in sequential circuits that
are time sensitive, such as asynchronous circuits. The Q_2 flip-flop is used as part of
the scan chain; i.e., by setting $T2 = 1$ and clocking $CK2$ the scan storage cells form a
shift register. By loading the Q_1 latch and setting $T1 = 1$, signals can be injected into
the circuit. Similarly, by setting $T2 = 0$ and clocking $CK2$, data can be loaded into the
scan register. One scenario for using this cell is outlined below.

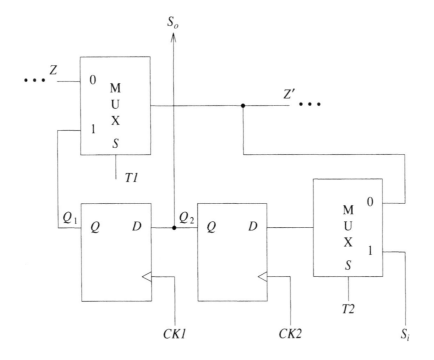

Figure 9.14   Complex scan storage cell

1. Load the scan register with test data by setting $T2 = 1$ and clocking $CK2$.

2. Drive the UUT to a predefined state with $T1 = 0$.

3. Load Q_2 into latch Q_1.

4. Optional

   a. Load $Z' = Z$ into Q_2 by setting $T2 = 0$ and clocking $CK2$ once.

   b. Scan-out the data in the Q_2 flip-flops.

5. Inject signals into the circuit by setting $T1 = 1$.

6. (Optional) Clock the UUT one or more times.

7. Observe the observation points by setting $T2 = 0$ and clock $CK2$ once.

8. Scan out these data by setting $T2 = 1$ and clocking $CK2$.

This last operation corresponds to step 1; i.e., data can be scanned-in and scanned-out of a scan register at the same time.

Observability Only

Figure 9.13(d) shows how an observation point can be tied to a storage cell in a scan register to gain observability. In Figure 9.13(e) the number of storage cells has been reduced by combining many observation points through a parity tree into a single scan cell. This technique detects single errors but may fail to detect faults affecting two signals.

Controllability Only

If only controllability is required, then the circuit of Figure 9.13(f) can be used.

Note that in all cases shown it was assumed that controllability required the ability to inject either a 0 or a 1 into a circuit. If this is not the case, then the MUX can be replaced by an AND, NAND, OR, or NOR gate.

Applications

Figure 9.15 shows a sequential circuit S having inputs X and outputs Z. To enhance controllability, control points have been added, denoted by X'. These can be driven by a scan register R_1. To enhance observability, observation points have been added, denoted by Z'. These can be tied to the data input ports of a scan register R_2. Thus X' act as pseudo-primary inputs and Z' as pseudo-primary outputs. Using X' and Z' can significantly simplify ATG. Assume one has run a sequential ATG algorithm on S and several faults remain undetected. Let f be such a fault. Then by simulating the test sequence generated for the faults that were detected, it is usually possible to find a state s_0 such that, if signals were injected at specific lines and a specific line were made observable, the fault could be detected. These points define where CPs and OPs should be assigned. Now to detect this fault the register R_1 is loaded with the appropriate scan data. An input sequence is then applied that drives the circuit to state s_0. Then the input X' is applied to S. The response Z and Z' can then be observed and the fault detected. By repeating this process for other undetected faults, the resulting fault coverage can be significantly increased. Of course, this increase is achieved at the expense of adding more scan storage cells. Note that the original storage cells (latches and flip-flops) in the circuit need not be modified. Finally, using the appropriate design for the scan storage cells, it is possible to combine registers R_1 and R_2 into a single register R where each scan storage cell in R is used both as a CP and an OP. The scan storage cell shown in Figure 9.14 is appropriate for this type of operation.

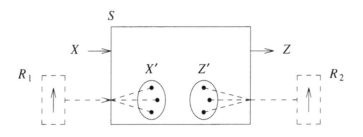

Figure 9.15   General architecture using test points tied to scan registers

Figure 9.16 illustrates a circuit that has been modified to have observation and control points to enhance its testability. The original circuit consists of those modules denoted by heavy lines. The circuits denoted by C_1, C_2, ..., are assumed to be complex logic blocks, either combinational or sequential. Most of their inputs and outputs are not shown. To inject a 0 into the circuit, an AND gate is used, e.g., G_1; to inject a 1, an OR gate is used, e.g., G_2. If such gates do not already exist in the circuit, they can be added. To inject either a 0 or a 1, a MUX is used, e.g., G_3 and G_4. Assume that the scan register R is able to hold its contents by disabling its clock. One major function of the test hardware shown is to observe data internal to the circuit via the OP lines. This process is initiated by carrying out a parallel load on R followed by a scan-out operation. The other major function of this circuit is to control the CP lines. These data are loaded into R via a scan-in operation. The data in Q_4 and Q_5 do not affect the circuit until $T = 1$. Then if $Q_1 = 1$, the data in Q_4 and Q_5 propagate through the MUXs; if Q_2 is 1, a 0 is injected at the output of G_1; if $Q_3 = 1$, a 1 is injected at the output of G_2. Since signals $\overline{N}/T$, S_{in}, S_{out}, T, and CK each require a separate pin, five additional I/O pins are required to implement this scheme. The number of CPs and OPs dictates the length of the register R. Note that the same storage cell can be used both to control a CP and accept data from an OP. The problem with this scheme is that the test data as supplied by R cannot change each clock time. Normally the circuit is logically partitioned using the CPs. N test vectors are then applied to the circuit. Data are then collected in R via the OPs and shifted out for observations. If the register is of length n, then n clock periods occur before the next test vector is applied. This scheme is particularly good for PCBs, where it is not easy to obtain CPs.

Several variations of this type of design philosophy can be incorporated into a special chip designed specifically to aid in making PCBs more testable. One such structure is shown in Figure 9.17. The normal path from A to B is broken when $Bl = 1$ and the top row of MUXs is used to inject data into a circuit from the scan register. The lower row of MUXs is used for monitoring data within the circuit. This circuit can be further enhanced by using MUXs and DEMUXs as shown earlier to concentrate several observation points into one or to control many control points from one scan storage cell.

9.3.1 Generic Boundary Scan

In designing modules such as complex chips or PCBs, it is often useful for purposes of testing and fault isolation to be able to isolate one module from the others. This can be done using the concept of *boundary scan*, which is illustrated in Figure 9.18. Here, the original circuit S has n primary inputs X and m primary outputs Y (see Figure 9.18(a)). In the modified circuit, shown in Figure 9.18(b), R_1 can be now used to observe all input data to S, and R_2 can be used to drive the output lines of S. Assume all chips on a board are designed using this boundary scan architecture. Then all board interconnects can be tested by scanning in test data into the R_2 register of each chip and latching the results into the R_1 registers. These results can then be verified by means of a scan-out operation. A chip can be tested by loading a test vector into the R_2 registers that drive the chip. Thus normal ATE static chip tests can be reapplied for in-circuit testing of chips on a board. The clocking of S must be inhibited as each test is loaded into the scan path. Naturally, many test vectors need to be processed fully to test a board. There are many other ways to implement the concept of boundary scan. For example, the storage cell shown in Figure 9.14 can be

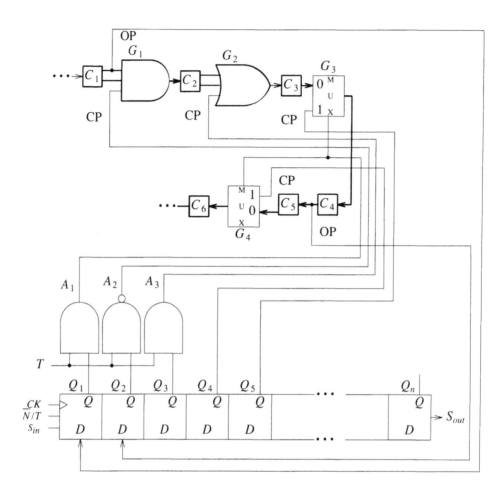

Figure 9.16   Adding controllability and observability to a circuit

inserted in every I/O line. The scan registers discussed so far fall into the category of being isolated scan or shadow registers. This is because they are not actually part of the functional circuitry itself. Scan registers that are part of the functional registers are referred to as integrated scan registers. They will be discussed next. Aspects of interconnect testing and related design-for-test techniques are discussed in [Goel and McMahon 1982].

9.4   Generic Scan-Based Designs

The most popular structured DFT technique used for external testing is referred to as *scan design* since it employs a scan register. We assume that the circuit to be made testable is synchronous. There are several forms of scan designs; they differ primarily in how the scan cells are designed. We will illustrate three generic forms of scan design and later go into the details for how the registers can be designed.

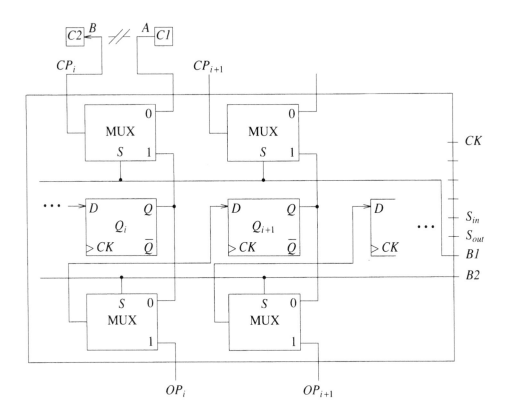

Figure 9.17 Controllability/observability circuitry with a scan chain

Usually these designs are considered to be *integrated* scan-based designs because all functional storage cells are made part of the scan registers. Thus the selection of what lines to be made observable or controllable becomes a moot point. So these techniques are referred to as *structured* rather than ad hoc.

9.4.1 Full Serial Integrated Scan

In Figure 9.19(a) we illustrate the classical Huffman model of a sequential circuit, and in Figure 9.19(b) the full scan version of the circuit. Structurally the change is simple. The normal parallel-load register R has been replaced by a scan register R_s. When $\overline{N}/T = 0$ (normal mode), R_s operates in the parallel-latch mode; hence both circuits operate the same way, except that the scan version may have more delay. Now Y becomes easily controllable and E easily observable. Hence test generation cost can be drastically reduced. Rather than considering the circuit S of Figure 9.19(a) as a sequential circuit, test generation can proceed directly on the combinational circuit C using any of a variety of algorithms, such as PODEM or FAN. The result is a series of test vectors (x_1,y_1), (x_2,y_2), ... and responses (z_1,e_1), (z_2,e_2), ... To test S^*, set $\overline{N}/T = 1$ and scan y_1 into R_s. During the k-th clock time, apply x_1 to X. Now the first test pattern, $t_1 = (x_1,y_1)$, is applied to C. During the $(k+1)st$ clock time, set

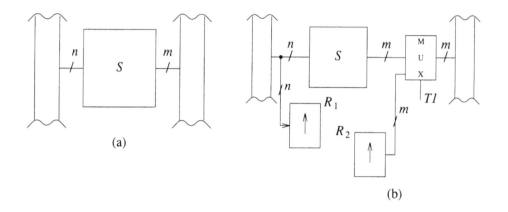

Figure 9.18   Boundary scan architecture (a) Original circuit (b) Modified
circuit

$\bar{N}/T = 0$ and load the state of E, which should be e_1, into R_s while observing the
response on Z, which should be z_1. This process is then repeated; i.e., while y_2 is
scanned into R_s, e_1 is scanned out and hence becomes observable. Thus the
response $r_1 = (z_1, e_1)$ can be easily observed. The shift register is tested by both the
normal test data for C and by a shift register test sequence, such as 01100xx ..., that
tests the setting, resetting, and holding of the state of a storage cell. It is thus seen that
the complex problem of testing a sequential circuit has been converted to a much
simpler one of testing a combinational circuit.

This concept is referred to as *full serial integrated scan* since all the original storage
cells in the circuit are made part of the scan register, and the scan register is used as a
serial shift register to achieve its scan function. Normally the storage cells in scan
designs do not have reset lines for global initialization; instead they are initialized by
means of shifting data into the SCAN register.

9.4.2 Isolated Serial Scan

Isolated serial scan designs differ from full serial integrated scan designs in that the
scan register is not in the normal data path. A common way of representing this
architecture is shown in Figure 9.20; it corresponds closely to that shown in
Figure 9.15.

This scan architecture is somewhat ad hoc since the selection of the CPs and OPs
associated with the scan register R_s is left up to the designer. Hence S may remain
sequential, in which case test generation may still be difficult. If R_s is used both to
observe and control all the storage cells in S, then the test generation problem again is
reduced to one of generating tests for combinational logic only. This design is shown
in Figure 9.21 and is referred to as *full isolated scan*. Here, S' consists of the circuit C
and register R', and R has been modified to have two data input ports. The testing of
this circuit is now similar to that for full serial scan designs. A test vector y_1 is
scanned (shifted) into R_s, loaded into R', and then applied to the circuit C. The

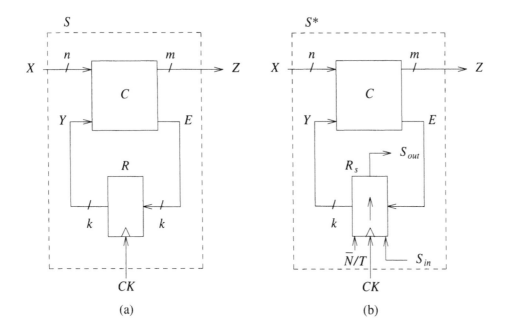

Figure 9.19 (a) Normal sequential circuit S (b) Full serial integrated scan version for circuit

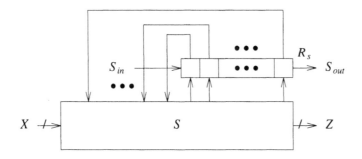

Figure 9.20 Isolated serial scan (scan/set)

response e_1 can be loaded into R', transferred to R_s, and then scanned out. The register R_s is said to act as a *shadow register* to R'. The overhead for this architecture is high compared to that for full serial integrated scan designs. Isolated scan designs have several useful features. One is that they support some forms of real-time and on-line testing. Real-time testing means that a single test can be applied at the operational clock rate of the system. In normal full serial scan, a test vector can only be applied at intervals of k clock periods. On-line infers that the circuit can be tested while in normal operation; i.e., a snapshot of the state of the circuit can be taken and loaded into R_s. This data can be scanned out while S continues normal operation. Finally this architecture supports latch-based designs; i.e., register R and hence R' can consist of just latches rather than flip-flops. It is not feasible to string these latches together to form a shift register; hence adding extra storage cells to form a scan register is required.

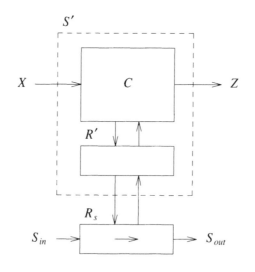

Figure 9.21   Full isolated scan

9.4.3 Nonserial Scan

Nonserial scan designs are similar to full serial integrated scan designs in that they aim to give full controllability and observability to all storage cells in a circuit. The technique differs from the previous techniques in that a shift register is not used. Instead the storage cells are arranged as in a random-access bit-addressable memory. (See Figure 9.22.) During normal operation the storage cells operate in their parallel-load mode. To scan in a bit, the appropriate cell is addressed, the data are applied to S_{in}, and a pulse on the scan clock SCK is issued. The outputs of the cells are wired-ORed together. To scan out the contents of a cell, the cell is addressed, a control signal is broadcast to all cells, and the state of the addressed cell appears at S_{out}. The major advantage of this design is that to scan in a new test vector, only bits in R that need be changed must be addressed and modified; also selected bits can be observed. This saves scanning data through the entire register. Unfortunately the overhead is high for this form of scan design. There is also considerable overhead associated with storing the addresses of the cells to be set and/or read.

9.5 Storage Cells for Scan Designs

Many storage cell designs have been proposed for use in scan cell designs. These designs have several common characteristics. Because a scan has both a normal data input and a scan data input, the appropriate input can be selected using a multiplexer controlled by a

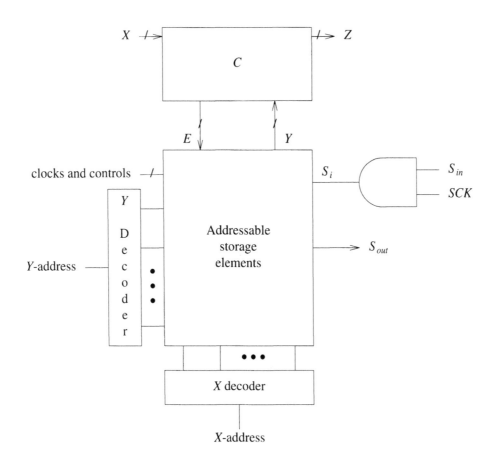

Figure 9.22   Random-access scan

normal/test ($\overline{N}/T$) input or by a two-clock system. Also, the cell can be implemented using a clocked edge-triggered flip-flop, a master-slave flip-flop, or level-sensitive latches controlled by clocks having two or more phases. Designs having up to four clocks have been proposed. In our examples master-slave rather than edge-triggered flip-flops will be used. Also only D flip-flops and latches will be discussed.

Several sets of notation will be used. Multiple data inputs will be denoted by $D1$, $D2$, ..., Dn, and multiple clocks by $CK1$, $CK2$, ..., CKm. In addition, if a latch or flip-flop has a single-system clock, the clock will be denoted by CK, a single-scan clock by SK, a scan data input by S_i, and a scan data output by S_o. Also, sometimes the notation used in some of the published literature describing a storage cell will be indicated.

Figure 9.23(a) shows a NAND gate realization of a clocked D-latch, denoted by L. This is the basic logic unit in many scan storage cell designs.

Figure 9.23(b) shows a two-port clocked master-slave flip-flop having a multiplexer on its input and denoted by (MD-F/F). Normal data (D) enter at port $1D$ when $\overline{N}/T = 0$. The device is in the scan mode when $\overline{N}/T = 1$, at which time scan data (S_i) enter at port $2D$.

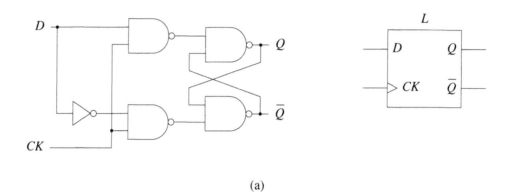

(a)

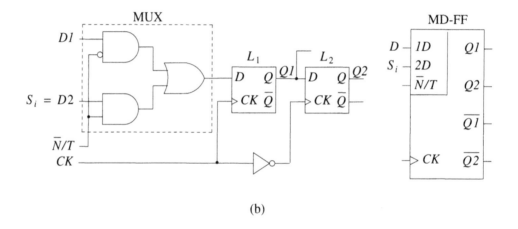

(b)

Figure 9.23   Some storage cell designs (a) Clocked D-latch and its symbol
(b) Multiplexed data flip-flop and its symbol (MD-FF)

Sometimes it is useful to separate the normal clock from the clock used for scan purposes. Figure 9.24(a) shows a two-port clocked flip-flop (2P-FF), which employs two clocks and a semimultiplexed input unit.

It is often desirable to insure race-free operation by employing a two-phase nonoverlapping clock. Figure 9.24(b) shows a two-port shift register latch consisting of latches L_1 and L_2 along with a multiplexed input (MD-SRL). This cell is not considered a flip-flop, since each latch has its own clock. Also, since this type of design is used primarily in scan paths, the term "shift" is used in its name.

To avoid the performance degradation (delay) introduced by the MUX in an MD-SRL, a two-port shift register latch (2P-SRL) can be used, as shown in Figure 9.25. This circuit is the NAND gate equivalent to the shift register latch used in a level-sensitive scan design (LSSD) methodology employed by IBM [Eichelberger and Williams 1977]. Note that three clocks are used. The inputs have the following functions (the notation in parenthesis is used by IBM):

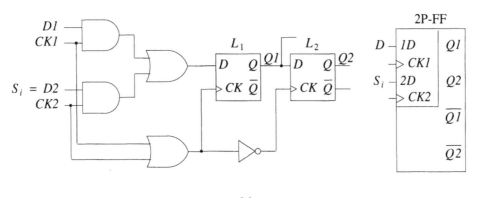

(a)

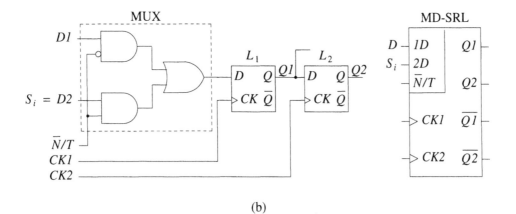

(b)

Figure 9.24   (a) Two-port dual-clock flip-flop and its symbol (2P-FF)
(b) Multiplex data shift register latch and its symbol (MD-SRL)

$D1(D)$ is the normal data input.
$D2$ or $S_i(I)$ is the scan data input.
$CK1(C)$ is the normal system clock.
$CK2(A)$ is the scan data input clock.
$CK3(B)$ is the L_2 latch clock.

If $CK1$ and $CK2$ are NORed together and used to clock L_2, then $CK3$ can be deleted. The result would be a two-port flip-flop.

Figure 9.26 shows a raceless master-slave D flip-flop. (The concept of races in latches and ways to eliminate them will be covered when Figure 9.28 is discussed.) Two clocks are used, one (CK) to select and control normal operation, the other (SK) to select scan data and control the scan process. In the normal mode, $SK = 1$ blocks scan data on S_i from entering the master latch; i.e., $G_1 = 1$. Also $G_7 = 1$. $CK = 0$ enables the value of the data input D to be latched into the master latch. When CK goes to 1, the state of the

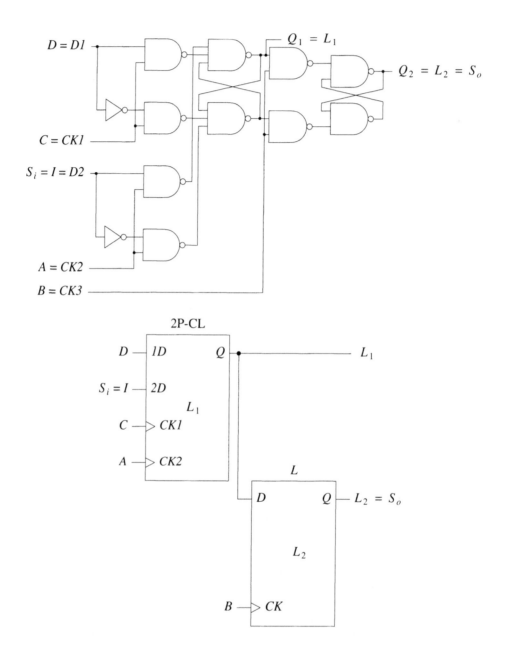

Figure 9.25 Two-port shift register latch and its symbol (2P-SRL)

master is transferred to the slave. Similarly, when $CK = 1$ and a pulse appears on SK, scan data enter the master and are transferred to the slave.

The random-access scan design employs an addressable polarity-hold latch. Several latch designs have been proposed for this application, one of which is shown in Figure 9.27.

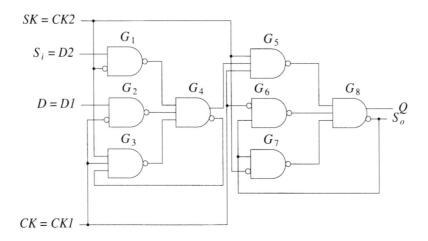

Figure 9.26 Raceless two-port D flip-flop

Since no shift operation occurs, a single latch per cell is sufficient. Note the use of a wired-AND gate. Latches that are not addressed produce a scan-out value of $S_o = 1$. Thus the S_o output of all latches can be wired-ANDed together to form the scan-out signal S_{out}.

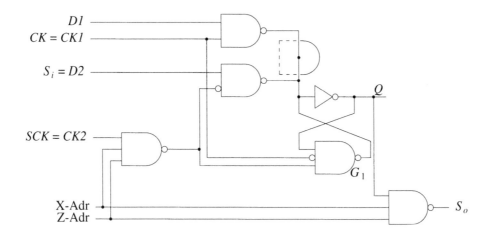

Figure 9.27 Polarity-hold addressable latch

The design of the storage cell is a critical aspect of a scan-based design. It is clearly important to achieve reliable operation. Hence a scan cell must be designed so that races and hazards either do not affect it or do not occur. In practice, storage cells are designed at the transistor level, and the gate-level circuits shown represent approximations to the actual operation of a cell. We will next consider one cell in somewhat more detail. Figure 9.28(a) shows a simple latch design. Figure 9.28(b) shows a logically equivalent design where G_1 has been replaced by a wired-AND and an inverter. The resulting latch requires only two NAND gates and two inverters, rather than the latch shown in Figure 9.23(a), which has four NAND gates and one inverter. Q is set to the value of D when $CK = 0$. Consider the case when $CK = 0$ and $D = 1$ (see Figure 9.28(b)). Let CK now go from 0 to 1. Because of the reconvergent fanout between CK and G_1', a race exists. If the effect of the clock transition at G_5 occurs before that at G_3, then $G_5 = 0$ and G_6 remains stable at 1. If, however, the effect at G_3 occurs first, then $G_3 = 1$, G_6 goes to 0, and this 0 keeps G_5 at a 1. One can attempt to rectify this situation by changing the threshold level of G_2 so that G_5 changes before G_2. This solution works as long as the fabrication process is reliable enough accurately to control the transition times of the gates. A somewhat simpler solution is to add a gate G_4 to eliminate the inherent hazard in this logic (see Figure 9.28(c)). Now, $G_4 = 0$ keeps G_6 at 1 as the clock drops from 1 to 0. The problem with this design is that G_4 is redundant; i.e., there is no static test for G_4 s-a-1.

Many other storage cells have been devised to be used in scan chains. Some will be discussed later in this chapter. In the next section we will discuss some specific scan approaches proposed by various researchers.

9.6 Classical Scan Designs

Scan Path

One of the first full serial integrated scan designs was called Scan Path [Kobayashi *et al.* 1968, Funatsu *et al.* 1975]. This design employs the generic scan architecture shown in Figure 9.19(b) and uses a raceless master-slave D flip-flop, such as the one shown in Figure 9.26.

Shift Register Modification

The scan architecture shown in Figure 9.19(b) using a MD-FF (see Figure 9.23(b)) was proposed by Williams and Angell [1973] and is referred to as Shift Register Modification.

Scan/Set

The concept of Scan/Set was proposed by Stewart [1977, 1978] and uses the generic isolated scan architectures shown in Figures 9.20 and 9.21. Stewart was not specific in terms of the types of latches and/or flip-flops to use.

Random-Access Scan

The Random-Access Scan concept was introduced by Ando [1980] and uses the generic nonserial scan architecture shown in Figure 9.22 and an addressable storage cell such as the one shown in Figure 9.27.

Level-Sensitive Scan Design (LSSD)

IBM has developed several full serial integrated scan architectures, referred to as Level-Sensitive Scan Design (LSSD), which have been used in many IBM systems [Eichelberger and Williams 1977, 1978, DasGupta *et al.* 1981]. Figure 9.25 shows one

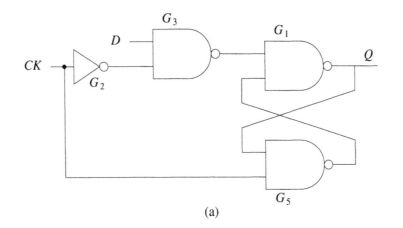

(a)

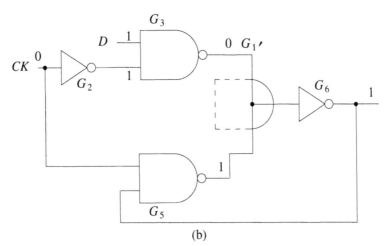

(b)

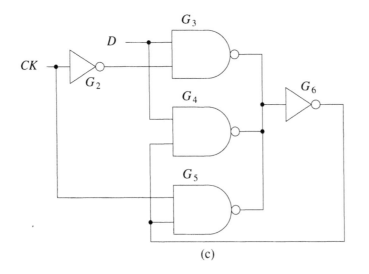

(c)

Figure 9.28   Analysis of a latch

design that uses a polarity-hold, hazard-free, and level-sensitive latch. When a clock is enabled, the state of a latch is sensitive to the level of the corresponding data input. To obtain race-free operation, clocks C and B as well as A and B are nonoverlapping.

Figure 9.29 shows the general structure for an LSSD *double-latch design*. The scan path is shown by the heavy dashed line. Note that the feedback Y comes from the output of the L_2 latches. In normal mode the C and B clocks are used; in test mode the A and B clocks are used.

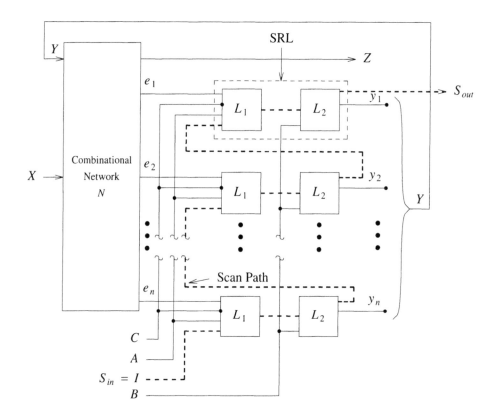

Figure 9.29 LSSD double-latch design

Sometimes it is desired to have combinational logic blocks separated by only a single latch rather than two in sequence. For such cases a *single-latch design* can be used. Figure 9.30 shows such a design. Here, during normal operation two-system clocks C_1 and C_2 are used in a nonoverlapping mode. The A and B clocks are not used; hence the L_2 latches are not used. The output Y_1 is an input to logic block N_2, whose outputs Y_2 are latched by clock C_2 producing the output at Y_2. The L_2 latches are used only when in the scan test mode; then the SRLs are clocked by A and B.

The LSSD latch design shown in Figure 9.25 has two problems. One is logic complexity. Also, when it is used in a single-latch design (see Figure 9.30), only L_1 is used during

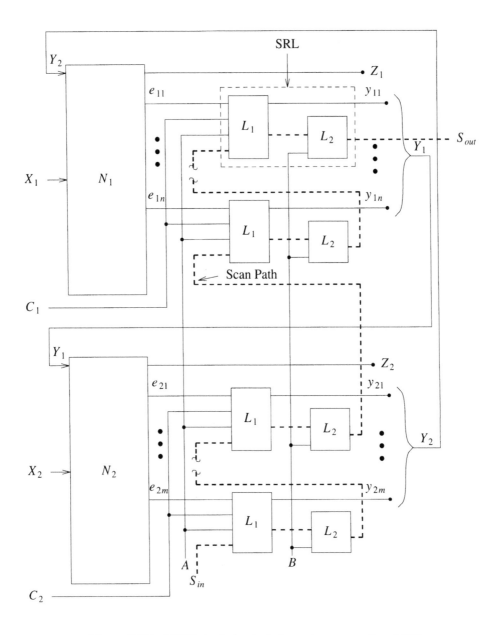

Figure 9.30 LSSD single-latch design using conventional SRLs

normal operation; L_2 is used only for shifting test data. A variation on this latch, known as the L_2^* latch, which reduces gate overhead, was reported in [DasGupta *et al.* 1982] and is shown in Figure 9.31. The primary difference between the *2P-SRL* of Figure 9.25 and L_2^* of Figure 9.31 is that L_2^* employs an additional clocked data port D^* and an additional clock C^*.

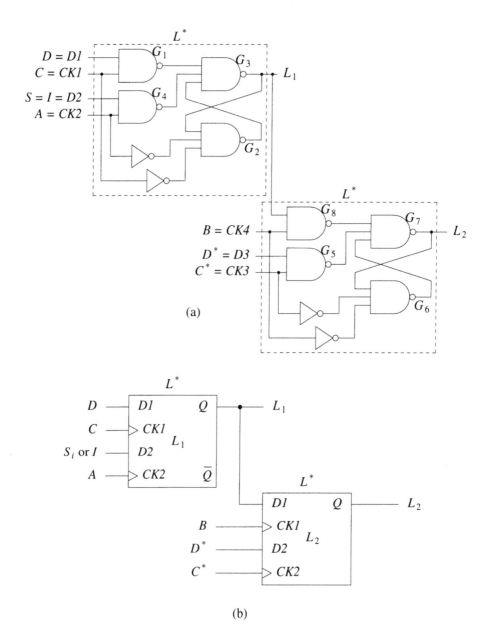

Figure 9.31   SRL using L_2^* latch with two data ports (a) Gate model
(b) Symbol

Using this latch in a single-latch SRL design, the outputs of N_1 are the D inputs to an SRL latch, while the outputs of N_2 are the D^* inputs (see Figure 9.32).

It is important to note that for the design shown in Figure 9.32, it is not possible to test N_1 and N_2 at the same time. That is, since an L_1 and L_2 pair of latches make up one

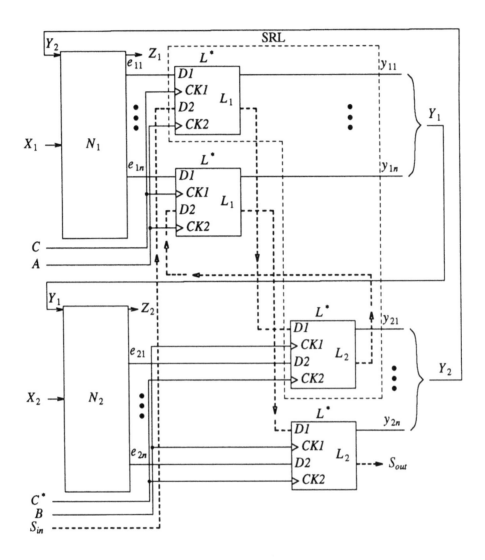

Figure 9.32 Single-latch scan design using SRLs with the L_2^* latch

SRL, only a test vector corresponding to Y_1 or Y_2 can be shifted through the scan chain, but not both Y_1 and Y_2.

LSSD Gate Overhead

Gate overhead is one important issue related to scan design. We will next compute this overhead for the three LSSD schemes presented. Let K be the ratio of combinational logic gates to latches in a nonscan design, such as that of Figure 9.30 without the L_2 latches. Assume a race-free single-latch architecture; i.e., phase 1 clocked latches feed a logic network N_2, which feeds phase 2 clocked latches, which feed a logic network N_1, which feeds the phase 1 clocked latches, again as in Figure 9.30.

Referring to Figure 9.31, a latch requires gates G_1, G_2, and G_3. Gates G_4, G_6, G_7, and G_8 are required to form an SRL. Thus for every latch in the original design, four extra gates are required to produce an SRL, resulting in the gate overhead of $(4/(K+3)) \times 100$ percent. For the L_2^* design, the L_1 and L_2 latches are both used during system operation; hence, the only extra gates are G_4 and G_5, leading to an overhead of $(1/(K+3)) \times 100$ percent; i.e., there is only one extra gate per latch. If the original design uses a double-latch storage cell with three gates per latch, then the overhead required to produce a double-latch scan design is $\left| \dfrac{1/2}{K+3} \right| \times 100$ percent. Figure 9.33 shows a plot of overhead as a function of K for these three scan designs.

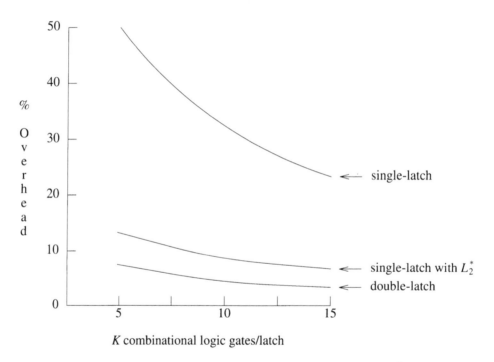

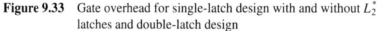

Figure 9.33   Gate overhead for single-latch design with and without L_2^* latches and double-latch design

LSSD Design Rules

In addition to producing a scan design, the design rules that define an LSSD network are intended to insure race-free and hazard-free operation. This is accomplished in part by the use of a level-sensitive rather than an edge-sensitive storage cell, as well as by the use of multiple clocks.

A network is said to be *level-sensitive* if and only if the steady state response to any of the allowed input changes is independent of the transistor and wire delays in that network.

Hence races and hazards should not affect the steady-state response of such a network. A level-sensitive design can be achieved by controlling when clocks change with respect to when input data lines change.

The rules for level-sensitive scan design are summarized as follows:

1. All internal storage elements must consist of polarity-hold latches.

2. Latches can be controlled by two or more nonoverlapping clocks that satisfy the following conditions.

    a. A latch X may feed the data port of another latch Y if and only if the clock that sets the data into latch Y does not clock latch X.

    b. A latch X may gate a clock C_i to produce a gated clock C_{ig}, which drives another latch Y if and only if clock C_{ig}, or any clock C_{ig} produced from C_{ig}, does not clock latch X.

3. There must exist a set of clock primary inputs from which the clock inputs to all SRLs are controlled either through (1) single-clock distribution trees or (2) logic that is gated by SRLs and/or nonclock primary inputs. In addition, the following conditions must hold:

    a. All clock inputs to SRLs must be at their "off" states when all clock primary inputs are held to their "off" states.

    b. A clock signal at any clock input of an SRL must be controlled from one or more clock primary inputs. That is, the clock signal must be enabled by one or more clock primary inputs as well as setting the required gating conditions from SRLs and/or nonclocked primary inputs.

    c. No clock can be ANDed with either the true or the complement of another clock.

4. Clock primary inputs cannot feed the data inputs to latches, either directly or through combinational logic. They may only feed clock inputs to latches or primary outputs.

    A network that satisfies these four rules is *level-sensitive*. The primary rule that provides for race-free operation is rule 2a, which does not allow one latch clocked by a given clock to feed another latch driven by the same clock. Rule 3 allows a test generation system to turn off system clocks and use the shift clocks to force data into and out of the scan latches. Rule 4 is also used to avoid races.

    The next two rules are used to support scan.

5. Every system latch must be part of an SRL. Also, each SRL must be part of some scan chain that has an input, output, and shift clocks available as primary inputs and/or outputs.

6. A scan state exists under the following conditions:

    a. Each SRL or scan-out primary output is a function of only the preceding SRL or scan-in primary input in its scan chain during the scan operation

    b. All clocks except the shift clocks are disabled at the SRL inputs

    c. Any shift clock to an SRL can be turned on or off by changing the corresponding clock primary input.

9.7 Scan Design Costs

Several attributes associated with the use of scan designs are listed below.

1. Flip-flops and latches are more complex. Hence scan designs are expensive in terms of board or silicon area.

2. One or more additional I/O are required. Note that some pins can be multiplexed with functional pins. In LSSD, four additional pins are used (S_{in}, S_{out}, A, and B).

3. With a given set of test patterns, test time per pattern is increased because of the need to shift the pattern serially into the scan path. The total test time for a circuit also usually increases, since the test vector set for a scan-path design is often not significantly smaller than for a nonscan design.

4. A slower clock rate may be required because of the extra delay in the scan-path flip-flops or latches, resulting in a degradation in performance. This performance penalty can be minimized by employing storage cells that have no additional delay introduced in series with the data inputs, such as the one shown in Figure 9.26.

5. Test generation costs can be significantly reduced. This can also lead to higher fault coverage.

6. Some designs are not easily realizable as scan designs.

9.8 Board-Level and System-Level DFT Approaches

By a *system* we mean a collection of modules, such as PCBs, which consist of collections of ICs. Many of the ad hoc DFT techniques referred to earlier apply to the board level. Structural techniques, such as scan, can also be applied at the board and system levels, assuming chips are designed to support these techniques. The primary system-level DFT approaches use existing functional busses, scan paths, and boundary scan.

9.8.1 System-Level Busses

This DFT approach makes use of a module's or system's functional bus to control and observe signals during functional level testing. A test and/or maintenance processor, such as the ATE, appears as another element attached to the system's busses. Figure 9.34 shows a simple bus-oriented, microprocessor-based system. During testing, the ATE can take control of the system busses and test the system. This is often done by emulating the system's processing engine. The ATE can also emulate the various units attached to the bus, monitor bus activity, and test the processing engine. In general, complex, manually generated functional tests are used.

9.8.2 System-Level Scan Paths

Figure 9.35 shows how the concept of scan can be extended to the board and system levels. Here the scan path of each chip on a board is interconnected in a daisy chain fashion to create one long scan path on each board. The boards all share a common S_{in}, $\overline{N/T}$, and CK input. Their S_{out} lines are wired-ORed together. The testing of such a system is under the control of a system-maintenance processor, which selects that board to be attached to the board-level scan line S_{out}. The interconnect that is external to the boards can be considered to be a test bus. It is seen that starting with a structured DFT approach at the lowest level of the design, i.e., at the chip level, leads to a hierarchical

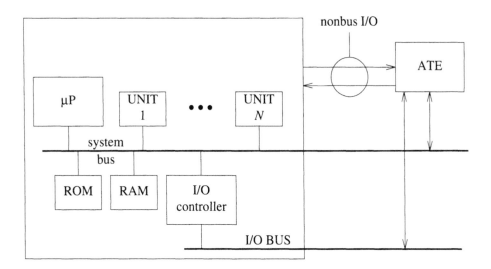

Figure 9.34   System-level test using system bus

DFT methodology. Assuming chips have boundary scan, tests developed for chips at the lowest level can be used for testing these same components at the higher levels. These tests must be extended to include tests for the interconnect that occurs at these higher levels.

9.9 Some Advanced Scan Concepts

In this section several advanced concepts related to scan-based designs are presented. The topics include the use of multiple test sessions and partial scan. *Partial scan* refers to a scan design in which a subset of the storage cells is included in the scan path. These techniques address some of the problems associated with full-scan designs discussed in Section 9.7.

9.9.1 Multiple Test Session

A *test session* consists of configuring the scan paths and other logic for testing blocks of logic, and then testing the logic using the scan-test methodology. Associated with a test session are several parameters, including the number of test patterns to be processed and the number of shifts associated with each test pattern. Often test application time can be reduced by using more than one test session. Consider the circuit shown in Figure 9.36, where C_1 has 8 inputs and 4 outputs and can be tested by 100 test patterns, and C_2 has 4 inputs and 8 outputs and can be tested by 20 test patterns. Thus the entire circuit can be tested by 100 test patterns, each of length 12, and the total test time is approximately $100 \times 12 = 1200$ clock cycles, where we have ignored the time required to load results into R_2 and R_4 as well as scan out the final result. Note that when a new test pattern is scanned into R_1 and R_3, a test result is scanned out of R_2 and R_4. We have just described one way of testing C_1 and C_2, referred to as the *together mode*. There are two other ways for testing C_1 and C_2, referred to as *separate mode* and *overlapped mode*.

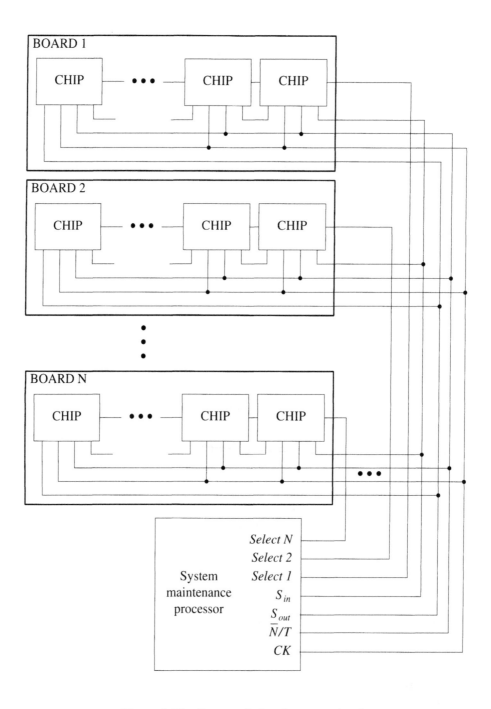

Figure 9.35   Scan applied to the system level

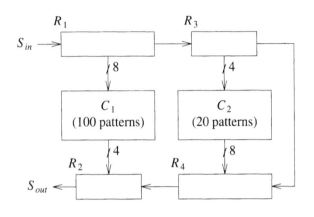

Figure 9.36   Testing using multiple test sessions

Separate Mode

The blocks of logic C_1 and C_2 can be tested *separately*. In this mode, while C_1 is being tested, C_2, R_3, and R_4 are ignored. To test C_1 it is only necessary to load R_1 with a test pattern, and capture and scan out the result in R_2. Let $|R_i|$ be the length of register R_i. Since max $\{|R_1|,|R_2|\} = 8$, only 8×100 = 800 clock cycles are required to test C_1. To test C_2 in a second test session the scan path needs to be reconfigured by having the scan output of R_4 drive a primary output. This can also be accomplished by disabling the parallel load inputs from C_1 to R_2. We will assume that the latter choice is made. Now C_2 can be tested by loading test patterns into the scan path formed by R_1 and R_3. Each test pattern can be separated by four don't-care bits. Thus each test pattern appears to be eight bits long. When a new test is loaded into R_3, a test result in R_4 must pass through R_2 before it is observed.

The result in R_4, when shifted eight times, makes four bits of the result available at S_{out} and leaves the remaining four bits in R_2. To test C_2 requires 20×8=160 clock cycles. The total time to test C_1 and C_2 is thus 960 clock cycles, compared to the previous case, which required 1200 clock cycles.

Overlapped Mode

C_1 and C_2 can be tested in an *overlapped mode*, that is, partly as one block of logic, and partly as separate blocks. Initially C_1 and C_2 can be combined and tested with 20 patterns, each 12 bits wide. This initial test requires 12×20=240 clock cycles. Now C_2 is completely tested and C_1 can be tested with just 80 of the remaining 100 test patterns. To complete the test of C_1, the scan path need not be reconfigured, but the length of each test pattern is now set to 8 rather than 12. The remaining 80 test patterns are now applied to C_1 as in the separate mode; this requires 80×8=640 clock cycles for a total of 880 clock cycles.

Therefore, by testing the various partitions of logic either together, separately, or in an overlapped mode, and by reorganizing the scan path, test application time can be reduced. No one technique is always better than another. The test time is a function of the scan-path configuration and relevant parameters such as number of test patterns, inputs, and outputs for each block of logic. More details on the efficient testing of scan-path designs can be found in [Breuer *et al.* 1988a].

Another way to reduce test application time and the number of stored test patterns is embodied in the method referred to as *scan path with look-ahead shifting* (SPLASH) described in [Abadir and Breuer 1986] and [Abadir 1987].

9.9.2 Partial Scan Using I-Paths

I-Modes

Abadir and Breuer [1985 a,b] introduced the concept of I-modes and I-paths to efficiently realize one form of partial scan. A module S with input port X and output port Y is said to have an *identity mode* (I-mode) between X and Y, denoted by IM $(S:X \to Y)$, if S has a mode of operation in which the data on port X is transferred (possibly after clocking) to port Y. A time tag t and activation-condition tags C and D are associated with every I-mode, where t is the time (in clock cycles or gate delays) for the data to be transferred from X to Y, C denotes the values required on the input control lines of S to activate the mode, and D denotes any values required on data inputs to ensure I-mode operation.

Latches, registers, MUXs, busses, and ALUs are examples of modules with I-modes. There are two I-modes associated with the multiplexer shown in Figure 9.37(a), denoted by $[\text{IM}(\text{MUX}:A \to C); x = 0; t = 10ns]$, and $[\text{IM}(\text{MUX}:B \to C); x = 1; t = 10ns]$. There are several I-modes associated with the ALU shown in Figure 9.37(b); one is denoted by $[\text{IM}(\text{ALU}:A \to C); x_1 x_2 = 00, t = 20ns]$, where $x_1 x_2 = 00$ is the condition code for the ALU to pass data from A to C; another I-mode is denoted by $[\text{IM}(\text{ALU}:A \to C); x_1 x_2 = 01; B = 0; C_{in} = 0]$, where $x_1 x_2 = 01$ is the condition code for the ALU to operate as an adder. The I-mode for the register shown in Figure 9.37(c) is denoted by $[\text{IM}(\text{Register}:A \to B); t = 1$ clock cycle].

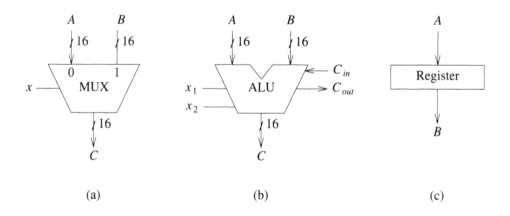

(a)   (b)   (c)

Figure 9.37   Three structures having I-modes

I-Paths

An *identity-transfer path* (I-path) exists from output port X of module $S1$ to input port Y of module $S2$, denoted by IP($S1$:$X \rightarrow S2$:Y), if data can be transferred unaltered, but possibly delayed, from port X to port Y. Every I-path has a time tag and activation plan. The time tag indicates the time delay for the data to be transferred from X to Y; the activation plan indicates the sequence of actions that must take place to establish the I-path. An I-path consists of a chain of modules, each of which has an I-mode.

Testing Using I-paths

Example 9.1: Consider the portion of a large circuit shown in Figure 9.38. All data paths are assumed to be 32 bits wide. Note that from the output port of the block of logic C, I-paths exist to the input ports of R_1, R_2, and R_3. Also, I-paths exist from the output ports of R_1, R_2, R_3, and R_4 to the input port of C. Register R_2 has a hold-enable line $\overline{H}/L$, where if $\overline{H}/L = 0$ R_2 holds its state, and if $\overline{H}/L = 1$ the register loads. Let $T = \{T_1, T_2, ..., T_n\}$ be a set of test patterns for C. Then C can be tested as shown in Figure 9.39. We assume that only one register can drive the bus at any one time, and a tristate driver is disabled (output is in the high impedance state) when its control line is high.

This process can now be repeated for test pattern T_{i+1}. While T_{i+1} is shifted into R_1, the response Z_i to T_i is shifted out over the S_{out} line.

In this partial scan design not only is the hardware overhead much less than for a full-scan design, but also the scan register has no impact on the signal delay in the path from R_2 to R_3.   □

This partial-scan approach leads to the following design problems.

1. Identifying a subset of registers to be included in the scan path.

2. Scheduling the testing of logic blocks. Since hardware resources, such as registers, busses, and MUXs are used in testing a block of logic it is usually impossible to test all the logic at one time. Hence after one block of logic is tested, another can be tested. As an example, for the circuit shown in Figure 9.38, R_1 along with other resources are first used to test C. This represents one test session. Later, R_1 and other resources can be used to test some other block of logic.

3. Determining efficient ways to activate the control lines when testing a block of logic.

4. Determining ways of organizing the scan paths to minimize the time required to test the logic.

The solution to some of these problems are discussed in [Breuer *et al.* 1988a,b].

More Complex Modes and Paths

The I-mode discussed previously is a parallel-to-parallel (P/P) I-mode, since data enter and exit modules as n-bit blocks of information. Other types of I-modes exist, such as serial-to-serial (S/S), serial-to-parallel (S/P), and parallel-to-serial (P/S). An example of a P/S I-mode would be a scan register which loads data in parallel and transmits them serially using its shift mode. Concatenating structures having various types of I-modes produces four types of I-paths, denoted by P/P, P/S, S/P, and S/S.

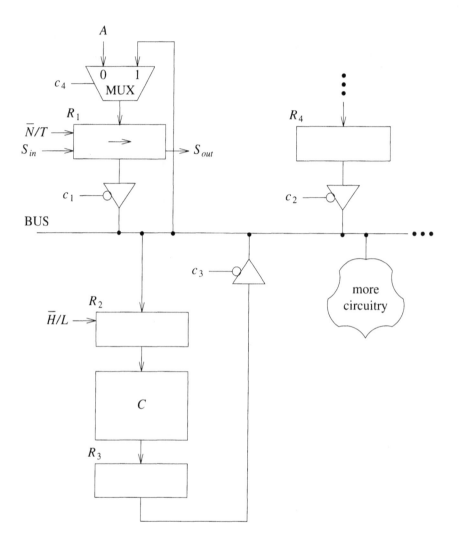

Figure 9.38   Logic block to be tested using I-path partial scan

In addition to I-modes, several other modes can be defined to aid in testing and in reducing the number of scan registers. A module S is said to have a *transfer-mode* (T-mode) if an onto mapping exists between input port X of S and output port Y of S. A trivial example of a structure having a T-mode is an array of inverters that maps the input vector X into NOT(X). A T-path consists of a chain of modules having zero or more I-modes and at least one T-mode. I-paths and T-paths are used primarily for transmitting data from a scan register to the input port of a block of logic to be tested.

A module V having an input port X and an output port Y is said to have a *sensitized mode* (S-mode) if V has a mode of operation such that an error in the data at port X produces an error in the data at port Y. An example is a parallel adder defined by the equation

Time	Controls	Operation
$t_1 \ldots t_{32}$	$\overline{N}/T = 1$	Scan T_i into R_1
t_{33}	$c_1 = 0$ $\overline{H}/L = 1$	Contents of R_1 are loaded onto the bus; data on bus are loaded into R_2
t_{34}	—	Test pattern T_i is applied to C; Response Z_i from C is loaded into R_3
t_{35}	$c_3 = 0$ $c_4 = 1, \overline{N}/T = 0$	Z_i is loaded onto bus; Z_i passes from bus through MUX and is loaded into R_1

Figure 9.39   Test process for block C

$SUM=A+B$. If B is held at any constant value, then an error in A produces an error in SUM. An S-path consists of a chain of structures having zero or more I-modes and at least one S-mode. S-modes correspond to one-to-one mappings.

When testing a block of logic, its response must be transmitted to a scan register or primary outputs. To transmit these data, I-paths and S-paths can be used.

More details on I-paths, S-paths, and T-paths can be found in [Breuer *et al.* 1988a]. Freeman [1988] has introduced the concept of F-paths, which correspond to "one-to-one" mappings, and S-paths, which correspond to "onto" mappings, and has shown how these paths can be effectively used in generating tests for data-path logic. He assumes that functional tests are used and does not require scan registers.

9.9.3 BALLAST — A Structured Partial Scan Design

Several methods have been proposed for selecting a subset of storage cells in a circuit to be replaced by scan-storage cells [Trischler 1980, Agrawal *et al.* 1987, Ma *et al.* 1988, Cheng and Agrawal 1989].

The resulting circuit is still sequential and in most cases sequential ATG is still required. But the amount of computation is now reduced. It is difficult to identify and/or specify the proper balance between adding storage cells to the scan path and reducing ATG cost. However, it appears that any heuristic for selecting storage cells to be made part of a scan path will lead to reduce ATG computation.

BALLAST (Balanced Structure Scan Test) is a structured partial scan method proposed by Gupta *et al.* [1989a,b]. In this design approach, a subset of storage cells is selected and made part of the scan path so that the resulting circuit has a special *balanced* property. Though the resulting circuit is sequential, only combinational ATG is required, and complete coverage of all detectable faults can be achieved.

The test plan associated with BALLAST is slightly different from that employed in a full-scan design, in that once a test pattern is shifted into the scan path, more than one normal system clock may be activated before the test result is loaded into the scan path and subsequently shifted out. In addition, in some cases the test data must be held in the scan path for several clock cycles while test data propagate through the circuitry.

Circuit Model

In general a synchronous sequential circuit S consists of blocks of combinational logic connected to each other, either directly or through registers, where a register is a collection of one or more storage cells. The combinational logic in S can be partitioned into maximal regions of connected combinational logic, referred to as *clouds*. The inputs to a cloud are either primary inputs or outputs of storage cells; the outputs of clouds are either primary outputs or inputs to storage cells. A group of wires forms a vacuous cloud if (1) it connects the outputs of one register directly to the inputs of another, (2) it represents circuit primary inputs feeding the inputs of a register, or (3) it represents the outputs of a register that are primary outputs. Storage cells can be clustered into registers as long as all storage cells in the register share the same control and clock lines. Storage cells can also be grouped together so that each register receives data from exactly one cloud and feeds exactly one cloud. However, a cloud can receive data from more than one register and can feed more than one register.

Figure 9.40 illustrates these concepts. C_1, C_2, and C_3 are nonvacuous clouds; A_1, A_2, and A_3 are vacuous clouds; c_1, c_2, ..., c_5 are blocks of logic, and R_1, R_2, ..., R_5 are registers. In forming clouds and registers, it is important first to create the clouds and then cluster storage cells into registers. If all storage cells are first clustered into one register, it is possible that only one cloud will be identified. A circuit can be partitioned into clouds in linear time.

A synchronous sequential circuit S is said to be *balanced*, denoted as a B-structure, if for any two clouds v_1 and v_2 in S, all signal paths (if any) between v_1 and v_2 go through the same number of registers. This condition also implies that S has an acyclic structure. The structure shown in Figure 9.40 is balanced. The structures shown in Figure 9.41 are not balanced.

Let $S*$ be a sequential circuit. There always exists a subset of registers which, if replaced by PIs and POs, creates a new circuit S, which is balanced. Let this subset of registers be made part of a scan path. This scan path represents pseudo-primary inputs and outputs to S. S is said to be the *kernel* of logic that is to be tested.

Given a B-structure S^B, its *combinational equivalent* C^B is the combinational circuit formed from S^B by replacing each storage cell (assumed to be a D flip-flop) in S^B by a wire. For simplicity we assume that only the Q output of the cell is used. Define the *depth* d of S^B as the largest number of registers on any path in S^B between any two clouds.

Let $T = \{t_1, t_2, ..., t_n\}$ be a complete test set for all detectable stuck at faults in C^B. Then the circuit S is tested as follows. Each test pattern $t_i = (t_i^a, t_i^b)$ consists of two parts, where t_i^a is applied to the primary inputs to S, and t_i^b is applied to the pseudo-primary inputs to S, that is, t_i^b is the data in the scan path. Then S^B can be tested as follows.

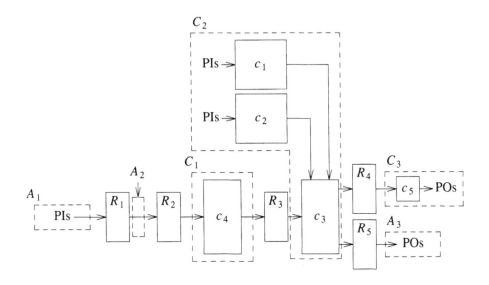

Figure 9.40   A partitioned circuit showing clouds and registers

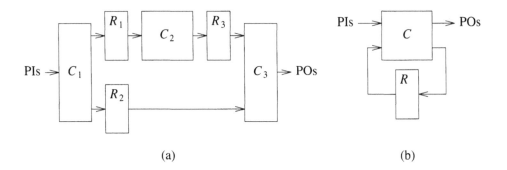

(a)	(b)

Figure 9.41   Nonbalanced structures (a) Unequal paths between C_1 and C_3
(b) A self-loop (unequal paths between C and itself)

Step 1:   Scan in the test pattern t_i^b.

Step 2:   Apply t_i^a to the primary inputs to S.

Step 3:   While holding t_i^a at the primary inputs and t_i^b in the scan path, clock the registers in S d times.

Step 4:   Place the scan path in its normal mode and clock it once.

Step 5:   Observe the value on the primary outputs.

Step 6:   Simultaneously, scan out the results in the scan paths and scan in t^b_{i+1}.

Note that in this test plan the scan path has three modes of operation, namely normal (parallel load), hold, and shift.

Example 9.2: Consider the circuit S shown in Figure 9.42(a). Selecting R_3 and R_6 to be scan registers produces the partial scan design shown in Figure 9.42(b). Replacing the scan path by PIs and POs results in the B-structure S^B, shown in Figure 9.42(c), which has depth $d = 2$.

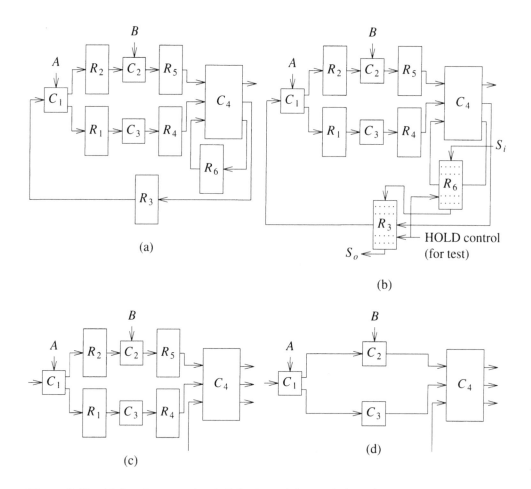

Figure 9.42   (a) Synchronous circuit S (b) A partial scan design of S (c) Kernel
K of S (d) Combinational equivalent of K

The combinational equivalent C^B is shown in Figure 9.42(d). To test S^B, a pattern is shifted into the scan path R_3 and R_6 and held there two clock periods while a test pattern is also applied to the primary inputs A and B. After one clock period test results are captured in registers R_1 and R_2. After the second clock period test results are captured in R_4 and R_5. Finally test results are captured in the scan path R_3 and R_6.   □

It has been shown that any single stuck fault that is detectable in the combinational logic in S is detectable by a test vector for C^B using the procedure just described [Gupta *et al.* 1989]. Thus, restricting one's attention to single stuck faults in the clouds, no fault coverage is sacrificed by employing the BALLAST test method. However, some additional test may be required to detect shorts between the I/O of storage cells.

A circuit can be transformed into a B-structure by removing an appropriate set of registers, which then become part of the scan path. Several criteria can be used to select these registers, such as information on critical paths in the circuit or the number of cells in the registers. For example, one criterion is to select a minimal number of storage cells to be made part of the scan path so that the resulting circuit is a B-structure.

Unfortunately, the problem just cited is NP-complete. A heuristic procedure for its solution is presented in [Gupta *et al.* 1989].

Figure 9.43 shows several different scan designs for the circuit shown in Figure 9.42(a), other than the one shown in Figure 9.42(b). Let n_i be the number of storage cells in register R_i. The design shown in Figure 9.43(a) can be considered superior to the one shown in Figure 9.42(b) if $n_1 + n_2 < n_3$, since fewer storage cells exist in the scan path. Assume that a scan storage cell introduces delay into a path terminating at its output port. Then the designs shown in Figure 9.43(c) and (d) may be well suited for the case where critical paths exist between R_5 and R_3 and between R_3 and R_1. Note that the depth for the design in Figure 9.43(d) is 3. Finally, Figure 9.43(e) shows a full-scan design. In all these designs, R_6 was made part of the scan path. For some partial scan designs it may not be necessary to transform each storage cell in R_6 into a scan-storage cell. The sequential circuit consisting of C_4 and R_6 can be replaced by a lower-level design description, where R_6 is replaced by its individual storage cells and C_4 is segmented into the cones of logic that feed each storage cell. The resulting circuit can then be made into a partial scan design using the BALLAST design methodology.

The importance of using a balanced structure is illustrated in the next example.

Example 9.3: Consider the circuit shown in Figure 9.44(a). This is a nonbalanced circuit having the structure shown in Figure 9.41(a). Consider the fault b s-a-1. The test pattern for this fault at time t within cloud C_3 consists of $(Q_3, Q_4, Q_5) = (0,0,1)$. This implies that at time $(t-1)$, $(Q_1, Q_2) = (1,0)$ and $A = 0$. This in turn implies that at time $(t-2)$, $(A,B) = (1,0)$. One test sequence for this fault is shown in Figure 9.44(a). The combinational equivalent of this circuit is shown in Figure 9.44(b), where now the fault b s-a-1 is redundant.   □

It is seen that some nonbalanced circuits may require sequential ATG.

In some cases, it is not necessary that the scan registers have a hold mode. Methods for removing hold modes as well as dealing with designs where the registers in the original circuit have hold modes are dealt with in [Gupta *et al.* 1989b]. For example, for the situation shown in Figure 9.42(b), let $t(3)$ and $t(6)$ be the test patterns to be loaded into R_3 and R_6, respectively. Consider the test pattern $t(3)$ *xx* $t(6)$ where two don't-care bits have been placed between $t(3)$ and $t(6)$. Now the scan path can continue shifting while the test data propagate through the circuit. Assume at time t that $t(3)$ is in R_3. At time $(t + 1)$, test results are latched into R_2 and R_3, at time $(t + 2)$ test pattern $t(6)$ is in R_6 and test results are latched into R_4 and R_5. At time $(t + 3)$ test results are latched into R_3 and R_6.

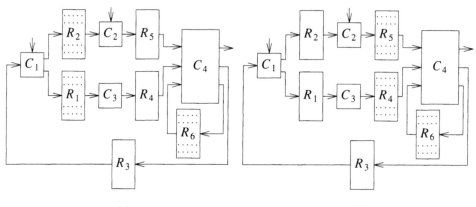

(a)   (b)

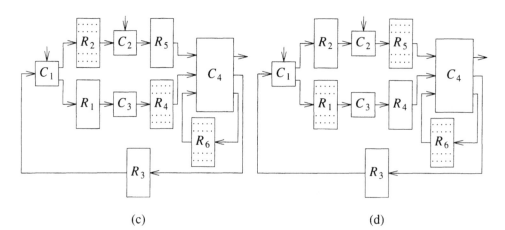

(c)   (d)

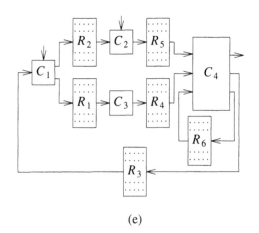

(e)

Figure 9.43   Different partial and full scan designs for S

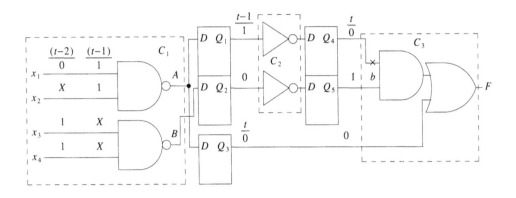

(a)

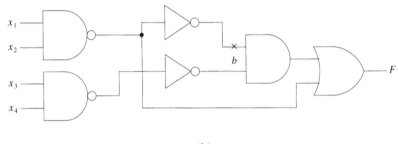

(b)

Figure 9.44   (a) A nonbalanced circuit (b) Its combinational equivalent

9.10 Boundary-Scan Standards

9.10.1 Background

To better address problems of board-level testing, several DFT standards have been and are currently being developed. The primary goal of these proposed standards is to ensure that chips of LSI and VLSI complexity contain a common denominator of DFT circuitry that will make the test development and testing of boards containing these chips significantly more effective and less costly. Some of these initiatives are known as the Joint Test Action Group (JTAG) Boundary Scan standard [JTAG 1988], the VHSIC Element-Test and Maintenance (ETM) — Bus standard [IBM *et al.* 1986a], the VHSIC Test and Maintenance (TM) — Bus standard [IBM *et al.* 1986b], and the IEEE 1149.1 Testability Bus Standard [IEEE 1989].

These standards deal primarily with the use of a test bus which will reside on a board, the protocol associated with this bus, elements of a bus master which controls the bus, I/O ports that tie a chip to the bus, and some control logic that must reside on a chip to interface the test bus ports to the DFT hardware residing on the application portion of the chip. In addition, the JTAG Boundary Scan and IEEE 1149.1 standards also require that a boundary-scan register exist on the chip.

The primary reasons for using boundary scan are to allow for efficient testing of board interconnect and to facilitate isolation and testing of chips either via the test bus or by built-in self-test hardware. With boundary scan, chip-level tests can be reused at the board level.

The description of a board-level test bus presented in this section is based on IEEE 1149.1. There are several major components associated with IEEE 1149.1, namely (1) the physical structure of the test bus and how it can be interconnected to chips, (2) the protocol associated with the bus, and (3) the on-chip test bus circuitry associated with a chip. The latter includes the boundary-scan registers and the test access port (TAP) controller which is a finite-state machine that decodes the state of the bus.

Figure 9.45 shows a general form of a chip which supports IEEE 1149.1. The application logic represents the normal chip design prior to the inclusion of logic required to support IEEE 1149.1. This circuitry may include DFT or BIST hardware. If so, the scan paths are connected via the test-bus circuitry to the chip's scan-in and scan-out ports. This is illustrated by the connection from the test data input line TDI to S_{in}, and S_{out} to the test data output line TDO. The normal I/O terminals of the application logic are connected through boundary-scan cells to the chips I/O pads.

The test-bus circuitry, also referred to as the bus slave, consists of the boundary-scan registers, a 1-bit bypass register, an instruction register, several miscellaneous registers, and the TAP.

The boundary-scan bus consists of four lines, namely a test clock (TCK), a test mode signal (TMS), the TDI line, and the TDO line.

Test instructions and test data are sent to a chip over the TDI line. Test results and status information are sent from a chip over the TDO line to whatever device is driving the bus. This information is transmitted serially. The sequence of operations is controlled by a bus master, which can be either ATE or a component that interfaces to a higher-level test bus that is part of a hierarchical test and maintenance system. Control of the test-bus circuitry is primarily carried out by the TAP, which responds to the state transitions on the TMS line.

Briefly, the test bus and associated logic operates as follows.

1. An instruction is sent serially over the TDI line into the instruction register.
2. The selected test circuitry is configured to respond to the instruction. In some cases this may involve sending more data over the TDI line into a register selected by the instruction.
3. The test instruction is executed. Test results can be shifted out of selected registers and transmitted over the TDO line to the bus master. It is possible to shift new data into registers using the TDI line while results are shifted out and transmitted over the TDO line.

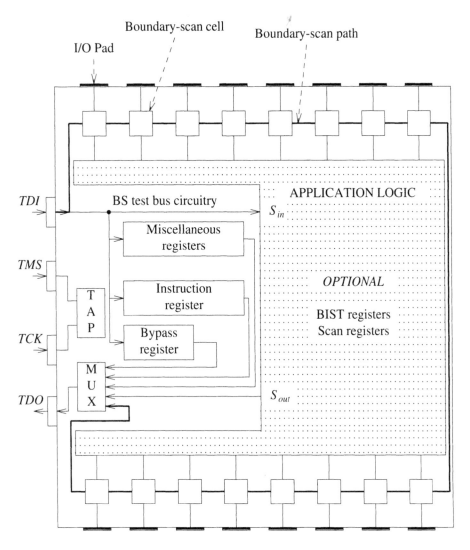

Figure 9.45   Chip architecture for IEEE 1149.1

9.10.2 Boundary-Scan Cell

Two possible boundary-scan cell designs are shown in Figure 9.46. These cells can be used as either output or input cells. Other cell designs exist for bidirectional I/O ports and tristate outputs.

As an input boundary-scan cell, *IN* corresponds to a chip input pad, and *OUT* is tied to a normal input to the application logic. As an output cell, *IN* corresponds to the output of the application logic, and *OUT* is tied to an output pad. The cell has several modes of operations.

Normal Mode: When *Mode_Control*=0, data passes from port *IN* to port *OUT*; then the cell is transparent to the application logic.

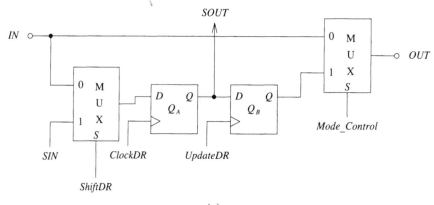

(a)

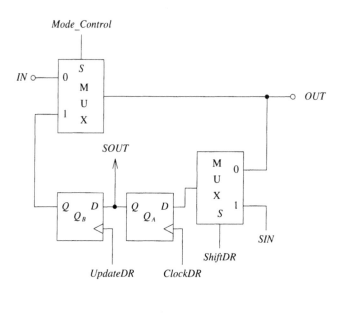

(b)

Figure 9.46   Two example boundary-scan cells

Scan Mode: The boundary-scan cells are interconnected into a scan path, where the *SOUT* terminal of one cell is connected to the *SIN* terminal of the next cell in the path. The first cell is driven by *TDI*, and the last one drives *TDO*. In the scan mode, *ShiftDR*=1 and clock pulses are applied to *ClockDR*. *DR* denotes a data register; *IR* denotes the instruction register.

Capture Mode: The data on line *IN* can be loaded into the scan path by setting *ShiftDR*=0 and applying one clock pulse to *ClockDR*. Thus the Q_A storage cells act as a shadow or snapshot register. The *OUT* terminal can be driven by either *IN* or the output of Q_B.

Update Mode: Once the Q_A storage cell is loaded, either by a capture or scan operation, its value can be applied to the *OUT* port by setting *Mode_Control* = 1 and applying a clock pulse to *UpdateDR*.

If a boundary-scan cell drives an output pad, then the select line of the multiplexer driving the output is driven by a signal denoted by *Output_Mode_Control*; if the cell is driven by an input pad the select line is driven by a signal labeled *Input_Mode_Control*.

It is easy to extend these design concepts to handle both tristate and bidirectional I/O pads.

9.10.3 Board and Chip Test Modes

Figure 9.47 shows a board containing four chips which support IEEE 1149.1. The boundary-scan cells are interconnected into a single scan path, where the *TDO* of one chip is tied to the *TDI* of another chip, except for the initial *TDI* and *TDO* ports which are tied to distinct terminals of the board. Some of the normal interconnect between chip pads is also shown. Using this configuration various tests can be carried out, including (1) interconnect test, (2) snapshot observation of normal system data, and (3) testing of each chip. To implement these tests, three test modes exist, namely external test, sample test, and internal test.

External Test Mode

To test the interconnect and/or logic external to chips supported by IEEE 1149.1 the circuit configuration shown in Figure 9.48 is used. Here a test pattern can be loaded into the Q_A cells of chips 1 and 2, for example, Q_A(chip 1) = 1 and Q_A(chip 2) = 0. Chip 1 can carry out an Update operation, where the data in Q_A(chip 1) are to drive the output pad. Chip 2 can carry out a Capture operation, where the data on its input pad are loaded into Q_A(chip 2). The data in the scan path can be shifted out to see if the correct response was received. By selecting test data appropriately, tests for shorts, opens, and stuck-at faults can be carried out. By using appropriate boundary-scan cells and test data, interconnect tests can be carried out for tristate logic and bidirectional pads.

Sample Test Mode

Figure 9.49 shows how the I/O data associated with a chip can be sampled during normal system operation. This sampled data can be scanned out while the board remains in normal operation. The boundary-scan circuitry must be designed so that the boundary-scan cells in the signal path between a chip I/O pins and the application logic do not interfere with the operation of the board logic.

Internal Test Mode

Figure 9.50 shows the configuration of the boundary-scan cells when in the internal test mode. In this configuration the inputs to the application logic are driven by the input boundary-scan cells, and the response can be captured in the output boundary-scan cells. Since these cells are part of a scan path, this mode of operation gives complete controllability and observability of the I/O pads of a chip. The chip may also have internal scan paths and in fact be designed to have built-in self-test (BIST) capability. While in the internal test mode, the internal scan paths and BIST operations can be activated to test the chip. This configuration replaces some forms of testing normally carried out by bed-of nails test equipment.

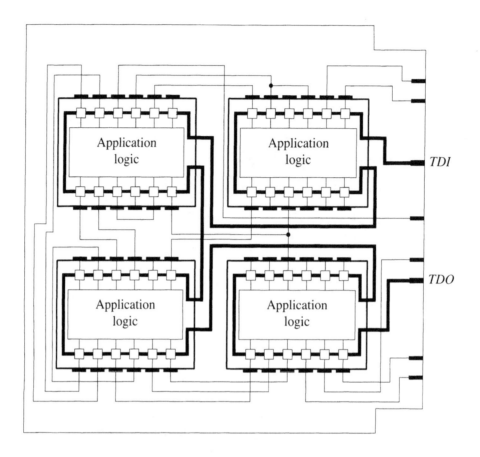

Figure 9.47   A printed circuit board with a IEEE 1149.1 test bus

9.10.4 The Test Bus

A board supporting IEEE 1149.1 contains a test bus consisting of at least four signals. (A fifth signal can be used to reset the on-chip test-bus circuitry.) These signals are connected to a chip via its test-bus ports. Each chip is considered to be a bus slave, and the bus is assumed to be driven by a bus master.

The minimum bus configuration consists of two broadcast signals (*TMS* and *TCK*) driven by the master, and a serial path formed by a "daisy-chain" connection of serial scan data pins (*TDI* and *TDO*) on the master and slave devices. Figure 9.51(a) shows a ring configuration; Figure 9.51(b) shows a star configuration, where each chip is associated with its own *TMS* signal. Star and ring configurations can be combined into hybrid configurations. The four bus signals associated with a slave TAP are defined as follows.

TCK — Test Clock. This is the master clock used during the boundary-scan process.

TDI — Test Data Input. Data or instructions are received via this line and are directed to an appropriate register within the application chip or test bus circuitry.

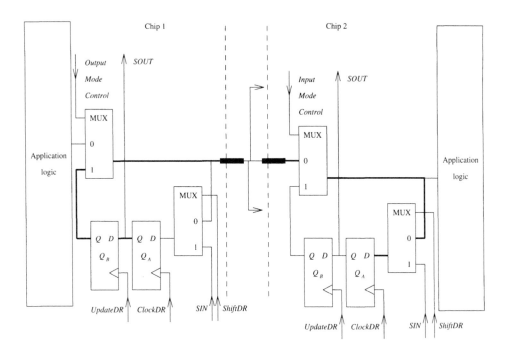

Figure 9.48   External test configuration

TDO — Test Data Output. The contents of a selected register (instruction or data) are shifted out of the chip over TDO.

TMS — Test Mode Selector. The value on TMS is used to control a finite state machine in a slave device so that this device knows when to accept test data or instructions.

Though the IEEE 1149.1 bus employs only a single clock, *TCK*, this clock can be decoded on-chip to generate other clocks, such as the two-phase nonoverlap clocks used in implementing LSSD.

9.10.5 Test-Bus Circuitry

The on-chip test-bus circuitry allows access to and control of the test features of a chip. A simplified version of this circuitry is shown in Figure 9.45, and more details are shown in Figure 9.52. This circuitry consists of four main elements, namely (1) a test access port (TAP) consisting of the ports associated with *TMS, TCK, TDI,* and *TDI,* (2) a TAP controller, (3) a scannable instruction register and associated logic, and (4) a group of scannable test data registers. We will next elaborate on some of these features.

9.10.5.1 The TAP Controller

The TAP controller is a synchronous finite-state machine whose state diagram is shown in Figure 9.53. It has a single input, labeled *TMS,* and its outputs are signals corresponding to a subset of the labels associated with the various states, such as Capture-IR. The state diagram shows that there are two parallel and almost identical

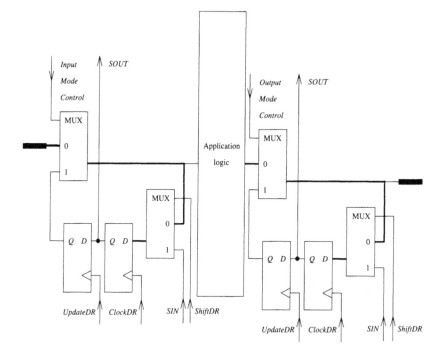

Figure 9.49   Sample test configuration

subdiagrams, one corresponding to controlling the operations of the instruction register, the other to controlling the operation of a data register. The controller can change state only when a clock pulse on *TCK* occurs; the next state is determined by the logic level of line *TMS*. The function of some of these control states is described below.

Test-Logic-Reset: In this state the test logic (boundary-scan) is disabled so that the application logic can operate in its normal mode.

Run-Test/Idle: This is a control state that exists between scan operations, and where an internal test, such as a built-in self-test, can be executed (see instruction RUNBIST).

A test is selected by first setting the instruction register with the appropriate information. The TAP controller remains in this state as long as *TMS*=0.

Select-DR-Scan: This is a temporary controller state. If *TMS* is held low then a scan-data sequence for the selected test-data register is initiated, starting with a transition to the state Capture-DR.

Capture-DR: In this state data can be loaded in parallel into the test-data registers selected by the current instruction. For example, the boundary-scan registers can be loaded while in this state. Referring to Figure 9.49, to capture the data on the input pad shown and the output of the application logic, it is necessary to set *ShiftDR* low and to activate the clock line *ClockDR*.

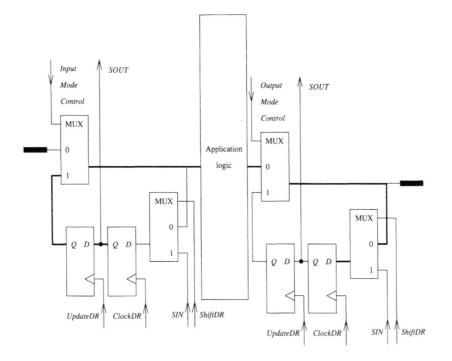

Figure 9.50   Internal test configuration

Shift-DR: In this state test-data registers specified by the data in the instruction register, and which lie between *TDI* and *TDO*, are shifted one position. A new data value enters the scan path via the *TDI* line, and a new data value now is observed at the *TDO* line. Other registers hold their state.

Exit1-DR: In this state all test-data registers selected by the current instruction hold their state. Termination of the scan process can be achieved by setting *TMS* high.

Pause-DR: In this control state the test-data registers in the scan path between *TDI* and *TDO* hold their state. The Pause-DR and Pause-IR states can be used to temporarily halt the scan operation and allow the bus master to reload data. This is often necessary during the transmission of long test sequences.

Update-DR: Some registers have a latched parallel output so that the output does not change during a scan process. In this control state test-data registers specified by the current instructions and having a latched parallel-output feature are loaded from their associated shift registers.

Referring to Figure 9.50, once the boundary-scan register has been loaded with scan data while the TAP controller is in state Shift-DR, these data can be applied to the inputs of the application logic when the controller is in state Update-DR by simply activating the clock *UpdateDR*.

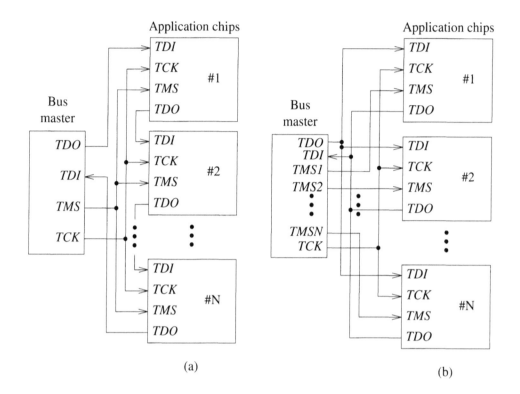

Figure 9.51   (a) Ring configuration (b) Star configuration

The states that control the instruction register operate similarly to those controlling the test-data registers. The instruction register is implemented using a latched parallel-output feature. This way a new instruction can be scanned into the scan path of the instruction register without affecting the output of the register. The Select-IR-Scan state is again a temporary transition state. In the Capture-IR state, the shift register associated with the instruction register is loaded in parallel. These data can be status information and/or fixed logic values. In the Shift-IR state this shift register is connected between *TDI* and *TDO* and shifts data one position. In the Update-IR control state the instruction shifted into the instruction register is latched onto the parallel output of the instruction register. This instruction now becomes the current instruction.

The TAP controller can be implemented using four flip-flops and about two dozen logic gates.

9.10.5.2 Registers

The Instruction Register and Commands

The instruction register has the ability to shift in a new instruction while holding the current instruction fixed at its output ports. The register can be used to specify operations to be executed and select test-data registers. Each instruction enables a single serial test-data register path between *TDI* and *TDO*. The instructions BYPASS, EXTEST, and

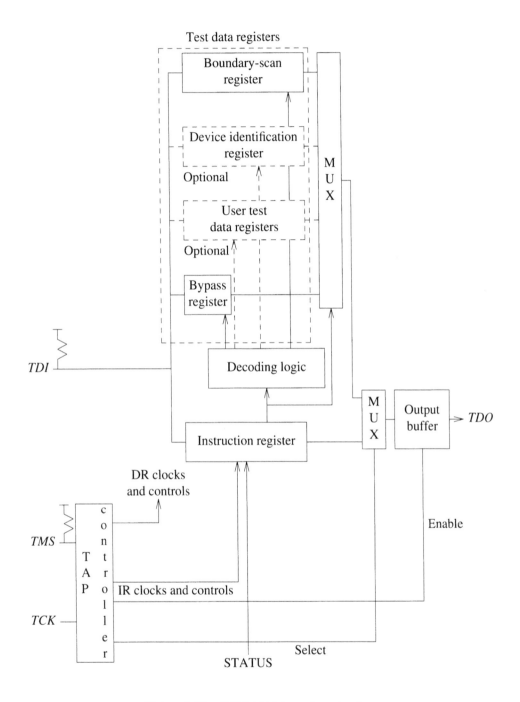

Figure 9.52   IEEE 1149.1 test bus circuitry

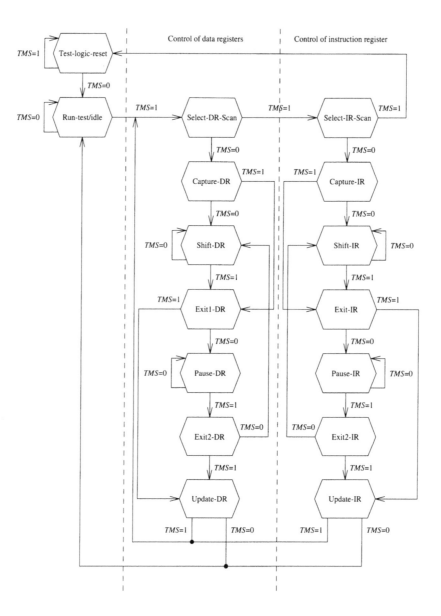

Figure 9.53   State diagram of TAP controller

SAMPLE are required; INTEST or RUNBIST are recommended. A brief description of these instructions follows.

BYPASS Instruction

Every chip must have a BYPASS register, which is a test-data register of length 1. Consider a board having 100 chips that are connected as shown in Figure 9.51(a). Test data must pass through all the chips when testing any one chip, say the ith chip. To reduce the length of the scan path, all chips except for the ith chip can be put into the bypass mode, where now only a single storage cell lies between the *TDI* and *TDO* ports of these chips. The BYPASS instruction is used to configure a chip in this bypass mode.

EXTEST Instruction

EXTEST is primarily used to test circuitry external to a chip, such as the board interconnect. When this instruction is executed, boundary-scan cells at output pads are used to drive the pads. Those cells at input pads capture test results when the TAP controller enters the Capture-DR state (see Figure 9.48).

SAMPLE Instruction

The SAMPLE instruction allows the data on the I/O pads of a chip to be sampled and placed in the boundary-scan register during normal board operation (see Figure 9.49).

INTEST Instruction

The INTEST instruction is used to apply a test vector to the application logic via the boundary-scan path, and to capture the response from this logic. First the INTEST instruction is loaded into the instruction register. Then the boundary-scan register is loaded with test data. Because of the INTEST instruction, the inputs to the application logic are driven by the input boundary-scan cells, and the output pads of the chip are driven by the output boundary-scan cells. The scan path is loaded while in the control state Shift-DR. In the Update-DR state these data get applied to the inputs of the application logic. Repeating a load-test-data register cycle, the test results are captured when in state Capture-DR. After this, another test pattern can be loaded into the boundary-scan path while the results are sent back to the bus master. This cycle is repeated for each test pattern.

RUNBIST Instruction

The RUNBIST instruction allows for the execution of a self-test process of the chip. This test is executed while the TAP controller is in the Run-Test/Idle state. The RUNBIST instruction must select the boundary-scan register to be connected between *TDI* and *TDO*. All inputs to the application logic are driven by the boundary-scan register during the execution of this instruction.

Test-Data Registers

The test-bus circuitry contains at least two test-data registers, namely the bypass and the boundary-scan registers. In addition, other scan registers can be accessed and connected between *TDI* and *TDO*, such as device identification registers and registers that are part of the application logic itself. Thus if the application logic is designed according to the full-scan DFT methodology, then this logic can be completely tested by the IEEE 1149.1 architecture.

More information on the use of boundary-scan and related interface logic can be found in [Avra 1987], [Beenker 1985], [Breuer *et al.* 1988], [Breuer and Lien 1988a, b], [Lagemaat and Bleeker 1987], [Lien and Breuer 1989], [Maunder and Beenker 1987], [Wagner 1987], and [Whetsel 1988a, b].

REFERENCES

[Abadir and Breuer 1985a] M. S. Abadir and M. A. Breuer, "Constructing Optimal Test Schedules for VLSI Circuits Having Built-in Test Hardware," *Proc. 15th Intn'l. Fault-Tolerant Computing Conf.*, pp. 165-170, June, 1985.

[Abadir and Breuer 1985b] M. S. Abadir and M. A. Breuer, "A Knowledge Based System for Designing Testable VLSI Chips," *IEEE Design & Test of Computers*, Vol. 2, No. 4, pp. 56-68, August, 1985.

[Abadir and Breuer 1986a] M. S. Abadir and M. A. Breuer, "Scan Path With Look Ahead Shifting," *Proc. Intn'l. Test Conf.*, pp. 699-704, September, 1986.

[Abadir and Breuer 1986b] M. S. Abadir and M. A. Breuer, "Test Schedules for VLSI Circuits," *IEEE Trans. on Computers*, Vol. C-35, No. 5, pp. 361-367, April, 1986.

[Abadir and Breuer 1987] M. S. Abadir and M. A. Breuer, "Test Schedules for VLSI Circuits Having Built-in Test Hardware," *Intn'l. Journal of Computers and Mathematics with Applications*, Vol. 13, No. 5/6, pp. 519-536, 1987.

[Abadir 1987] M. S. Abadir, "Efficient Scan Path Testing Using Sliding Parity Response Compaction," *Proc. Intn'l. Conf. on Computer Aided Design,* pp. 332-335, November, 1987.

[Agrawal and Mercer 1982] V. D. Agrawal and M. R. Mercer, "Testability Measures — What Do They Tell Us?," *Digest of Papers 1982 Intn'l. Test Conf.*, pp. 391-396, November, 1982.

[Agrawal *et al.* 1987] V. D. Agrawal, K.-T. Cheng, D. D. Johnson, and T. Lin, "A Complete Solution to the Partial Scan Problem," *Proc. Intn'l. Test Conf.*, pp. 44-51, September, 1987.

[Ando 1980] H. Ando, "Testing VLSI with Random Access Scan," *Proc. COMPCON*, pp. 50-52, 1980.

[Avra 1987] L. Avra, "A VHSIC ETM-BUS Compatible Test and Maintenance Interface," *Proc. Intn'l. Test Conf.*, pp. 964-971, September, 1987.

[Beenker 1985] F. Beenker, "Systematic and Structured Methods for Digital Board Testing," *Proc. Intn'l. Test Conf.*, pp. 380-385, November, 1985.

[Bennetts 1984] R. G. Bennetts, *Design of Testable Logic Circuits*, Addison-Wesley, Reading, Massachusetts, 1984.

[Bennetts *et al.* 1981] R. G. Bennetts, C. M. Maunder, and G. D. Robinson, "CAMELOT: A Computer-Aided Measure for Logic Testability," *IEE Proc.*, Vol. 128, Part E, No. 5, pp. 177-189, 1981.

[Breuer 1978] M. A. Breuer, "New Concepts in Automated Testing of Digital Circuits," *Proc. EEC Symp. on CAD of Digital Electronic Circuits and Systems*, Brussels, pp. 69-92, 1978.

[Breuer *et al.* 1988a] M. A. Breuer, R. Gupta, and R. Gupta, "AI Aspects of TEST: A System for Designing Testable VLSI Chips," *IFIP Workshop on*

Knowledge-Based Systems for Test and Diagnosis, pp. 29-75, September 27-29, 1988.

[Breuer *et al.* 1988b] M. A. Breuer, R. Gupta, and J. C. Lien, "Concurrent Control of Multiple BIT Structures," *Proc. Intn'l. Test Conf.*, pp. 431-442, September, 1988.

[Breuer and Lien 1988a] M. A. Breuer and J. C. Lien, "A Test and Maintenance Controller for a Module Containing Testable Chips," *Proc. Intn'l. Test Conf.*, pp. 502-513, September, 1988.

[Breuer and Lien 1988b] M. A. Breuer and J. C. Lien, "A Methodology for the Design of Hierarchically Testable and Maintainable Digital Systems," *Proc. 8th Digital Avionics Systems Conf. (DASC)*, pp. 40-47, October 17-20, 1988.

[Chandra and Patel 1989] S. J. Chandra and J. H. Patel, "Experimental Evaluation of Testability Measures for Test Generation," *IEEE Trans. on Computer-Aided Design*, Vol. CAD-8, No. 1, pp. 93-97, January, 1989.

[Chen and Breuer 1985] T-H. Chen and M. A. Breuer, "Automatic Design for Testability Via Testability Measures," *IEEE Trans. on Computer-Aided Design*, Vol. CAD-4, pp. 3-11, January, 1985.

[Cheng and Agrawal 1989] K.-T. Cheng and V. D. Agrawal, "An Economical Scan Design for Sequential Logic Test Generation," *Proc. 19th Intn'l. Symp. on Fault-Tolerant Computing*, pp. 28-35, June, 1989.

[DasGupta *et al.* 1981] S. DasGupta, R. G. Walther, and T. W. Williams, "An Enhancement to LSSD and Some Applications of LSSD in Reliability, Availability, and Serviceability," *Digest of Papers 11th Annual Intn'l. Symp. on Fault-Tolerant Computing*, pp. 32-34, June, 1981.

[DasGupta *et al.* 1982] S. DasGupta, P. Goel, R. G. Walther, and T. W. Williams, "A Variation of LSSD and Its Implications on Design and Test Pattern Generation in VLSI," *Proc. Test Conf.*, pp. 63-66, November, 1982.

[Davidson 1979] R. P. Davidson, "Some Straightforward Guidelines Help Improve Board Testability," *Electronic Design News*, pp. 127-129, May 5, 1979.

[Eichelberger 1983] E. B. Eichelberger, "Latch Design Using 'Level Sensitive Scan Design,'" *Proc. COMPCON*, pp. 380-383, February, 1983.

[Eichelberger and Williams 1977] E. B. Eichelberger and T. W. Williams, "A Logic Design Structure for LSI Testing," *Proc. 14th Design Automation Conf.*, pp. 462-468, June, 1977.

[Eichelberger and Williams 1978] E. B. Eichelberger and T. W. Williams, "A Logic Design Structure for LSI Testability," *Journal Design Automation & Fault-Tolerant Computing*, Vol. 2, No. 2, pp. 165-178, May, 1978.

[Eichelberger *et al.* 1978] E. B. Eichelberger, T. W. Williams, E. I. Muehldorf, and R. G. Walther, "A Logic Design Structure for Testing Internal Array," *3rd USA-JAPAN Computer Conf.*, pp. 266-272, October, 1978.

[Freeman 1988] S. Freeman, "Test Generation for Data-path Logic: The F-Path Method," *IEEE Journal of Solid-State Circuits*, Vol. 23, No. 2, pp. 421-427, April, 1988.

[Funatsu *et al.* 1978] S. Funatsu, N. Wakatsuki,, and A. Yamada, "Designing Digital Circuits with Easily Testable Consideration," *Proc. Test Conf.*, pp. 98-102, September, 1978.

[Goel and McMahon 1982] P. Goel and M. T. McMahon, "Electronic Chip-in-Place Test," *Digest of Papers 1982 Intn'l. Test Conf.*, pp. 83-90, November, 1982.

[Goldstein 1979] L. H. Goldstein, "Controllability/Observability Analysis of Digital Circuits," *IEEE Trans. on Circuits and Systems*, Vol. CAS-26, No. 9, pp. 685-693, September, 1979.

[Goldstein and Thigen 1980] L. M. Goldstein and E. L. Thigen, "SCOAP: Sandia Controllability/Observability Analysis Program," *Proc. 17th Design Automation Conf.*, pp. 190-196, June, 1980.

[Grason 1979] J. Grason, "TMEAS -- A Testability Measurement Program," *Proc. 16th Design Automation Conf.*, pp. 156-161, June, 1979.

[Grason and Nagel 1981] J. Grason and A. W. Nagel, "Digital Test Generation and Design for Testability," *Journal Digital Systems*, Vol. 5, No. 4, pp. 319-359, 1981.

[Gupta *et al.* 1989a] R. Gupta, R. Gupta, and M. A. Breuer, "BALLAST: A Methodology for Partial Scan Design," *Proc. 19th Intn'l. Symp. on Fault-Tolerant Computing*, pp. 118-125, June, 1989.

[Gupta *et al.* 1989b] R. Gupta, R. Gupta, and M. A. Breuer, "An Efficient Implementation of the BALLAST Partial Scan Architecture," *Proc. IFIP Intn'l. Conf. on Very Large Scale Integration (VLSI 89)*, pp. 133-142, August 16-18, 1989.

[Hayes and Friedman 1974] J. P. Hayes and A. D. Friedman, "Test Point Placement to Simplify Fault Detection," *IEEE Trans. on Computers*, Vol. C-33, pp. 727-735, July, 1974.

[IBM *et al.* 1986a] IBM, Honeywell, and TRW, "VHSIC Phase 2 Interoperability Standards," ETM-BUS Specification, December, 1986.

[IBM *et al.* 1986b] IBM, Honeywell and TRW, "VHSIC Phase 2 Interoperability Standards," TM-BUS Specification, December, 1986.

[IEEE 1989] "Standard Test Access Port and Boundary-Scan Architecture," Sponsored by Test Technology Technical Committee of the IEEE Computer Society, Document P1149.1/D5 (Draft), June 20, 1989.

[JTAG 1988] Technical Subcommittee of Joint Test Action Group (JTAG), "Boundary-Scan Architecture Standard Proposal," Version 2.0, March, 1988.

[Kobayashi *et al.* 1968] T. Kobayashi, T. Matsue, and H. Shiba, "Flip-Flop Circuit with FLT Capability," *Proc. IECEO Conf.*, in Japanese, p. 692, 1968.

[Kovijanic 1979] P. G. Kovijanic, "Computer Aided Testability Analysis," *Proc. IEEE Automatic Test Conf.*, pp. 292-294, 1979.

[Kovijanic 1981] P. G. Kovijanic, "Single Testability Figure of Merit," *Proc. Intn'l. Test Conf.*, pp. 521-529, October, 1981.

[Lagemaat and Bleeker 1987] D. Van de Lagemaat and H. Bleeker, "Testing a Board with Boundary-Scan," *Proc. Intn'l. Test Conf.*, pp. 724-729, September, 1987.

[Lien and Breuer 1989] J. C. Lien and M. A. Breuer, "A Universal Test and Maintenance Controller for Modules and Boards," *IEEE Trans. on Industrial Electronics*, Vol. 36, No. 2, pp. 231-240, May, 1989.

[Lioy and Mezzalama 1978] A. Lioy and M. Mezzalama, "On Parameters Affecting ATPG Performance," *Proc. COMPEURO Conf.*, pp. 394-397, May, 1987.

[Lippman and Donn 1979] M. D. Lippman and E. S. Donn, "Design Forethought Promotes Easier Testing of Microcomputer Boards," *Electronics International*, Vol. 52, No. 2, pp. 113-119, January, 1979.

[Ma et al. 1988] H.-K. T. Ma, S. Devadas, A. R. Newton, and A. Sangiovanni-Vincentelli, "An Incomplete Scan Design Approach to Test Generation for Sequential Machines," *Proc. Intn'l. Test Conf.*, pp. 730-734, September, 1988.

[Maunder and Beenker 1987] C. Maunder and F. Beenker, "Boundary Scan: A Framework for Structured Design-for-Test," *Proc. Intn'l. Test Conf.*, pp. 714-723, September, 1987.

[Ratiu et al. 1982] I. M. Ratiu, A. Sangiovanni-Vincentelli, and D. O. Peterson, "VICTOR: A Fast VLSI Testability Analysis Program," *Proc. Intn'l. Test Conf.*, pp. 397-401, November, 1982.

[Rutman 1972] R. A. Rutman, "Fault Detection Test Generation for Sequential Logic Heuristic Tree Search," *IEEE Computer Repository Paper* No. R-72-187, 1972.

[Savir 1983] J. Savir, "Good Controllability and Observability Do Not Guarantee Good Testability," *IEEE Trans. on Computers*, Vol. C-32, pp. 1198-1200, December, 1983.

[Stephenson and Grason 1976] J. E. Stephenson and J. Grason, "A Testability Measure for Register Transfer Level Digital Circuits," *Proc. Intn'l. Symp. on Fault-Tolerant Computing*, pp. 101-107, June, 1976.

[Stewart 1977] J. H. Stewart, "Future Testing of Large LSI Circuit Cards," *Digest of Papers 1977 Semiconductor Test Symp.*, pp. 6-15, October, 1977.

[Stewart 1978] J. H. Stewart, "Application of Scan/Set for Error Detection and Diagnostics," *Digest of Papers 1978 Semiconductor Test Conf.*, pp. 152-158, 1978.

[Trischler 1980] E. Trischler, "Incomplete Scan Path with an Automatic Test Generation Methodology," *Digest of Papers 1980 Test Conf.*, pp. 153-162, November, 1980.

[Wagner 1987] P. T. Wagner, "Interconnect Testing with Boundary-Scan," *Proc. Intn'l. Test Conf.*, pp. 52-57, September, 1987.

[Whetsel 1988a] L. Whetsel, "A Proposed Standard Test Bus and Boundary-Scan Architecture," *Proc. Intn'l. Conf. on Computer Design*, pp. 330-333, 1988.

[Whetsel 1988b] L. Whetsel, "A View of the JTAG Port and Architecture," *ATE Instrumentation Conf. West*, pp. 385-410, January, 1988.

[Williams and Angell 1973] M. J. Y. Williams and J. B. Angell, "Enhancing Testability of Large Scale Integrated Circuits Via Test Points and Additional Logic," *IEEE Trans. on Computers*, Vol. C-22, pp. 46-60, January, 1973.

[Writer 1975] P. L. Writer, "Design for Testability," *Proc. IEEE Automated Support Systems Conf.*, pp. 84-87, 1975.

PROBLEMS

9.1   Assume that it costs $1.00 to test an IC chip, $10.00 to locate a faulty IC chip on a PCB, the test sets provide 100 percent fault coverage, and a board has 100 ICs. Option 1 is to test every chip, and thus all assembled boards have only good ICs. Option 2 is not to test any ICs before assembly but to test and locate faulty ICs at the board level. At what value of IC yield is it cost-effective to switch from option 1 to option 2?

Hint: This is a hard problem. Note that under option 2 PCBs may contain several bad ICs, and a bad IC may be replaced by another bad IC. Make simplifying assumptions if necessary.

9.2   Partition the circuit shown in Figure 9.54 using the partitioning scheme illustrated in Figure 9.11. Attempt to keep the number of inputs to C_1 and C_2 close to the same, and try to minimize the number of signals between the two partitions.

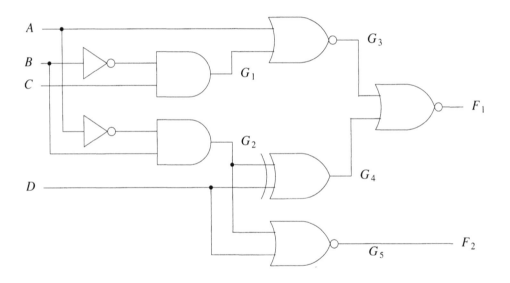

Figure 9.54

9.3   The Am 29818 is a general-purpose pipeline register with parallel load and an on-board shadow register for performing serial shadow register diagnostics and/or writable control store loading. The block diagram for this device is shown in Figure 9.55, and its function table is shown in Figure 9.56. The functional I/Os are defined below. Show how this device can be used in making circuits scan testable. Which generic scan architecture is best suited for implementation by this device, and why?

$D_7 - D_0$ parallel data input to the pipeline register or parallel data output from the shadow register

DCLK   diagnostic clock for loading shadow register (serial or parallel modes)

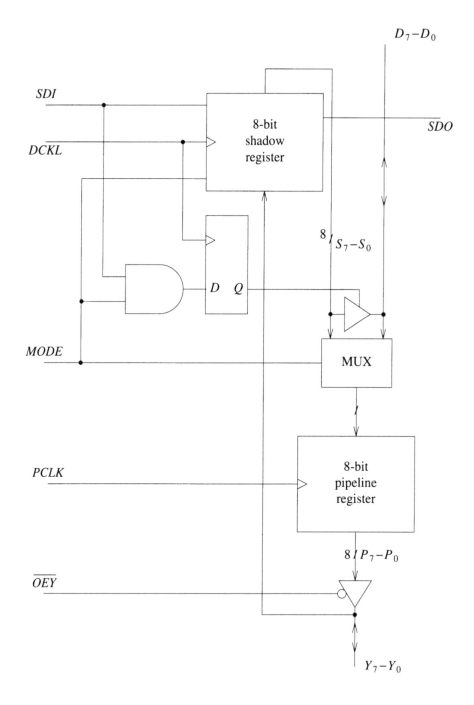

Figure 9.55   The Am 29818

Inputs				Outputs			
SDI	MODE	DCLK	PCLK	SDO	Shadow register	Pipeline register	Operation
X	L	↑	X	S_7	$S_i \leftarrow S_{i-1}$ $S_o \leftarrow SDI$	NA	Serial shift; D_{7-D_0} disabled
X	L	X	↑	S_7	NA	$P_i \leftarrow D_i$	Normal load pipeline register
L	H	↑	X	L	$S_i \leftarrow Y_i$	NA	Load shadow register from Y; $D_7 - D_0$ disabled
X	H	X	↑	SDI	NA	$P_i \leftarrow S_i$	Load pipeline register from shadow register
H	H	↑	X	H	Hold	NA	Hold shadow register; $D_7 - D_0$ enabled

NOTATION

INPUTS		OUTPUTS	
H	= high	$S_7 - S_o$	= shadow register outputs
L	= low	$P_7 - P_0$	= pipeline register outputs
X	= don't care	$D_7 - D_0$	= data I/O port
↑	= low-to-high	$Y_7 - Y_0$	= Y I/O port
	clock transition	NA	= not applicable; output is not a function of the specified input combinations

Figure 9.56 Am 29818 function table description

MODE control input for pipeline register multiplexer and shadow register control

$\overline{OEY}$ active LOW output enable for Y-port

PCLK pipeline register clock input loads D-port or shadow register contents on low to high transition

SDI serial data input to shadow register

SDO serial data output from shadow register

$Y_7 - Y_0$ data outputs from the pipeline register and parallel inputs to the shadow register

9.4 Consider an asynchronous circuit having many inputs, outputs, and feedback lines. Assume that during testing, line 1 requires 0-injection only; line 2 requires 1-injection only; line 3 requires both 0 and 1 injection; lines 4 and 5 require observation only.

  a. Modify the design so that it has the desired test features. Minimize the amount of hardware you add.

b.   Reduce your I/O requirements by sharing functional and test I/O pins.

c.   Modify your results to part a. by using a scan register so that observation points and control points do not have to be tied to I/O pins. Assume that during the scan operation the normal state of the circuit should not be disturbed.

9.5   Consider IC1 and IC2, where each IC contains boundary scan registers as shown in Figure 9.18(b). Assume the outputs of IC1 drive the inputs to IC2, and the outputs of IC2 drive the inputs to IC1. Show how the scan registers of these ICs can be interconnected. Describe how the interconnections between these ICs can be tested by means of the boundary scan registers. Explain the control sequencing required for the signal T1 as well as the signal $\overline{N/T}$ that controls the scan registers.

Are these control signals also tied to scan flip-flops? If not, how are these lines tested? If so, does the scan operation interfere with their values?

9.6   Consider a scan register implemented using MD-FF elements as shown in Figure 9.23(b). Show how clock skew can lead to erroneous operation during shift operations.

9.7   Consider a random-access scan architecture. How would you organize the test data to minimize the total test time? Describe a simple heuristic for ordering these data.

9.8   Figure 9.29 shows the double-latch LSSD design. Assume there are 100 SRLs and the circuit requires 1000 test vectors. How many clock cycles are required to test this circuit?

9.9   Consider the single-latch LSSD design shown in Figure 9.30. For each case listed below, determine the best way to organize the testing of the blocks N_1 and N_2 to minimize test time. Let $n = 10$ and $m = 30$.

a.   Assume N_1 requires 10 test vectors, and N_2 requires 30.

b.   Assume N_1 requires 30 test vectors and N_2 requires only 10.

9.10   Single-latch LSSD designs appear to require a significant amount of overhead with respect to other types of LSSD designs (see Figure 9.33). What advantages, if any, do such designs have over the other designs?

9.11   For each of the circuit configurations shown in Figure 9.57 state whether it satisfies the LSSD design rules and which rules, if any, it violates.

9.12   Assume that the normal function of a register in a large circuit is only to act as a shift register. To make this circuit scan testable by either the LSSD design methodology or the shift register modification technique requires some extra logic and interconnect since both a normal-mode and test-mode shift register will exist in the final design. Comment on the pros and cons of such an approach, and suggest how the normal-mode shift register can be used in the scan mode to reduce hardware overhead. What problems may this create?

9.13   Assume that the normal function of a register (made up of D flip-flops) in a large circuit is both to load data in parallel as well as shift. A multiplexer is used to select the input appropriate to each flip-flop. Show how in a scan design it may be possible to use

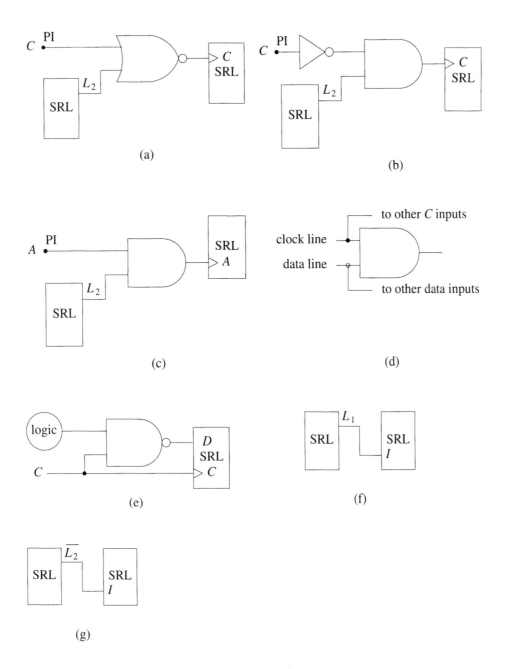

Figure 9.57   Legal and illegal LSSD circuit configurations

the scan operation for both functional and test mode operation and thus eliminate the multiplexers referred to above. What problems may exist because of this dual use of the scan registers?

9.14 Describe and illustrate a linear time algorithm for partitioning a combinational circuit into clouds. Assume no storage cells have been clustered into registers.

9.15 Given a clouded version of a circuit S, describe an algorithm for generating the minimal number of registers from the storage cells consistent with the existing set of clouds.

9.16 Argue for or against the following proposition. Randomly selecting a subset of flip-flops to be made part of a scan path in a partial scan design will reduce ATG cost compared to a nonscan design.

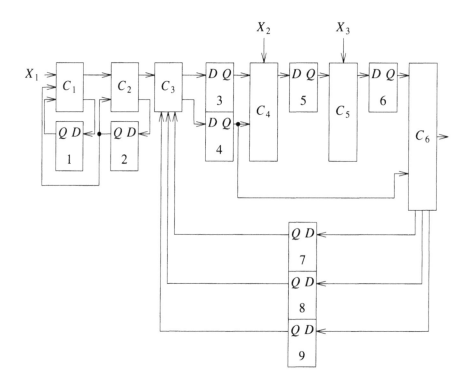

Figure 9.58   Circuit to be made balanced

9.17 Cloud the circuit shown in Figure 9.58, and identify a minimum number of flip-flops to be made part of the scan path so the resulting circuit is a balanced structure.

9.18 Show a design for the boundary-scan cell circuitry for a bidirectional I/O pin in the IEEE 1149.1 methodology.

9.19 Two boundary-scan cells are needed to implement a tristate output pad, one to specify whether or not the pad should be in the high impedance state, and the other to determine the logic value of the pad when it is not in the high impedance state. Indicate the logic associated with such an output pad.

9.20 The following questions deal with IEEE 1149.1.

a) Discuss various applications for the capability to load the instruction register with system status information.

b) Discuss how the on-chip test bus circuitry can be tested.

c) Suggest a format for the information in the instruction register, including fields, their width and functionality.

d) Suggest several uses for an identification test data register.

9.21 Describe a generic test program, consisting of instructions and test data, for use in testing the interconnect between two chips which support IEEE 1149.1. At each step, indicate the level of the *TMS* line and the state of the TAP controller.

10. COMPRESSION TECHNIQUES

About This Chapter

In previous chapters we have considered a conventional testing approach, which involves a bit-by-bit comparison of observed output values with the corrected values as previously computed and saved. This approach requires a significant amount of memory storage for saving the correct outputs associated with all test vectors. In this chapter we consider an alternative approach, which is simpler and requires less memory storage. In this approach the information saved is a compressed form of the observed test outcome, called a *signature*. A circuit is tested by comparing the observed signature with the correct computed signature. The process of reducing the complete output response to a signature is referred to as *response compacting* or *compressing*.

This concept is illustrated in Figure 10.1. A fault is detected if the signature $S(R')$ obtained from the circuit under test (CUT) differs from the precomputed signature $S(R_0)$ of a fault-free circuit.

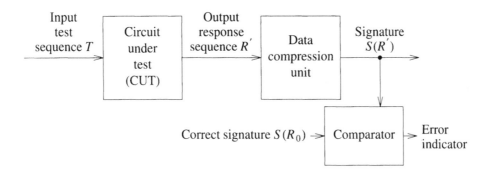

Figure 10.1   Testing using test-response compression

By making the compression circuitry simple, it is possible to embed this function within the circuit being tested, thus realizing one important aspect of *built-in self-test* (BIST), namely detection of response errors.

This chapter describes five compression techniques, namely *ones counting*, *transition counting*, *parity checking*, *syndrome checking*, and *signature analysis*. Since signature-analysis techniques have become extremely popular, they will be discussed in detail. Chapter 11 deals with the use of data-compression techniques in BIST circuits.

10.1 General Aspects of Compression Techniques

A reliable and practical compression technique should be easily implemented by a simple circuit, which, if desired, can be included in the CUT as part of its BIST logic. Furthermore the compression procedure should not introduce signal delays that affect either the normal behavior or the test execution time of the CUT. In addition the length of the test signature should be a logarithmic factor of the length of the output-response data so that the amount of storage is significantly reduced. And for any fault and associated output response containing one or more errors, the signatures of the faulty and fault-free cases should not be the same. This is required to ensure that the compression method does not lose information.

There is no known compression procedure that satisfies all these requirements. Of particular difficulty is insuring that the faulty and fault-free signatures are different, since a fault may produce offsetting errors and hence go undetected. This situation is referred to as *error masking*, and the erroneous output response is said to be an *alias* of the correct output response. There are three common ways to measure the masking characteristics associated with a compression technique. Some also provide information on the associated fault coverage.

The first way is to simulate the circuit and compression technique and determine which faults are detected. This method requires fault simulation and hence is computationally expensive, especially if the test sequence is long.

A second method classifies the output-response sequences from faulty circuits into categories, such as single-bit error or burst errors, which are errors that lie within a fixed number of patterns of one another. This set of error patterns is then analyzed to determine the degree of masking associated with various compression procedures.

A third method of assessing masking is to compute the fraction of all possible erroneous response sequences that will cause masking. To obtain a realistic result one needs to know the distribution of erroneous sequences. This often is impractical to determine. Usually all possible output sequences are assumed to be equally likely. The problem with this technique is that it is not possible to correlate the probability of obtaining an erroneous signature with fault coverage.

Because the degree of masking is so hard to measure in arbitrary circuits with arbitrary test sets, some compression techniques restrict the configuration of the CUT or require special test sets or both. Most common BIST compression methods fall into one of the following categories:

1. Those that do not require special test sets or circuit designs.

2. Those that require special test sets.

3. Those that require special test sets and circuit designs. Examples of these methods can be found in Chapter 14 dealing with PLAs.

To get the same signature for multiple copies of a fault-free circuit, all sequential circuits are assumed to be initialized to some fixed state before applying a test sequence.

One further problem associated with response-compression techniques is that of calculating the good signature. One way to obtain this signature is to identify a good part, apply the actual test to the part, and have the compression hardware generate the signature. Another approach is to simulate the CUT and compression procedure using the actual test patterns. If the number of test vectors is large, this process can be computationally expensive. Still another technique is to produce many copies of a CUT and attempt to deduce the correct signature by finding a subset of circuits that produce the same signature. These circuits are assumed to be fault-free.

10.2 Ones-Count Compression

Assuming a single-output circuit C, let the output response of C be $R = r_1, r_2, ..., r_m$. In *ones counting*, the signature $IC(R)$ is the number of 1s appearing in R, i.e.

$$IC(R) = \sum_i r_i$$

where

$$0 \le IC(R) \le m.$$

The compressor is simply a counter and the degree of compression is $\lceil \log_2(m+1) \rceil$. Figure 10.2 shows a simple circuit, an exhaustive test set, and the response data corresponding to the fault-free and two faulty circuits. Figure 10.3 shows how this circuit can be tested via ones-count compression and the corresponding signatures. Here R_0 refers to the fault-free response and R_i, $i > 0$, to the response corresponding to fault f_i.

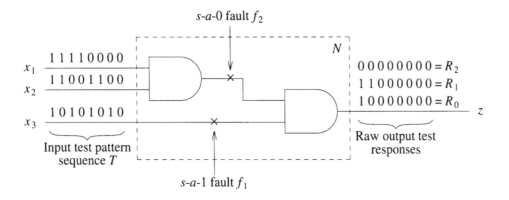

Figure 10.2   Input/output response data

Consider a circuit tested with m random input vectors and let $IC(R_0) = r$, $0 \le r \le m$. The number of m-bit sequences having r 1s is $\binom{m}{r}$. Thus $\binom{m}{r} - 1$ such sequences are aliases. The ratio of masking sequences to all possible erroneous sequences, given that $IC(R_0) = r$, is

$$P_{IC}(M \mid m,r) = \frac{\binom{m}{r} - 1}{2^m - 1}.$$

If all $2^m - 1$ error sequences are equally probable, which is usually not a realistic assumption, then $P_{IC}(M \mid m,r)$ can be considered to be the probability of masking. Ones-count testing has the following additional attributes.

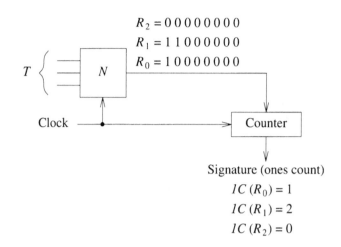

$$R_2 = 0\,0\,0\,0\,0\,0\,0\,0$$
$$R_1 = 1\,1\,0\,0\,0\,0\,0\,0$$
$$R_0 = 1\,0\,0\,0\,0\,0\,0\,0$$

T

N

Clock

Counter

Signature (ones count)

$$1C\,(R_0) = 1$$
$$1C\,(R_1) = 2$$
$$1C\,(R_2) = 0$$

Figure 10.3   Testing via ones counting

1. Because of the bell-shaped form of the binomial coefficient $\binom{m}{r}$, the probability of masking is low when the signature lies near the extreme values of its range, but increases rapidly as it approaches the midpoint, i.e., $\lfloor m/2 \rfloor$.

2. When $1C(R_0) = 0$ or m, then no masking can occur (see Problem 10.1).

3. A fault that creates an odd number of errors in the output response is always detected; if the number of errors is even, it may be detected.

4. Inverting the output does not change the statistical properties of this test technique.

5. If the circuit under test is combinational, then the test vectors in T can be permuted without affecting the fault coverage. This permutation will not affect the signature or the masking characteristics.

Multiple-output circuits can be tested using a separate counter on each of the k outputs. To reduce area overhead, a parallel-to-serial conversion of the outputs can be used along with a single counter. Here each output is loaded into a register, and the test process stops while the k-bit response is shifted out of the register into the ones counter, which must have $\lceil \log_2(km+1) \rceil$ flip-flops.

Combinational Circuits

For the following result we assume that the CUT is tested by random test vectors and that all error patterns are equally likely to occur.

Theorem 10.1: The masking probability for ones-count compression for a combinational circuit asymptotically approaches $(\pi m)^{-1/2}$.

Proof: As shown previously, when $1C(R_0) = r$ the masking probability is

$$P_{IC}(M \mid m,r) = \frac{\binom{m}{r} - 1}{2^m - 1} \qquad (10.1)$$

Consider the set S of all possible n-input functions. Assuming random test vectors, the probability that the random test vectors will produce r ones at the output F of these functions over this set S is

$$P(r) = \frac{\binom{m}{r}}{2^m} \qquad (10.2)$$

The masking probability is therefore

$$P(M) = \sum_{r=0}^{m} P_{IC}(M \mid m,r) P(r) \qquad (10.3)$$

Substituting (10.1) and (10.2) into (10.3) yields

$$P(M) = \frac{\binom{2m}{m} - 2^m}{2^m(2^m - 1)} \qquad (10.4)$$

Using Stirling's formula $n! \approx (2\pi n)^{1/2} e^{-n} n^n$ in (10.4) produces the result $P(M) \approx (\pi m)^{-1/2}$ □

Next we show that for any combinational circuit C there exists a test $T'(1C)$ which detects all faults of interest, even when the results are compressed using ones counting. Let $T = \{T^0, T^1\}$ be a set of m test vectors that detect a set of faults F in C, and where for all $t \in T^0$, the fault-free response is 0, and for all $t \in T^1$, the fault-free response is 1. Let the test set $T'(1C)$ consist of one copy of every test pattern in T^0, and $|T^0| + 1$ copies of every test pattern in T^1. In the worst case, $T'(1C)$ consists of the order of m^2 test vectors.

Theorem 10.2: When applying ones counting and the test set $T'(1C)$ to C, no error masking occurs for any fault in F.

Proof: See Problem 10.3.

Further details on ones counting can be found in [Barzilai *et al.* 1981], [Hayes 1976], [Losq 1978], and [Parker 1976].

10.3 Transition-Count Compression

In *transition-count* (TC) *testing*, the signature is the number of 0-to-1 and 1-to-0 transitions in the output data stream. Thus the transition count associated with a sequence $R = r_1, r_2, ..., r_m$ is

$$TC(R) = \sum_{i=1}^{m-1} (r_i \oplus r_{i+1})$$

where $\sum$ denotes ordinary arithmetic addition and $\oplus$ is modulo-2 addition. Since $0 \le TC(R) \le (m-1)$, the response-compression circuitry consists of a transition detector and a counter with $\lceil \log_2 m \rceil$ stages. Figure 10.4 illustrates this concept. Note that the output of the transition detector in Figure 10.4(b) is a function of the initial state of the flip-flop.

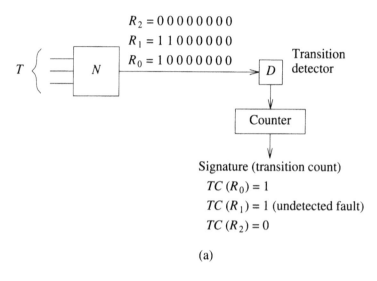

$$R_2 = 0\,0\,0\,0\,0\,0\,0\,0$$
$$R_1 = 1\,1\,0\,0\,0\,0\,0\,0$$
$$R_0 = 1\,0\,0\,0\,0\,0\,0\,0$$

Transition detector

Counter

Signature (transition count)

$$TC\,(R_0) = 1$$
$$TC\,(R_1) = 1 \text{ (undetected fault)}$$
$$TC\,(R_2) = 0$$

(a)

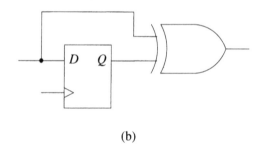

(b)

Figure 10.4   (a) Transition-count testing (b) A transition detector

Let T be a test sequence of length m for a circuit and R_0 be the fault-free response, where $TC(R_0) = r$. Let R' be an arbitrary binary sequence of length m. R' has $(m-1)$ boundaries between bits where a transition can occur. There are $\binom{m-1}{r}$ ways of assigning r transitions to $m-1$ boundaries so that R' will also have a transition count of r. Since the sequence $\overline{R'}$ obtained by complementing every bit of R' has the same transition count as R', there are $2\binom{m-1}{r}$ possible sequences that have transition count r, only one of which is the response of the fault-free circuit. Thus there are $2\binom{m-1}{r} - 1$ possible error sequences that lead to aliasing. If all faulty sequences are equally likely to occur as the response of a faulty circuit, then the probability of masking is given by

$$P_{TC}(M \mid m,r) = \frac{2\binom{m-1}{r} - 1}{2^m - 1}.$$

This function has similar properties to that derived for the case of ones counting.

Unlike ones counting, transition counting is sensitive to the order of the bits in the response vector. Also transition counting does not guarantee detecting all single-bit errors. (See Figure 10.4). The single-bit error masking capability of a TC test is summarized in the following result.

Theorem 10.3: In an arbitrary m-bit sequence the probability of a single-bit error being masked is $(m-2)/2m$.

Proof: Note that for the sequence of bits r_{j-1}, r_j, r_{j+1}, where $r_{j-1} \neq r_{j+1}$, if r_j is in error, then this error will not be detected. Consider bit r_j. Selecting a value for r_{j-1} also determines the value of r_{j+1}. Thus there are 2^{m-1} sequences in which an error on bit r_j is TC-undetectable. Also there are $(m-2)$ choices for j, since r_j cannot be the first or last bit in R. Thus there are $(m-2)\,2^{m-1}$ TC-undetectable single-bit errors in an arbitrary sequence of length m. Since there are 2^m possible m-bit sequences and each can be associated with m single-bit errors, the probability of a TC-undetectable single-bit error occurring is

$$P_r = \frac{(m-2)\,2^{m-1}}{m\,2^m} = \frac{(m-2)}{2m}.$$

This value approaches $1/2$ for large values of m. □

For multiple-output circuits, a transition detector is needed for each output pin. Each detector can drive a counter, or a single counter can be used to compress a weighted sum of each of the detectors.

Combinational Circuits

Again assume that the CUT is tested by random test vectors and that all error patterns are equally likely to occur.

Theorem 10.4: The masking probability for transition-count compression for a combinational circuit asymptotically approaches $(\pi m)^{-1/2}$. □

The proof is similar to that of Theorem 10.1, but now

$$P(M) = \frac{4\binom{2m-2}{m-1} - 2^m}{2^m\,(2^m-1)}.$$

We next focus on the problem of testing a combinational circuit using a deterministic test set where faults are detected by means of transition-count testing.

Let X be an input test sequence and $F = \{\,f_1,...,f_n\,\}$ be a set of faults in a combinational circuit C. Let R_0 be the response to X from C, and R_i be the response to X from C when fault f_i is present. For simplicity we assume that C is a single-output circuit. Then X is a transition-count test for C with respect to F if $TC(R_0) \neq TC(R_i)$ for $1 \leq i \leq n$. Here, every fault is said to be TC-detectable. Note that in general not all logic faults are TC-detectable. For example, consider a fault that causes the output function realized by C to be changed from $f(x_1,...,x_n)$ to $\overline{f}(x_1,...,x_n)$. For this case the response to X will be

$\overline{R}_0$; that is, each bit in the output sequence will be complemented. But $TC(R_0) = TC(\overline{R}_0)$; hence this fault is TC-undetectable even though every response bit is in error (see Problem 10.11).

Faults f_i and f_j are said to be TC-*distinguishable* if there exists an input sequence X such that $TC(R_i) \neq TC(R_j)$. Transition-count testing is weaker than the conventional testing procedure in the following sense.

1. A fault in C is TC-detectable only if it is detectable; the converse is not true.

2. Two faults in C are TC-distinguishable only if they are distinguishable; the converse is also false.

We will restrict out attention to TC testing of single and multiple stuck-type faults. Note that in an irredundant single-output combinational circuit C no single fault can change the output function $f(x_1,...,x_n)$ to $\overline{f}(x_1,...,x_n)$. It is conjectured that this result also holds true for multiple faults. However, in a single-output redundant circuit it is possible for a multiple fault to complement the output.

The following theorem specifies a constructive procedure for generating a complete TC test for a combinational circuit C.

Theorem 10.5: Let T be a single-fault test set for an irredundant single-output combinational circuit C. Let $T^0(T^1)$ be all tests in T producing output 0(1). Construct a test sequence $X^* = t(1)t(2)...t(p)$ as follows:

1. X^* contains every element in T.

2. X^* is an alternating sequence of tests from T^0 and T^1. If $|T^0| \geq |T^1|$, let $t(1) \in T^0$, otherwise $t(1) \in T^1$. If $t(i) \in T^d$, then select $t(i+1) \in T^d$ for $1 \leq i \leq p - 1$. The resulting sequence X^* is a single-fault TC test for C.

Proof: The response R^* to X^* is an alternating sequence of 0s and 1s. The only other sequence with this same transition count is $\overline{R}_0$. But X^* detects all stuck-at faults, and none of these faults can produce $\overline{R}_0$. Note that the length p of X^* is bounded by the inequalities $p \leq 2 \ \max \ \{ |T^0|, |T^1| \} < 2(|T|-1)$. In addition, if T is a minimal-length test for C, and if the difference D between $|T^0|$ and $|T^1|$ is at most 1, then X^* is a minimal-length single-fault TC test for C.   □

Thus, every single stuck fault in a single-output irredundant combinational circuit is TC-detectable using less than twice the number of tests required by conventional testing. For $D > 1$, the problem of finding a minimal-length TC test is unsolved. An extension to this analysis dealing with both single and multiple faults in two-level sum of product circuits is given in Problem 10.12.

Further material on transition count testing can be found in [Hayes 1976a, 1976b], [Reddy 1977], and [Savir and McAnney 1985].

10.4 Parity-Check Compression

Parity-check compression is carried out using a simple circuit such as that shown in Figure 10.5. As will be shown later, the compression circuit consisting of the D flip-flop and the XOR gate implements a linear feedback shift register whose primitive polynomial is $G(x) = x + 1$. If the initial state of the flip-flop is 0, the signature S is the parity of the circuit response, namely it is 0 if the parity is even and 1 if the parity is odd. This scheme detects all single-bit errors and all multiple-bit errors consisting of an odd number of error bits in the response sequence. Faults that create an even number of errors are not detected. Assuming all faulty bit streams are equally likely, the probability of masking approaches 1/2 as m increases.

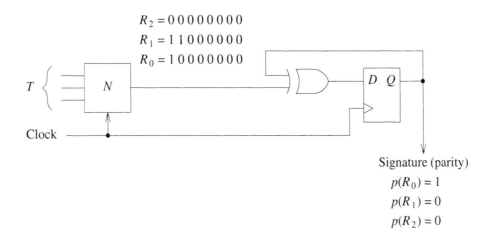

$$R_2 = 0\,0\,0\,0\,0\,0\,0\,0$$
$$R_1 = 1\,1\,0\,0\,0\,0\,0\,0$$
$$R_0 = 1\,0\,0\,0\,0\,0\,0\,0$$

$$p(R_0) = 1$$
$$p(R_1) = 0$$
$$p(R_2) = 0$$

Figure 10.5   Parity-check compression

This technique can be extended to multiple-output circuits in several ways. One way is to replace the input exclusive-or gate by a multiple-input gate or exclusive-or network and have all outputs of the CUT drive this network. Unfortunately, an error internal to a circuit may affect more than one output line. If such an error propagates simultaneously to an even number of output lines, then it will have no effect on the signature.

A second method is to use a separate parity-check compressor in each output, though this requires additional hardware.

More information on parity-check compression can be found in [Tzidon *et al.* 1978] and [Carter 1982a, 1982b].

10.5 Syndrome Testing

Syndrome testing relies on exhaustive testing, i.e., on applying all 2^n test vectors to an n-input combinational circuit. First consider a single-output circuit implementing a function f. The *syndrome S* (or signature) is the normalized number of 1s in the resulting bit stream; i.e., $S = K/2^n$, where K is the number of minterms in the function f. Thus syndrome testing is a special case of ones counting. Clearly $0 \le S \le 1$. The syndrome of a 3-input AND gate is 1/8 and that of a 3-input OR gate is 7/8. Also the syndrome of f is a functional property of f independent of its implementation.

Syndrome testing is of interest because of the concept of *syndrome testability*; i.e., any function f can be realized in such a way that all single stuck-at faults are syndrome detectable.

Consider the circuit configuration shown in Figure 10.6. Assume circuits C_1 and C_2 have no inputs in common, and let $S(F) = S_1$, and $S(G) = S_2$. The input-output syndrome relation for this circuit as a function of the type of gate used to realize the circuit C is given in Figure 10.7.

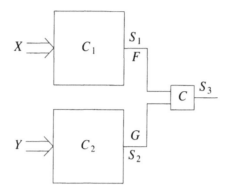

Figure 10.6   Circuit for calculating syndromes

Gate type for C	Syndrome S_3
OR	$S_1 + S_2 - S_1 S_2$
AND	$S_1 S_2$
NAND	$1 - S_1 S_2$
NOR	$1 - (S_1 + S_2 - S_1 S_2)$
XOR	$S_1 + S_2 - 2 S_1 S_2$

Figure 10.7   Syndromes for circuits having nonreconvergent fanout

Let C_i have n_i inputs and K_i minterms. For the case where C is an OR gate, the result in Figure 10.7 can be derived as follows. Note that $K = K_1 2^{n_2} + K_2 2^{n_1} - K_1 K_2$. Clearly each minterm of C_1 can be associated with 2^{n_2} inputs to C_2, each of which produces an output of 1. The last term is required so that no minterms are counted twice. Then

$$S = \frac{K}{2^{n_1 + n_2}} = \frac{K_1}{2^{n_1}} + \frac{K_2}{2^{n_2}} - \frac{K_1}{2^{n_1}} \frac{K_2}{2^{n_2}} = S_1 + S_2 - S_1 S_2 .$$

The reader can verify the other entries in Figure 10.7 in a similar way. (See Problem 10.17.)

Assuming C_1 and C_2 share inputs, i.e., there is reconvergent fanout, the output syndrome can be expressed as shown in Figure 10.8.

Gate type for C	Syndrome S_3
OR	$S_1 + S_2 - S(FG)$
AND	$S_1 + S_2 + S(\overline{F}\ \overline{G}) - 1$
XOR	$S(F\overline{G}) + S(\overline{F}G)$

Figure 10.8   Syndromes for circuits having reconvergent fanout

Some basic results dealing with the testability of combinational circuits using syndrome testing are given below. Further details can be found in [Savir 1980], [Barzilai *et al.* 1981], and [Markowsky 1981].

A realization C of a function f is said to be *syndrome-testable* if no single stuck-at fault causes the circuit to have the same syndrome as the fault-free circuit.

A function $f(x_1, x_2, ..., x_n)$ is said to be *unate* in x_i if there exists a sum of product expression for f where variable x_i appears only in uncomplemented (x_i) or complemented $(\overline{x}_i)$ form. For example, $f = x_1\overline{x}_2x_3 + x_1\overline{x}_2x_4 + \overline{x}_4\overline{x}_5$ is unate in x_1, x_2, x_3, and x_5, but not in x_4.

Lemma 10.1:   A two-level irredundant circuit that realizes a unate function in all its variables is syndrome-testable.   □

There exist two-level irredundant circuits that are not syndrome-testable. For example, for the function $f = x\overline{z} + yz$, the faults z s-a-0 and z s-a-1 are not syndrome-testable. Note that z is not unate in f. By adding an extra input, this function can be converted into a function g, which is syndrome-testable. For example, $g = wx\overline{z} + yz$ is syndrome-testable, and by setting $w = 1$ during normal operation, g reduces to f. Using this technique leads to the following result:

Lemma 10.2:   Every two-level irredundant combinational circuit can be made syndrome-testable by adding control inputs to the AND gates.   □

Lemma 10.3:   Every fanout-free irredundant combinational circuit composed of AND, OR, NAND, NOR, and NOT gates is syndrome-testable.   □

Lemma 10.4:   Let g be an internal line in an arbitrary combinational circuit. The output f can be expressed as $f = Ag + B\overline{g} + C$ where A, B, C, and g are functions of some or all of the variables of f. Then g s-a-0 is syndrome-untestable iff $S(A\overline{C}g) = S(B\overline{C}g)$, and g s-a-1 is syndrome-untestable iff $S(A\overline{C}\ \overline{g}) = S(B\overline{C}\ \overline{g})$.   □

Note that if only one path exists from g to the output f, then f is syndrome-testable with respect to the faults g s-a-0 and g s-a-1.

10.6   Signature Analysis

Signature analysis is a compression technique based on the concept of cyclic redundancy checking (CRC) and realized in hardware using linear feedback shift registers (LFSRs). Before we discuss the details of signature analysis, some background on the theory and operation of LFSRs is necessary. General information on this subject can be found in [Golomb 1982] and [Peterson and Weldon 1972].

10.6.1 Theory and Operation of Linear Feedback Shift Registers

In this section we present some of the formal properties associated with linear feedback shift registers. These devices, as well as modified versions of LFSRs, are used extensively in two capacities in DFT and BIST designs. One application is as a source of pseudorandom binary test sequences; the other is as a means to carry out response compression — known as *signature analysis*.

Consider the feedback shift registers shown in Figure 10.9. These circuits are all autonomous; i.e., they have no inputs except for clocks. Each cell is assumed to be a clocked D flip-flop. It is well known that such circuits are cyclic in the sense that when clocked repeatedly, they go through a fixed sequence of states. For example, a binary counter consisting of n flip-flops would go through the states 0, 1, ..., 2^{n-1}, 0, 1, ... The maximum number of states for such a device is 2^n. The shift register shown in Figure 10.9(a) cycles through only two states. If the initial state were 00 or 11, it would never change state. An n-bit shift register cycles through at most n states. Notice that the output sequence generated by such a device is also cyclic. The circuit of Figure 10.9(b) starting in the initial state 111 (or 000) produces a cyclic sequence of states of length 1. In Figure 10.9(c) we show the sequence generated for the circuit of Figure 10.9(b) if the initial state is 011. (The reader should analyze the case if the initial state is 101.)

In Figure 10.9(d) we illustrate the case where the state sequence generated by the feedback shift register is of length $2^3 - 1$. Note that for the class of circuits being illustrated, the all-0 state leads to a state sequence of length 1, namely the all-0 state itself. The circuit of Figure 10.9(d) is said to be a maximal-length shift register, since it generated a cyclic state sequence of length $2^n - 1$, as long as its initial state is not all-0s. Note also that if one of these circuits generates a cyclic state sequence of length k, then the output sequence also repeats itself every k clock cycles.

A *linear circuit* is a logic network constructed from the following basic components:

- unit delays or D flip-flops;

- modulo-2 adders;

- modulo-2 scalar multipliers.

In the analysis of such circuits, all operations are done modulo 2. The truth table for modulo-2 addition and subtraction is shown below.

$\pm$	0	1
0	0	1
1	1	0

Thus $x + x = -x - x = x - x = 0$.

Such a circuit is considered to be linear since it preserves the principle of superposition; i.e., its response to a linear combination of stimuli is the linear combination of the responses of the circuit to the individual stimuli.

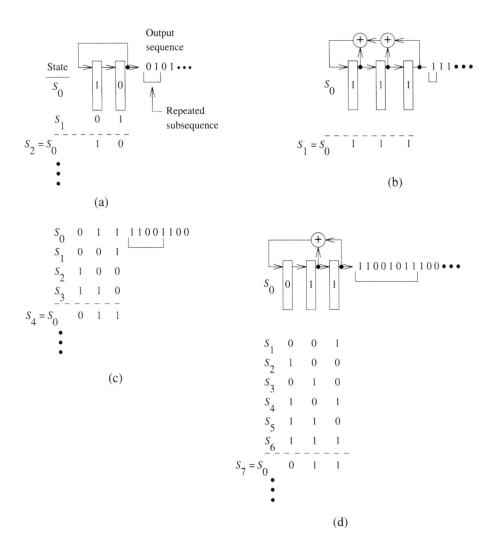

Figure 10.9   Feedback shift registers

In this section we will deal primarily with a class of linear circuits, known as *autonomous linear feedback shift registers*, that have the canonical form shown in Figures 10.10 and 10.11. Here c_i is a binary constant, and $c_i = 1$ implies that a connection exists, while $c_i = 0$ implies that no connection exists. When $c_i = 0$ the corresponding XOR gate can be replaced by a direct connection from its input to its output.

Characteristic Polynomials

A sequence of numbers $a_0, a_1, a_2, ..., a_m, ...$ can be associated with a polynomial, called a *generating function $G(x)$*, by the rule

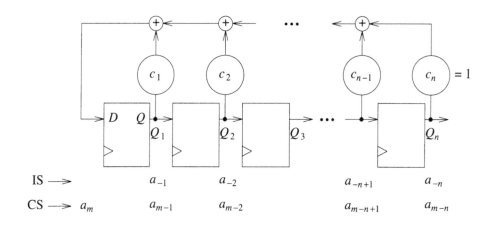

Figure 10.10   Type 1 (external-XOR) LFSR

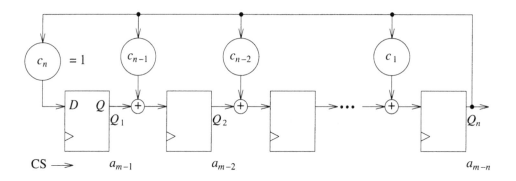

Figure 10.11   Type 2 (internal-XOR) LFSR

$$G(x) = a_0 + a_1 x + a_2 x^2 + \ldots + a_m x^m \ldots .$$

Let $\{ a_m \} = a_0, a_1, a_2, \ldots$ represent the output sequence generated by an LFSR, where $a_i = 0$ or 1. Then this sequence can be expressed as

$$G(x) = \sum_{m=0}^{\infty} a_m x^m \tag{10.5}$$

Recall that polynomials can be multiplied and divided (modulo 2). An example is shown below.

Example 10.1:

(a)

$$
\begin{array}{r}
x^2 + x\ + 1 \\
\times\quad x^2 + 1 \\
\hline
x^2 + x\ + 1 \\
(+)\ \ x^4 + x^3 + x^2 \\
\hline
x^4 + x^3 \qquad\ + x\ + 1
\end{array}
$$

since $x^2 + x^2 = 0$.

(b)

$$
\begin{array}{r}
x^2 +\ x\ +\ 1 \\
x^2 + 1\ \overline{\big)\ x^4 +\ x^3 \qquad\quad + x + 1} \\
(-)\ x^4 \qquad\quad +\ x^2 \\
\hline
(-)\ x^3 +\ x^2 + x + 1\ \ (\text{note } x^2 = -x^2) \\
x^3 \qquad\quad + x \\
\hline
(-)\ x^2 \qquad + 1 \\
x^2 \qquad + 1 \\
\hline
0
\end{array}
$$

There is no remainder; hence $x^2 + 1$ is said to *divide* $x^4 + x^3 + x + 1$. □

From the structure of the type 1 LFSR it is seen that if the current state (CS) of Q_i is a_{m-i}, for $i = 1, 2, ..., n$, then

$$
a_m = \sum_{i=1}^{n} c_i\, a_{m-i} \tag{10.6}
$$

Thus the operation of the circuit can be defined by a recurrence relation. Let the initial state (IS) of the LFSR be $a_{-1}, a_{-2}, ..., a_{-n+1}, a_{-n}$. The operation of the circuit starts n clock periods before generating the output a_0. Since

$$
G(x) = \sum_{m=0}^{\infty} a_m\, x^m
$$

substituting for a_m we get

$$
G(x) = \sum_{m=0}^{\infty}\sum_{i=1}^{n} c_i a_{m-i} x^m = \sum_{i=1}^{n} c_i x^i \sum_{m=0}^{\infty} a_{m-i} x^{m-i}
$$

$$
= \sum_{i=1}^{n} c_i x^i [a_{-i} x^{-i} + ... + a_{-1} x^{-1} + \sum_{m=0}^{\infty} a_m x^m]
$$

$$
= \sum_{i=1}^{n} c_i x^i [a_{-i} x^{-i} + ... + a_{-1} x^{-1} + G(x)].
$$

Hence

$$
G(x) = \sum_{i=1}^{n} c_i x^i G(x) + \sum_{i=1}^{n} c_i x^i (a_{-i} x^{-i} + ... + a_{-1} x^{-1})
$$

or

$$G(x) = \frac{\sum\limits_{i=1}^{n} c_i x^i (a_{-i} x^{-i} + \dots + a_{-1} x^{-1})}{1 + \sum\limits_{i=1}^{n} c_i x^i} \qquad (10.7)$$

Thus $G(x)$ is a function of the initial state $a_{-1}, a_{-2}, \dots, a_{-n}$ of the LFSR and the feedback coefficients $c_1, c_2, \dots, c_n$. The denominator in (10.7), denoted by

$$P(x) = 1 + c_1 x + c_2 x^2 + \dots + c_n x^n$$

is referred to as the *characteristic polynomial* of the sequence $\{a_m\}$ and of the LFSR. For an n-stage LFSR, $c_n = 1$. Note that $P(x)$ is only a function of the feedback coefficients. If we set $a_{-1} = a_{-2} = \dots = a_{1-n} = 0$, and $a_{-n} = 1$, then (10.7) reduces to $G(x) = 1 / P(x)$.

Thus the characteristic polynomial along with the initial state characterizes the cyclic nature of an LFSR and hence characterizes the output sequence. For $a_{-1} = a_{-2} = \dots = a_{1-n} = 0$, and $a_{-n} = 1$, then

$$G(x) = \frac{1}{P(x)} = \sum_{m=0}^{\infty} a_m x^m \qquad (10.8)$$

and since the sequence $\{a_m\}$ is cyclic with period p, (10.8) can be rewritten as

$$1/P(x) = (a_0 + a_1 x + \dots + a_{p-1} x^{p-1})$$

$$+ x^p (a_0 + a_1 x + \dots + a_{p-1} x^{p-1})$$

$$+ x^{2p} (a_0 + a_1 x + \dots + a_{p-1} x^{p-1}) + \dots$$

$$= (a_0 + a_1 x + \dots + a_{p-1} x^{p-1}) (1 + x^p + x^{2p} + \dots)$$

$$= \frac{a_0 + a_1 x + \dots + a_{p-1} x^{p-1}}{1 - x^p}.$$

Thus it is seen that $P(x)$ evenly divides into $1 - x^p$.

The analysis of the type 1 LFSR can also proceed as follows.

Recall that

$$a_m(t) = \sum_{i=1}^{n} c_i a_{m-i}(t)$$

and note that $a_i(t) = a_{i+1}(t-1)$ Let x be a "shift" operator such that

$$x^k a_i(t) = a_i(t-k) \qquad (10.9)$$

Then

$$a_m(t) = \sum_{i=1}^{n} c_i a_{m-i}(t) = \sum_{i=1}^{n} c_i x^i a_m(t).$$

Note, for example, that $x a_m(t) = a_m(t-1) = a_{m-1}(t)$.

Thus

$$a_m + c_1 x a_m + c_2 x^2 a_m + \ldots + c_n x^n a_m = 0$$

or equivalently

$$[1 + c_1 x + c_2 x^2 + \ldots + c_n x^n] a_m = 0.$$

The term in brackets is again the characteristic polynomial associated with the LFSR.

Given a characteristic polynomial, it is easy to implement a type 1 LFSR to realize it. Conversely, given a type 1 LFSR, it is a simple process to determine its corresponding characteristic polynomial. For the circuit of Figure 10.9(b), $P(x) = 1 + x + x^2 + x^3$, and for Figure 10.9(d), $P(x) = 1 + x^2 + x^3$.

Referring to (10.9), let $y^{-k} a_i(t) = a_i(t-k)$. Then, carrying out the same algebraic manipulation as before, we obtain

$$[1 + c_1 y^{-1} + c_2 y^{-2} + \ldots + c_n y^{-n}] a_m = 0$$

or equivalently

$$[y^n + c_1 y^{n-1} + c_2 y^{n-2} + \ldots + c_n] a_m y^{-n} = 0.$$

Again the term in the brackets can be considered to be a characteristic polynomial of the LFSR. Replacing y by x we obtain

$$P^*(x) = c_n + c_{n-1} x + c_{n-2} x^2 + \ldots + c_1 x^{n-1} + x^n.$$

$P^*(x)$ is said to be the *reciprocal polynomial* of $P(x)$, since $P^*(x) = x^n P(1/x)$.

Thus every LFSR can be associated with two characteristic polynomials. Referring to Figures 10.9(c) and 10.9(d), if $P(x) = 1 + x + x^2 + x^3$, then $P^*(x) = P(x)$, and if $P(x) = 1 + x^2 + x^3$, then $P^*(x) = 1 + x + x^3$.

If in Figure 10.10 one associates x^i with Q_i, then $P(x)$ can be read off directly from the figure. If, however, one associates x^i with Q_{n-i} and labels the input to the first flip-flop Q_0, then $P^*(x)$ can be read off directly from the figure. Figure 10.12 illustrates these two labelings.

Finally, note that a given characteristic polynomial, say $Q(x)$, can be realized by two different LFSRs depending on the labeling used.

Referring to Figure 10.11, we see that for $i = 2, 3, \ldots, n$

$$a_{m-i}(t+1) = a_{m-i+1}(t) + c_{n-i+1} a_{m-n}(t) \tag{10.10}$$

If we define $a_m(t) = 0$, then (10.10) is also true for $i = 1$. Let x be a "shift" operator such that $x^k a_i(t) = a_i(t-k)$. Then the equations (10.10) can be written as

$$x^{-1} a_{m-i}(t) = a_{m-i+1}(t) + c_{n-i+1} a_{m-n}(t) \tag{10.11}$$

for $i = 1, 2, \ldots, n$. Multiplying the i-th equation by x^{-i+1} we get

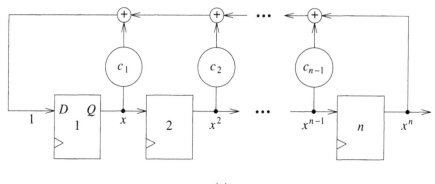

(a)

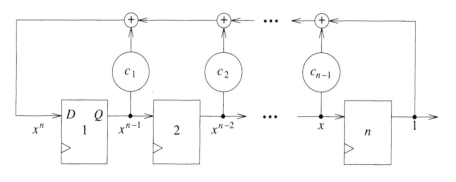

(b)

Figure 10.12 Reciprocal characteristic polynomials
(a) $P(x) = 1 + c_1 x + c_2 x_2 + \ldots + c_{n-1}x^{n-1} + x^n$
(b) $P^*(x) = 1 + c_{n-1} x + c_{n-2} x^2 + \ldots + c_1 x^{n-1} + x^n$

$$x^{-1}a_{m-1} = c_n a_{m-n}$$

$$x^{-2}a_{m-2} = x^{-1}a_{m-1} + c_{n-1}x^{-1}a_{m-n}$$

$$x^{-3}a_{m-3} = x^{-2}a_{m-2} + c_{n-2}x^{-2}a_{m-n}$$

$$\vdots$$

$$x^{-n}a_{m-n} = x^{-n+1}a_{m-n+1} + c_1 x^{-n+1}a_{m-n}.$$

Summing these n equations and canceling terms that appear on both sides of the equation
yield

$$x^{-n}a_{m-n} = [c_1 x^{-n+1} + \ldots + c_{n-2}x^{-2} + c_{n-1}x^{-1} + c_n]a_{m-n}$$

or

$$[x^{-n} + c_1 x^{-n+1} + \ldots + c_{n-1} x^{-1} + c_n] a_{m-n} = 0 .$$

Multiplying by x^n we get

$$[1 + c_1 x + \ldots + c_{n-1} x^{n-1} + c_n x^n] a_{m-n} = 0 .$$

The term in the brackets is the characteristic polynomial for the type 2 LFSR.

Again it can be shown that a characteristic polynomial $P(x)$ has two type 2 realizations, or conversely, a type 2 LFSR can be associated with two characteristic polynomials that are reciprocals of each other.

Periodicity of LFSRs

We have seen that an LFSR goes through a cyclic or periodic sequence of states and that the output produced is also periodic. The maximum length of this period is $2^n - 1$, where n is the number of stages. In this section we consider properties related to the period of an LFSR. Most results will be presented without proof. Details can be found in [Bardell *et al.* 1987], [Golomb 1982], and [Peterson and Weldon 1972].

Theorem 10.6:   If the initial state of an LFSR is $a_{-1} = a_{-2} = \ldots = a_{1-n} = 0$, $a_{-n} = 1$, then the LFSR sequence $\{a_m\}$ is periodic with a period that is the smallest integer k for which $P(x)$ divides $(1-x^k)$. $\qquad\qquad\qquad\qquad\qquad\qquad\qquad\qquad\qquad\qquad\qquad\qquad\qquad\quad$ □

Definition: If the sequence generated by an n-stage LFSR has period $2^n - 1$, then it is called a *maximum-length sequence*.

Definition: The characteristic polynomial associated with a maximum-length sequence is called a *primitive polynomial*.

Definition: An *irreducible polynomial* is one that cannot be factored; i.e., it is not divisible by any other polynomial other than 1 and itself.

Theorem 10.7:   An *irreducible polynomial* $P(x)$ of degree n satisfies the following two conditions:

  1.   For $n \geq 2$, $P(x)$ has an odd number of terms including the 1 term.

  2.   For $n \geq 4$, $P(x)$ must divide (evenly) into $1 + x^k$, where $k = 2^n - 1$. $\qquad\qquad$ □

The next result follows from Theorems 10.6 and 10.7.

Theorem 10.8:   An irreducible polynomial is primitive if the smallest positive integer k that allows the polynomial to divide evenly into $1 + x^k$ occurs for $k = 2^n - 1$, where n is the degree of the polynomial. $\qquad\qquad\qquad\qquad\qquad\qquad\qquad\qquad\qquad\qquad\qquad\qquad\qquad$ □

The number of primitive polynomials for an n-stage LFSR is given by the formula

$$\lambda_2(n) = \Phi(2^n - 1) / n$$

where

$$\Phi(n) = n \prod_{p \mid n} (1 - 1/p)$$

and p is taken over all primes that divide n. Figure 10.13 shows some values of $\lambda_2(n)$.

n	$\lambda_2(n)$
1	1
2	1
4	2
8	16
16	2048
32	67108864

Figure 10.13   Number of primitive polynomials of degree n

Figure 10.14 gives one primitive polynomial for every value of n between 1 and 36. A shorthand notation is employed. For example, the polynomial $x^{12} + x^7 + x^4 + x^3 + 1$ is represented by the entry 12:7 4 3 0, listing the exponents of those x^i associated with a "1" coefficient.

1:	0				13:	4	3	1	0	25:	3	0		
2:	1	0			14:	12	11	1	0	26:	8	7	1	0
3:	1	0			15:	1	0			27:	8	7	1	0
4:	1	0			16:	5	3	2	0	28:	3	0		
5:	2	0			17:	3	0			29:	2	0		
6:	1	0			18:	7	0			30:	16	15	1	0
7:	1	0			19:	6	5	1	0	31:	3	0		
8:	6	5	1	0	20:	3	0			32:	28	27	1	0
9:	4	0			21:	2	0			33:	13	0		
10:	3	0			22:	1	0			34:	15	14	1	0
11:	2	0			23:	5	0			35:	2	0		
12:	7	4	3	0	24:	4	3	1	0	36:	11	0		

Figure 10.14   Exponents of terms of primitive polynomials

Characteristics of Maximum-Length Sequences

Sequences generated by LFSRs that are associated with a primitive polynomial are called *pseudorandom sequences*, since they have many properties like those of random sequences. However, since they are periodic and deterministic, they are pseudorandom, not random. Some of these properties are listed next.

In the following, any string of $2^n - 1$ consecutive outputs is referred to as an *m-sequence*.

Property 1.   The number of 1s in an *m*-sequence differs from the number of 0s by one.

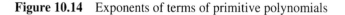

Property 2. An *m*-sequence produces an equal number of runs of 1s and 0s.

Property 3. In every *m*-sequence, one half the runs have length 1, one fourth have length 2, one eighth have length 3, and so forth, as long as the fractions result in integral numbers of runs.

These properties of randomness make feasible the use of LFSRs as test sequence generators in BIST circuitry.

10.6.2 LFSRs Used as Signature Analyzers

Signature analysis is a compression technique based on the concept of *cyclic redundancy checking* (CRC) [Peterson and Weldon 1972]. In the simplest form of this scheme, the signature generator consists of a single-input LFSR. The signature is the contents of this register after the last input bit has been sampled. Figure 10.15 illustrates this concept.

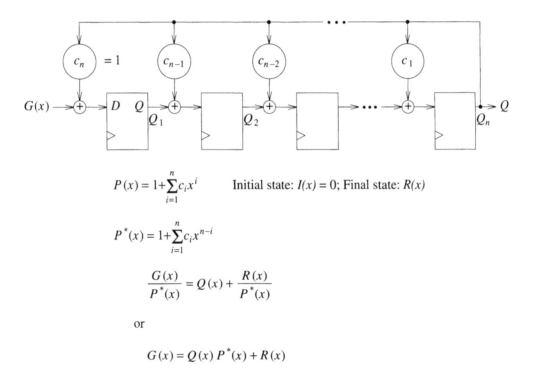

$$P(x) = 1 + \sum_{i=1}^{n} c_i x^i \qquad \text{Initial state: } I(x) = 0; \text{ Final state: } R(x)$$

$$P^*(x) = 1 + \sum_{i=1}^{n} c_i x^{n-i}$$

$$\frac{G(x)}{P^*(x)} = Q(x) + \frac{R(x)}{P^*(x)}$$

or

$$G(x) = Q(x) P^*(x) + R(x)$$

Figure 10.15   A type 2 LFSR used as a signature analyzer

The proportion of error streams that mask to the correct signature $S(R_0)$ is independent of the actual signature. For a test bit stream of length *m*, there are 2^m possible response streams, one of which is correct. It will be shown later that the structure of an LFSR distributes all possible input bit streams evenly over all possible signatures; i.e., the number of bit streams that produce a specific signature is

$$\frac{2^m}{2^n} = 2^{m-n} \tag{10.12}$$

where the LFSR consists of n stages, and the all-0 state is now possible because of the existence of an external input. For a particular fault-free response, there are $2^{m-n} - 1$ erroneous bit streams that will produce the same signature. Since there are a total of $2^m - 1$ possible erroneous response streams, the proportion of masking error streams is

$$P_{SA}(M \mid m,n) = \frac{2^{m-n} - 1}{2^m - 1} \simeq 2^{-n} \tag{10.13}$$

where the approximation holds for $m \gg n$.

If all possible error streams are equally likely, which is rarely the case, then $P_{SA}(M \mid m,n)$ is the probability that an incorrect response will go undetected; i.e., the probability of no masking is $1 - 2^{-n}$. This is a somewhat strange result since it is only a function of the length of the LFSR and not of the feedback network. Increasing the register length by one stage reduces the masking probability by a factor of 2. Note that because of the feedback network, all single-bit errors are detectable. However, there is no direct correlation between faults and error masking. Thus a 16-bit signature analyzer may detect $100(1-2^{-16}) = 99.9984$ percent of the erroneous responses but not necessarily this same percentage of faults.

Signature analysis is the most popular method employed for test data compression because it usually produces the smallest degree of masking. This results from the signature being sensitive to the number of 1s in the data stream as well as to their positions. Because of the widespread use of signature analyzers in both built-in test circuitry as well as in ATE equipment for PCB, the rest of this chapter will deal primarily with both practical and theoretical aspects of their design.

Shift Register Polynomial Division

The theory behind the use of an LFSR for signature analysis is based on the concept of polynomial division, where the "remainder" left in the register after completion of the test process corresponds to the final signature.

Consider the type 2 (internal-XOR) LFSR shown in Figure 10.15. The input sequence $\{a_m\}$ can be represented by the polynomial $G(x)$ and the output sequence by $Q(x)$. The highest degree of the polynomials $G(x)$ and $Q(x)$ correspond, respectively, to the first input bit to enter the LFSR and the first output bit produced n clock periods later, where n is the degree of the LFSR. If the initial state of the LFSR is all 0s, let the final state of the LFSR be represented by the polynomial $R(x)$. Then it can be shown that these polynomials are related by the equation

$$\frac{G(x)}{P^*(x)} = Q(x) + \frac{R(x)}{P^*(x)}$$

where $P^*(x)$ is the reciprocal characteristic polynomial of the LFSR. The reciprocal characteristic polynomial is used because a_m corresponds to the first bit of the input stream rather than the last bit. Hence an LFSR carries out (polynomial) division on the input stream by the characteristic polynomial, producing an output stream corresponding to the quotient $Q(x)$ and a remainder $R(x)$. We illustrate this in the next example.

Example 10.2: Figure 10.16(a) shows a single-input signature analyzer where $P^*(x) = 1 + x^2 + x^4 + x^5$. Let the input sequence be 1 1 1 1 0 1 0 1 where the data are entered in the order shown. The first bit corresponds to a_m in the sequence $a_m a_{m-1} \ldots a_2 a_1$. Figure 10.16(c) shows a simulation of the processing of this input by the LFSR. The first five bits of the output response are ignored since they are independent of the input sequence. It is seen that $R(x) = x^4 + x^2$ and $Q(x) = x^2 + 1$.

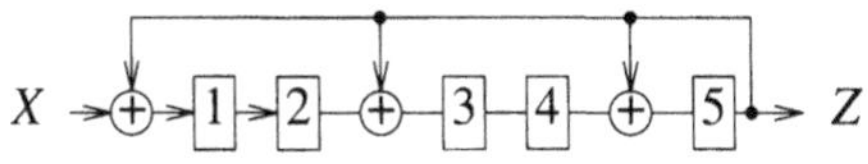

$X \longrightarrow \oplus \rightarrow \boxed{1} \rightarrow \boxed{2} \oplus \boxed{3} \boxed{4} \oplus \boxed{5} \longrightarrow Z$

$P^*(x) = 1 + x^2 + x^4 + x^5$

(a)

Input sequence: 1 1 1 1 0 1 0 1 (8 bits)

$G(x) = x^7 + x^6 + x^5 + x^4 + x^2 + 1$

(b)

Time	Input stream	Register contents	Output stream
		1 2 3 4 5	
0	1 0 1 0 1 1 1 1	0 0 0 0 0 ← Initial state	
1	1 0 1 0 1 1 1	1 0 0 0 0	
⋮	⋮	⋮	
5	1 0 1	0 1 1 1 1	
6	1 0	0 0 0 1 0	1
7	1	0 0 0 0 1	0 1
8	Remainder → 0 0 1 0 1		1 0 1

Remainder   Quotient

$R(x) = x^2 + x^4$ $1 + x^2$

(c)

Figure 10.16   Polynomial division

To check this result we have

$$
\begin{aligned}
P^*(x)&: \; x^5 + x^4 + x^2 + 1 \\
\times Q(x)&: \; x^2 + 1 \\
\hline
& x^7 + x^6 + x^4 + x^2 + x^5 + x^4 + x^2 + 1 \\
&= x^7 + x^6 + x^5 + 1
\end{aligned}
$$

Thus

$$P^*(x)Q(x) + R(x) = x^7 + x^6 + x^5 + x^4 + x^2 + 1 = G(x).$$

□

Type 1 (external-XOR) LFSRs also carry out polynomial division and produce the correct quotient. However, the contents of the LFSR is not the remainder as is the case for type 2 LFSRs. But it can be shown that all input sequences which are equal to each other modulo $P(x)$ produce the same remainder.

Error Polynomials and Masking

Let $E(x)$ be an *error polynomial*; i.e., each non-0 coefficient represents an error occurring in the corresponding bit position. As an example, let the correct response be $R_0 = 10111$ and the erroneous response be $R' = 11101$. Then the difference, or error polynomial, is 01010. Thus $G_0(x) = x^4 + x^2 + x + 1$, $G'(x) = x^4 + x^3 + x^2 + 1$, and $E(x) = x^3 + x$. Clearly $G'(x) = G(x) + E(x)$ (modulo 2). Since $G(x) = Q(x)P^*(x) + R(x)$, an undetectable response sequence is one that satisfies the equation $G'(x) = G(x) + E(x) = Q'(x)P^*(x) + R(x)$; i.e., $G'(x)$ and $G(x)$ produce the same remainder. From this observation we obtain the following well-known result from algebraic coding theory.

Theorem 10.9: Let $R(x)$ be the signature generated for an input $G(x)$ using the characteristic polynomial $P(x)$ as a divisor in an LFSR. For an error polynomial $E(x)$, $G(x)$ and $G'(x) = G(x) + E(x)$ have the same signature $R(x)$ if and only if $E(x)$ is a multiple of $P(x)$. □

Thus both type 1 and type 2 LFSRs can be used to generate a signature $R(x)$. Henceforth, the final contents of the LFSR will be referred to as the *signature*, because sometimes, depending on the initial state and polynomial $P(x)$, the final state does not correspond to the remainder of $G(x)/P(x)$.

Theorem 10.10: For an input data stream of length m, if all possible error patterns are equally likely, then the probability that an n-bit signature generator will not detect an error is

$$P(M) = \frac{2^{m-n} - 1}{2^m - 1}$$

which, for $m \gg n$, approaches 2^{-n}. □

This result follows directly from the previous theorem because $P(x)$ has $2^{m-n} - 1$ non-0 multiples of degree less than m. It also corresponds to the same result given earlier but based on a different argument. Note that this result is independent of the polynomial $P(x)$. This includes $P(x) = x^n$, which has no feedback, i.e., is just a shift register. For this case, the signature is just the last n bits of the data stream. In fact one can use the first n bits and truncate the rest of the test sequence and obtain the same results. These strange conclusions follow from the assumption that all error patterns are equally likely. If this were the case, long test sequences would certainly not be necessary.

To see why this assumption is flawed, consider a minimal-length test sequence of length m for a combinational circuit. Clearly the i-th test vector t_i detects some fault f_i not detected by t_j, $j = 1, 2, ..., i - 1$. Thus if f_i is present in the circuit, the error pattern is of the form 00...01xx...; i.e., the first $i-1$ bits must be 0. Several other arguments can be made to show that all error patterns are not equally likely.

Theorem 10.11: An LFSR signature analyzer based on any polynomial with two or more non-0 coefficients detects all single-bit errors.

Proof: Assume $P(x)$ has two or more non-0 coefficients. Then all non-0 multiples of $P(x)$ must have at least two non-0 coefficients. Hence an error pattern with only one non-0 coefficient cannot be a multiple of $P(x)$ and must be detectable.   □

As an example, $P(x) = x + 1$ has two non-0 coefficients and thus detects all single errors. (See Figure 10.5.)

Definition: A (k,k) *burst error* is one where all erroneous bits are within k consecutive bit positions, and at most k bits are in error.

Theorem 10.12:   If $P(x)$ is of degree n and the coefficient of x^0 is 1, then all (k,k) burst errors are detected as long as $n \geq k$.   □

Rather than assuming that all error patterns are equally likely, one can assume that the probability that a response bit is in error is p. Then for $p = 0.5$, the probability of masking is again 2^{-n}. For very small or very large values of p, the probability of masking approaches the value

$$2^{-n} + (2^n - 1)(1 - 2p)^{m(1 - 1/(2^n - 1))}$$

where m is the length of the test sequence [Williams *et al.* 1987].

Experimental results also show that using primitive polynomials helps in reducing masking.

In conclusion, the bound of 2^{-n} on error masking is not too useful since it is based on unrealistic assumptions. However, in general, signature analysis gives excellent results. Results are sensitive to $P(x)$ and improve as n increases. Open problems deal with selecting the best $P(x)$ to use, characterizing error patterns, correlating fault coverage to $P(x)$, and determining the probability of masking.

Further results on signature analysis can be found in [Frohwerk 1977], [Chan 1977], [Smith 1980], [Bhaskar 1982], [Sridhar *et al.* 1982], and [Carter 1982]. McAnney and Savir [1988] describe a procedure for determining the initial state of a signature register so that the final signature in a fault-free circuit is a given constant, say all zeros. Sometimes this can simplify the checking of the final signature, such as when more than one test sequence is used.

10.6.3 Multiple-Input Signature Registers

Signature analysis can be extended to testing multiple-output circuits. Normally a single-input signature analyzer is not attached to every output because of the resulting high overhead. A single signature analyzer could be time-multiplexed, but that would require repeating the test sequence for each output, resulting in a potentially long test time. The most common technique is to use a multiple-input signature register (MISR), such as the one shown in Figure 10.17 [Hassan *et al.* 1983, David 1984]. Here we assume that the CUT has n (or less) outputs. It is seen that this circuit operates as n single-input signature analyzers. For example, by setting $D_i = 0$ for all $i \neq j$, the circuit computes the signature of the data entering on line D_j. The mathematical theory associated with MISRs will not be presented here, but follows as a direct extension of the results presented previously [Bardell *et al.* 1987]. One can again associate an error pattern with each input D_i. These error patterns are merged within the LFSR. Again, assuming all error patterns are equally likely, the probability that a MISR will not detect an error is approximately 2^{-n}.

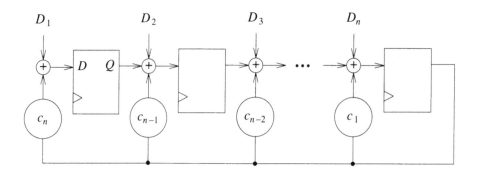

Figure 10.17   Multiple-input signature register

Selection of the Polynomial P(x)

As stated previously, a MISR having n stages has a masking probability approximately equal to 2^{-n} for equally likely error patterns and long data streams. Also, this result is independent of $P(x)$. Let the error bit associated with D_i at time j be denoted by e_{ij}, where $i = 1,2,...,n$, and $j = 1,2,...,m$. Then the error polynomial associated with D_i is $E_i = \sum_{j=1}^{m} e_{ij}x^{j-1}$. Then the *effective error polynomial* is $E(x) = \sum_{i=1}^{n} E_i x^{i-1}$ assuming that the initial state of the register is all zeros. The error polynomial $E(x)$ is masked if it is a multiple of $P(x)$. So a complex-feedback LFSR structure is typically used on the assumption that it will reduce the chances of masking an error. For example, the Hewlett-Packard 5004A signature analyzer employs the characteristic polynomial $P(x) = x^{16} + x^9 + x^7 + x^4 + 1$ [Frohwerk 1977]. $P(x)$ is often selected to be a primitive polynomial.

When the characteristic polynomial is the product of the parity generator polynomial $g(x) = x + 1$ and a primitive polynomial of degree $(n-1)$, an n-stage MISR has the property that the parity over all the bits in the input streams equals the parity of the final signature. Hence masking will not occur for an odd number of errors.

Increasing the Effectiveness of Signature Analysis

There are several ways to decrease the probability of masking. Based on the theory presented, the probability of masking can be reduced by increasing the length of the LFSR. Also a test can be repeated using a different feedback polynomial. When testing combinational circuits, a test can be repeated after first changing the order of the test vectors, thus producing a different error polynomial. This technique can also be used for sequential circuits, but now the fault-free signature also changes.

Masking occurs because once an error exists within an LFSR, it can be canceled by new errors occurring on the inputs. Inspecting the contents of the signature analyzer several times during the testing process decreases the chance that a faulty circuit will go undetected. This technique is equivalent to periodically sampling the output of the signature analyzer. The degree of storage compression is a function of how often the output is sampled.

Implementation Issues

It is often desirable to modify a functional register in a circuit so that it can also operate as a signature analyzer. Figure 10.18(a) shows the original circuit where the output of network N feeds a register R. In the modified circuit, shown in Figure 10.18(b), R is modified to be both a signature analyzer (MISR) as well as a scan register. The MISR is used to generate a signature when N is tested. The scan aspect of R^* is used first to initialize R^*, referred to as seeding R^*, and for scanning out the final signature. Figure 10.19 shows one design for an appropriate cell for R^*, based on the LSSD double latch SRL, where R^* is a type 1 LFSR. Referring to Figure 10.19, when clock pulses are applied to CK and then B, the register functions in a normal parallel-load mode, accepting data from D_i. When $S/T = 0$ and clocks A and B are alternately activated, the register acts as a scan register, each cell accepting the data bit from S_i and transferring it to Q^+. When $S/T = 1$, the scan input is $I = D_i \oplus S_i$, which is the condition required for the MISR. Again A and B are alternately activated. Usually these design changes have a small impact on the normal performance of the circuit. However, during the test mode, the extra gates in the path D_i to I may require that the clock rate be reduced.

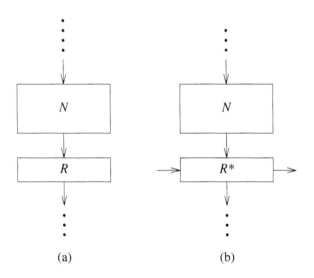

(a)   (b)

Figure 10.18   (a) Original circuit (b) Modified circuit

10.7 Concluding Remarks

Compression techniques are widely used since they are easy to implement, can be used for field test and self-testing, and can provide high fault coverage, though the correlation between error coverage and fault coverage is hard to predict.

Transition-count testing can provide good fault coverage with short, deterministic test sequences. This technique can be used to monitor asynchronous line activity if the

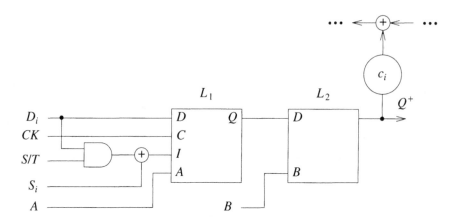

Figure 10.19   Storage cell for a signature analyzer

circuitry driving the line is race-free and hazard-free. The results depend on the order of the test patterns.

Ones counting can provide adequate fault coverage, though longer test sequences may be required. For combinational circuits the results are independent of the order of the test patterns.

All Boolean switching functions can be implemented by a circuit that is syndrome-testable. Since syndrome testing requires an exhaustive test set, it is not a practical technique for circuits having many inputs.

Signature analysis is widely used because it provides excellent fault and error coverage, though fault coverage must be determined using a fault simulator or a statistical fault simulator. Unlike the other techniques, several means exist for improving the coverage without changing the test, such as by changing the characteristic polynomial or increasing the length of the register.

REFERENCES

[Abadir 1987] M. S. Abadir, "Efficient Scan Path Testing Using Sliding Parity Response Compaction," *Proc. Intn'l. Conf. on Computer-Aided Design*, pp. 332-335, November, 1987.

[Agarwal 1983] V. K. Agarwal, "Increasing Effectiveness of Built-In Testing by Output Data Modification," *Digest of Papers 13th Annual Intn'l. Symp. Fault-Tolerant Computing*, pp. 227-233, June, 1983.

[Bardell and McAnney 1982] P. H. Bardell and W. H. McAnney, "Self-Testing of Multichip Logic Modules," *Digest of Papers 1982 Intn'l. Test Conf.*, pp. 200-204, November, 1982.

[Bardell *et al.* 1987] P. H. Bardell, W. H. McAnney, and J. Savir, *Built-in Test for VLSI: Pseudorandom Techniques,* John Wiley and Sons, New York, 1987.

[Barzilai *et al.* 1981]   Z. Barzilai,   J. Savir,   G. Markowsky,   and   M. G.   Smith,   "The Weighted   Syndrome   Sums   Approach   to   VLSI   Testing,"   *IEEE   Trans.   on Computers*, Vol. C-30, No. 12, pp. 996-1000, December, 1981.

[Beauchamp 1975]   K. G.   Beauchamp,   *Walsh   Functions   and   Their   Applications*, Academic Press, New York, 1975.

[Bhaskar 1982] K. S. Bhaskar, "Signature Analysis: Yet Another Perspective," *Digest of Papers 1982 Intn'l. Test Conf.*, pp. 132-134, November, 1982.

[Bhavsar and Heckelman 1981]   D. K.   Bhavsar   and   R. W.   Heckelman,   "Self-Testing   by Polynomial Division," *Digest of Papers 1981 Intn'l. Test Conf.*, pp. 208-216, October, 1981.

[Bhavsar and Krishnamurthy 1984]   D. K.   Bhavsar   and   B.   Krishnamurthy,   "Can   We Eliminate   Fault   Escape   in   Self   Testing   by   Polynomial   Division   (Signature Analysis)?," *Proc. Intn'l. Test Conf.*, pp. 134-139, October, 1984.

[Brillhart *et al.* 1983] J. Brillhart, D. H. Lehmer, J. L. Selfridge, B. Tuckerman, and S. S. Wagstaff, Jr., *Factorization of $b^n \pm 1$, $b = 2,3,5,6,7,10,11,12$, Up to High Powers*, American Mathematical Society, Providence, 1983.

[Carter 1982] J. L. Carter, "The Theory of Signature Testing for VLSI," *14th ACM Symp. on the Theory of Computing*, pp. 66-76, May, 1982.

[Carter 1982a] W. C. Carter, "The Ubiquitous Parity Bit," *Digest of Papers 12th Annual Intn'l. Symp. Fault-Tolerant Computing*, pp. 289-296, June, 1982.

[Carter 1982b]   W. C.   Carter,   "Signature   Testing   with   Guaranteed   Bounds   for   Fault Coverage,"   *Digest   of   Papers   1982   Intn'l.   Test   Conf.*,   pp. 75-82,   November, 1982.

[Chan 1977]   A. Y. Chan,   "Easy-to-use   Signature   Analysis   Accurately   Troubleshoots Complex   Logic   Circuits,"   *Hewlett-Packard   Journal*,   Vol. 28,   pp. 9-14,   May, 1977.

[Daniels and Bruce 1975]   R. G.   Daniels   and   W. C.   Bruce,   "Built-In   Self   Test   Trends   in Motorola Microprocessors," *IEEE Design & Test of Computers*, Vol. 2, No. 2, pp. 64-71, April, 1985.

[David 1984] R. David, "Signature Analysis of Multi-Output Circuits," *Digest of Papers 14th   Annual   Intn'l.   Symp.   on   Fault-Tolerant   Computing*,   pp. 366-371,   June, 1984.

[Frohwerk 1977]   R. A.   Frohwerk,   "Signature   Analysis: A   New   Digital   Field   Service Method," *Hewlett-Packard Journal*, Vol. 28, pp. 2-8, September, 1977.

[Golomb 1982] S. W. Golomb, *Shift Register Sequences*, rev. ed., Aegean Park Press, Laguna Hills, California, 1982.

[Gupta and Pradhan 1988] S. K. Gupta and D. K. Pradhan, "A New Framework for Designing and Analyzing BIST Techniques: Computation of Exact Aliasing Probability," *Proc. Intn'l. Test Conf.*, pp. 329-342, September, 1988.

[Hassan *et al.* 1983] S. Z. Hassan, D. J. Lu, and E. J. McCluskey, "Parallel Signature Analyzers — Detection Capability and Extensions," *26th IEEE Computer Society Intn'l. Conf.*, COMPCON, Spring 1983, pp. 440-445, February-March, 1983.

[Hassan and McCluskey 1983] S. Z. Hassan and E. J. McCluskey, "Increased Fault Coverage Through Multiple Signatures," *Digest of Papers 14th Annual Intn'l. Fault-Tolerant Computing Symp.*, pp. 354-359, June, 1984.

[Hayes 1976a] J. P. Hayes, "Transition Count Testing of Combinational Logic Circuits," *IEEE Trans. on Computers*, Vol. C-25, pp. 613-620, June, 1976.

[Hayes 1976b] J. P. Hayes, "Check Sum Methods for Test Data Compression," *Journal of Design Automation & Fault-Tolerant Computing*, Vol. 1, No. 1, pp. 3-17, 1976.

[Hsiao and Seth 1984] T-C. Hsiao and S. C. Seth, "An Analysis of the Use of Rademacher-Walsh Spectrum in Compact Testing," *IEEE Trans. on Computers*, Vol. C-33, No. 10, pp. 934-937, October, 1984.

[Hurst *et al.* 1985] S. L. Hurst, D. M. Miller, and J. C. Muzio, *Spectral Techniques in Digital Logic*, Academic Press, New York, 1985.

[Karpovsky and Nagvajara] M. Karpovsky and P. Nagvajara, "Optional Time and Space Compression of Test Responses for VLSI Devices," *Proc. Intn'l. Test Conf.*, pp. 523-529, September, 1987.

[Losq 1978] J. Losq, "Efficiency of Random Compact Testing," *IEEE Trans. on Computers*, Vol. C-27, No. 6, pp. 516-525, June, 1978.

[Markowsky 1981] G. Markowsky, "Syndrome-Testability Can Be Achieved by Circuit Modification," *IEEE Trans. on Computers*, Vol. C-30, No. 8, pp. 604-606, August, 1981.

[McAnney and Savir 1986] W. H. McAnney and J. Savir, "Built-In Checking of the Correct Self-Test Signature," *Proc. Intn'l. Test Conf.*, pp. 54-58, September, 1986.

[McAnney and Savir 1988] W. H. McAnney and J. Savir, "Built-in Checking of the Correct Self-Test Signature," *IEEE Trans. on Computers*, Vol. C-37, No. 9, pp. 1142-1145, September, 1988.

[Parker 1976] K. P. Parker, "Compact Testing: Testing with Compressed Data," *Proc. 1976 Intn'l. Symp. on Fault-Tolerant Computing*, pp. 93-98, June, 1976.

[Peterson and Weldon 1972] W. W. Peterson and E. J. Weldon, Jr., *Error-Correcting Codes*, 2nd ed., MIT Press, Cambridge, Massachussetts, 1972.

[Reddy 1977] S. M. Reddy, "A Note on Testing Logic Circuits by Transition Counting," *IEEE Trans. on Computers*, Vol. C-26, No. 3, pp. 313-314, March, 1977.

[Robinson 1985] J. P. Robinson, "Segmented Testing," *IEEE Trans. on Computers*, Vol. C-34, No. 5, pp. 461-471, May, 1985.

[Robinson and Saxena 1987] J. P. Robinson and N. R. Saxena, "A Unified View of Test Compression Methods," *IEEE Trans. on Computers*, Vol. C-36, No. 1, pp. 94-99, January, 1987.

[Saluja and Karpovsky 1983] K. K. Saluja and M. Karpovsky, "Test Compression Hardware Through Data Compression in Space and Time," *Proc. Intn'l. Test Conf.*, pp. 83-88, 1983.

[Savir 1980] J. Savir, "Syndrome-Testable Design of Combinational Circuits," *IEEE Trans. on Computers*, Vol. C-29, No. 6, pp. 442-451, June 1980; and No. 11, pp. 1012-1013, November, 1980.

[Savir 1981] J. Savir, "Syndrome-Testing of Syndrome-Untestable Combinational Circuits," *IEEE Trans. on Computers*, Vol. C-30, No. 8, pp. 606-608, August, 1981.

[Savir and Bardell 1984] J. Savir and P. H. Bardell, "On Random Pattern Test Length," *IEEE Trans. on Computers*, Vol. C-33, No. 6, pp. 467-474, June, 1984.

[Savir et al. 1984] J. Savir, G. S. Ditlow, and P. H. Bardell, "Random Pattern Testability," *IEEE Trans. on Computers*, Vol. C-33, No. 1, pp. 79-90, January, 1984.

[Savir and McAnney 1985] J. Savir and W. H. McAnney, "On the Masking Probability with Ones Count and Transition Count," *Proc. Intn'l. Conf. on Computer-Aided Design*, pp. 111-113, November, 1985.

[Segers 1981] M. T. M. Segers, "A Self-Test Method for Digital Circuits," *Digest of Papers 1981 Intn'l. Test Conf.*, pp. 79-85, October, 1981.

[Shanks 1969] J. L. Shanks, "Computation of the Fast Walsh-Fourier Transform," *IEEE Trans. on Computers*, Vol. C-18, No. 5, pp. 457-459, May, 1969.

[Smith 1980] J. E. Smith, "Measures of the Effectiveness of Fault Signature Analysis," *IEEE Trans. on Computers*, Vol. C-29, No. 6, pp. 510-514, June, 1980.

[Sridhar *et al.* 1982] T. Sridhar, D. S. Ho, T. J. Powell, and S. M. Thatte, "Analysis and Simulation of Parallel Signature Analyzers," *Digest of Papers 1982 Intn'l. Test Conf.*, pp. 661-665, November, 1982.

[Susskind 1983] A. K. Susskind, "Testing by Verifying Walsh Coefficients," *IEEE Trans. on Computers*, Vol. C-32, No. 2, pp. 198-201, February, 1983.

[Tzidon *et al.* 1978] A. Tzidon, I. Berger, and M. Yoeli, "A Practical Approach to Fault Detection in Combinational Networks," *IEEE Trans. on Computers*, Vol. C-27, No. 10, pp. 968-971, October, 1978.

[Williams *et al.* 1986] T. W. Williams, W. Daehn, M. Gruetzner, and C. W. Starke, "Comparison of Aliasing Errors for Primitive and Non-Primitive Polynomials," *Proc. Intn'l. Test Conf.*, pp. 282-288, September, 1986.

[Williams *et al.* 1987] T. W. Williams, W. Daehn, M. Gruetzner, and C. W. Starke, "Aliasing Errors in Signature Analysis Registers," *IEEE Design & Test of Computers*, Vol. 4, pp. 39-45, April, 1987.

[Zorian and Agarwal 1984] Y. Zorian and V. K. Agarwal, "Higher Certainty of Error Coverage by Output Data Modification," *Proc. Intn'l. Test Conf.*, pp. 140-147, October, 1984.

PROBLEMS

10.1 For the ones-count compression technique, show that if $1C(R) = 0$ or m, then no error masking can occur.

10.2 The number of binary sequences of length m having a transition count of r is $2 \begin{bmatrix} m-1 \\ r \end{bmatrix}$. Based on the concept of the transition count of a sequence, show that $\begin{bmatrix} m \\ r \end{bmatrix} = \begin{bmatrix} m-1 \\ r \end{bmatrix} + \begin{bmatrix} m-1 \\ r-1 \end{bmatrix}$. Don't use the fact that $\begin{bmatrix} m \\ r \end{bmatrix} = \dfrac{m!}{r!(m-r)!}$.

10.3 Prove Theorem 10.2.

10.4 Assume that a single-output combinational circuit C is tested using the parity-check compression technique along with an exhaustive test set, and that the output stream has odd parity. Show that this scheme will detect all primary input and output stuck-at faults associated with C.

10.5 Prove that any two-level (AND-OR) irredundant circuit that realizes a unate function in all its variables is syndrome-testable.

10.6 The function $F = x_1 x_2 + \bar{x}_2 \bar{x}_3$ is not syndrome-testable. Determine which stuck-at fault associated with x_2 is not syndrome-detectable, and show that the modified function $F' = c x_1 x_2 + x_2 x_3$ is syndrome-testable, where c is a new primary input. Note that for $c = 1$, $F' = F$.

10.7 For an autonomous LFSR, show that if its initial state is not the all-0 state, then it will never enter the all-0 state.

10.8 Consider an autonomous LFSR whose characteristic polynomial is primitive.

    a. Show that if at time t its state is in error, then at any time after t its state will always be in error.

    b. Does the result for part (a) hold true when $P(x)$ is not a primitive polynomial? If so, prove it; otherwise show an example.

10.9 Consider an n-input NAND gate, denoted by G_n. All single and multiple stuck faults associated with G_n can be detected by the unique minimal test set T consisting of $n + 1$ test vectors shown below:

$$
\begin{aligned}
u &= (1,1,\ldots\ldots,1) \\
e_1 &= (0,1,1,\ldots,1,1) \\
e_2 &= (1,0,1,\ldots,1,1) \\
&\vdots \\
e_n &= (1,1,1,\ldots 1,0).
\end{aligned}
$$

The correct response to u and e_i is 0 and 1 respectively. Every complete TC test sequence X_n for G_n must include every test pattern in T, hence $|X_n| \geq n + 1$.

    a. Show that $X_1 = u e_1$ and $X_2 = e_1 u e_2$ are minimal TC tests for $n = 1$ and 2 respectively.

    b. For $n > 2$, the minimal value of $|X_n|$ is $n + 2$, as stated in the following theorem.

Theorem [Hayes 1976a]: $X_n = ue_1e_2...e_{n-1}e_ne_1$ is a minimal TC test of length $n + 2$ for both single and multiple stuck-type faults for an n-input NAND gate G_n. Prove this theorem.

10.10 For an n-input NAND gate G_n, how many single and multiple stuck faults are distinguishable and how many have a unique transition count?

10.11 Prove the following theorem [Hayes 1976a]:

Let T be any test set for a two-level circuit C. Let R_0 denote the fault-free response of C to any sequence X containing every member of T. No single or multiple fault can change the response of C to X from R_0 to $\bar{R}_0$.

10.12 Let $T = [T^0, T^1]$ be an arbitrary single-fault test set for a two-level sum of product irredundant circuit C containing $r = p + q$ vectors, where $T^0 = \{t_1^0, t_2^0, ..., t_p^0\}$ and $T^1 = \{t_1^1, t_2^1, ..., t_q^1\}$. It is well known that T also detects all multiple stuck-type faults.

a. Show that the sequence $X^2 = t_q^1 t_1^1 t_2^1 ... t_q^1 t_p^0 t_1^0 t_2^0 ... t_p^0$ of length $r + 2$ is a TC test for C with respect to both single and multiple faults.

b. Assume T is a minimal test set. Show that (1) if $D = 0$ or 1, the sequence X^* defined in Theorem 10.5 is a minimal length TC test; (2) for $D = 2$, both X^* and X^2 are of the same length; and (3) for $D > 2$, X^2 is shorter than X^*.

10.13 Consider the circuit shown in Figure 10.20. Construct a minimal-length test set T for all stuck-at faults. Construct X^* and X^2 (defined in Problem 10.12). Is X^* a TC test for all multiple faults?

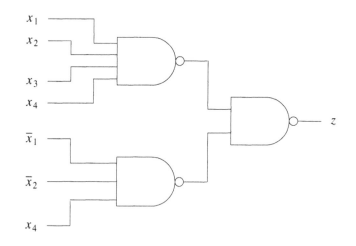

Figure 10.20

10.14 Construct a minimal-length test sequence T for all single stuck-at faults in a 3-input NAND gate and identify one fault not TC-detectable by T.

10.15 A circuit implementing the function $f = xy + \bar{y}z$ is to be tested using the syndrome-test method. Show that the single faults z s-a-0 and z s-a-1 are not detected, while all other single stuck-at faults are detected.

10.16 Consider a circuit implementing the function $f = x_1 x_3 + x_2 \bar{x}_3$ and an input sequence consisting of all eight input patterns generated by a 3-bit binary counter that starts in the state $(x_3, x_2, x_1) = (0,0,0)$ and produces the sequence $(0,0,0)$, $(0,0,1)$, $(0,1,0)$, ..., $(1,1,1)$. Assume the response is compressed using a type 2 signature analyzer that has a characteristic polynomial $x^4 + x^3 + 1$ and whose initial state is all 0s. Verify that the following faults are detected.

   a. x_3 *s-a*-0

   b. x_3 *s-a*-1

   c. x_1 *s-a*-0

   d. x_1 *s-a*-1

10.17 Verify the entries in Figures 10.7 and 10.8.

11. BUILT-IN SELF-TEST

About This Chapter

In previous chapters we have discussed algorithmic methods for test generation and techniques for design for testability (DFT). These methods are primarily used when external testing is employed. Built-in self-test (BIST) is a design technique in which parts of a circuit are used to test the circuit itself. The first part of this chapter covers the basic concepts associated with BIST. We then focus on the problem of built-in generation of test patterns. Various ways of partitioning a circuit for self-testing are described, as are ways of generating test patterns. Test-pattern generation techniques discussed include exhaustive testing, pseudorandom testing, and pseudoexhaustive testing; the latter includes the concepts of verification and segmentation testing.

Generic BIST architectures are described, including the major ways of characterizing such architectures in terms of centralized versus distributed BIST hardware and internal versus external BIST hardware. Next many specific BIST architectures are presented.

Finally several advanced BIST techniques are discussed, including the identification of a minimal number of test sessions, the control of BIST structures, and the notion of partial-intrusion BIST designs.

11.1 Introduction to BIST Concepts

Built-in self-test is the capability of a circuit (chip, board, or system) to test itself. BIST represents a merger of the concepts of *built-in test* (BIT) and *self-test*, and has come to be synonymous with these terms. The related term *built-in-test equipment* (BITE) refers to the hardware and/or software incorporated into a unit to provide DFT or BIST capability.

BIST techniques can be classified into two categories, namely on-line BIST, which includes concurrent and nonconcurrent techniques, and off-line BIST, which includes functional and structural approaches (see Figure 11.1). These terms were mentioned in Chapter 1 but will be briefly discussed here for ease of reference.

In *on-line* BIST, testing occurs during normal functional operating conditions; i.e., the circuit under test (CUT) is not placed into a test mode where normal functional operation is locked out. *Concurrent on-line* BIST is a form of testing that occurs simultaneously with normal functional operation. This form of testing is usually accomplished using coding techniques or duplication and comparison; these techniques will be described in more detail in Chapter 13. In *nonconcurrent on-line* BIST, testing is carried out while a system is in an idle state. This is often accomplished by executing diagnostic software routines (macrocode) or diagnostic firmware routines (microcode). The test process can be interrupted at any time so that normal operation can resume.

Off-line BIST deals with testing a system when it is not carrying out its normal functions. Systems, boards, and chips can be tested in this mode. This form of testing is also applicable at the manufacturing, field, depot, and operational levels. Often Off-line testing is carried out using on-chip or on-board test-pattern generators (TPGs) and output response analyzers (ORAs) or microdiagnostic routines. Off-line testing does not detect

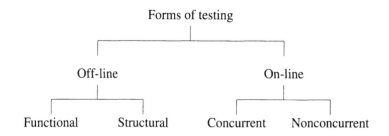

Figure 11.1   Forms of testing

errors in real time, i.e., when they first occur, as is possible with many on-line concurrent BIST techniques.

Functional off-line BIST deals with the execution of a test based on a functional description of the CUT and often employs a functional, or high-level, fault model. Normally such a test is implemented as diagnostic software or firmware.

Structural off-line BIST deals with the execution of a test based on the structure of the CUT. An explicit structural fault model may be used. Fault coverage is based on detecting structural faults. Usually tests are generated and responses are compressed using some form of an LFSR. Figure 11.2 lists several types of test structures used in BIST circuits. Two common TPG circuits exist. A *pseudorandom pattern generator* (PRPG) is a multioutput device normally implemented using an LFSR, while a *shift register pattern generator* (SRPG) is a single-output autonomous LFSR. For simplicity the reader can consider a PRPG to represent a "parallel random-pattern generator," and a SRPG to be a "serial random-pattern generator." Two common ORA circuits also exist. One is a *multiple-input signature register* (MISR), the other a *single-input signature register* (SISR). Both are implemented using an LFSR.

This chapter deals primarily with structural off-line BIST. Thus LFSRs will be used extensively. The reader is referred to Chapter 10 for a review of the theory related to LFSRs. Before discussing the structure of off-line BIST circuitry, some basic concepts will be reviewed.

11.1.1 Hardcore

Some parts of a circuit must be operational to execute a self-test. This circuitry is referred to as the *hardcore*. At a minimum the hardcore usually includes power, ground, and clock distribution circuitry. The hardcore is usually difficult to test explicitly. If faulty, the self-test normally fails. Thus detection is often easy to achieve, but little if any diagnostic capability exists. If a circuit fails during self-test, the problem may be in the hardcore rather than in the hardware presumably being tested. The hardcore is normally tested by external test equipment or is designed to be self-testable by using various forms of redundancy, such as duplication or self-checking checkers (see Chapter 13). Normally a designer attempts to minimize the complexity of the hardcore.

BILBO — built-in logic block observer (register)

LFSR — linear feedback shift register

MISR — multiple-input signature register

ORA — (generic) output response analyzer

PRPG — pseudorandom pattern generator, often referred to as a pseudorandom number generator

SISR — single-input signature register

SRSG — shift-register sequence generator; also a single-output PRPG

TPG — (generic) test-pattern generator

Figure 11.2 Glossary of key BIST test structures

11.1.2 Levels of Test

Production Testing

We refer to the testing of newly manufactured components as production testing. Production testing can occur at many levels, such as the chip, board, or system levels. Using BIST at these levels reduces the need for expensive ATE in go/no-go testing and simplifies some aspects of diagnostic testing. For example, the Intel 80386 microprocessor employs about 1.8 percent area overhead for BIST to test portions of the circuit that would be difficult to test by other means [Gelsinger 1987]. The BIST aspects of several other chips and/or boards are discussed in [Benowitz *et al.* 1975] [Boney and Rupp 1979], [Fasang 1982], [Kuban and Bruce 1984], [Beenker 1985], and [Karpovsky and Nagvajara 1989].

When implemented at the chip level along with boundary scan, BIST can be used effectively at all levels of a system's hierarchy. Since many BIST techniques can be run in real time, this method is superior to many non-BIST approaches and to some extent can be used for delay testing. It is not applicable, however, to parametric testing.

Field Testing

BIST can be used for field-level testing, eliminating the need for expensive special test equipment to diagnose faults down to field-replaceable units. This can have a great influence on the maintainability and thus life-cycle costs of both commercial and military hardware. For example, the U.S. military is attempting to implement the concept of *two-level maintenance*. Here a system must carry out a self-test and automatically diagnose a fault to a field-replaceable unit, such as a printed circuit board. This board is then replaced "in the field" and the faulty board is either discarded or sent to a depot for further testing and repair.

11.2 Test-Pattern Generation for BIST

In Chapter 10 the design of a pseudorandom pattern generator based on the use of an LFSR was described. In this section various TPG designs will be described. We assume that the unit being tested is an n-input, m-output combinational circuit. The various forms of testing and related TPGs are summarized next.

> Exhaustive testing
>> Exhaustive test-pattern generators

> Pseudorandom testing
>> Weighted test generator
>> Adaptive test generator

> Pseudoexhaustive testing
>> Syndrome driver counter
>> Constant-weight counter
>> Combined LFSR and shift register
>> Combined LFSR and XOR gates
>> Condensed LFSR
>> Cyclic LFSR

11.2.1 Exhaustive Testing

Exhaustive testing deals with the testing of an n-input combinational circuit where all 2^n inputs are applied. A binary counter can be used as TPG. If a maximum-length autonomous LFSR is used, its design can be modified to include the all-zero state. Such an LFSR is referred to as a *complete LFSR*, and its design is described in [McCluskey 1981] and [Wang and McCluskey 1986d].

Exhaustive testing guarantees that all detectable faults that do not produce sequential behavior will be detected. Depending on the clock rate, this approach is usually not feasible if n is larger than about 22. Other techniques to be described are more practical when n is large. The concept of exhaustive testing is not generally applicable to sequential circuits.

11.2.2 Pseudorandom Testing

Pseudorandom testing deals with testing a circuit with test patterns that have many characteristics of random patterns but where the patterns are generated deterministically and hence are repeatable. Pseudorandom patterns can be generated with or without replacement. Generation with replacement implies that a test pattern may be generated more than once; without replacement implies that each pattern is unique. Not all 2^n test patterns need be generated. Pseudorandom test patterns without replacement can be generated by an autonomous LFSR. Pseudorandom testing is applicable to both combinational and sequential circuits. Fault coverage can be determined by fault simulation. The test length is selected to achieve an acceptable level of fault coverage. Unfortunately, some circuits contain random-pattern-resistant faults and thus require long test lengths to insure a high fault coverage. Methods for estimating the probability of detecting a fault with random tests are described in Chapter 6. Further information on estimating test length as a function of fault coverage can be found in [Savir and Bardell 1984], [Williams 1985], [Chin and McCluskey 1987], and [Wagner *et al.* 1987].

The inherent attributes of an LFSR tend to produce test patterns having equal numbers of 0s and 1s on each output line. For many circuits it is better to bias the distribution of 0s and 1s to achieve a higher fault coverage with fewer test vectors. Consider, for example, a 4-input AND gate. When applying unbiased random inputs, the probability of applying at least one 0 to any input is 15/16. A 0 on any input makes it impossible to test any other input for *s-a*-0 or *s-a*-1. Thus there is a need to be able to generate test patterns having different distributions of 0s and 1s.

Some results relating the effectiveness of testing in terms of test length and fault coverage to the distribution characteristics of the test patterns have been reported in [Kumar 1980], [Archambeau and McCluskey 1984], [Lisanke *et al.* 1987], and [Wunderlich 1988].

Weighted Test Generation

A *weighted test generator* is a TPG where the distribution of 0s and 1s produced on the output lines is not necessarily uniform. Such a generator can be constructed using an autonomous LFSR and a combinational circuit. For example, the probability distribution of 0.5 for a 1 that is normally produced by a maximal-length LFSR can be easily changed to 0.25 or 0.75 to improve fault coverage. When testing a circuit using a weighted test generator, a preprocessing procedure is employed to determine one or more sets of weights. Different parts of a circuit may be tested more effectively than other parts by pseudorandom patterns having different distributions. Once these weights are determined, the appropriate circuitry can be designed to generate the pseudorandom patterns having the desired distributions. Further information on these test generators can be found in [Schnurmann *et al.* 1975], [Chin and McCluskey 1984], [Ha and Reddy 1986], and [Wunderlich 1987].

Adaptive Test Generation

Adaptive test generation also employs a weighted test-pattern generator. For this technique the results of fault simulation are used to modify the weights, thereby resulting in one or more probability distributions for the test patterns. Once these distributions are determined, an appropriate TPG can be designed. Tests generated by such a device tend to be efficient in terms of test length; however, the test-pattern-generation hardware can be complex. Further information can be found in [Parker 1976] and [Timoc *et al.* 1983].

11.2.3 Pseudoexhaustive Testing

Pseudoexhaustive testing achieves many of the benefits of exhaustive testing but usually requires far fewer test patterns. It relies on various forms of circuit segmentation and attempts to test each segment exhaustively. Because of the many subjects that are associated with pseudoexhaustive testing, we will first briefly outline the main topics to be discussed in this section.

A segment is a subcircuit of a circuit C. Segments need not be disjoint. There are several forms of segmentation, a few of which are listed below:

1. Logical segmentation

    a. Cone segmentation (verification testing)

    b. Sensitized path segmentation

2. Physical segmentation

When employing a pseudoexhaustive test to an n-input circuit, it is often possible to reconfigure the input lines so that tests need only be generated on m lines, where $m < n$, and these m lines can fanout and drive the n lines to the CUT. These m signals are referred to as *test signals*. A procedure for identifying these test signals will be presented.

Pseudoexhaustive testing can often be accomplished using constant-weight test patterns. Some theoretical results about such patterns will be presented. We will also describe several circuit structures that generate these patterns as well as other patterns used in pseudoexhaustive testing.

11.2.3.1 Logical Segmentation

In this section we will briefly describe two logical-segmentation techniques.

Cone Segmentation

In *cone segmentation* an m output circuit is logically segmented into m cones, each cone consisting of all logic associated with one output. Each cone is tested exhaustively, and all cones are tested concurrently. This form of testing was originally suggested by McCluskey [1984] and is called *verification testing*.

Consider a combinational circuit C with inputs $X = \{x_1, x_2, ..., x_n\}$ and outputs $Y = \{y_1, y_2, ..., y_m\}$. Let $y_i = f_i(X_i)$, where $X_i \subseteq X$. Let $w = \max_i \{|X_i|\}$. One form of a verification test produces all 2^w input patterns on all $\binom{n}{w}$ subsets of w inputs to C.

The circuit under test is denoted as an (n,w)-CUT, where $w < n$. If $w = n$, then pseudoexhaustive testing simply becomes exhaustive testing. Figure 11.3 shows a $(4,2)$-CUT.

Sensitized-Path Segmentation

In Chapter 8 a segmentation technique based on circuit partitioning was presented. Some circuits can be segmented based on the concept of path sensitization. A trivial example is shown in Figure 11.4. To test C_1 exhaustively, 2^{n_1} patterns are applied to A while B is set to some value so that $D = 1$. Thus a sensitized path is established from C to F. C_2 is tested in a similar manner. By this process, the AND gate is also completely tested. Thus this circuit can be effectively tested using only $2^{n_1} + 2^{n_2} + 1$ test patterns, rather than $2^{n_1 + n_2}$. More details on this form of testing can be found in [McCluskey and Bozorgui-Nesbat 1981], [Chandra *et al.* 1983], [Patashnik 1983], [Udell 1986], [Chen 1987], [Shperling and McCluskey 1987], and [Udell and McCluskey 1989].

11.2.3.2 Constant-Weight Patterns

Consider two positive integers n and k, where $k \leq n$. Let T be a set of binary n-tuples. Then T is said to *exhaustively cover all k-subspaces* if for all subsets of k bit positions, each of the 2^k binary patterns appears at least once among the $|T|$ n-tuples. For example, the set T shown below exhaustively covers all 2-spaces.

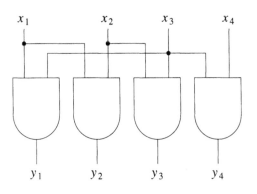

Figure 11.3   A (4,2)-CUT

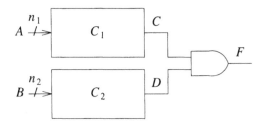

Figure 11.4   Segmentation testing via path sensitization

$$T = \begin{bmatrix} 0 & 0 & 0 \\ 0 & 1 & 1 \\ 1 & 1 & 0 \\ 1 & 0 & 1 \end{bmatrix}$$

If $|T|_{\min}$ is the smallest possible size for such a set T, then clearly $2^k \leq |T|_{\min} \leq 2^n$.

A binary n-tuple is said to be of weight k if it contains exactly k 1s. There are $\binom{n}{k}$ binary n-tuples having weight k.

The following results have been derived by [Tang and Woo 1983] and will be presented here without proof.

Theorem 11.1:   Given n and k, then T exhaustively covers all binary k-subspaces if it contains all binary n-tuples of weight(s) w such that $w = c \bmod (n-k+1)$ for some integer constant c, where $0 \leq c \leq n-k$.   □

Let T_c denote the set produced to using Theorem 11.1 for a specific value of c.

Example 11.1: $n = 20, k = 2, n - k + 1 = 19.$

Case 1: $c = 0$

Setting $w = 0$ mod 19 produces $w = 0$ and 19. Thus T_0 consists of the all-0 patterns and 20 patterns of weight 19. Hence

$$|T_0| = \begin{pmatrix} 20 \\ 0 \end{pmatrix} + \begin{pmatrix} 20 \\ 19 \end{pmatrix} = 1 + 20 = 21$$

and

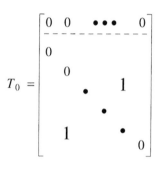

Case 2: $c = 1$

Setting $w = 1$ mod 19 results in $w = 1, 20$. Therefore

$$|T_1| = \begin{pmatrix} 20 \\ 1 \end{pmatrix} + \begin{pmatrix} 20 \\ 20 \end{pmatrix} = 20 + 1 = 21$$

and

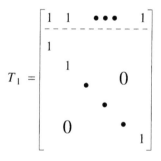

Case 3: $2 \le c \le 18$

For $2 \le c \le 18$, $w = c$ mod 19 implies that $w = c$. Thus in each case T_c consists of all weight-c binary n-tuples, and $|T_c| = \begin{pmatrix} n \\ c \end{pmatrix}$. Note that T_0 and T_1 are the smallest among the 19 sets; in fact T_1 and T_0 are complements of each other.   □

Example 11.2: $n = 20, k = 3, n - k + 1 = 18$

Case 1: $c = 0$

For $w = 0$ mod 18, $w = 0$ and 18. Thus

$$|T_0| = \begin{pmatrix} 20 \\ 0 \end{pmatrix} + \begin{pmatrix} 20 \\ 18 \end{pmatrix} = 1 + 190 = 191.$$

Case 2: $c = 1$

For $w = 1$ mod 18, $w = 1$, 19 and $|T_1| = \begin{pmatrix} 20 \\ 1 \end{pmatrix} + \begin{pmatrix} 20 \\ 19 \end{pmatrix} = 20 + 20 = 40.$

Case 3: $c = 2$

For $w = 2$ mod 18, $w = 2$, 20 and T_2 is the complement of T_0.

Case 4: $3 \leq c \leq 17$

For $w = c$ mod 8, $3 \leq c \leq 17$, we have $w = c$.   □

Note that for both examples, for any value of w, $0 \leq w \leq 20$, all n-tuples of weight w exist in exactly one case considered.

The general situation is covered by the following corollary.

Corollary 11.1:   There are $(n-k+1)$ solution sets T_i, $0 \leq i \leq n - k$ obtained from Theorem 11.1, and these solution sets are disjoint and partition the set of all 2^n n-tuples into disjoint classes.   □

Since these are $(n-k+1)$ solution sets that partition the set of 2^n distinct n-tuples into $(n-k+1)$ disjoint sets, and the smallest set cannot be larger than the average set, then an upper bound on the size of $|T|_{min}$ is

$$|T|_{min} \leq \frac{2^n}{n - k + 1} = B_n .$$

Theorem 11.2:   Let T_c be a set generated according to Theorem 11.1. Then T_c is minimal; i.e., no proper subset of T_c also exhaustively covers all k-subspaces, if $c \leq k$ or $c = n - k$.   □

Example 11.3: $n = 6$, $k = 2$, $n - k + 1 = 5$.

For $c = 3$, $w = 3$ mod 5, thus $w = 3$ and $|T_3| = \begin{pmatrix} 6 \\ 3 \end{pmatrix} = 20$. One subset of T_3 that exhaustively covers all 2-subspaces is shown below as T'_3.

$$T'_3 = \begin{bmatrix} 1 & 1 & 0 & 0 & 1 & 0 \\ 1 & 0 & 1 & 1 & 0 & 0 \\ 1 & 0 & 0 & 1 & 0 & 1 \\ 0 & 1 & 1 & 0 & 0 & 1 \\ 0 & 1 & 0 & 1 & 1 & 0 \\ 0 & 0 & 1 & 0 & 1 & 1 \end{bmatrix}$$

T'_3 is minimal and $|T'_3| = 6$.   □

The upper bound of $|T_{min}| \leq B_n$ is tight when k is close to n and is loose when k is small. B_n grows exponentially as n increases, independent of the value of k.

For $k \leq n/2$, the size of the minimum test set occurs when $w = \lfloor k/2 \rfloor$ and $\lfloor k/2 \rfloor + (n-k+1)$, resulting in

$$|T_{\min}| = \binom{n}{\lfloor k/2 \rfloor} + \binom{n}{k - \lfloor k/2 \rfloor - 1}$$

We will now consider two special cases.

Case 1: $n = k$

For this case $w = 0$ mod 1, hence $w = 0, 1, 2, ..., n$. Thus T_0 consists of the set of all 2^n binary n-tuples.

Case 2: $n = k + 1$

For this case $w = c$ mod 2, and T_0 consists of the set of all binary n-tuples having odd parity, and T_1 consists of the set of all n-tuples having even parity. This situation is shown below for the case of $n = 4$ and $k = 3$.

$$T_0 = \begin{bmatrix} 0 & 0 & 0 & 0 \\ 0 & 0 & 1 & 1 \\ 0 & 1 & 0 & 1 \\ 1 & 0 & 0 & 1 \\ 0 & 1 & 1 & 0 \\ 1 & 0 & 1 & 0 \\ 1 & 1 & 0 & 0 \\ 1 & 1 & 1 & 1 \end{bmatrix} \text{even parity} \qquad T_1 = \begin{bmatrix} 1 & 0 & 0 & 0 \\ 0 & 1 & 0 & 0 \\ 0 & 0 & 1 & 0 \\ 0 & 0 & 0 & 1 \\ 1 & 1 & 1 & 0 \\ 1 & 1 & 0 & 1 \\ 1 & 0 & 1 & 1 \\ 0 & 1 & 1 & 1 \end{bmatrix} \text{odd parity}$$

The following theorem verifies the correctness of this last case.

Theorem 11.3: In a matrix with 2^k distinct rows of $(k+1)$ binary values, where every row has the same parity, every set of k columns has all 2^k combinations of k values.

Proof: A set of k columns is obtained by removing one column of the matrix. We want to show that the 2^k rows of the remaining matrix are distinct. Let us assume the contrary, namely that two of the remaining rows — say, i and j — are identical. Because of the constant parity assumption, it follows that the bits of the removed column in the rows i and j of the original matrix must be equal. Hence rows i and j of the original matrix were also identical, which contradicts the assumption that all 2^k rows of $(k+1)$ bits are distinct. Therefore every set of k columns has all 2^k combinations of k values.   □

11.2.3.3 Identification of Test Signal Inputs

Consider an n-input circuit. During testing it may be possible to apply inputs to p *test signal lines*, and have these p lines drive the n lines, where $p < n$. Clearly some of these p lines must fanout to two or more of the normal input lines. This section deals with the identification of these test signal lines and the tests associated with these lines.

Consider the circuit and test patterns shown in Figure 11.5. Note that f is a function of x and y, while g is a function of y and z. The four test vectors shown in Figure 11.5 test the individual functions f and g exhaustively and concurrently, even though to test the multiple-output function (f,g) exhaustively requires eight test vectors. Note that since no output is a function of both x and z, the same test data can be applied to both of these lines. Thus this circuit can be tested with only two test signals. A circuit is said to be a *maximal-test-concurrency* (MTC) *circuit*, if the minimal number of required test signals for the circuit is equal to the maximum number of inputs upon which any output depends. The circuit shown in Figure 11.5 is a MTC circuit.

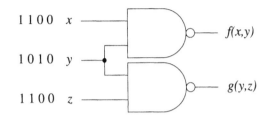

Figure 11.5 A maximal-test-concurrency circuit with verification test inputs
$x = z$

Figure 11.6 shows a non-MTC circuit. Here every output is a function of only two inputs, but three test signals are required. However, each output can still be tested exhaustively by just four test patterns.

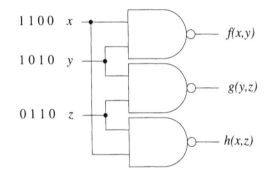

Figure 11.6 A nonmaximal-test-concurrency circuit with verification test inputs

Figure 11.7 shows a non-MTC circuit that requires four test signals; each output is a function of only two inputs, but five test patterns are required to exhaustively test all six outputs.

We next present a procedure for partitioning the inputs of a circuit to determine (1) the minimal number of signals required to test a circuit and (2) which inputs to the CUT can share the same test signal. We will also show how constant-weight patterns can be used to test the circuit. The various steps of the procedure will be illustrated as they are presented using the circuit C^* shown in functional form in Figure 11.8. From these results it will be possible to identify MTC circuits and to construct tests for MTC and non-MTC circuits.

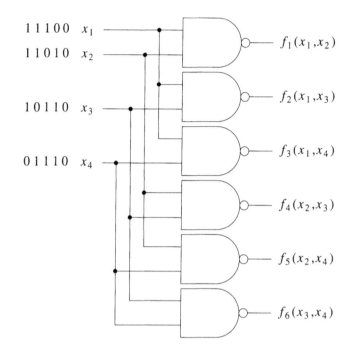

Figure 11.7   A nonmaximal-test-concurrency circuit with verification test inputs

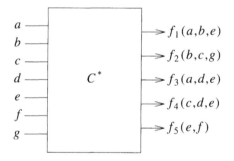

Figure 11.8   Circuit C^*

Procedure 11.1: Identification of Minimal Set of Test Signals

Step 1: Partition the circuit into disjoint subcircuits.

C^* consists of only one partition.

Step 2: For each disjoint subcircuit, carry out the following steps.

    a.   Generate a dependency matrix.

    b.   Partition the matrix into groups of inputs so that two or more inputs in a group do not affect the same output.

    c.   Collapse each group to form an equivalent input, called a test signal input.

For an n-input, m-output circuit, the *dependency matrix* $D = [d_{ij}]$ consists of m rows and n columns, where $d_{ij} = 1$ if output i depends on input j; otherwise $d_{ij} = 0$. For circuit C^*, we have

$$D = \begin{matrix} & \begin{matrix} a & b & c & d & e & f & g \end{matrix} \\ \begin{bmatrix} 1 & 1 & 0 & 0 & 1 & 0 & 0 \\ 0 & 1 & 1 & 0 & 0 & 0 & 1 \\ 1 & 0 & 0 & 1 & 1 & 0 & 0 \\ 0 & 0 & 1 & 1 & 1 & 0 & 0 \\ 0 & 0 & 0 & 0 & 1 & 1 & 0 \end{bmatrix} & \begin{matrix} f_1 \\ f_2 \\ f_3 \\ f_4 \\ f_5 \end{matrix} \end{matrix}$$

Reordering and grouping the inputs produce the modified matrix D_g.

$$D_g = \begin{matrix} & \begin{matrix} & \textit{Group} & \\ I\; & II & III & IV \\ a \;\; c & b \;\; d & e & f \;\; g \end{matrix} \\ \begin{bmatrix} 1 & 0 & 1 & 0 & 1 & 0 & 0 \\ 0 & 1 & 1 & 0 & 0 & 0 & 1 \\ 1 & 0 & 0 & 1 & 1 & 0 & 0 \\ 0 & 1 & 0 & 1 & 1 & 0 & 0 \\ 0 & 0 & 0 & 0 & 1 & 1 & 0 \end{bmatrix} & \begin{matrix} f_1 \\ f_2 \\ f_3 \\ f_4 \\ f_5 \end{matrix} \end{matrix}$$

In each group there must be less than two 1s in each row and the number of groups should be minimal. This insures that no output is driven by more than one input from each group.

Procedures for finding such a partition, which is a NP-complete problem, can be found in [McCluskey 1984] and [Barzilai *et al.* 1981].

The collapsed equivalent matrix D_c is obtained by ORing each row within a group to form a single column. The result for circuit C^* is

$$\begin{array}{cccc} I & II & III & IV \end{array}$$

$$D_c = \begin{bmatrix} 1 & 1 & 1 & 0 \\ 1 & 1 & 0 & 1 \\ 1 & 1 & 1 & 0 \\ 1 & 1 & 1 & 0 \\ 0 & 0 & 1 & 1 \end{bmatrix} \begin{array}{l} f_1 \\ f_2 \\ f_3 \\ f_4 \\ f_5 \end{array}$$

Step 3: Characterize the collapsed matrix D_c in terms of two parameters p and w, where p is the number of partitions in D_c, called the *width* of D_c, and w is the maximum number of 1s in any row, and is the *weight* of D_c. Note that p represents the maximum number of input signals required to test a disjoint subcircuit, and w represents the maximum number of signals upon which any output depends. Hence a pseudoexhaustive test must be of length 2^w or greater but need not exceed 2^p.

For D_c, we have $p = 4$ and $w = 3$.

A *universal minimal pseudoexhaustive test set* having parameters (p,w) is a minimal set of test patterns that contains, for all $\begin{pmatrix} p \\ w \end{pmatrix}$ subsets consisting of w of the p signal lines, all 2^w test patterns. The properties of these test sets are determined by the relative values of p and w, where by definition, $p \geq w$. For a specific circuit, if all outputs are a function of w inputs, then this test set is minimal in length. Otherwise it may not be.

Step 4: Construct the test patterns for the circuit based upon the following three cases.

$$\begin{array}{ll} \textit{Case 1:} & p = w \\ \textit{Case 2:} & p = w + 1 \\ \textit{Case 3:} & p > w + 1 \end{array}$$

Case 1: $p = w$

This case corresponds to MTC circuits and the test consists of all 2^p test patterns of p bits. The test can be easily generated by a counter or a complete LFSR. Clearly this is a universal minimal pseudoexhaustive test set. This case applies to Figure 11.5, where $p = w = 2$. Referring to the prior discussion on constant weight patterns, this case corresponds to the previous case where $k = n$ resulting in the test set T_0.

Case 2: $p = w + 1$

This case corresponds to the case where $n = k + 1$ in the previous discussion on constant weight patterns. The minimal test set consists of all possible patterns of p bits with either odd or even parity. There are $2^{p-1} = 2^w$ such patterns. Selecting odd parity, the tests for C^* are listed next.

$$
\begin{array}{cccc}
A & B & C & D \\
\hline
0 & 0 & 0 & 1 \\
0 & 0 & 1 & 0 \\
0 & 1 & 0 & 0 \\
1 & 0 & 0 & 0
\end{array}
\left. \right\} parity = 1
$$

$$
\begin{array}{cccc}
0 & 1 & 1 & 1 \\
1 & 0 & 1 & 1 \\
1 & 1 & 0 & 1 \\
1 & 1 & 1 & 0
\end{array}
\left. \right\} parity = 3
$$

Note that for any subset of three columns, all 2^3 binary triplets occur.

This pseudoexhaustive test set consists of 8 patterns, while an exhaustive test would consist of 128 patterns. Each column represents an input to each line in a group; e.g., column A can be the input to lines a and c in the circuit shown in Figure 11.8.

This case also applies to the circuit shown in Figure 11.6, where $p = 3$ and test patterns of even parity are selected.

Case 3: $p > w + 1$

For this case, the test set consists of two or more pattern subsets, each of which contains all possible patterns of p bits having a specific constant weight.   □

The total number of test patterns T is a function of p and w. Figure 11.9 shows the value of the constant weights and T for various values of p and w. Unfortunately, constant weights do not exist for all pairs of p and w. For such cases, w can be increased so as to achieve a constant-weight pseudoexhaustive test, but it may not be minimal in length.

The minimal test set for $p = 5$ and $w = 3$ corresponding to the constant-weight pair (1,4) (see Figure 11.9) is shown in Figure 11.10.

Unfortunately it is not always easy to construct a circuit to generate a pseudoexhaustive test set for $p > w + 1$, and the hardware overhead of some of these circuits is sometimes quite high. In the next subsection several techniques for designing circuits that generate pseudoexhaustive tests will be briefly described. Many of these designs do not generate minimal test sets, but the techniques lead to efficient hardware designs. Because most TPGs use some form of an LFSR, and since more than one test sequence is sometimes needed, often more than one seed value is required for initializing the state of the LFSR.

11.2.3.4 Test-Pattern Generators for Pseudoexhaustive Tests

Syndrome-Driver Counter

For the circuit shown in Figure 11.3, $y_1 = f_1(x_1,x_3)$, $y_2 = f_2(x_1,x_2)$, $y_3 = f_3(x_2,x_3)$, and $y_4 = f_4(x_3,x_4)$. Thus no output is a function of both x_1 and x_4, or of x_2 and x_4. (The width of this circuit is 3.) Hence x_1 and x_4 (or x_2 and x_4) can share the same input during testing. Thus this circuit can be pseudoexhaustively tested by applying all 2^3 input to x_1, x_2, and x_3, and by having the line driving x_1 or x_2 also drive x_4.

p	w	T	Constant weights
$p > 3$	2	$p + 1$	$(0,p{-}1)$ or $(1,p)$
$p > 4$	3	$2p$	$(1,p{-}1)$
$p > 5$	4	$\frac{1}{2}p(p{+}1)$	$(1,p{-}2)$ or $(2,p{-}1)$
$p > 6$	5	$p(p{-}1)$	$(2,p{-}2)$
$p = 8$	6	$\frac{1}{2}p(p{+}1)$	$(1,4,7)$
$p > 8$	6	$B(p{+}1,3)$	$(2,p{-}3)$ or $(3,p{-}2)$
$p = 9$	7	170	$(0,3,6,9)$
$p > 9$	7	$2B(p,3)$	$(3,p{-}3)$
$p = 10$	8	341	$(0,3,6,9)$ or $(1,4,7,10)$
$p = 11$	8	496	$(0,4,8)$ or $(3,7,9)$
$p > 11$	8	$B(p{+}1,4)$	$(3,p{-}4)$ or $(4,p{-}3)$
$p = 11$	9	682	$(1,4,7,10)$
$p = 12$	9	992	$(0,4,8,12)$
$p > 12$	9	$2B(p,4)$	$(4,p{-}4)$
$p = 12$	10	1365	$(1,4,7,10)$ or $(2,5,8,11)$
$p = 13$	10	2016	$(0,4,8,12)$ or $(1,5,9,13)$
$p = 14$	10	3004	$(0,5,10)$ or $(4,9,14)$
$p > 14$	10	$B(p{+}1,5)$	$(4,p{-}5)$ or $(5,p{-}4)$

Figure 11.9   Constant weights for pseudoexhaustive tests. Note: $(\alpha_1, \alpha_2, ...)$ represents a constant-weight vector, where α_i is the constant weight of one subset of patterns. $B(n,m)$ represents the binomial coefficient.

```
0  0  0  0  1 ⎞                    0  1  1  1  1 ⎞
0  0  0  1  0 ⎪                    1  0  1  1  1 ⎪
0  0  1  0  0 ⎬ Weight of 1        1  1  0  1  1 ⎬ Weight of 4
0  1  0  0  0 ⎪                    1  1  1  0  1 ⎪
1  0  0  0  0 ⎠                    1  1  1  1  0 ⎠
```

Figure 11.10   Constant-weight test set for (1,4)

In general if $(n{-}p)$ inputs, $p < n$, can share test signals with the p other inputs, then the circuit can be exhaustively tested using these p inputs. If $p = w$, only 2^p tests are required.

If $p > w$, $T = 2^p - 1$ tests are enough to test the circuit. T can be much greater than the lower-bound value of 2^w. For the example shown in Figure 11.3, $p = 3$ and

$2^3 - 1 = 7$. The reader can verify that the (1,1,1) test pattern is not required. In general the test patterns consisting of all 0s and all 1s are not required; hence $2^p - 2$ tests can be used.

The major problem with this approach is that when p is close in value to n a large number of test patterns are still required.

This testing scheme was proposed by Barzilai *et al.* [1981] and uses a *syndrome-driver counter* (SDC) to generate test patterns. The SDC can be either a binary counter or an LFSR and contains only p storage cells.

Constant-Weight Counter

A *constant-weight code* (CWC), also known as an *N-out-of-M* code, consists of the set of codewords of M binary bits, where each codeword has exactly N 1s. A constant-weight code is analogous to the constant-weight patterns discussed previously. The 2-out-of-4 CWC is shown below.

$$\begin{array}{cccc} 1 & 1 & 0 & 0 \\ 1 & 0 & 1 & 0 \\ 1 & 0 & 0 & 1 \\ 0 & 1 & 1 & 0 \\ 0 & 1 & 0 & 1 \\ 0 & 0 & 1 & 1 \end{array}$$

Note that for any two columns all possible pairs of binary values appear. The same is not true of a 2-out-of-3 CWC.

Any (n,w)-circuit can be pseudoexhaustively tested by a constant-weight counter implementing a w-out-of-K code, for an appropriate value of K. For large values of N and M, the complexity of the associated constant-weight counter is often high. More information on CWCs and the design of constant-weight counters can be found in [Ichikawa 1982], [Tang and Woo 1983], and [McCluskey 1984].

Combined LFSR/SR

Another way to produce a pseudoexhaustive test, proposed in [Barzilai *et al.* 1983] and [Tang and Chen 1984a], is to employ a combination of an LFSR and a shift register (SR), as shown in Figure 11.11. A 4-stage combined LFSR/SR for testing a (4,2)-CUT is shown in Figure 11.12. The implementation cost for this design approach is less than for a constant-weight counter, but the resulting circuit usually generates more test vectors. Also, the LFSR usually requires at least two seed values. The number of test patterns generated is near minimal when w is much less than n; e.g., $w < n/2$. LFSRs are usually designed to have a shift mode of operation. This mode is used for loading seed values into the registers. If the register is also used as a signature analyzer, the shift mode is also used for scanning out the final signature.

Combined LFSR/XOR Gates

Still another way to produce a pseudoexhaustive test, shown in Figure 11.13, uses a combination of an LFSR and a XOR (linear) network. The design of these networks is based on the use of linear sums [Akers 1985] or linear codes [Vasanthavada and Marinos

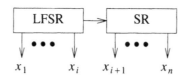

Figure 11.11   An LFSR/SR verification test generator

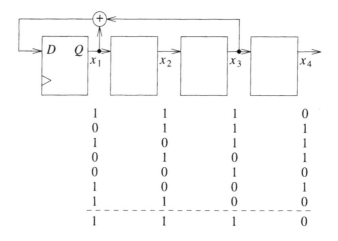

Figure 11.12   A 4-stage LFSR/SR for a (4,2)-CUT

1985]. These designs require at most two seeds, and the number of test patterns needed to ensure pseudoexhaustive testing is close to that required for LFSR/SR designs.

Figure 11.14 shows a combined LFSR/XOR TPG along with the patterns it produced. This device can test a (4,2)-CUT.

Condensed LFSR

Another design approach, proposed by Wang and McCluskey [1984, 1986b] and referred to as *condensed LFSR*, uses at most two seeds, leads to simple designs, and produces a very efficient test set when $w \geq n/2$. When $w < n/2$ this technique uses more tests than the LFSR/SR approach. Condensed LFSRs are based on the concept of linear codes [Peterson and Weldon 1972, Lin and Costello 1983]. An *(n,k)-linear code* over a Galois field of 2 generates a set S of n-tuples containing 2^k distinct code words, where if $c_1 \in S$ and $c_2 \in S$, then $c_1 \oplus c_2 \in S$.

Using a type 2 LFSR having a characteristic polynomial $p(x)$, a condensed LFSR for a (n,w)-CUT can be constructed as follows. Let k be the smallest integer such that

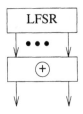

Figure 11.13   A combined LFSR/XOR verification-test generator

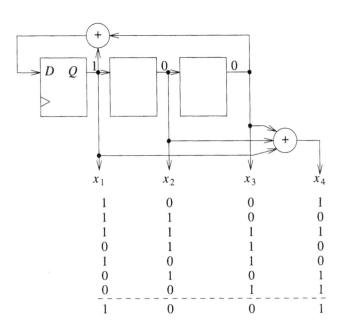

Figure 11.14   A combined LFSR/XOR TPG for a (4,2)-CUT

$$w \leq \lceil k/(n-k+1) \rceil + \lfloor k/(n-k+1) \rfloor$$

Then a type 2 LFSR realizes a condensed LFSR if

$$p(x) = (1+x+x^2+ \ ... \ +x^{n-k})q(x)$$

where $q(x)$ is a primitive polynomial of degree k. Also the seed polynomial $S_0(x)$ must be divisible by $(1+x+x^2+ \ ... \ +x^{n-k})$.

Again consider the case $(n,w)=(4,2)$: then $k = 3$ and $(n-k) = 1$. Selecting $q(x) = 1 + x + x^3$, we obtain

$$p(x) = (1+x)(1+x+x^3) = 1 + x^2 + x^3 + x^4$$

Figure 11.15 shows the resulting design and initial seed. Although a condensed LFSR has n stages, the feedback circuitry is usually simple.

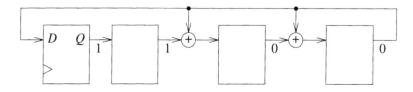

Figure 11.15   A condensed LFSR for a (4,2)-CUT

Cyclic LFSR

When $w < n/2$, condensed LFSR designs produce long tests for (n,w)-CUTs. LFSR/XOR designs reduce this test length but have a high hardware overhead. For $w < n/2$, *cyclic LFSRs* lead to both efficient tests and low hardware overhead. Cyclic LFSRs are based on cyclic codes [Peterson and Weldon 1972, Lin and Costello 1983]. An *(n,k)-cyclic code* over the Galois field of 2 contains a set of 2^k distinct codewords, each of which is an n-tuple satisfying the following property: if c is a codeword, then the n-tuple obtained by rotating c one place to the right is also a code word. Cyclic codes are a subclass of linear codes. The design of cyclic LFSRs and details for obtaining the characteristic polynomial for a cyclic LFSR are presented in [Wang 1982] and [Wang and McCluskey 1986f, 1986g, 1987a, 1987c].

11.2.3.5 Physical Segmentation

For very large circuits, the techniques described for pseudoexhaustive testing often lead to large test sets. In these cases, pseudoexhaustive testing can still be achieved by employing the concept of *physical segmentation*. Here a circuit is divided or partitioned into subcircuits by employing hardware-segmentation techniques.

One such technique is shown in Figure 9.11. Various ways for segmenting a circuit based on this type of structure are presented in [Patashnik 1983], [Archambeau 1985], and [Shperling and McCluskey 1987].

More details on this form of testing can be found in [McCluskey and Bozorgui-Nesbat 1981], [Chandra *et al.* 1983], [Udell 1986], [Chen 1987], and [Udell and McCluskey 1989].

Physical segmentation can also be achieved by inserting bypass storage cells in various signal lines. A *bypass storage cell* is a storage cell that in normal mode acts as wire, but in the test mode can be part of an LFSR circuit. It is similar to a cell used in boundary-scan designs, such as the one shown in Figure 9.14. If inserted into line x, then the associated LFSR can be used as a MISR and hence to detect errors occurring on line x, or it can be used as a PRPG and hence to generate test patterns on line x.

Example 11.4:   Consider the circuit C shown in Figure 11.16(a), where the logic blocks G_i, $i = 1, 2, ..., 9$ are represented by circles. Next to each block is an integer indicating the number of primary inputs that can affect the output of the block. Assume it is desired to segment this circuit using bypass storage cells so that no signal is a function of more than w variables. Figure 11.16(b) shows how three bypass storage cells can be added to this circuit to achieve a value of $w = 4$.

Figure 11.17 shows the four segments of C created by the bypass storage cells. Figure 11.17(a) depicts the subcircuit C_1; here G_3 is tested along with its associated interconnecting lines, and the storage cells associated with x_4, x_5, x_6 and x_7, labeled P, are part of a TPG; the bypass storage cell on the output of G_3, labeled S, is part of a ORA. Similarly, Figure 11.17(b) shows the subcircuit C_2 where G_4 and G_7 are tested. G_1 and G_5 are tested as part of C_3 (see Figure 11.17(c)); G_2, G_6, and G_8 are tested as part of C_4 (see Figure 11.17(d)). One additional subcircuit C_5 (not shown) is needed to test G_9, along with G_2 and G_6.   □

Procedures for this form of segmentation and that also take into account the design of the corresponding TPG and ORA circuitry are discussed in [Jone and Papachristou 1989].

11.3 Generic Off-Line BIST Architectures

In the previous section we considered BIST approaches for testing a single block of combinational logic. In this section we will consider general off-line BIST structures that are applicable to chips and boards consisting of blocks of combinational logic interconnected by storage cells.

Off-line BIST architectures at the chip and board level can be classified according to the following criteria:

1. centralized or distributed BIST circuitry;

2. embedded or separate BIST elements.

BIST architectures consist of several key elements, namely

1. test-pattern generators;

2. output-response analyzers;

3. the circuit under test;

4. a distribution system (DIST) for transmitting data from TPGs to CUTs and from CUTs to ORAs;

5. a BIST controller for controlling the BIST circuitry and CUT during self-test.

Often all or parts of the controller are off-chip. The distribution system consists primarily of direct interconnections (wires), busses, multiplexers, and scan paths.

The general form of a centralized BIST architecture is shown in Figure 11.18. Here several CUTs share TPG and ORA circuitry. This leads to reduced overhead but increased test time.

During testing, the BIST controller may carry out one or more of the following functions:

1. Single-step the CUTs through some test sequence.

2. Inhibit system clocks and control test clocks.

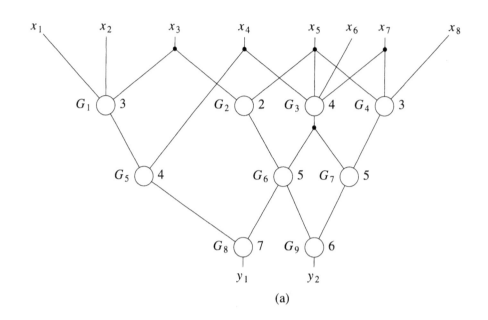

(a)

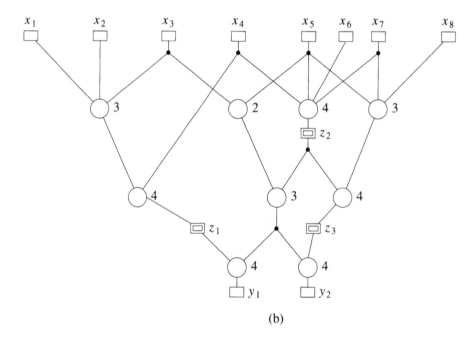

(b)

Key

☐ — normal I/O storage cell   ▭ — bypass storage cell   ○ — a logic block

Figure 11.16   Inserting bypass storage cells to achieve $w = 4$

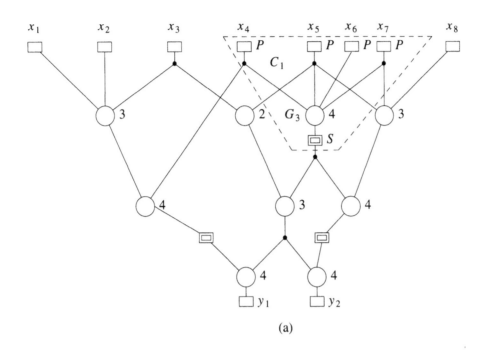

(a)

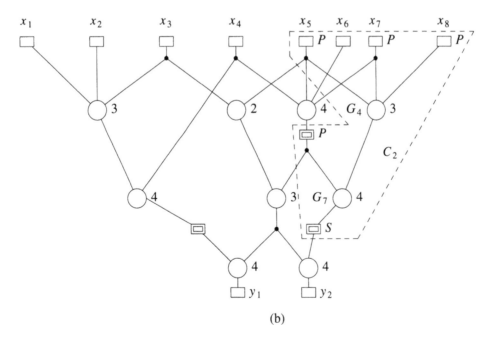

(b)

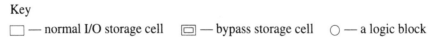

Key

□ — normal I/O storage cell   ▣ — bypass storage cell   ○ — a logic block

Figure 11.17   Four segments formed by the bypass storage cells

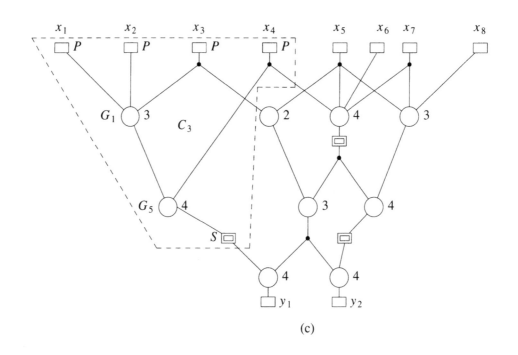

(c)

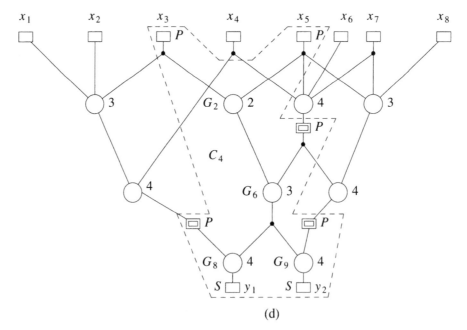

(d)

Key

 — normal I/O storage cell   ⊡ — bypass storage cell   ○ — a logic block

Figure 11.17   (Continued)

Chip, board, or system

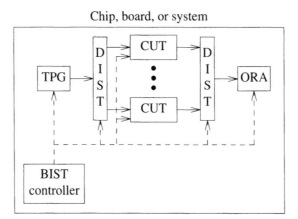

Figure 11.18   Generic form of centralized and separate BIST architecture

3.   Communicate with other test controllers, possibly using test busses.

4.   Control the operation of a self-test, including seeding of registers, keeping track of the number of shift commands required in a scan operation, and keeping track of the number of test patterns that have been processed.

Further information on the design of controllers for BIST circuitry can be found in [Breuer *et al.* 1988].

The distributed BIST architecture is shown in Figure 11.19. Here each CUT is associated with its own TPG and ORA circuitry. This leads to more overhead but less test time and usually more accurate diagnosis. The BIST control circuitry is not shown. The designs shown in Figures 11.18 and 11.19 are examples of the separate BIST architecture, since the TPG and ORA circuitry is external to the CUT and hence not part of the functional circuitry.

Chip, board, or system

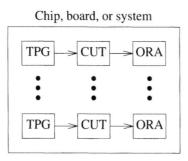

Figure 11.19   Generic form of distributed and separate BIST architecture

Figure 11.20 shows the general form of the distributed and embedded BIST architecture. Here the TPG and ORA elements are configured from functional elements within the CUT, such as registers. This leads to a more complex design to control, but has less hardware than distributed and separate architectures have.

Chip, board, or system

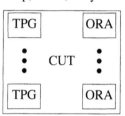

Figure 11.20   Generic form of distributed and embedded BIST architecture

The choice of a BIST architecture is a function of several factors, some of which are listed next.

1. *Degree of test parallelism:* Distributed BIST provides a higher degree of test parallelism, since several CUTs can be tested at the same time.

2. *Fault coverage:* Distributed BIST usually leads to a higher fault coverage since TPG and ORA circuits can be customized to each CUT. For example, a BIST technique for a block of combinational logic may not be suitable for a RAM.

3. *Level of packaging:* At higher levels, centralized BIST becomes more natural. For example, microdiagnostics testing is only applicable at the level where a microprogrammable controller exists. This controller can then be used to test many components of a system.

4. *Test time:* Distributed BIST usually leads to a reduction in test time.

5. *Physical constraints:* Size, weight, power, cooling costs, and other factors influence the design. Often embedded and separate BIST architectures require more hardware and degrade performance.

6. *Complexity of replaceable units:* If a board is a replaceable unit and is to be self-testable, then it must contain TPG and ORA circuitry. If a system is the lowest level of replaceable unit, then its constituent boards need not have TPG and ORA circuitry and a more centralized BIST architecture can be used.

7. *Factory and field test-and-repair strategy:* The type of ATE and degree to which it is used to test and diagnose failures influences and is influenced by BIST. For example, because of the increasing use of surface-mounted devices, in-circuit or bed-of-nails testing is becoming more difficult to use and hence the need for BIST and the use of boundary scan.

8. *Performance degradation:* Adding BIST hardware in critical timing paths of a circuit may require a reduction in the system clock rate.

11.4 Specific BIST Architectures

In this section several BIST architectures proposed by various research and development groups will be described. We will denote sequential blocks of logic by S and combinational blocks by C.

11.4.1 A Centralized and Separate Board-Level BIST Architecture (CSBL)

Figure 11.21 illustrates a *centralized and separate board-level* (CSBL) BIST architecture proposed by Benowitz *et al.* [1975]. It has the following attributes:

- centralized and separate BIST architecture;

- no boundary scan;

- combinational or sequential CUT.

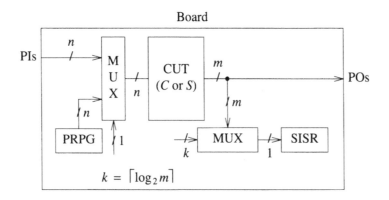

Figure 11.21   A centralized and separate BIST architecture (CSBL)

During the off-line test mode, the inputs are driven by a PRPG and the outputs are monitored using a single-input signature analyzer. To reduce hardware costs, the test is repeated m times, once for each output; hence only one signature analyzer is required.

This method is best suited for pipeline circuits with limited feedback. General finite-state controllers can be more complex to test. Microprocessors represent a very complex test situation. Extensive fault simulation is required to determine the number of test vectors required to achieve an adequate level of fault coverage.

11.4.2 Built-In Evaluation and Self-Test (BEST)

The *built-in evaluation and self-test* (BEST) architecture is an application of the CSBL design to chips (see Figure 11.22). In general the logic being tested is a sequential circuit. The inputs to the CUT are driven by a PRPG and the outputs are compressed using a MISR. The details of switching between the primary inputs and the output of the PRPG when applying the normal or the test inputs to the CUT are not shown. Either a MUX can be used, or the primary inputs can first be loaded into the PRPG and then applied to the CUT. The same concepts apply to the outputs. Both an embedded and a separate version of this architecture exist. For example, if the primary inputs to the CUT go directly to registers, and the primary outputs are driven by registers, then an embedded BIST design can be used. That is, these I/O registers can be modified for use as the PRPG and MISR. Otherwise the PRPG and MISR need to be added to the CUT, which results in a separate architecture. In the latter case, the PRPG and MISR can be made part of the boundary-scan registers.

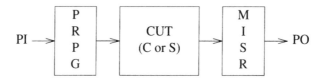

Figure 11.22   The built-in evaluation and self-test (BEST) BIST architecture

The hardware overhead for the BEST architecture is low. Again, extensive fault simulation is required to determine an acceptable balance between fault coverage and test length. For some circuits this technique can be ineffective in achieving an acceptable level of fault coverage. Further information can be found in [Resnick 1983] and [Lake 1986].

The previous BIST architectures often result in low fault coverage because they rely on the use of pseudorandom patterns for testing a sequential circuit. To circumvent this problem, an internal scan path can be used within the CUT so that the testing of the CUT can be reduced to the problem of testing combinational logic. The next few BIST architectures will illustrate this concept.

11.4.3 Random-Test Socket (RTS)

The *random-test socket* (RTS) [Bardell and McAnney 1982] is not a true BIST architecture because the test circuitry is external to the CUT. The RTS architecture, shown in Figure 11.23, has the following attributes:

- distributed and separate test hardware;
- no boundary scan;
- scan path (LSSD) CUT architecture.

The CUT is tested as follows:

1. Initialize the LFSRs.

2. Load a pseudorandom test pattern into the scan path using R_2.

3. Generate a new pseudorandom test pattern using R_1. This takes one clock cycle. The vector is applied to the primary inputs of the CUT.

4. Capture the response on the primary outputs of the CUT by applying one clock pulse to R_3.

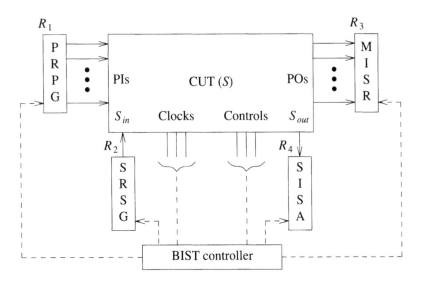

Figure 11.23   Random-test socket (RTS)

5. Execute a parallel-load operation on the system storage cells to capture the response to the random test pattern.

6. Scan out the data in the internal scan path of the CUT and compress these data in R_4.

Steps 2-6 are repeated until either an adequate fault coverage is achieved or the maximum allowable test time is reached. Also, steps 2 and 6 can be carried out simultaneously. The circuit is considered to be fault-free if the final values in R_3 and R_4 are correct.

The RTS test approach can suffer from problems related to fault coverage and test time. Testing is inherently slow, since for each test pattern the entire scan path must be loaded. Test time can be reduced by partitioning the storage cells into several scan paths, but this requires additional I/O pins.

Testing with pseudorandom test patterns requires more patterns than testing with deterministic test patterns. The number of test patterns is determined by the desired fault coverage, the circuit itself, and the characteristics of the LFSRs used to generate the test data and produce the signatures. For a high fault coverage, the population of random-pattern-resistant faults will dictate the number of tests required. Test points can be effectively used to reduce this number.

Fault coverage can be adversely affected by linear correlations among the bits generated by LFSRs and by the periodicity of their sequences [Bardell and McAnney 1986, Chen 1986]. Assume, for example, that the LFSR R_2 has a period p and that the length of the scan path is k, where either p divides k, or k divides p. Then, in the former case only k/p unique patterns can be loaded in the scan path; in the latter case only p/k unique patterns exist. Thus it is important that neither of these situations occur. Also, if a primitive polynomial is not used, then all possible patterns are not generated. Some of these patterns may be required to detect specific faults.

Even when $p \gg k$, a complete load of the scan path between each test vector implies that large blocks of combinational logic in the CUT will probably not be tested exhaustively.

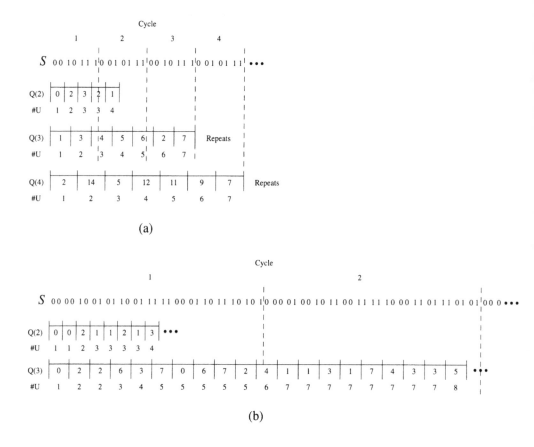

Figure 11.24   Test patterns generated by LFSRs

Figure 11.24 illustrates the test patterns generated by two LFSRs for different values of k and p. S refers to the sequence of bits generated by the LFSR. $Q(k)$ indicates the integer (base 10) value of the k-bit test pattern generated, and # U indicates the number of unique test patterns generated. For example, Figure 11.24(a) shows the result for a type 1 LFSR with characteristic polynomial $P(x) = 1 + x^2 + x^3$ and initial state 100. The sequence generated is 0010111 and $p = 7$. For $k = 2$, five test patterns must be generated before all four unique test patterns are produced. For $k = 3$ after three cycles seven unique test patterns have been generated. The test patterns now repeat, hence the (000) pattern is never generated. For $k = 4$ the patterns repeat after four cycles and only 7 of the 16 possible test patterns are generated. Figure 11.24(b) shows comparable results for the case $P(x) = 1 + x^3 + x^5$, where $p = 31$. For $k = 3$, 20 test patterns must be generated before all 8 possible patterns are produced.

By incorporating the LFSRs and controller within the CUT, and by multiplexing between test inputs and normal system inputs, the RTS architecture can be transformed into a BIST architecture.

11.4.4 LSSD On-Chip Self-Test (LOCST)

Figure 11.25 shows the general form of the *LSSD on-chip self-test* (LOCST) architecture, which is an extension of a BIST version of the RTS design. This architecture has the following attributes:

- centralized and separate BIST architecture;
- scan path (LSSD) CUT architecture;
- boundary scan;
- on-chip test controller.

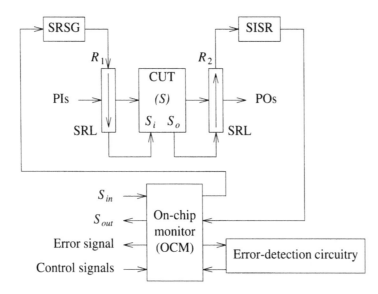

Figure 11.25 The LOCST architecture

The inputs and outputs are buffered through two boundary scan registers, denoted by R_1 and R_2 in Figure 11.25. The design includes an on-chip controller, referred to as an on-chip monitor (OCM), which is driven by a test bus and is used to partially control the test process. One implementation of LOCST uses a 20-bit SRSG having the characteristic polynomial $x^{20} + x^{17} + 1$, and a 16-bit SISR with characteristic polynomial $x^{16} + x^9 + x^7 + x^4 + 1$.

The test process is as follows:

1. *Initialize:* The scan path is loaded with seed data via the S_{in} line.

2. *Activate self-test mode:* Disable system clocks on R_1, R_2; enable LFSR operation.

3. *Execute self-test operation:*

    a. Load the scan path with a pseudorandom test pattern. Data in the scan path is compressed in the SISR. This step requires multiple clock cycles and is controlled by off-chip signals.

    b. Activate the system clocks for one cycle; register R_2 and the internal scan path will load data.

    c. Steps a and b are repeated until an adequate level of fault coverage is achieved.

4. *Check result:* Compare the final value in the SISR with the known good signature.

Data in the scan path can also be scanned out of the chip using the S_{out} line. The design can be embedded rather than separate by implementing the TPG and ORA using the system latches. For this case, the SRSG becomes a PRPG, and the SISR a MISR.

Some versions of this form of BIST architecture do not employ boundary-scan registers. Here some logic, referred to as *external logic*, does not get fully tested unless external test equipment is used. The external logic consists of (a) those blocks of logic driven by primary inputs and (b) blocks of logic that drive primary outputs.

More details on this architecture can be found in [Eichelberger and Lindbloom 1983] and [LeBlanc 1984].

11.4.5 Self-Testing Using MISR and Parallel SRSG* (STUMPS)

The *self-testing using MISR and parallel SRSG* (STUMPS) architecture is shown in Figure 11.26 [Bardell and McAnney 1982, 1984]. It was originally applied at the board level, and subsequently at the chip level. It has the following attributes:

- centralized and separate BIST architecture;
- multiple scan paths;
- no boundary scan.

The scan paths are driven in parallel by a PRPG, and the signature is generated in parallel from each scan path using a MISR. At the board level, each scan path corresponds to the scan path in a separate chip; at the chip level each scan path is just one segment of the entire scan path of a chip.

The use of multiple scan paths leads to a significant reduction in test time. Since the scan paths may be of different lengths, the PRPG is run for K clock cycles to load up the scan paths, where K is the length of the longest scan path. For short scan paths, some of the data generated by the PRPG flow over into the MISR.

When this approach is applied at the board level to chips designed with a scan path, then the PRPG and the MISR can be combined into a special-purpose test chip, which must be added to the board.

---

* A SRSG is equivalent to a PRPG.

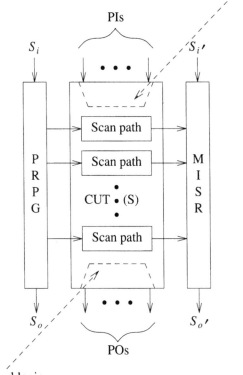

Figure 11.26   Self-test using MISR and parallel SRSG (STUMPS)

As before, problems related to linear correlation of data and periodicity can adversely affect the performance of this architecture. Note that if a type 1 LFSR is used in the PRPG, then $Q_i(t) = Q_{i-1}(t-1)$. Hence data in one scan path are a shifted version of data in another scan path. To avoid this situation, a type 2 LFSR can be used. The external logic must be tested via ATE or by adding boundary-scan registers.

11.4.6 A Concurrent BIST Architecture (CBIST)

Saluja *et al.* [1988] have extended the concepts of off-line BIST to include on-line BIST. One version of their proposed *concurrent BIST* (CBIST) architecture is shown in Figure 11.27. The CUT must be combinational logic. The BIST hardware is separate from the normal hardware. The technique can be employed in sequential circuits, if the circuitry is partitioned into blocks of combinational logic that can be tested as separate entities. No boundary or internal scan paths need to be used. To reduce hardware overhead, a centralized BIST architecture can be employed. During off-line testing, the PRPG drives the CUT, whose response is compressed in the MISR. For on-line testing, the PRPG and MISR are first initialized and held in their initial state until enabled by the *EN* signal. Normal inputs are applied to the CUT. When a match exists between the normal inputs and the state of the PRPG, an enable signal is generated allowing the PRPG to advance to its next state, while the MISR samples the output and also advances to its next state. This process is repeated whenever a match occurs between the normal input data and the current state of the PRPG. When the state of the PRPG reaches some prespecified final state, the signature in the MISR is verified.

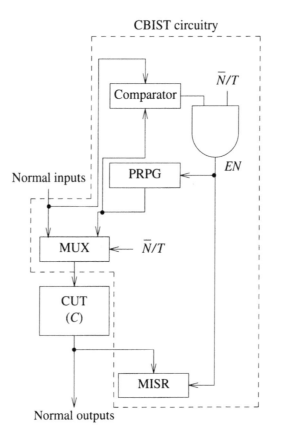

Figure 11.27   Concurrent BIST for combinational logic

The *test latency* for a circuit is the time required for the PRPG to go through all states corresponding to the desired test sequence. The expected value of the test latency can be estimated if certain assumptions on the distribution of inputs are made. For example, assume that the CUT has n inputs and that all input patterns are equally likely. Then the probability that an input will match the state of the PRPG, denoted by p, is $1/2^n$. Let $q = 1 - p$. Assume that L test patterns are applied to the CUT, that the CUT is to be tested exhaustively, and that $L \geq 2^n$. Let P_s denote the probability that the PRPG reaches its final state during the time these patterns are being applied. Then

$$P_s = 1 - \sum_{k=0}^{2^n - 1} \binom{L}{k} p^k q^{L-k}$$

Let the concurrent test latency for exhaustive testing, denoted by $C(\alpha)$, be the time (in clock cycles) during normal operation that it takes the PRPG to go through all 2^n states with a probability α. Figure 11.28 shows a few values of $C(\alpha)$ for various values of α and n. Here $C(\alpha)$ is expressed in terms of both L and time, where a 10 MHz system clock is assumed.

n	C(0.90)		C(0.99)	
	L	sec.	L	sec.
10	109089	0.11	1127044	0.11
12	17114557	1.71	17395982	1.74
14	271129177	27.11	273349519	27.33

Figure 11.28 $C(\alpha)$ for different values of n and α

11.4.7 A Centralized and Embedded BIST Architecture with Boundary Scan (CEBS)

Figure 11.29 shows a *centralized and embedded BIST architecture with boundary scan* (CEBS); this is a version of the architecture depicted in Figure 11.25. Here the first r bits of the input boundary scan register labeled RPG act as a PRPG and a SRSG, and the last s bits of the output boundary scan register act as both a MISR and a SISR. Testing proceeds as follows. A test mode signal is set and the scan registers are seeded. Then the RPG loads the scan path with pseudorandom test data. A system clock is issued and the scan-path registers are loaded in parallel with system data, except for the signature register SR, which operates in the MISR mode. The scan path is again loaded with pseudorandom data while the signature register operates in the SISR mode, compressing data that were in the scan path.

More details on this type of design can be found in [Komanytsky 1982, 1983] and [Butt and El-Ziq 1984].

11.4.8 Random Test Data (RTD)

One major problem with many of the techniques presented previously is that to apply a single test pattern to the CUT requires that the entire scan path be loaded with new data. This problem can be eliminated by using the *random test data* (RTD) BIST architecture, shown in Figure 11.30, which has the following attributes:

- distributed and embedded BIST architecture;

- boundary scan.

In this architecture, signature analyzers are used for both compression and test pattern generation [Bardell *et al.* 1987].

The test process operates as follows. The registers R_1, R_2, and R_3 are first set to their scan mode so that they can be loaded with a seed pattern. R_1 cannot be loaded with the all-0 pattern. The registers are then put into their test mode and held in this state while the circuit is tested. For each clock cycle, R_1 and R_2 generate a new test pattern, and R_2 and R_3 operate as a MISR.

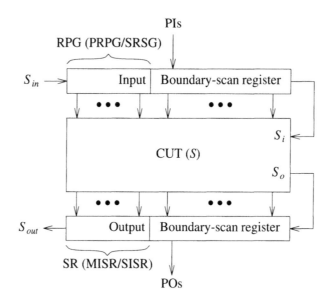

Figure 11.29   A centralized and embedded architecture with boundary scan (CEBS)

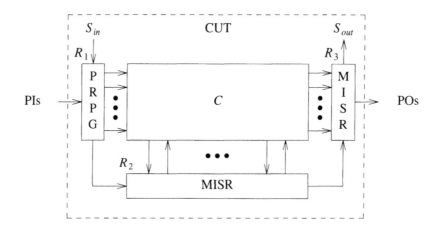

Figure 11.30   Random test data (RTD) BIST architecture

Little can be said for the tests generated by R_2 since these patterns are the state of a MISR. Clearly the patterns generated are a function of the logic C being tested. One problem with this technique is that some binary patterns may be repeated, and others may never be generated. Fault simulation is required to determine the number of clock cycles to run in the test mode to achieve a desired level of fault coverage.

Most of the previous BIST approaches use some form of LFSRs for generating test patterns and a signature. The RTD technique assumes that the normal data in a circuit, if suitably combined with some form of error capturing mechanism, is sufficient to test a circuit. This concept is the basis of the next three BIST architectures.

11.4.9 Simultaneous Self-Test (SST)

Simultaneous self-test (SST) is a BIST approach with the following attributes:

- distributed and embedded BIST architecture;

- scan CUT architecture;

- no boundary scan;

- no LFSRs used.

In this approach each storage cell in the CUT is modified to be a *self-test storage cell*. Such a storage cell has three modes of operation, namely normal mode, scan mode and test mode. A self-test SRL storage cell is shown in Figure 11.31. During normal operation, the value of the system data signal D is loaded into the flip-flop by $CK1$. To scan data into or out of the scan path, T is set to 0 and clock $CK2$ is used. During self-test, T is set to 1 and again clock $CK2$ is used. The new state of the flip-flop is equal to $D \oplus S_i$. A part of a circuit employing simultaneous self-test is shown in Figure 11.32. Note that the self-test scan register is not explicitly configured as an LFSR. Feedback can exist from the output to the input of this register by way of the combinational logic block C.

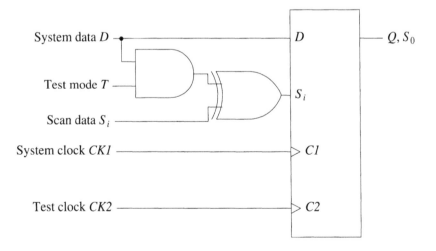

Figure 11.31   A self-test storage cell

To test a SST design, the scan path is first loaded with seed data. The circuit is then put into the test mode and clocked a sufficient number of times (N) to be adequately tested. After N clock cycles, the circuit is put back into the scan mode and the contents of the scan path shifted out. Faulty operation is detected if during the test process an output

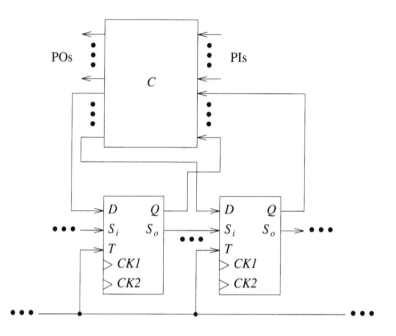

Figure 11.32   BIST architecture using self-test storage cells

error occurs, or if the final contents of the scan path is incorrect. During the test process the self-test scan path is simultaneously collecting test results from C and supplying test values to C. This procedure leads to a very fast test process since test data need not be shifted into the scan path.

Unfortunately, there are several problems associated with this test method. One problem again deals with testing external logic. For production testing, this can be solved by means of the ATE. For field test, some other means is required, such as the use of boundary-scan registers. Another solution is to connect the outputs of the CUT back to its inputs during self-test. In general, the number of POs and PIs are not the same. If there are more POs than PIs, then groups of POs can be XORed together to drive a single PI; if there are more PIs than POs, then some PIs can be driven by a unique PO, while other PIs are driven by the XOR of several POs. Another alternative is to let the ICs connected to the CUT drive (observe) inputs (outputs) which cannot be driven (observed) by the chip's own scan path.

Another problem deals with characterizing the quality of this test process. Note that though the scan path acts as both a random-pattern generator and as a data-compression circuit, there is no way of predicting *a priori* information related to the randomness of the test patterns generated or the degree of masking that may occur. Note that if an error is generated at the output of C (see Figure 11.32), then an error will occur in the scan path, say in cell Q_j. On the one hand, during the next clock cycle this error may propagate through C and produce several errors in the scan path. On the other hand, possibly no sensitized paths exist through C and this initial error just propagates to the next cell Q_{j+1}

in the scan path. However, if the data input D_{j+1} to this cell also is in error during this clock cycle, then the two errors *cancel* and masking occurs. Though errors frequently are canceled in the scan path, usually not all errors are canceled. For more information on simultaneous self-test see [DasGupta *et al.* 1982] and [Bardell and McAnney 1985].

11.4.10 Cyclic Analysis Testing System (CATS)

Figure 11.33 shows a BIST architecture referred to as *cyclic analysis testing systems* (CATS) [Burkness 1987]. The test aspects of this architecture are based on the fact that sequential circuits, by their very nature, are nonlinear binary-sequence generators. Figure 11.33(a) shows a sequential circuit that is the CUT. The CATS architecture is shown in Figure 11.33(b). Here during test mode, the outputs of S drive the inputs. If there are more outputs than inputs, then extra outputs can be combined using XOR gates. If the number of outputs is less than the number of inputs, then some outputs can drive more than one input. Thus all outputs and inputs can be exercised during testing. The analysis of this type of BIST architecture is complex. Different permutations of interconnections of outputs to inputs clearly lead to different test effectiveness. Also asynchronous feedback loops can be created, including ones having critical races. Hence care must be exercised in forming the feedback.

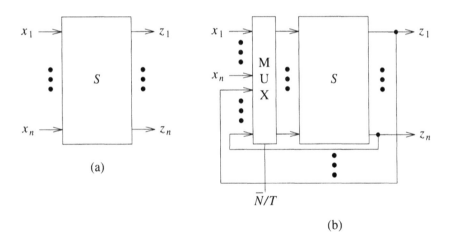

Figure 11.33   The CATS BIST architecture

To test the CUT, the outputs are connected to the inputs and the CUT clocked a predetermined number of times. For large circuits this can be tens of thousands of steps. The actual number is a function of the desired fault coverage and must be determined by fault simulation. The area overhead for this technique is low. Its effectiveness is clearly circuit dependent.

11.4.11 Circular Self-Test Path (CSTP)

The *circular self-test path* (CSTP) BIST architecture is similar to the SST design. It differs primarily in three aspects. First, it employs a self-test cell design shown in Figure 11.34(a), rather than the one shown in Figure 11.31. Second, it is a register-based architecture; i.e., self-test cells are grouped into registers. Third, it employs partial self-test; i.e., not all registers must consist of self-test cells. Some necessary features of this architecture are (1) all inputs and outputs must be associated with boundary scan cells and (2) all storage cells must be initializable to a known state before testing.

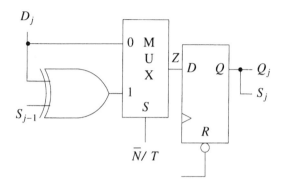

$\overline{N}/T$	Z	Mode
0	D_j	System
1	$D_j \oplus S_{j-1}$	Test

(a)

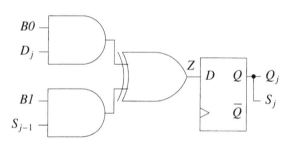

$B0$	$B1$	Z	Mode
0	0	0	Reset
0	1	S_j	Scan
1	0	D_j	System
1	1	$D_j \oplus S_{j-1}$	Test

(b)

Figure 11.34 Storage cell designs for use in circular self-test path BIST architectures

Consider the circuit shown in Figure 11.35. The registers R_1, R_2, R_3, R_7, and R_8 are part of the circular self-test path. Note that the self-test cells form a circular path. If this circular path contains m cells, then it corresponds to a MISR having the characteristic polynomial $(1 + x^m)$. Registers R_4, R_5, and R_6 need not be in the self-test path, nor do they require reset or set lines, since they can be initialized based on the state of the rest of the circuit. That is, once R_1, R_2, and R_3 are initialized, if R_4, R_5, and R_6 are issued two system clocks, then they too will be initialized to a known state.

The same cannot be said of R_3 because of the feedback loop formed by the path $R_3 - C_3 - R_6 - C_6 - R_3$. However, if R_3 has a reset line, then it need not be in the self-test path. Increasing the number of cells in the self-test path increases both the BIST hardware overhead and the fault coverage for a fixed test length.

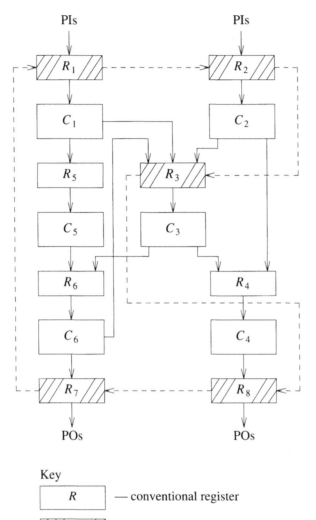

Figure 11.35   A design employing the circular self-test path architecture

The test process requires three phases.

1. *Initialization:*   All registers are placed into a known state.

2. *Testing of CUT:*   The circuit is run in the test mode; registers that are not in the self-test path operate in their normal mode.

3. *Response evaluation.*

During phase 2 the self-test path operates as both a random-pattern generator and response compactor. During phase 3 the circuit is again run in the test mode. But now the sequences of outputs from one or more self-test cells are compared with precomputed fault-free values. This comparison can be done either on-chip or off-chip.

More information on this approach can be found in [Krasniewski and Pilarski 1989].

A similar BIST approach has been reported by Stroud [1988]. Here a signature analyzer is incorporated in the BIST circuitry to simplify identification of faulty circuit operation. Also a self-test cell having a scan mode is used, as shown in Figure 11.34(b).

Performance Analysis

Figure 11.36 shows the general form of a circular self-test path design. The circular path corresponds to an LFSR having the primitive polynomial $p(x) = 1 + x^m$. In this section we will briefly present some theoretical and experimental results about certain performance aspects of this class of design. These results are applicable to many designs where a MISR is used as a PRPG.

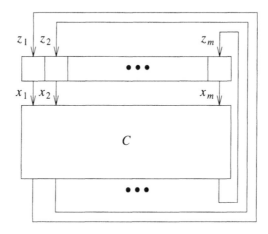

Figure 11.36   General form of a circular self-test path design

Let $z_i(t)$ and $x_i(t)$ be the input and output, respectively, to the ith cell in a circular self-test path. Then assume that the sequences of bits applied to each cell in the path are independent and that each sequence is characterized by a constant (in time) probability of a 1; e.g., for input z_i, $p_i = Prob\{z_i(t) = 1\}$, $t = 1,2,...$.

Theorem 11.4: [Krasniewski and Pilarski 1989]. If there exists an input to the circular path, z_i, such that $0 < p_i < 1$, then, independent of the initial state of the path, $\lim_{t \to \infty} Prob\{x_j(t) = 1\} = 0.5$ for $j = 1, 2, ..., m$. $\qquad \square$

Thus if the response of the circuit to the initial state of the circular path is neither the all-0 nor the all-1 pattern, then some time after initiatization the probability of a 1 at any bit position of the circular path is close to 0.5.

By carrying out extensive simulations, Krasniewski and Pilarski have made the following observations.

Observation 1: The number of clock cycles required for $x_j(t)$ to converge to 0.5 (the equiprobable steady state) is a function of the length of the circular path, and is usually small compared to the number of test patterns normally applied to the circuit.

The *pattern coverage* (also known as the *state coverage*) is denoted by $C_{n,r}$ and is defined as the fraction of all 2^n binary patterns occurring during r clock cycles of the self-testing process at n arbitrary selected outputs of the circular path. These outputs can be the n inputs to a block of logic C.

Observation 2: As the length of the circular path increases, the impact of the value of p_i on the pattern coverage decreases.

Observation 3: The circular path provides a block C with an almost exhaustive test for test lengths a few (four to eight) times longer than an exhaustive test.

Observation 4: When the number of clock cycles associated with a test exceeds the length of the circular path, the impact of the location of the n cells feeding the block C on the pattern coverage is negligible. For long test times, the pattern coverage associated with n cells is almost independent of the length of the path; it is close to the pattern coverage of an n-bit path.

More recently, Sastry and Majumdar [1989] have obtained the following theoretical results dealing with the distribution of pattern coverage in r cycles, as well as the distribution of r, the number of clock cycles required to obtain a pattern coverage of j (out of $N = 2^n$ of the possible inputs to an n-input circuit). In the results presented here we assume that all cells in the circular path are self-test scan cells.

$E[C_{n,r}]$ is the expected value of the random variable $C_{n,r}$. Let Y be a random variable representing the number of distinct patterns applied at time r, where we assume that at time 0 no patterns have yet been applied. Let $P_{j,r}$ denote the probability that $Y = j$, where $j \leq r$ and $j \leq 2^n$.

Definition 11.1: $\left\{ {z \atop k} \right\}$ are Stirling's numbers of the second kind, where

$$x^z = \sum_{k=0}^{z} \left\{ {z \atop k} \right\} \binom{x}{k} k! \text{ and } \binom{x}{y} = x!/y!(x-y)!.$$

Theorem 11.5: Once the circular path has reached the steady (equiprobable) state, then

$$P_{j,r} = \frac{j! \binom{N}{j} \left\{ {r \atop j} \right\}}{N^r} \tag{11.1}$$

The computation of $P_{j,r}$ is complex. When r and N are both large such that $0 < \lambda = r/N < \infty$, we can approximate $P_{j,r}$ by

$$P_{j,r} = \binom{N}{j} (1-e^{-\lambda})^j e^{-\lambda(N-j)} \tag{11.2}$$

For the case where $j = r$, from (11.1) we obtain

$$P_{r,r} = \frac{N!}{(N-r)! N^r} = \prod_{k=0}^{r-1} \left(1 - \frac{k}{N} \right)$$

Let $\mu_{k,N,r}$ be the k-th moment of Y conditioned on N and r; i.e.,

$$\mu_{k,N,r} = E[Y^k|N,r].$$

Then

$$\mu_{k,N,r} = N\left[\mu_{k-1,N,r+1} - \left(1-\frac{1}{N}\right)^r \sum_{m=0}^{k-1} \binom{k-1}{m} \mu_{m,N-1,r}\right].$$

From this result we can determine the expected value and variance of $C_{n,r}$

$$E[Y|N,r] = \mu_{1,N,r} = N\left(1-\left(1-\frac{1}{N}\right)^r\right)$$

hence

$$E[C_{n,r}] = \left(1-\left(1-\frac{1}{2^n}\right)^r\right).$$

This same result was obtained by Kim *et al.* [1988].

The variance on the random variable Y can be expressed as $Var[Y] = E[Y^2] - E^2[Y]$. For Y conditioned on N and r we have

$$Var[Y|N,r] = N\left[\left[\frac{N-1}{N}\right]^r + (N-1)\left[\frac{N-2}{N}\right]^r - N\left[\frac{N-1}{N}\right]^{2r}\right]$$

Thus the variance of Y approaches zero as r approaches infinity, and therefore the accuracy of $E[Y]$ as an estimate of the actual pattern coverage increases as r increases.

Let R be the random variable representing the number of clock cycles required to obtain a pattern coverage of j (out of N). Then the probability of requiring r clock cycles to achieve the pattern coverage of j ($r \geq j$) is given by

$$P\{R = r|N,j\} = \frac{(N-j+1)}{N} P_{j-i,r-1}$$

where $P_{j,r}$ is given in equation (11.1). Then

$$E[R|N,j] = j + \sum_{k=1}^{j-1} \frac{k/N}{(1-k/N)}$$

11.4.12 Built-In Logic-Block Observation (BILBO)

One major problem with several of the designs presented previously is that they deal with an unpartitioned version of a CUT; i.e., all primary inputs are grouped together into one set, all primary outputs into a second set, and all storage cells into a third set. These sets are then associated with PRPGs and MISRs. Since the number of cells in these registers is usually large, it is not feasible to consider exhaustive or pseudoexhaustive test techniques. For example, a chip can easily have over 100 inputs and several hundred storage cells. To circumvent this problem one can attempt to cluster storage cells into groups, commonly called registers. In general these groups correspond to the functional registers found in many designs, such as the program counter, the instruction register, and the accumulator found in a microprocessor design. Some BIST architectures take

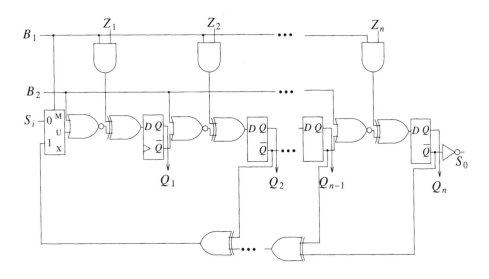

(a)

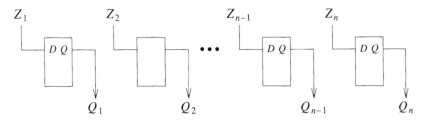

(b)

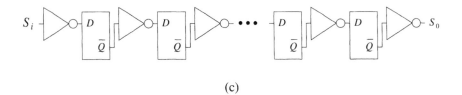

(c)

Figure 11.37   (a) n-bit BILBO register (b) normal mode ($B_1 = B_2 = 1$) (c) shift
register mode ($B_1 = B_2 = 0$) (d) LFSR (test) mode ($B_1 = 1, B_2 = 0$)

advantage of the register aspects of many designs to achieve a more effective test
methodology.

One such architecture employs *built-in logic-block observation* (BILBO) registers
[Koenemann *et al.* 1979, 1980], shown in Figure 11.37(a). In this register design the
inverted output $\overline{Q}$ of a storage cell is connected via a NOR and an XOR gate to the data
input of the next cell. A BILBO register operates in one of four modes, as specified by

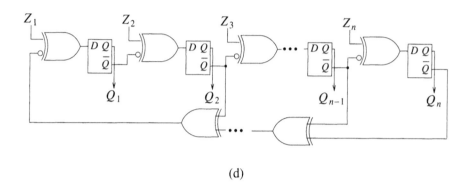

(d)

Figure 11.37   (Continued)

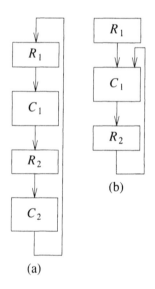

(b)

(a)

Figure 11.38   BIST designs with BILBO registers

the control inputs B_1 and B_2. When $B_1 = B_2 = 1$, the BILBO register operates in its normal parallel load mode (see Figure 11.37(b)). When $B_1 = B_2 = 0$, it operates as a shift register with scan input S_i (see Figure 11.37(c)). Note that data is complemented as it enters the scan register. When $B_1 = 0$ and $B_2 = 1$, all storage cells are reset. When $B_1 = 1$ and $B_2 = 0$, the BILBO register is configured as an LSFR (see Figure 11.37(d)), or more accurately the register operates as a MISR. If the Z_is are the outputs of a CUT, then the register compresses the response to form a signature. If the inputs $Z_1, Z_2, ..., Z_n$ are held at a constant value of 0, and the initial value of the register is not all 0s, then the LFSR operates as a pseudorandom-pattern generator.

A simple form of a BILBO BIST architecture consists of partitioning a circuit into a set of registers and blocks of combinational logic, where the normal registers are replaced by BILBO registers. In addition, the inputs to a block of logic C are driven by a BILBO register R_i, and the outputs of C drive another BILBO register R_j.

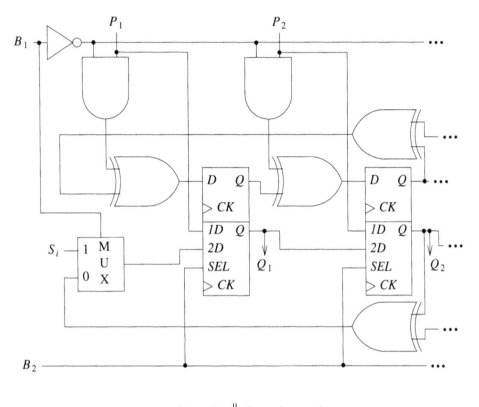

B_1	B_2	Operation mode
—	0	Normal
1	1	Scan
0	1	PRPG/MISR

Figure 11.39   Concurrent BILBO register

Consider the circuit shown in Figure 11.38(a), where the registers are all BILBOs. To test C_1, first R_1 and R_2 are seeded, and then R_1 is put into the PRPG mode and R_2 into the MISR mode. Assume the inputs of R_1 are held at the value 0. The circuit is then run in this mode for N clock cycles. If the number of inputs of C_1 is not too large, C_1 can even be tested exhaustively, except for the all-zero pattern. At the end of this test process, called a *test session*, the contents of R_2 can be scanned out and the signature checked. Similarly C_2 can be tested by configuring R_1 to be a MISR and R_2 to be a PRPG. Thus the circuit is tested in two test sessions.

Figure 11.38(b) shows a different type of circuit configuration, one having a self-loop around the R_2 BILBO register. This design does not conform to a normal BILBO architecture. To test C_1, R_1 must be in the PRPG mode. Also, R_2 should be in both the MISR mode and the PRPG mode. This is not possible for the design shown in Figure 11.37(a). What can be done is to place R_2 in the MISR mode. Now its outputs are essentially random vectors that can be used as test data to C_1. One feature of this scheme is that errors in the MISR produce "erroneous" test patterns that are applied to C_1, which in turn tend to produce more errors in R_2. The bad aspect of this approach is that there may exist faults that are never detected. This could occur, for example, if the input data to C_1 never propagate the effect of a fault to the output of C_1.

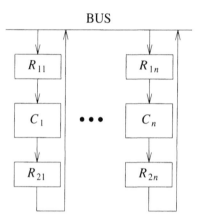

Figure 11.40   Bus-oriented BILBO architecture

This situation can be rectified by using a *concurrent built-in logic-block observation* (CBILBO) register [Wang and McCluskey 1986c]. This register, shown in Figure 11.39, operates simultaneously as a MISR and a PRPG. The top row of D flip-flops and associated logic form a MISR. The bottom row of dual-port flip-flops and associated logic form a register that can operate either in the normal parallel-load mode or the scan mode or as a PRPG.

Recall that when a BILBO register is in the PRPG mode, its inputs need to be held at some constant value. This can be achieved in several ways. Often the BILBO test methodology is applied to a modular and bus-oriented system in which functional modules, such as ALUs, RAMs, and ROMs, are connected via a register to a bus (see Figure 11.40). By disabling all bus drivers and using pull-up or pull-down circuitry, the register inputs can be held in a constant state.

However, some architectures have a pipeline structure as in Figure 11.41. To deactivate the inputs to a BILBO register during its PRPG mode, a modified BILBO register design having three control signals (B_1, B_2, B_3) can be used, as shown in Figure 11.42. There are now eight possible control states, and one can be used to specify the MISR mode and another the PRPG mode.

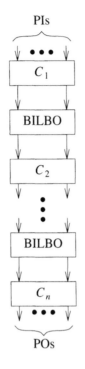

Figure 11.41   Pipeline-oriented BILBO architecture

One aspect that differentiates the BILBO architecture from the previously discussed BIST architectures is the partitioning of storage cells to form registers and the partitioning of the combinational logic into blocks of logic. Once such a partitioning is achieved, then many of the techniques discussed previously dealing with pseudoexhaustive testing can be employed. The BILBO registers can be replaced by other types of registers, such as constant-weight counters or more complex forms of LFSRs.

The *built-in digital-circuit observer* (BIDCO) is an extension of the BILBO design approach to board-level testing [Fasang 1980]. Related work can be found in [Beucler and Manner 1984].

Several techniques, such as RTD, CSTP, and BILBO, use a MISR as a PRPG. Previously we have presented some theoretical results pertaining to the effectiveness of using a circular self-test path as a pattern generator. We next present some additional results related to the use of a MISR for test-pattern generation.

Analysis of Patterns Generated by a MISR

Background

In this section we will analyze some of the properties of patterns generated by a MISR. The results of this analysis are useful when a MISR is used as a TPG. Consider an n-bit MISR whose characteristic polynomial is primitive. Then if the input is held constant, the MISR behaves as a maximum-length LFSR and generates $2^n - 1$ patterns. However,

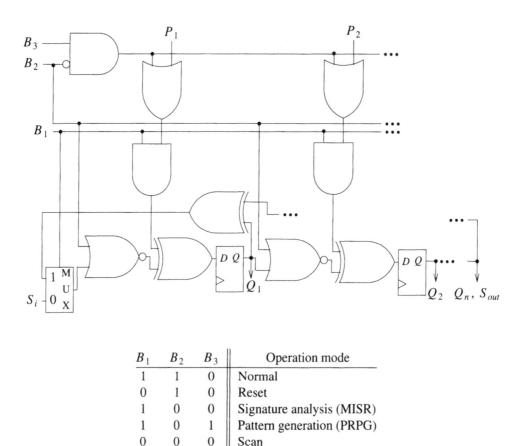

B_1	B_2	B_3	Operation mode
1	1	0	Normal
0	1	0	Reset
1	0	0	Signature analysis (MISR)
1	0	1	Pattern generation (PRPG)
0	0	0	Scan

Figure 11.42   BILBO register with input disable control

even if a primitive polynomial is not employed, all states can still be generated by changing the inputs to the MISR.

Figure 11.43 shows a 3-bit MISR and the state transitions that occur under two different inputs.

The predecessor state of state 101 under input pattern 010 is 110; its successor state is 000. Because of the linear nature of LFSRs, the following results can be shown [Golomb 1982].

Property 1:   The number of successor states of a given state of a MISR is equal to the number of unique input patterns of the MISR.

Property 2:   The number of predecessor states of a given state of a MISR is equal to the number of unique input patterns of the MISR.

Property 3:   If the number of input patterns of an n-bit MISR is greater than 1, the MISR can possibly generate all 2^n patterns.

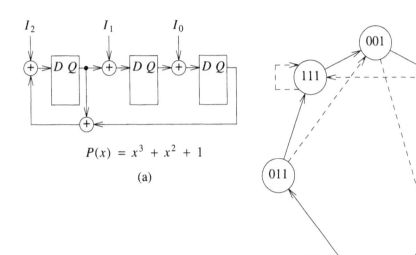

$$P(x) = x^3 + x^2 + 1$$

(a)

Input pattern

$\longrightarrow$ $I_2 I_1 I_0$ = 010

$- - \!\!\! \to$ $I_2 I_1 I_0$ = 100

(b)

Figure 11.43 (a) A 3-bit MISR (b) State transition diagram for two different inputs

Completeness of State Space

The behavior of a MISR can be modeled by a *Markov chain*. Let p_{ij} be the probability of a state transition from state s_i to state s_j. Then the state-transition diagram of an n-bit MISR can be described by a $2^n \times 2^n$ matrix, called the *transition-probability matrix*, and denoted by $P = [p_{ij}]$, $i, j = 1, 2, ..., 2^n$.

For the example shown in Figure 11.43, let the probability of occurrence of input pattern 010 be 1/3 and that of 100 be 2/3. The transition-probability matrix P is shown in Figure 11.44.

Let π_j^k be the probability that a MISR is in state s_j after the application of k input patterns. The state-probability vector is denoted by $\pi(k) = (\pi_1^k, \pi_2^k, ..., \pi_N^k)$ where $N = 2^n$. The $\pi(k)$ is related to the initial-probability vector $\pi(0)$ and P by the equation $\pi(k) = \pi(0) P^k$.

$$P = \begin{bmatrix} 0 & 0 & 1/3 & 0 & 2/3 & 0 & 0 & 0 \\ 2/3 & 0 & 0 & 0 & 0 & 0 & 1/3 & 0 \\ 0 & 0 & 0 & 1/3 & 0 & 2/3 & 0 & 0 \\ 0 & 2/3 & 0 & 0 & 0 & 0 & 0 & 1/3 \\ 0 & 0 & 2/3 & 0 & 1/3 & 0 & 0 & 0 \\ 1/3 & 0 & 0 & 0 & 0 & 0 & 2/3 & 0 \\ 0 & 0 & 0 & 2/3 & 0 & 1/3 & 0 & 0 \\ 0 & 1/3 & 0 & 0 & 0 & 0 & 0 & 2/3 \end{bmatrix}$$

Figure 11.44

For example, several of the state-probability vectors for the circuit of Figure 11.43, given the initial state (111) and the P matrix given above are listed below:

$\pi(0) = (00000001)$
$\pi(1) = \pi(0)\,P = (0\ 1/3\ 00000\ 2/3)$
$\pi(2) = \pi(1)\,P = \pi(0)\,P^2 = (0.22\ 0.22\ 0\ 0\ 0\ 0\ 0.11\ 0.44)$
$\vdots$
$\pi(7) = (0.1235\ 0.1235\ 0.1235\ 0.1235\ 0.1235\ 0.1235\ 0.1235\ 0.1358)$
$\pi(8) = (0.1235\ 0.1235\ 0.1235\ 0.1235\ 0.1235\ 0.1235\ 0.1235\ 0.1317)$

Note that as k increases, each element of $\pi(k)$ approaches 1/8. We will show that in general as k increases each state becomes equally probable.

A Markov chain is said to be *regular* if every state can be reached from every other state. A transition-probability matrix P corresponding to a Markov chain is *doubly stochastic* if each row and column sum is 1.

A Markov process having these properties satisfies the following theorem [Taylor and Karlin 1984]:

Theorem 11.6: If a Markov chain is regular and the transition-probability matrix P is doubly stochastic, then $\lim_{k\to\infty} \pi(k) = (1/N, 1/N, ..., 1/N)$, where N is the number of states.

$\square$

From properties 1, 2, and 3, the state diagram of a MISR is regular if the number of input patterns is greater than 1. Also, the transition-probability matrix P associated with a MISR is doubly stochastic. Based on the law of large numbers [Taylor and Kalin 1984] and Theorem 11.6, we have the following result:

Theorem 11.7: For a given n-bit MISR the probability of appearance of each pattern becomes $1/2^n$ after a sufficient number of clock cycles, provided that the number of different input patterns to the MISR is greater than 1. $\square$

Let x_i be the output of one the i-th bit of the MISR, and let $p_i = Prob\{x_i = 1\}$. Then from Theorem 11.7 it follows that $p_i = 0.5$.

Effectiveness as a PRPG

In this section we will show that a MISR can be effectively used as a PRPG. Let $\{v_1, v_2, ..., v_N\}$ be the set of all binary n-tuples. Each v_i represents a state of an n-bit MISR. Assume a MISR operates for m clock cycles and generates a sequence of patterns $S = S_1, S_2, ..., S_m$. Let D_m be the number of distinct patterns in this sequence. Let

$$
x_{i,m} = \begin{cases} 1 & \text{if } v_i \text{ occurs in the sequence } S \\ 0 & \text{otherwise.} \end{cases}
$$

Then $D_m = x_{1,m} + x_{2,m} + ... + x_{N,m}$, and the expected value of D_m is

$$
E[D_m] = E\left[\sum_{i=1}^{N} x_{i,m}\right] = \sum_{i=1}^{N} E(x_{i,m}).
$$

Since $Prob(x_{i,m}) = 1 - (1 - \frac{1}{2^n})^m$, then $E[D_m] = N(1 - (1 - \frac{1}{2^n})^m)$.

When $m \ll 2^n$, then

$$
E[D_m] = 2^n(1 - (1 - \frac{1}{2^n})^m) \cong 2^n(1 - (1 - \frac{m}{2^n})) = m.
$$

Thus for a small number of patterns, the chances of generating repeated patterns is small.

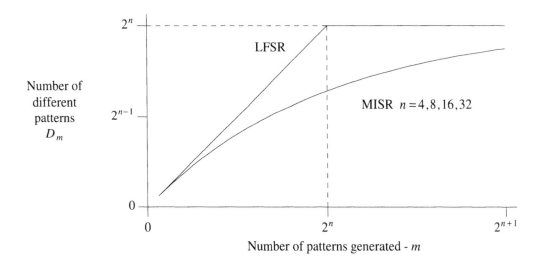

Figure 11.45

Some results relating the value of D_m to m for various values of n are shown in Figure 11.45. Since the results for $n = 4$, 8, 16, and 32 are similar, only one curve is shown. It is seen that for $m \ll 2^n$, the MISR acts as a PRPG; but for larger values of m it acts more as a random-pattern generator. Kim *et al.* [1988] have also shown that the properties pertaining to the randomness of the patterns generated by a MISR are true even if the inputs are nonequiprobable.

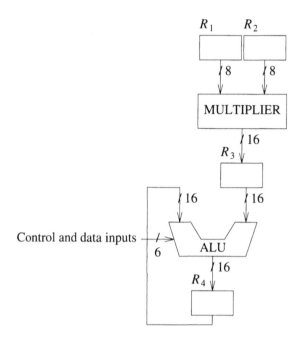

Figure 11.46   Portion of the TMS32010 signal-processing chip

11.4.12.1 Case Study

Figure 11.46 shows a portion of the TMS32010 signal-processing chip. Various BIST architectures were applied to this chip, and the results derived by Kim *et al.* [1988] are presented here. The 8×8-bit multiplier and 16-bit ALU were modeled using 880 and 354 gates respectively. The numbers of single stuck-fault classes in these circuits are 1255 and 574 respectively.

Four BIST designs are shown in Figure 11.47.

Design 1:   Figure 11.47(a) shows a BILBO-type BIST architecture that requires two test sessions, one to test the multiplier, at which time the BILBO register operates as a MISR; the second to test the ALU, at which time the BILBO register operates as a PRPG. This design requires the addition of a 16-bit and a 6-bit PRPG to drive the leftmost 16-bits of the ALU, and the five control inputs and carry-in of the ALU.

Design 2:   Figure 11.47(b) shows a BIST design where R_3 is unmodified and again a 22-bit PRPG is added to the circuit. The response data produced by the multiplier are used as test data to the ALU. Now only one test session is required.

Design 3:   Figure 11.47(c) shows a BIST design where R_3 is extended to be a 38-bit MISR. The inputs to the first 22-bits of this MISR are held constant at 0. Again only one test session is required.

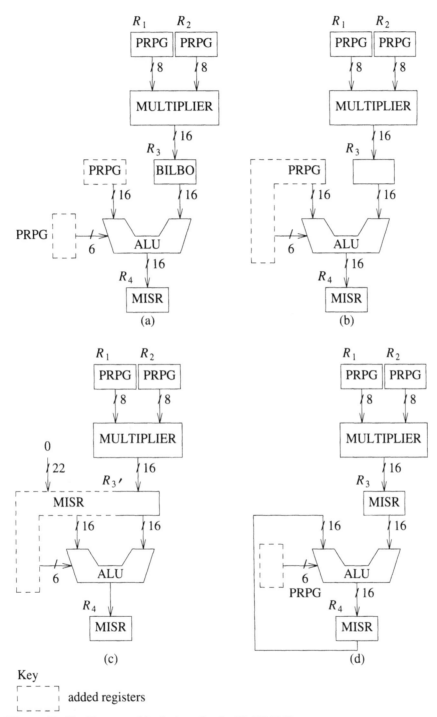

Figure 11.47   Four testable designs for the TMS32010
(a) BILBO solution (b) Solution where R_3 remains
unmodified (c) Solution where ALU is tested by separate
MISRs (d) Solution where ALU is tested by sharing a
MISR

Design 4: Figure 11.47(d) shows still another BIST design. Here R_3 and R_4 are MISRs and are used to drive the data inputs to the ALU. Of the four designs, this one uses the least amount of test hardware. Again only one test session is required.

Simulation results based on 20 different runs are shown in Figure 11.48, where each run corresponds to a different selection of seed values in the LFSRs.

Number of test patterns	Design			
	1	2	3	4
Average	2,177	$\geq 3,000$	1,457	1378
Minimum	830	—	634	721
Maximum	3,619	—	2,531	2,121
Fault coverage (%)	100	64.5	100	100

Figure 11.48   Fault-simulation results for various BIST designs

For designs 1, 3, and 4, simulation terminated as soon as 100 percent fault coverage was achieved. For design 2, fault coverage saturated at 64.5 percent. The number of test patterns for BILBO design 1 is the sum of the test patterns for each of the two test sessions. Note that designs 3 and 4 lead to a substantial reduction in test length. For design 4, the input patterns to the ALU are not necessarily uniformly distributed since they are dependent on the current state of the MISR R_3; i.e., the probability-transition matrix P is not doubly stochastic.

In summary, it appears that a MISR can be used effectively as a source of test patterns. Also, it is usually not necessary to test circuits exhaustively or even to apply a verification test.

11.4.13 Summary

BIST represents an active research area, and new BIST architectures are continually being developed. In this section we have shown how some of these architectures have evolved. Initially, applying techniques such as CSBL and BEST, sequential circuits were tested using pseudorandom test patterns, and results compressed using signature analyzers. To obtain better access to internal circuitry, scan paths can be used so that the kernel being tested is combinational. The next step was to embed the PRPG and MISR into the CUT and employ boundary scan to test isolated logic. To reduce test time, approaches such as SST and CSTP were developed where a scan path does not have to be serially loaded before the application of each test pattern. Here a more complex pseudo-scan-path design is used. Finally, a register-based BIST architecture (BILBO) was presented. Note that for some BIST approaches, such as CSBL, BEST, and CSSP, the circuit under test operates as a sequential circuit during the test process; hence some aspects of dynamic testing are present. For those approaches that use a scan operation between test patterns, such as RTS, LOCST, and STUMPS, testing is more of a static operation.

One problem associated with several BIST designs such as RTD, SST, and CSTP, is that during self-test, the D inputs to the storage cells are not tested. Hence some additional tests need to be run while in the normal mode. This can be done by loading the scan path with a test pattern, applying a normal clock, and then checking the contents of the scan path. Usually a few patterns are enough to test the logic not tested by the self-test process.

Figure 11.49 shows some of the main attributes of the architectures presented.

Architecture (section)	Centralized (C) or distributed (D)	Separated (S) or embedded (E)	Combinational (C) or sequential (S) kernels	Boundary scan	Chip (C) or board (B) level	
CSBL (11.4.1)	C	S	C or S	N	B	
BEST (11.4.2)	C	S or E	C or S	Y	C	
RTS (11.4.3)	D	S	C	N	C	(1)
LOCST (11.4.4)	C	S	C	Y	C	
STUMPS (11.4.5)	C	S	C	N	B or C	
CBIST (11.4.6)	C or D	S	C	Optional	C	(2)
CEBS (11.4.7)	C	E	C	Y	C	
RTD (11.4.8)	D	E	C	Y	C	
SST (11.4.9)	D	E	C	N	C	
CATS (11.4.10)	NA	NA	S	N	C	(3)
CSTP (11.4.11)	D	E	C or S	Y	C	
BILBO (11.4.12)	D	E	C	N	C	(4)

(1) A non-BIST architecture
(2) A concurrent BIST architecture
(3) In a minimal configuration there is no BIST hardware except for MUXs
(4) Can be extended to sequential kernels

Figure 11.49   Summary of BIST architecture

11.5 Some Advanced BIST Concepts

In this section we will consider three advanced concepts related to some BIST architectures: (1) scheduling of test sessions, (2) control of multiple test sessions when a distributed and embedded BIST architecture is employed, and (3) partial BIST designs. For simplicity we will assume a BILBO architecture is employed. First, some basic concepts will be presented.

A register R_i is said to be a *driver* of a block of logic C if some outputs of R_i are inputs to C. A register R_j is said to be a *receiver* of C if some outputs of C are inputs to R_j. R_i is said to be *adjacent* to R_j if there exists a block of logic C such that R_i is a driver of C, and R_j is a receiver of C. If R_i is both a receiver and a driver of C, then it is *self-adjacent*.

Example 11.5: Consider the portion of the circuit shown in Figure 11.50(a). R_1, R_3, and R_4 are drivers of C, R_2 and R_3 are receivers of C, and R_3 is a self-adjacent register. Also R_1 and R_2 are adjacent, but R_1 and R_4 are not adjacent.

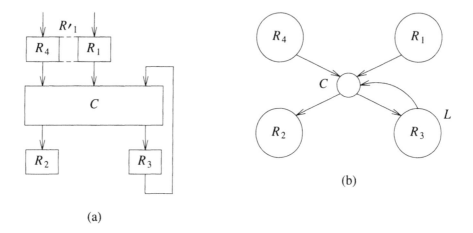

(a)   (b)

Figure 11.50   (a) Part of a circuit (b) Its register adjacency graph

To test C, R_1 and R_4 should be placed in the PRPG mode, and R_2 in the MISR mode. R_3 should be in both the PRPG mode and the MISR mode. This can be accomplished by using a CBILBO register, such as the one shown in Figure 11.39. To reduce hardware overhead, we will operate self-adjacent registers in the MISR mode. Note that storage cells and functional registers can be clustered into new registers in order to form a BILBO register. For example, it may be feasible to cluster R_1 and R_4 into a single LFSR R'_1 during the test mode. However, it would not be beneficial to cluster R_2 and R_4 together since they operate in different ways during the test mode.   □

11.5.1 Test Schedules

A *test session* is defined as an assignment of test modes to BILBO registers to test one or more blocks of logic. A block of logic is considered to be tested if its driver registers are in the PRPG mode, and its receivers and self-adjacent registers are in the MISR mode.

A circuit can be modeled by a bipartite graph $G = (N_A, N_B, E)$, referred to as a *register adjacency graph* (RAG), where N_A is a set of type A nodes, each of which represents a register, N_B is a set of type B nodes, each of which represents a block of combinational logic, and E is a set of edges between type A and B nodes. A directed edge exists from a type A node (register R_i) to a type B node (block of logic C_k) if R_i is a driver of C_k, and a directed edge exists from a type B node C_k to a type A node R_j if R_j is a receiver of C_k. In addition, a type A node corresponding to a self-adjacent register is flagged with an L. Figure 11.50(b) shows the RAG associated with Figure 11.50(a).

The *test-scheduling* problem is defined as follows: determine the minimal number of test sessions required to test all blocks of combinational logic. This problem is related to determining the *chromatic number* of a graph G^*, which is the minimal number of colors that can be assigned to the nodes of G^* such that no edge connects two nodes of the same color. Methods for determining the minimal number of test sessions have been presented in [Craig *et al.* 1988].

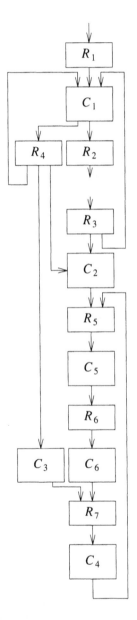

Figure 11.51   Example circuit

Example 11.6:   Figure 11.51 shows a circuit to be made testable using the BILBO methodology. Figure 11.52 shows its corresponding RAG. The circuit can be completely tested with just three test sessions, as shown in Figure 11.53. Figure 11.53(a) shows a test session where C_1 and C_5 are tested simultaneously.

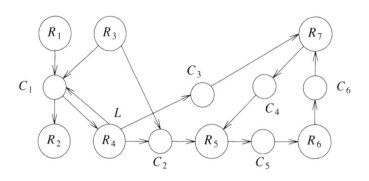

Figure 11.52   Register adjacency graph for the circuit in Figure 11.51

The test-scheduling problem becomes more complex when the test time for each block of logic is considered. Now the objective is to minimize the test time for the entire chip. A test session can now be more complex; for example, while C_i is being tested, C_j is also tested, and when the testing of C_j is completed, C_k is then tested. The analysis of these time-dependent test schedules is considered in [Jone *et al.* 1989].

11.5.2 Control of BILBO Registers

When multiple test sessions exist, the efficient control of the BILBO registers becomes an important issue. Each BILBO register requires three bits of control information to specify in which of five possible modes to operate, namely, normal, scan, reset, PRPG, or MISR. It would be desirable if these control bits were common to all registers and could then be broadcast using three common control signals. However, since in each test session the role of a register can be either a MISR or a PRPG, it is only possible to broadcast two signals. These two signals can be used to specify one of four modes, namely, normal, scan, reset, or test. Each register can then have one unique control signal to indicate if it should operate as a PRPG or a MISR while in the test mode. The distribution and control of these individual signals could require a significant area overhead. An alternative approach is to distribute the control. This can be done by associating with each BILBO register R_i a test storage cell T_i. If $T_i = 1$, then R_i operates in the MISR mode when in the test mode; otherwise it operates in the PRPG mode. The test storage cells can be organized into a separate scan path or made part of the same scan path that incorporates the BILBO registers. Before executing a test session the test cells can be initialized so that each BILBO register operates in its appropriate mode. Figure 11.54 shows a BILBO register design that incorporates a test mode storage cell [Hudson and Peterson 1987]. A basic cell in the BILBO register is shown in Figure 11.54(a). The first cell of the BILBO register and the control lines to all the cells are driven by the logic shown in Figure 11.54(b). Figure 11.54(c) shows the truth table for the control circuitry C_1. The test mode storage cell T_i is in the same scan path as the BILBO registers. The output of T_i is T2, which is an input to C_1. The signal $S*$ drives the input S_0 to the first cell in the BILBO register corresponding to $j = 1$. S_i is the normal scan-in input used to scan data into the BILBO register; *FB* is the feedback signal from the linear feedback network. Setting $T0 = 1$ disables the clock to T_i, allowing T_i to hold its state during the test mode.

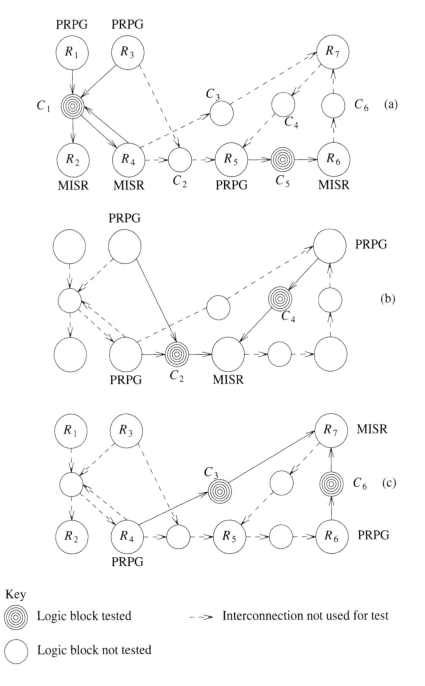

Figure 11.53   Three test sessions for the circuit in Figure 11.51

To control a test session completely, a controller is required to keep count of the number of test-patterns processed during each test session and the number of shift operations required to load and read data from the various scan paths. The design of these test

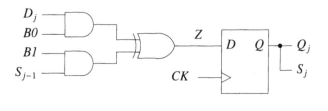

(a)

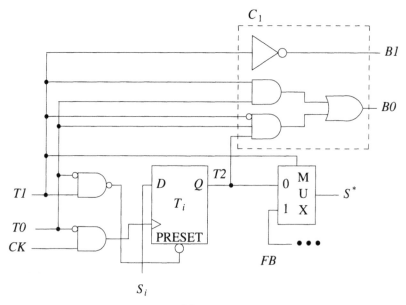

(b)

$T0$	$T1$	$T2$	Mode	$B0$	$B1$	S^*
1	0	0	PRPG	0	1	FB
1	0	1	MISR	1	1	FB
0	0	Q	SHIFT	0	1	Q
1	1	x	LATCH	1	0	x
0	1	1	RESET	0	0	x

(c)

Figure 11.54   Control of a BILBO register

controllers is discussed in [Kalinowski *et al.* 1986] and [Breuer *et al.* 1988]. Most control sequences for the various BIST architectures are similar and consist of the following major steps:

1.   Inhibit system clocks and enter the test mode.

2. Initialize control registers with data specific to a test session, such as seed data, scan-path lengths, identification of desired scan paths, and number of test patterns to be processed.

3. Seed the LFSRs and scan paths.

4. Initiate the test process.

5. Process the final signature to determine if an error has been detected.

11.5.3 Partial-Intrusion BIST

Partial-intrusion BIST refers to the concept where only a subset of the registers and/or storage cells in a circuit are made part of an LFSR or shift-register path. The motivation for partial-intrusion BIST is to reduce circuit complexity and enhance performance. Partial-intrusion BIST can be implemented in several ways. We will focus on those techniques where the kernel is combinational logic.

A partial-intrusion BIST design can be achieved using *I*-paths [Abadir and Breuer 1985], which were discussed in Chapter 9.

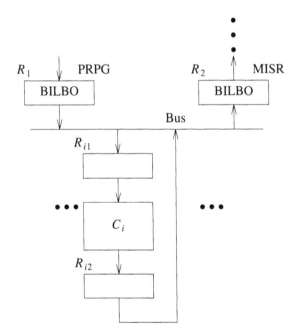

Figure 11.55   A design employing partial BIST

Example 11.7: A portion of a complex design is shown in Figure 11.55. Using the classical BILBO methodology to test C_i, R_{i1} and R_{i2} must be BILBO registers; R_{i1} must be put into the PRPG mode and R_{i2} into the MISR mode. Using a partial-intrusion BIST/BILBO methodology, it is possible to keep R_{i1} and R_{i2} as normal registers and only convert R_1 and R_2 into BILBO registers. To test C_i, R_1 is configured as a PRPG and R_2 as a MISR. Figure 11.56 illustrates a *test plan* which specifies how C_i is tested by a single test pattern.

Time	Operation	
t	OP_1:	R_1 generates a new test pattern.
$t + 1$	OP_2:	R_1 drives bus and R_{i1} loads from bus.
$t + 2$	OP_3:	Test data propagate through C_i and response is loaded into R_{i2}.
$t + 3$	OP_4:	R_{i2} drives bus and R_2 loads from bus (while in the MISR mode).

Figure 11.56 Initial test plan for C_i

This plan can be repeated until C_i is completely tested. The plan can be optimized so that C_i is tested by a new test pattern every other clock cycle rather than every four clock cycles. To achieve this new plan the design will be considered to be a pipeline. Since both operations OP_2 and OP_4 use the bus, these operations cannot occur simultaneously. To process a new test pattern every other clock cycle, the original test plan must be modified and a hold operation inserted between OP_3 and OP_4. The new test plan is shown in Figure 11.57.

Time	Operation	
t	OP_1:	R_1 generates a new test pattern.
$t + 1$	OP_2:	R_1 drives bus and R_{i1} loads from bus.
$t + 2$	OP_3:	Test data propagate through C_i and response is loaded into R_{i2}.
$t + 3$	OP_4:	R_{i2} holds its state.
$t + 4$	OP_5:	R_{i2} drives bus and R_2 loads from bus.

Figure 11.57 Modified test plan for C_i

The resulting pipelined test plan, shown in Figure 11.58, consists of a sequence of parallel operations, where now OP_2 and OP_5 cannot occur simultaneously.

It is seen that two clock cycles after the initiation of test pattern $(j-1)$ occurs, the j test pattern is initiated. The test process consists of two phases that are repeatedly executed. During phase 1 operations OP_4 and OP_2 occur simultaneously, while during phase 2

Time	Operations	Phase
t	OP_1 ⎫	
$t+1$	OP_2 ⎬ Test pattern $j-1$	
$t+2$	OP_3 ⎭ OP_1 ⎫	
$t+3$	OP_4 OP_2 ⎬ Test pattern j	1
$t+4$	OP_5 OP_3 ⎭ OP_1 ⎫	2
$t+5$	OP_4 OP_2 ⎬ Test pattern $j+1$	1
$t+6$	OP_5 OP_3 • • •	2
$t+7$	OP_4 ⎫	
$t+8$	OP_5 ⎭	

Figure 11.58 Pipelined test plan

operations OP_5, OP_3, and OP_1 are executed. It is assumed that when R_1 and R_2 are not executed in the MISR or PRPG mode, they remain in the hold mode. This is an optimal plan in that it is not possible to process a new test pattern any faster than every other clock cycle using the given hardware configuration. This is because both the driver and receiver must use the bus.

This design approach leads to a reduction in test hardware overhead. Also the delay between R_{i1} and R_{i2} is not increased, which it would be if R_{i1} and R_{i2} were replaced by BILBO registers. This is important if a critical timing path exists between registers R_{i1} and R_{i2}. In addition, R_1 and R_2 can be used to test kernels other than C_i. □

Algorithms for constructing these forms of optimal test plans that are executed with a minimal delay between consecutive test patterns are discussed in [Abadir and Breuer 1985, 1986]. Using this partial-intrusion BIST approach, the blocks of logic in a circuit can be tested sequentially. This is an example of a centralized and embedded version of the original distributed BILBO concept.

Another partial-intrusion BIST approach has been suggested by Krasniewski and Albicki [1985a, 1985b]; an example is shown in Figure 11.59. Here no I-paths are used, and R_2 is a normal latch register. Assume that a BILBO-type test method is to be used, that all registers are of width 16, that R_1 is configured as a PRPG and R_3 as a MISR, and that C_1 is to be tested exhaustively. Then C_2 is said to be tested *functionally exhaustively*. That is, even though all 2^{16} patterns are probably not generated at the output of R_2, all possible patterns that can occur in this circuit under normal operation are generated. Thus some input patterns corresponding to don't-care minterms associated with C_2 may not be applied to C_2. This approach can be extended to more complex architectures having both feedforward and feedback interconnections.

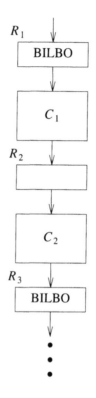

Figure 11.59   A partial BILBO pipeline architecture

11.6 Design for Self-Test at Board Level

Most PCBs consist of commercial chips that do not support IEEE 1149.1. Figure 11.60 shows part of a typical microprocessor-based design. Here the normal buffers, latch registers, and transceivers have been replaced by functionally equivalent ICs that have added capabilities including compatibility with IEEE 1149.1, a pseudorandom-pattern-generation mode, and a multiple-input signature-register mode. This design can now be configured, under control of the test bus, to be put in a BIST mode, where the microprocessor, glue logic, and memory are tested. For example, to test the glue logic, the input buffers can be configured to be in the PRPG mode, while the output buffer and latch register are placed in the MISR mode. Once the test has been run, the test registers can be placed into their scan mode and the resulting signature verified to see if it is correct. A family of circuitry is emerging that supports IEEE 1149.1 and includes test-related functions and can be easily incorporated into board designs to enhance their testability [TI 1989].

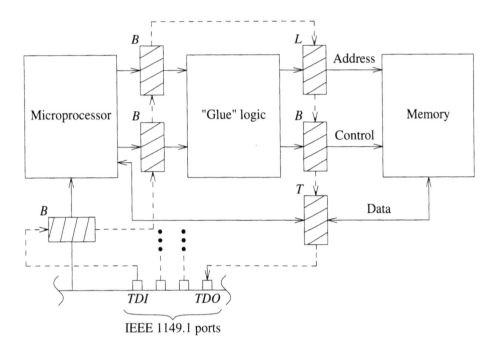

B — buffers   L — latch register   T — transceiver

Figure 11.60   PCB with IEEE 1149.1 test chips

REFERENCES

[Abadir and Breuer 1985] M. Abadir and M. A. Breuer, "A Knowledge Based System for Designing Testable VLSI Chips," *IEEE Design & Test of Computers*, Vol. 2, No. 4, pp. 55-68, August, 1985.

[Abadir and Breuer 1986] M. Abadir and M. A. Breuer, "Test Schedules for VLSI Circuits," *IEEE Trans. on Computers*, Vol. C-35, No. 4, pp. 361-367, April, 1986.

[Akers 1985] S. B. Akers, "On the Use of Linear Sums in Exhaustive Testing," *Digest of Papers 15th Annual Intn'l. Fault-Tolerant Computing Symp.*, pp. 148-153, June, 1985.

[Archambeau 1985] E. C. Archambeau, "Network Segmentation for Pseudoexhaustive Testing," Center for Reliable Computing Technical Report No. 85-10, Stanford University, 1985.

[Archambeau and McCluskey 1984] E. C. Archambeau and E. J. McCluskey, "Fault Coverage of Pseudoexhaustive Testing," *Digest of Papers 14th Annual Intn'l. Fault-Tolerant Computing Symp.*, pp. 141-145, June, 1984.

[Bardell *et al.* 1987] P. H. Bardell, W. H. McAnney, and J. Savir, *Built-in Test for VLSI: Pseudorandom Techniques*, John Wiley and Sons, New York, 1987.

[Bardell and McAnney 1982] P. H. Bardell and W. H. McAnney, "Self-Testing of Multichip Logic Modules," *Digest of Papers 1982 Intn'l. Test Conf.*, pp. 200-204, November, 1982.

[Bardell and McAnney 1984] P. H. Bardell and W. H. McAnney, "Parallel Pseudorandom Sequences for Built-In Test," *Proc. Intn'l. Test Conf.*, pp. 302-308, October, 1984.

[Bardell and McAnney 1985] P. H. Bardell and W. H. McAnney, "Simultaneous Self-Testing System," U.S. Patent No. 4,513,418, April 23, 1985.

[Bardell and McAnney 1986] P. H. Bardell and W. H. McAnney, "Pseudorandom Arrays for Built-In Tests," *IEEE Trans. on Computers*, Vol. C-35, No. 7, pp. 653-658, July, 1986.

[Bardell and Spencer 1984] P. H. Bardell and T. H. Spencer, "A Class of Shift-Register Sequence Generators: Hurd Generators Applied to Built-In Test," IBM Corp. Technical Report No. TR00.3300, IBM, Poughkeepsie, NY, September, 1984.

[Barzilai *et al.* 1981] Z. Barzilai, J. Savir, G. Markowsky, and M. G. Smith, "The Weighted Syndrome Sums Approach to VLSI Testing," *IEEE Trans. on Computers*, Vol. C-30, No. 12, pp. 996-1000, December, 1981.

[Barzilai *et al.* 1983] Z. Barzilai, D. Coppersmith, and A. L. Rosenberg, "Exhaustive Generation of Bit Patterns with Applications to VLSI Self-Testing," *IEEE Trans. on Computers*, Vol. C-32, No. 2, pp. 190-194, February, 1983.

[Beenker 1985] F. P. M. Beenker, "Systematic and Structured Methods for Digital Board Testing," *Proc. Intn'l. Test Conf.*, pp. 380-385, November, 1985.

[Benowitz *et al.* 1975] N. Benowitz, D. F. Calhoun, G. E. Alderson, J. E. Bauer, and C. T. Joeckel, "An Advanced Fault Isolation System for Digital Logic," *IEEE Trans. on Computers*, Vol. C-24, No. 5, pp. 489-497, May, 1975.

[Beucler and Manner 1984] F. P. Beucler and M. J. Manner, "HILDO: The Highly Integrated Logic Device Observer," *VLSI Design*, Vol. 5, No. 6, pp. 88-96, June, 1984.

[Boney and Rupp 1979] J. Boney and E. Rupp, "Let Your Next Microcomputer Check Itself and Cut Down Your Testing Overhead," *Electronic Design*, pp. 101-106, 1979.

[Breuer *et al.* 1988] M. A. Breuer, R. Gupta, and J. C. Lien, "Concurrent Control of Multiple BIT Structures," *Proc. Intn'l. Test Conf.*, pp. 431-442, September, 1988.

[Buehler and Sievers 1982] M. G. Buehler and M. W. Sievers, "Off-Line, Built-In Test Techniques for VLSI Circuits," *Computer*, Vol. 18 pp. 69-82, 1982.

[Burkness 1987] D. C. Burkness, "Self Diagnostic Cyclic Analysis Testing System (CATS) for LSI/VLSI," U.S. Patent No. 4,680,761, July 14, 1987.

[Butt and El-Ziq 1984] H. H. Butt and Y. M. El-Ziq, "Impact on Mixed-Mode Self-Test of Life Cycle Cost of VLSI Based Designs," *Proc. Intn'l. Test Conf.*, pp. 338-347, October, 1984

[Chandra *et al.* 1983] A. K. Chandra, L. T. Kou, G. Markowsky, and S. Zaks, "On Sets of Boolean n-Vectors with All k-projections Surjective," *Acta Inform.*, Vol. 19, pp. 103-111, October, 1983.

[Chen 1986] C. L. Chen, "Linear Dependencies in Linear Feedback Shift Register," *IEEE Trans. on Computers*, Vol. C-35, No. 12, pp. 1086-1088, December, 1986.

[Chen 1987] C. L. Chen, "Exhaustive Test Pattern Generation Using Cyclic Codes," *IEEE Trans. on Computers*, Vol. 37, No. 3, pp. 329-338, March, 1987.

[Chin and McCluskey 1984] C. Chin and E. J. McCluskey, "Weighted Pattern Generation for Built-In Self-Test," Center for Reliable Computing Technical Report No. 84-7, Stanford University, 1984.

[Chin and McCluskey 1987] C. K. Chin and E. J. McCluskey, "Test Length for Pseudorandom Testing," *IEEE Trans. on Computers*, Vol. C-36, No. 2, pp. 252-256, February, 1987.

[Craig *et al.* 1988] G. L. Craig, C. R. Kime, and K. K. Saluja, "Test Scheduling and Control for VLSI Built-In Self-Test," *IEEE Trans. on Computers*, Vol. 37, No. 9, pp. 1099-1109, September, 1988.

[DasGupta *et al.* 1982] S. DasGupta, P. Goel, R. F. Walther, and T. W. Williams, "A Variation of LSSD and Its Implications on Design and Test Pattern Generation in VLSI," *Digest of Papers 1982 Intn'l. Test Conf.*, pp. 63-66, November, 1982.

[Eichelberger and Lindbloom 1983] E. B. Eichelberger and E. Lindbloom, "Random-Pattern Coverage Enhancement and Diagnosis for LSSD Logic Self-Test," *IBM Journal of Research & Development*, Vol. 27, No. 3, pp. 265-272, May, 1983.

[El-Ziq and Butt 1983] Y. M. El-Ziq and H. H. Butt, "A Mixed-Mode Built-In Self-Test Technique Using Scan Path and Signature Analysis," *Proc. Intn'l. Test Conf.*, pp. 269-274, October, 1983.

[Fasang 1980] P. P. Fasang, "BIDCO, Built-In Digital Circuit Observer," *Digest of Papers 1980 Test Conf.*, pp. 261-266, November, 1980.

[Fasang 1982] P. P. Fasang, "A Fault Detection and Isolation Technique for Microprocessors," *Digest of Papers 1982 Intn'l. Test Conf.*, pp. 214-219, November, 1982.

[Gelsinger 1987] P. P. Gelsinger, "Design and Test of the 80386," *IEEE Design & Test of Computers*, Vol. 4, No. 3, pp. 42-50, June, 1987.

[Gloster and Brglez 1989] C. Gloster, Jr., and F. Brglez, "Boundary Scan with Built-In Self Test," *IEEE Design & Test of Computers*, Vol. 6, No. 1, pp. 36-44, February, 1989.

[Golomb 1982] S. W. Golomb, *Shift Register Sequences*, Aegean Park Press, Laguna Hills, California, 1982.

[Ha and Reddy 1986] D. S. Ha and S. M. Reddy, "On the Design of Random Pattern Testable PLAs," *Proc. Intn'l Test Conf.*, pp. 688-695, September, 1986.

[Hassan 1986] S. Z. Hassan, "An Efficient Self-Test Structure for Sequential Machines," *Proc. Intn'l. Test Conf.*, pp. 12-17, September, 1986.

[Hortensius *et al.* 1989] P. D. Hortensius, R. D. McLeod, W. Pries, D. M. Miller, and H. C. Card, "Cellular Automata-Based Pseudorandom Number Generators for Built-In Self-Test," *IEEE Trans. on Computer-Aided Design*, Vol. 8, No. 8, pp. 842-859, August, 1989.

[Hudson and Peterson 1987] C. L. Hudson, Jr., and G. D. Peterson, "Parallel Self-Test with Pseudorandom Test Patterns," *Proc. Intn'l. Test Conf.*, pp. 954-963, September, 1987.

[Ichikawa 1982] M. Ichikawa, "Constant Weight Code Generators," Center for Reliable Computing Technical Report No. 82-7, Stanford University, June, 1982.

[Jone and Papachristou 1989] W. B. Jone and C. A. Papachristou, "A Coordinated Approach to Partitioning and Test Pattern Generation for Pseudoexhaustive Testing," *Proc. 26th Design Automation Conf.*, pp. 525-530, June, 1989.

[Jone *et al.* 1989] W. B. Jone, C. A. Papachristou, and M. Pereina, "A Scheme for Overlaying Concurrent Testing of VLSI Circuits," *Proc. 26th Design Automation Conf.*, pp. 531-536, June, 1989.

[Kalinowski *et al.* 1986] J. Kalinowski, A. Albicki, and J. Beausang, "Test Control Line Distribution in Self-Testable VLSI Circuits," *Proc. Intn'l. Conf. on Computer-Aided Design*, pp. 60-63, November, 1986.

[Karpovsky and Nagvajara 1989] M. G. Karpovsky and P. Nagvajara, "Design of Self-Diagnostic Boards by Signature Analysis," *IEEE Trans. on Industrial Electronics*, Vol. 36, No. 2, pp. 241-245, May, 1989.

[Kim *et al.* 1988] K. Kim, D. S. Ha, and J. G. Tront," On Using Signature Registers as Pseudorandom Pattern Generators in Built-in Self-testing," *IEEE Trans. on Computer-Aided Design*, Vol. 7, No. 8, pp. 919-928, August, 1988.

[Komonytsky 1982] D. Komonytsky, "LSI Self-Test Using Level-Sensitive Scan Design and Signature Analysis," *Digest of Papers 1982 Intn'l. Test Conf.*, pp. 414-424, November, 1982.

[Komonystky 1983] D. Komonytsky, "Synthesis of Techniques Creates Complete System Self-Test," *Electronics*, pp. 110-115, March, 1983.

[Konemann *et al.* 1979] B. Konemann, J. Mucha, and G. Zwiehoff, "Built-In Logic Block Observation Technique," *Digest of Papers 1979 Test Conf.*, pp. 37-41, October, 1979.

[Konemann *et al.* 1980] B. Konemann, J. Mucha, and G. Zwiehoff, "Built-In Test for Complex Digital Integrated Circuits," *IEEE Journal Solid State Circuits*, Vol. SC-15, No. 3, pp. 315-318, June, 1980.

[Krasniewski and Albicki 1985a] A. Krasniewski and A. Albicki, "Automatic Design of Exhaustively Self-Testing Chips with BILBO Modules," *Proc. Intn'l. Test Conf.*, pp. 362-371, November, 1985.

[Krasniewski and Albicki 1985b] A. Krasniewski and A. Albicki, "Self-Testing Pipelines," *Proc. Intn'l. Conf. on Computer Design*, pp. 702-706, October, 1985.

[Krasniewski and Pilarski 1989] A. Krasniewski and S. Pilarski, "Circular Self-Test Path: A Low-Cost BIST Technique for VLSI Circuits," *IEEE Trans. on Computer-Aided Design*, Vol. 8, No. 1, pp. 46-55, January, 1989.

[Kuban and Bruce 1984] J. R. Kuban and W. C. Bruce, "Self-Testing of the Motorola MC6804P2," *IEEE Design & Test of Computers*, Vol. 1, No. 2, pp. 33-41, May, 1984.

[Kumar 1980] S. K. Kumar, "Theoretical Aspects of the Behavior of Digital Circuits Under Random Inputs," Ph.D. thesis, University of Southern California, June, 1980.

[Lake 1986] R. Lake, "A Fast 20K Gate Array with On-Chip Test System," *VLSI System Design*, Vol. 7, No. 6, pp. 46-66, June, 1986.

[LeBlanc 1984] J. LeBlanc, "LOCST: A Built-In Self-Test Technique," *IEEE Design & Test of Computers*, Vol. 1, No. 4, pp. 42-52, November, 1984.

[Lin and Costello 1983] S. Lin and D. J. Costello, *Error Control Coding: Fundamentals and Applications*, Prentice-Hall, Englewood Cliffs, New Jersey, 1983.

[Lisanke *et al.* 1987] R. Lisanke, F. Brglez, A. J. deGeus, and D. Gregory, "Testability-Driven Random Test-Pattern Generation," *IEEE Trans. on Computer-Aided Design*, Vol. CAD-6, No. 6, pp. 1082-1087, November, 1987.

[McCluskey 1984] E. J. McCluskey, "Verification Testing — A Pseudoexhaustive Test Technique," *IEEE Trans. on Computers*, Vol. C-33, No. 6, pp. 541-546, June, 1984.

[McCluskey and Bozorgui-Nesbat 1981] E. L. McCluskey and S. Bozorgui-Nesbat, "Design for Autonomous Test," *IEEE Trans. on Computers*, Vol. C-30, No. 11, pp. 860-875, November, 1981.

[Parker 1976] K. P. Parker, "Adaptive Random Test Generation," *Journal of Design Automation and Fault-Tolerant Computing*, Vol. 1, No. 1, pp. 52-83, October, 1976.

[Patashnik 1983] O. Patashnik, "Circuit Segmentation for Pseudoexhaustive Testing," Center for Reliable Computing Technical Report No. 83-14, Stanford University, 1983.

[Peterson and Weldon 1972] W. W. Peterson and E. J. Weldon, Jr., *Error Correcting Codes*, 2nd ed., M.I.T. Press, Cambridge, Massachusetts, 1972.

[Resnick 1983] D. R. Resnick, "Testability and Maintainability with a New 6K Gate Array," *VLSI Design*, Vol. 4, No. 2, pp. 34-38, March/April, 1983.

[Saluja *et al.* 1988] K. K. Saluja, R. Sharma, and C. R. Kime, "A Concurrent Testing Technique for Digital Circuits," *IEEE Trans. on Computer-Aided Design*, Vol. 7, No. 12, pp. 1250-1259, December, 1988.

[Sastry and Marjumdar 1989] S. Sastry and A. Marjumdar, private communication, September, 1989.

[Savir and Bardell 1984] J. Savir and P. H. Bardell, "On Random Pattern Test Length," *IEEE Trans. on Computers*, Vol C-33, No. 6, pp. 467-474, June, 1984.

[Schnurmann *et al.* 1975] H. D. Schnurmann, E. Lindbloom, and R. F. Carpenter, "The Weighted Random Test-Pattern Generator," *IEEE Trans. on Computers*, Vol. C-24, No. 7, pp. 695-700, July, 1975.

[Sedmak 1979] R. M. Sedmak, "Design for Self-Verification: An Approach for Dealing with Testability Problems in VLSI-Based Designs," *Digest of Papers 1979 Test Conf.*, pp. 112-124, October, 1979.

[Shperling and McCluskey 1987] I. Shperling and E. J. McCluskey, "Circuit Segmentation for Pseudoexhaustive Testing via Simulated Annealing," *Proc. Intn'l. Test Conf.*, pp. 58-66, September, 1987.

[Stroud 1988] C. S. Stroud, "Automated BIST for Sequential Logic Synthesis," *IEEE Design & Test of Computers*, Vol. 5, No. 6, pp. 22-32, December, 1988.

[Tang and Woo 1983] D. T. Tang and L. S. Woo, "Exhaustive Test Pattern Generation with Constant Weight Vectors," *IEEE Trans. on Computers*, Vol. C-32, No. 12, pp. 1145-1150, December, 1983.

[Tang and Chen 1984a] D. T. Tang and C. L. Chen, "Logic Test Pattern Generation Using Linear Codes," *IEEE Trans. on Computers*, Vol. C-33, No. 9, pp. 845-850, September, 1984.

[Tang and Chen 1984b] D. T. Tang and C. L. Chen, "Iterative Exhaustive Pattern Generation for Logic Testing," *IBM Journal of Research & Development*, Vol. 28, pp. 212-219, March, 1984.

[Taylor and Karlin 1984] H. M. Taylor and S. Karlin, *An Introduction to Stochastic Modeling*, Academic Press, New York, 1984.

[TI 1989] Texas Instruments, "SCOPE Testability Octals," Preview Bulletin SCBT098, 1989.

[Timoc *et al.* 1983] C. Timoc, F. Stott, K. Wickman, and L. Hess, "Adaptive Self-Test for a Microprocessor," *Proc. Intn'l. Test Conf.*, pp. 701-703, October, 1983.

[Udell 1986] J. G. Udell, Jr., "Test Set Generation for Pseudoexhaustive BIST," *Proc. Intn'l. Conf. on Computer-Aided Design*, pp. 52-55, November, 1986.

[Udell and McCluskey 1989] J. G. Udell, Jr., and E. J. McCluskey, "Pseudoexhaustive Test and Segmentation: Formal Definitions and Extended Fault Coverage Results," *Digest of Papers 19th Intn'l. Symp. on Fault-Tolerant Computing*, pp. 292-298, June, 1989.

[Vasanthavada and Marinos 1985] N. Vasanthavada and P. N. Marinos, "An Operationally Efficient Scheme for Exhaustive Test-Pattern Generation Using Linear Codes," *Proc. Intn'l. Test Conf.*, pp. 476-482, November, 1985.

[Wagner *et al.* 1987] K. D. Wagner, C. K. Chin, and E. J. McCluskey, "Pseudorandom Testing," *IEEE Trans. on Computers*, Vol. C-36, No. 3, pp. 332-343, March, 1987.

[Wang 1982] L. T. Wang, "Autonomous Linear Feedback Shift Register with On-Line Fault-Detection Capability," *Digest of Papers 12th Annual Intn'l. Symp. Fault-Tolerant Computing*, pp. 311-314, June, 1982.

[Wang 1984] L. T. Wang and E. J. McCluskey, "A New Condensed Linear Feedback Shift Register Design for VLSI/System Testing," *Digest of Papers 14th Intn'l Symp. on Fault-Tolerant Computing*, pp. 360-365, June, 1984.

[Wang and McCluskey 1986a] L. T. Wang and E. J. McCluskey, "A Hybrid Design of Maximum-Length Sequence Generators," *Proc. Intn'l. Test Conf.*, pp. 38-47, September, 1986.

[Wang and McCluskey 1986b] L. T. Wang and E. J. McCluskey, "Condensed Linear Feedback Shift Register (LFSR) Testing — A Pseudoexhaustive Test Technique," *IEEE Trans. on Computers*, Vol. C-35, No. 4, pp. 367-370, April, 1986.

[Wang and McCluskey 1986c] L. T. Wang and E. J. McCluskey, "Concurrent Built-In Logic Block Observer (CBILBO)," *Intn'l. Symp. on Circuits and Systems*, Vol. 3, pp. 1054-1057, 1986.

[Wang and McCluskey 1986d] L. T. Wang and E. J. McCluskey, "Complete Feedback Shift Register Design for Built-In Self-Test," *Proc. Intn'l. Conf. on Computer-Aided Design*, pp. 56-59, November, 1986.

[Wang and McCluskey 1986e] L. T. Wang and E. J. McCluskey, "Feedback Shift Registers for Self-Testing Circuits," *VLSI Systems Design*, pp. 50-58, 1986.

[Wang and McCluskey 1986f] L. T. Wang and E. J. McCluskey, "Circuits for Pseudoexhaustive Test Pattern Generation," *Proc. Intn'l Test Conf.*, pp. 25-37, September, 1986.

[Wang and McCluskey 1986g] L. T. Wang and E. J. McCluskey, "Circuits for Pseudoexhaustive Test Pattern Generation Using Cyclic Codes," Center for

Reliable Computing Technical Report (CRC TR) No. 86-8, Stanford University, July, 1986.

[Wang and McCluskey 1987a] L. T. Wang and E. J. McCluskey, "Circuits for Pseudoexhaustive Test Pattern Generation," *IEEE Trans. on Computer-Aided Design*, Vol. 7, No. 1, pp. 91-99, January, 1987.

[Wang and McCluskey 1987b] L. T. Wang and E. J. McCluskey, "Built-In Self-Test for Sequential Machines," *Proc. Intn'l. Test Conf.*, pp. 334-341, September, 1987.

[Wang and McCluskey 1987c] L. T. Wang and E. J. McCluskey, "Linear Feedback Shift Register Design Using Cyclic Codes," *IEEE Trans. on Computers*, Vol. 37, No. 10, pp. 1302-1306, October, 1987.

[Williams 1985] T. W. Williams, "Test Length in a Self-Testing Environment," *IEEE Design & Test of Computers*, Vol. 2, No. 2, pp. 59-63, April, 1985.

[Wunderlich 1987] H. J. Wunderlich, "Self Test Using Unequiprobable Random Patterns," *Digest of Papers 17th Intn'l. Symp. on Fault-Tolerant Computing*, pp. 258-263, July, 1987.

[Wunderlich 1988] H. J. Wunderlich, "Multiple Distributions for Biased Random Test Patterns," *Proc. Intn'l. Test Conf.*, pp. 236-244, September, 1988.

PROBLEMS

11.1 Construct the test sequence generated by the LFSR shown in Figure 11.15.

11.2   a.   Give what you believe is a good definition of an exhaustive test for a sequential circuit.

   b.   Can the test described in part a. be realized by using a complete LFSR on the primary inputs?

   c.   What if the inputs and feedback lines were driven by a complete LFSR?

11.3 Show that by ANDing together pairs of outputs from a 4-stage maximal-length autonomous LFSR, a pseudorandom weighted-patterns generator can be constructed where the probability of generating a 1 is 0.25.

11.4 Show that the modified LFSR shown in Figure 11.61 generates all 2^4 states. The added OR and NOR gates force the LFSR into the (0000) state after the (1000) state.

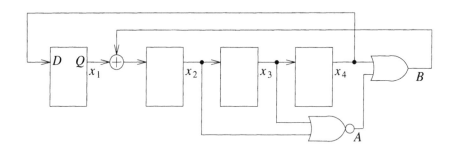

Figure 11.61   A complete LFSR

11.5   For the circuit of Figure 11.5

   a.   Find a fault that creates additional 1-entries in the dependency matrix but is detected by the given pseudoexhaustive test set.

   b.   Determine a pseudoexhaustive test set of four vectors that also detects the BF ($a.c$) (hint: modify the given test set).

11.6   Determine two partitions of the dependency matrix given in Figure 11.62.

11.7 Show that complementing one column of a matrix representing a pseudoexhaustive test set creates a test set that is also pseudoexhaustive.

11.8 Consider a binary matrix where any set of n columns has all 2^n combinations. Show that any set of $k < n$ columns has all 2^k combinations.

11.9 Construct a pseudoexhaustive test set of minimum length for a circuit whose dependency matrix is given in Figure 11.63.

	a	b	c	d	e	f
x	1	0	1	1	0	0
y	0	1	0	1	0	0
z	0	1	0	0	1	1

Figure 11.62

	a	b	c
x	1	1	0
y	0	1	1
z	1	0	1

Figure 11.63

11.10 Based on Theorem 11.1, construct the matrices T_0 and T_2 for the case $n=5$ and $k=3$. What relation can be observed between T_0 and T_2?

11.11 Determine if the circuit shown in Figure 11.64 is a maximal-test-concurrency circuit, and derive a minimal verification test for this circuit.

11.12 Apply Procedure 11.1 to the circuit of Figure 11.64.

11.13 Apply Procedure 11.1 to the circuit of Figure 11.65.

11.14 Prove that for the case where $p = w + 1$, applying all possible binary patterns of p bits with either odd or even parity will produce a test set where for every subset of w lines, all possible binary patterns will occur.

11.15 For the circuit shown in Figure 11.16, add a minimal number of bypass storage cells so that the resulting circuit can be tested by a verification test for the values of w specified below. Show the various segments produced by the addition of the bypass storage cells.

    a. $w = 3$

    b. $w = 5$

11.16 Consider a microprocessor chip that contains a PLA-based finite-state controller, a data path consisting of registers and logic, and a RAM and a ROM for memory. Describe the pros and cons of employing the following generic BIST architectures to this circuit.

    a.   central and separate;

    b.   distributed and separate;

    c.   distributed and embedded.

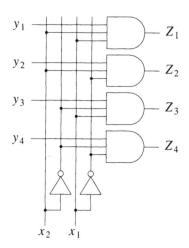

Figure 11.64   A data selector

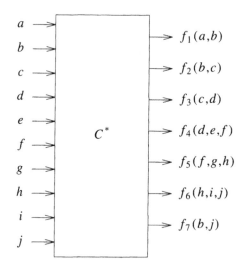

Figure 11.65

11.17 Give a logic design for the BIST hardware required to support the STUMPS methodology at the board level. Assume each chip is designed with a single scan path. Thus every LFSR must be placed into one or more "STUMPS" chips.

11.18 In the CATS BIST architecture why is it necessary that all storage cells be initialized to a known state before executing a test?

11.19 Show the complete circuitry required to test the circuit shown in Figure 11.66 using the following BIST architectures.

    a. CSBL;

    b. BEST.

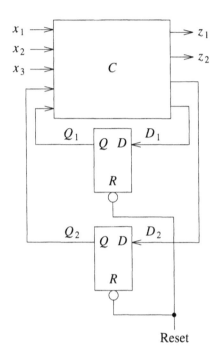

Figure 11.66

11.20 Replace the storage cells in the circuit shown in Figure 11.66 with a scan path. Show the complete circuitry required to test this circuit using the LOCST BIST architecture.

11.21 Consider the concurrent BIST architecture CBIST, where the normal data into the CUT correspond to the binary equivalent of the input sequence 7, 6, 5, 4, 3, 2, 1, 0, 7, 6, 5, ..., while the PRPG goes though the state sequence 0, 1, 2, ..., 7. Determine the number of clock cycles required to test the CUT.

11.22 Determine optimal test plans for the partial-intrusion BIST designs shown in Figures 11.67(a) and 11.67(b).

11.23 Consider the partial-intrusion BIST architecture shown in Figure 11.68, where R_1 generates all 2^{16} test patterns. Is C_2 tested functionally exhaustively? Justify your answer.

11.24 For the register adjacency graph shown in Figure 11.69, determine the minimal number of test sessions required so that each block of logic is tested. Assume that self-adjacent registers can operate as CBILBO registers.

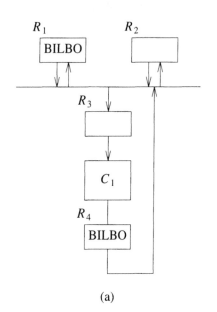

(a)

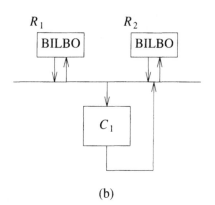

(b)

Figure 11.67

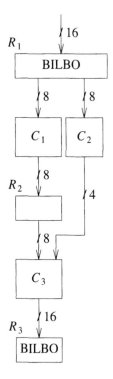

Figure 11.68

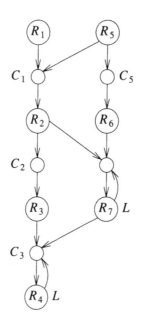

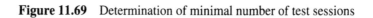

Figure 11.69   Determination of minimal number of test sessions

12. LOGIC-LEVEL DIAGNOSIS

About This Chapter

In this chapter we analyze logic-level fault-location techniques. After discussing the basic concepts of diagnosis, we review the fault-dictionary method, guided-probe testing, diagnosis by progressive reduction of the unit under test, methods specialized for combinational circuits, expert systems for diagnosis, effect-cause analysis, and a reasoning technique based on structure and behavior.

12.1 Basic Concepts

A unit under test (UUT) *fails* when its observed behavior is different from its expected behavior. If the UUT is to be repaired, the cause of the observed error(s) must be diagnosed. Diagnosis consists of locating the physical fault(s) in a structural model of the UUT. In other words, *diagnosis maps the observed misbehavior of the UUT into physical faults affecting its components or their interconnections.*

The degree of accuracy to which faults can be located is referred to as *diagnostic resolution*. No external testing experiment can distinguish among functionally equivalent faults. The partition of all the possible faults into distinct sets of functionally equivalent faults defines the *maximal fault resolution*, which is an intrinsic characteristic of the system. The *fault resolution of a test sequence* reflects its capability of distinguishing among faults, and it is bounded by the maximal fault resolution. This is illustrated in Figure 12.1(a). A, B, C, D, E, and F represent the sets of equivalent faults of a system. Consider a test sequence T that does not distinguish between (the faults in) A and (those in) B, or between E and F. The maximal fault resolution is the partition $\{A,B,C,D,E,F\}$, while the resolution of T is $\{A{\cup}B, C, D, E{\cup}F\}$. A test (sequence) that achieves the maximal fault resolution is said to be a *complete fault-location test.*

Repairing the UUT often consists of substituting one of its *replaceable units* (RUs) identified as containing some faults, and referred to as a *faulty RU*, by a good unit. Hence usually we are interested only in locating a faulty RU, rather than in an accurate identification of the fault inside an RU. This diagnosis process is characterized by the *RU resolution*. Figure 12.1(b) illustrates the relation between fault resolution and RU resolution. U_1, U_2, U_3, and U_4 are the RUs of the system, and the faults are physically partitioned as shown. If the actual fault belongs to A or B, in either case we can identify U_1 as the faulty RU. But if the actual fault belongs to C, then we cannot determine whether the faulty RU is U_2 or U_3. However, distinguishing between U_2 and U_3 is feasible when the fault belongs to D or E.

Clearly, the location of the faulty RU is more difficult when equivalent faults span different RUs. The RU resolution corresponding to a fault resolution defined by a partition $\{F_1, F_2, ..., F_k\}$ of the set of faults, is obtained by replacing every F_i by the set of RUs spanned by the faults in F_i. For our example, the maximal RU resolution, corresponding to the maximal fault resolution, is given by $\{U_1, U_2, U_3, \{U_2, U_3\}, U_4\}$. The *RU resolution* of any test is bounded by the maximal RU resolution. For example, the RU resolution of T is $\{U_1, U_2, \{U_2, U_3\}, \{U_3, U_4\}\}$. A test that

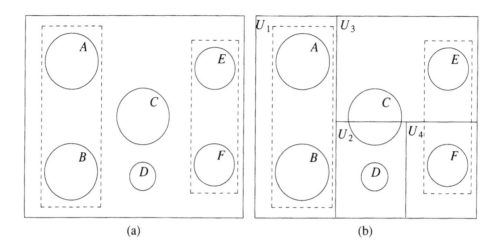

(a)   (b)

Figure 12.1   Fault resolution and RU resolution

achieves the maximal RU resolution (i.e., it distinguishes between every pair of
nonequivalent faults that reside in different RUs) is said to be a *complete RU-location
test*.

For the preceding example, suppose that the results of the test do not distinguish
between U_3 and U_4. In such a case, it is sometimes possible to replace one of the
suspected RUs, say U_3, with a good RU, and rerun the experiment. If the new results
are correct, the faulty RU is the replaced one; otherwise, it is the remaining one (U_4).
This type of approach is an example of a *sequential diagnosis procedure*, in which
diagnosis and repair are interleaved.

The diagnosis process is often *hierarchical* such that the faulty RU identified at one
level becomes the UUT at the next level. For example, to minimize the downtime of a
computing system, first-level diagnosis deals with "large" RUs, such as boards
containing many components; these are referred to as *field-replaceable units*. The
faulty board is then tested in a maintenance center, where the objective is to locate a
faulty component on the board; this is done to minimize the cost of the replaced unit.
A typical RU at this level is an IC. Although further repair beyond the IC level is not
possible, accurate location of faults inside a faulty IC may be useful for improving its
manufacturing process.

The hierarchical diagnosis process described above proceeds *top-down*, starting with a
system operating in the field. During the fabrication of a system, however, its testing
proceeds *bottom-up* (e.g., ICs → boards → system), such that a higher level is
assembled only from components already tested at a lower level. This is done to
minimize the cost of diagnosis and repair, which increases substantially with the level
at which the faults are detected. For example, if it costs $1 to test an IC, the cost of
locating the same defective IC when mounted on a board and of repairing the board
may be about $10; when the defective board is plugged into a system, the cost of
finding the fault and repairing the system may be about $100.

Note that in the bottom-up process, the faults most likely to occur are fabrication errors affecting the interconnections between components, while in the top-down process, the most likely faults are physical failures internal to components (because every UUT had been successfully tested in the past). Knowing the most likely class of faults is a definite help in fault location.

Fault diagnosis can be approached in two different ways. The first approach does most of the work before the testing experiment. It uses fault simulation to determine the possible responses to a given test in the presence of faults. The data base constructed in this step is called a *fault dictionary*. To locate faults, one tries to match the actual response obtained from the UUT with one of the precomputed responses stored in the fault dictionary. If this look-up process is successful, the dictionary indicates the corresponding fault(s) or faulty RU(s) in the UUT.

Fault diagnosis based on fault dictionaries can be characterized as a *cause-effect analysis* that starts with possible causes (faults) and determines their corresponding effects (responses). A second type of approach, employed by several diagnosis methods, relies on an *effect-cause analysis*, in which the effect (the actual response obtained from the UUT) is processed to determine its possible causes (faults).

12.2 Fault Dictionary

Example 12.1: To illustrate the concepts involved in building and using a fault dictionary, we will analyze the circuit shown in Figure 12.2(a). The circuit has 13 lines and 26 single stuck-at faults. Fault collapsing based on structural equivalence partitions the faults into the following 14 equivalence classes (x_i denotes x s-a-i):

1.	$\{a_0\}$	8.	$\{g_1\}$
2.	$\{a_1\}$	9.	$\{i_0, h_1, l_0, j_0, e_1\}$
3.	$\{b_1\}$	10.	$\{i_1, h_0\}$
4.	$\{c_1\}$	11.	$\{j_1, e_0\}$
5.	$\{d_1\}$	12.	$\{k_0, d_0, g_0\}$
6.	$\{f_0, b_0, c_0\}$	13.	$\{k_1, l_1, m_1\}$
7.	$\{f_1\}$	14.	$\{m_0\}$

We use the first fault in a class as the representative of its class. For convenience, we assume that the fault-free circuit has an empty fault Φ. Figure 12.2(b) shows all the possible responses of the circuit to the given test set under the SSF model. Error values are marked by "*". Note that the test set does not distinguish between d_1 and i_1, or between g_1 and j_1.

For this simple example, we can arrange the fault dictionary as a mapping between the 12 distinct responses and the faults that can produce them. Thus if we obtain the response 00001, the dictionary will point to the faults $\{k_0, d_0, g_0\}$.   □

To reduce the amount of data used for fault location, a fault dictionary does not store the entire response R_f caused by the fault f, but only a "signature" usually consisting of the list of errors contained in R_f. An error occurring in test t_i at output o_j is denoted by (i,j). In Example 12.1, the error caused by k_0 is (4,1). Other methods for reducing the size of a fault dictionary can be found in [Tulloss 1978, 1980].

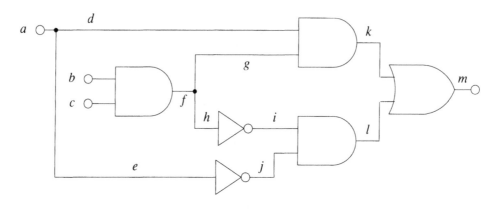

(a)

	a b c	Φ	a_0	a_1	b_1	c_1	d_1	f_0	f_1	g_1	i_0	i_1	j_1	k_0	k_1	m_0
t_1	0 1 1	0	0	1*	0	0	1*	1*	0	0	0	1*	0	0	1*	0
t_2	1 1 0	0	1*	0	0	1*	0	0	1*	1*	0	0	1*	0	1*	0
t_3	1 0 1	0	1*	0	1*	0	0	0	1*	1*	0	0	1*	0	1*	0
t_4	1 1 1	1	0*	1	1	1	1	0*	1	1	1	1	1	0*	1	0*
t_5	0 0 1	1	1	0*	0*	1	1	1	0*	1	0*	1	1	1	1	0*

(b)

Figure 12.2   (a) Circuit (b) Applied tests and responses in the presence of faults

To reduce the large computational effort involved in building a fault dictionary, in fault simulation the detected faults are dropped from the set of simulated faults. Hence all the faults detected for the first time by the same vector at the same output will produce the same signature and will be included in the same equivalence class. In Example 12.1, a fault dictionary constructed in this way will not distinguish among the faults detected by t_1, namely $\{a_1, d_1, f_0, i_1, k_1\}$, even if most of them are further distinguished by subsequent tests. In this case the testing experiment can stop after the first failing test, because the information provided by the following tests is not used. Such a testing experiment achieves a lower diagnostic resolution. (A trade-off between computation time and diagnostic resolution can be achieved by dropping faults after $k>1$ detections.)

Figure 12.3 shows the possible results for Example 12.1 in the form of a *diagnostic tree*. The results of a test are indicated as *pass* (P) or *fail* (F). Every test distinguishes between the faults it detects and the ones it does not. (In a multioutput circuit, faults detected by the same test can be distinguished if they are detected at different outputs.) The set of faults shown in a rectangle are equivalent under the currently applied test set. Note that faults remaining undetected are equivalent to Φ.

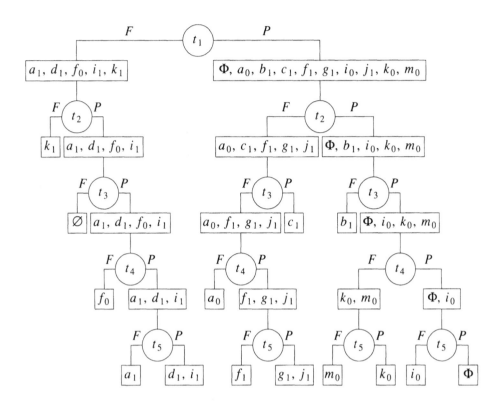

Figure 12.3   Diagnostic tree

From Figure 12.3 we can observe that some faults are uniquely identified before the entire test sequence is applied. For example, k_1 is the only fault detected in both t_1 and t_2; thus, if both t_1 and t_2 fail, the fault in the UUT is located to within the equivalence class $\{k_1, l_1, m_1\}$. Here the testing experiment can stop after the first two tests, because no more diagnostic information can be obtained from the following tests (according to the SSF model).

Rather than applying the entire test sequence in a fixed order, *adaptive testing* determines the next vector to be applied based on the results obtained by the preceding vectors. In our example, if t_1 fails and t_2 passes, the possible faults are $\{a_1, d_1, f_0, i_1\}$. At this point applying t_3 would be wasteful, because t_3 does not distinguish among these faults. The use of adaptive testing may substantially decrease the average number of tests required to locate a fault (see, for example, [Koren and Kohavi 1977]).

Generating Tests to Distinguish Faults

To improve the fault resolution of a test set T, it is necessary to generate tests to distinguish among faults equivalent under T. Note that the goal of having a high fault resolution is somehow at odds with the goal of minimizing the size of the test set, which requires every vector to detect as many faults as possible.

Consider the problem of generating a test to distinguish between two faults, f and g, in a combinational circuit. Such a test must detect f but not g on some output, or vice versa. The fault-oriented algorithms presented in Chapter 6 can be generalized to handle this problem, by using the following strategy:

Case 1: If f and g do not influence the same set of outputs, let O_f (O_g) be the set of outputs influenced by $f(g)$ and not by $g(f)$. Try to generate a test for f using only the circuit feeding the outputs O_f. If this fails, try to generate a test for g using only the circuit feeding the outputs O_g [Savir and Roth 1982].

Case 2: (Here $O_f = O_g = \varnothing$, or test generation for Case 1 has failed.) Try to generate a test for f without activating g (if g is the fault l s-a-v, this is simply done by imposing the additional constraint that the value of l should be v). If this fails, try to generate a test for g without activating f.

Case 3: (Here f and g are simultaneously activated.) Try to generate a test that propagates the effect of either f or g (but not both) to some primary output.

While Cases 1 and 2 require only minor changes to any fault-oriented test generation algorithm, Case 3 is more difficult, because the algorithm must process the fault effects of both f and g. We will present a generalization of the D-algorithm for this problem. Let us use the symbols D and $\overline{D}$ to represent effects of the fault f and the symbols E and $\overline{E}$ to represent effects of the fault g. Both these pairs of values are propagated in the same way when they interact with 0 and 1 signals and the objective is to propagate a D (or $\overline{D}$) or an E (or $\overline{E}$) but not both to some output. However, when these two values appear on different inputs to the same gate, some differences arise in defining how they propagate through the gate. Specifically it must be remembered that even though we are propagating two single faults, at most one fault is present. Thus if the inputs to a 2-input OR gate are D and $\overline{E}$, then the normal inputs are 1 and 0, the inputs with f present are 0 and 0, and the inputs with g present are 1 and 1. The normal gate output is 1, the output with f present is 0, and the output with g is 1. Hence only the D propagates to the output ($D + \overline{E} = D$). An AND gate with inputs D and E would have normal output 1 and output 0 with either f or g present, and hence both D and E propagate to the gate output; this is denoted by D,E ($D.E = D,E$). The tables of Figure 12.4 represent the signal propagations for AND and OR gates.

Note that when all the lines in the fault-propagation frontier have only D,E or $\overline{D},\overline{E}$ values (which means that the D-frontiers in the circuits N_f and N_g are identical), the algorithm must backtrack, because the fault effects cannot be propagated separately.

Example 12.2: For the circuit of Figure 12.5, we want to generate a test to distinguish the faults A s-a-1 and F s-a-0. These two faults influence the same output, so Case 1 does not apply. First we try to derive a test for A s-a-1 while setting $F=0$. By implication we obtain $A=0$, $B=1$, $C=\overline{D}$, $H=\overline{D}$. But if $G=0$, then $J=\overline{D}.D=0$; if $G=1$, then $I=0$ and $J=0$. Thus either case leads to a failure. Then we try to derive a test for F s-a-0 while setting $A=1$. By implication we obtain $F=1$, $C=0$, $B=0$, $H=D$, $I=1$, $G=0$, $J=D$. Thus the vector 1010 distinguishes the two faults. $\square$

Example 12.3: For the circuit of Figure 12.5, we want to generate a test to distinguish the faults C_1 s-a-0 and C_2 s-a-0. Any attempt to activate the two faults separately fails, so Case 2 does not apply. The only way to activate the faults is $C=1$, which implies $A=B=1$. Then $C_1=D$ and $C_2=E$. First we try to propagate the D by

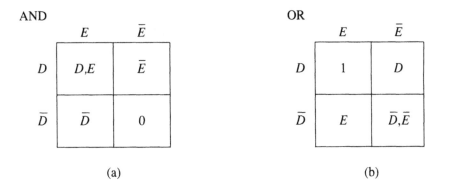

Figure 12.4

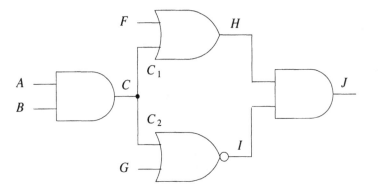

Figure 12.5

setting $F=0$. This implies $H=D$. Trying to inhibit the propagation of E by setting $G=1$ results in $I=0$ and $J=0$. We backtrack by setting $G=0$, which leads to $I=\bar{E}$ and $J=D.\bar{E}=\bar{E}$. Thus the vector 1100 distinguishes the two faults by propagating only one fault effect to the output.   □

Problems with Fault Dictionaries

One problem with the fault-dictionary approach is the large computational effort involved in building fault dictionaries for large circuits tested with long test sequences. Fault simulation consumes much CPU time and storage, even when used with early fault dropping (typically on first detection). Moreover, early fault dropping results in lower diagnostic resolution.

A fault dictionary is constructed only for a specific fault universe, usually consisting of SSFs. A fault that is not equivalent under the applied test sequence to any of the simulated faults cannot be located via the fault dictionary, because its corresponding response does not match any response computed by fault simulation.

Example 12.4: Consider the AND bridging fault $(a.c)$ in the circuit of Figure 12.2(a). The test set given in Figure 12.2(b) detects this fault in vectors t_1 and t_2. But none of the SSFs produces errors only in t_1 and t_2. Similarly, the multiple fault $\{b_1, i_1\}$ is detected in t_1 and t_3, but this does not match the signature of any SSF.   □

When the signature of the UUT does not exactly match any of the precomputed signatures stored in the fault dictionary, location is based on closest-match analysis techniques. As the signature S_i derived from a response R_i is the set of errors contained in R_i, we can measure the closeness c_{12} of two signatures S_1 and S_2 by the amount of overlap between the two sets, that is,

$$c_{12} = \frac{|S_1 \cap S_2|}{|S_1 \cup S_2|}$$

(Note that $c_{12}=1$ when $S_1=S_2$)

The diagnosis process selects as the most likely set of faults the one whose signature is closest to the signature of the UUT. While this heuristic may work in many cases, it is not guaranteed to produce a correct diagnosis. For example, the closest match for the signatures of both faults analyzed in Example 12.4 — the bridging fault $(a.c)$ and the multiple fault $\{b_1, i_1\}$ — would be the signature of k_1; but this fault is totally unrelated to either $(a.c)$ or $\{b_1, i_1\}$. (It is interesting to observe that the set of tests that detect $\{b_1, i_1\}$ is not the union of the sets of tests that detect b_1 and i_1 — $\{t_1, t_3, t_5\}$ — because i_1 masks b_1 under the test t_5.)

Another problem encountered by some testing systems based on the fault-dictionary approach is caused by potential detections. A potential detection occurs when the response generated by a fault contains an unknown (u) value, while the corresponding value in the fault-free circuit is binary. (This situation is typical for faults that prevent initialization). As the actual response of the UUT contains only binary values, an observed error should be interpreted as a match for a potential detection [Richman and Bowden 1985]. For example, consider a fault dictionary built from data including the values shown in Figure 12.6, and assume that the response of the UUT is 100. The signature of the UUT (with errors in t_1 and t_2) should match the signature of f_1 (error in t_1 and potential detection in t_2), but not that of f_2 (error only in t_1).

	Φ	f_1	f_2
t_1	0	1*	1*
t_2	1	u	1
t_3	0	0	u

Figure 12.6

12.3 Guided-Probe Testing

Guided-probe testing extends an edge-pin testing process by monitoring internal signals in the UUT via a *probe* which is moved (usually by an operator) following the guidance provided by the ATE. The principle of guided-probe testing is to backtrace an error from the primary output where it has been observed during edge-pin testing to its source (physical fault) in the UUT (see Figure 12.7). Typical faults located by guided-probe testing are opens and defective components. An open between two points A and B is identified by a mismatch between the error observed at B and the correct value measured at the signal source A. A faulty device is identified by detecting an error at one of its outputs, while only expected values are observed at its inputs.

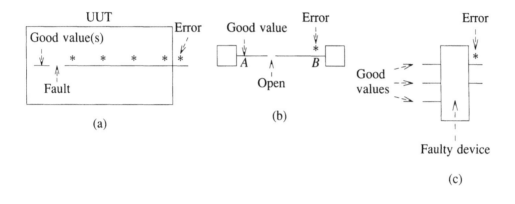

Figure 12.7   (a) Backtracing errors (b) Identifying an open (c) Identifying a faulty device

Unlike the fault-dictionary method, guided-probe testing is not limited to the SSF model. The concept of "defective component" covers any type of internal device fault detected by the applied test; this generality makes guided-probe testing independent of a particular fault model.

In addition to the applied stimuli and the expected response needed for edge-pin testing, ATE supporting guided-probe testing also needs the expected values of all the internal lines accessible for probing and the structural model of the UUT (see Figure 12.8). Because the backtrace is guided by the structural model, faults that create additional connections are inherently difficult to locate. For example, the faulty-device diagnosis illustrated in Figure 12.7(c) would be incorrect had the error on the device output been caused by a short.

The *Guided-probe* procedure outlined in Figure 12.9 recursively backtraces a given error along its propagation path and returns a diagnostic message when it reaches the location where the fault is activated. *Probe (j)* represents the action of moving the probe to line j. Initially, *Guided-probe* starts at a primary output and with the first test in which an error has been observed at that output. As long as the backtrace traverses a combinational region of the UUT, it would be sufficient to analyze only the values

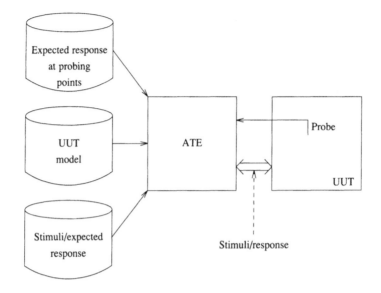

Figure 12.8   ATE with guided-probe testing

Guided-probe (i, t_r)
/* backtrace error observed at i in test t_r */
begin
    j = source of signal observed at i
    Probe (j)
    if no error at j in t_r **then return** ("open between j and i")
    /* error at j */
    if j is a PI **then return** ("faulty PI: j")
    m = the device whose output is j
    for every input k of m
       begin
          Probe (k)
          reapply $(t_1, t_2, ..., t_r)$
          if error at k in t_q $(q{\leq}r)$ **then return** *Guided-probe (k, t_q)*
       end
    /* no errors at the inputs of m */
    return ("faulty device: m")
end

Figure 12.9   Basic guided-probe procedure

existing after the last vector t_r has been applied, because in a combinational circuit the same vector activates the fault and propagates the resulting error. In a sequential circuit, however, fault activation and error propagation may be done by sequences.

Hence when probing an input k of a device whose output has errors in t_r, the entire sequence $(t_1, t_2, ..., t_r)$ must be reapplied, because the error could have been propagated to k in a vector t_q applied earlier $(q<r)$. In such a case, backtracing continues with the earliest vector t_q that propagates an error to k. Although the reapplication of the sequence can be avoided for a combinational device, the test application time is usually negligible compared to the time involved in moving the probe.

Example 12.5: Consider the circuit and the test sequence given in Figure 12.10. The errors shown are caused by D s-a-0. (Note that the F/F input called D and the primary input DT are different probing points.) We assume that every signal is accessible for probing. Backtracing the error observed at Z in t_4, *Guided-probe* reaches the output Q of the F/F. Note that none of the F/F inputs has an error in t_4. Next the F/F inputs are probed while the entire sequence is reapplied, and the earliest error is found on the D input in t_2. The last line probed is the primary input DT.   □

The procedure outlined in Figure 12.9 describes only the basic mechanism of guided-probe testing. In the following, we discuss the extensions it needs to handle feedback loops and wired logic.

Feedback Loops

The following example illustrates the problems encountered in circuits with feedback loops and the approach used to solve them.

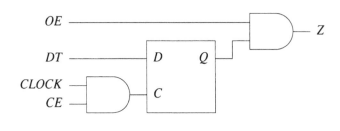

(a)

test	CLOCK	CE	DT	OE	Q	Z
t_1	↑	1	0	1	0	0
t_2	↑	0	1/0	1	0	0
t_3	↑	1	1/0	0	1/0	0
t_4	0	1	0	1	1/0	1/0

(b)

Figure 12.10

Example 12.6: Consider the circuit and the test sequence given in Figure 12.11. First assume that the fault is DT s-a-1, which causes errors in t_4 on DT, D, Z, and $\bar{Q}$. Backtracing the error observed at Z in t_4 leads to Q, D, and, assuming that the $\bar{Q}$ input of the AND gate is probed before DT, the procedure again reaches the F/F. Hence to identify loops, one should keep track of the previously encountered devices. An identified loop is then retraced probing all the inputs of the devices in the loop (rather than following only the first input with an error). In this way, the procedure can determine whether the error originates within or outside the loop. In our example, retracing the loop and probing all the inputs of the AND gate will find DT to be the source of the errors along the loop.

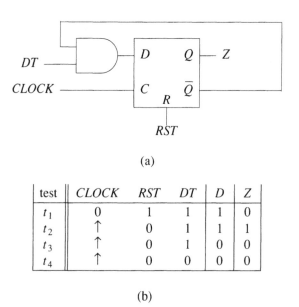

(a)

test	CLOCK	RST	DT	D	Z
t_1	0	1	1	1	0
t_2	↑	0	1	1	1
t_3	↑	0	1	0	0
t_4	↑	0	0	0	0

(b)

Figure 12.11

Now assume that the fault is D s-a-1, which causes D, Z and $\overline{Q}$ to have errors in t_3 and t_4. After the loop is identified, probing all the inputs of the devices in the loop shows that the error originates within the loop. To further improve the diagnostic resolution, one needs to observe the values of the probed lines both before and after the CLOCK pulse is applied. This will show that the error at D in t_3 precedes the CLOCK pulse, which propagates it around the loop. (Alternatively, we can consider that the clock and data changes define separate vectors.)   □

Wired Logic

One difficulty in backtracing errors through a "device" created by wired logic is that the inputs and the output of such a device form the same physical entity and hence cannot be separately probed. Also the previously used criterion for identifying a faulty device — error on output and good values on inputs — is no longer valid for a device driving wired logic, because the output error may propagate from another output connected to the same wire. For example, in Figure 12.12 it would be incorrect to identify gate E as faulty. The solution in such a case is to probe the inputs of all the driving devices as if they were inputs of the wired gate. In Figure 12.12, this strategy will result in continuing the backtrace from C.

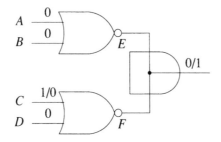

Figure 12.12

If no errors are found while probing the inputs of the driving devices, the conventional guided-probe testing strategy cannot distinguish among these devices. In the circuit of Figure 12.12, if $E=F=1$, while (A,B,C,D) have their expected values 0010, the obtained diagnostic resolution is $\{E,F\}$. This can be further improved by additional analysis. For example, we can observe that a failure of gate E (driving a 0 instead of 1) cannot produce the observed result, which can be explained only by a failure of gate F (outputting a 1 instead of 0).

Speed-up Techniques

The most time-consuming part of guided-probe testing is moving the probe. Hence to speed up the fault-diagnosis process, one needs to reduce the number of probed lines. Here are some techniques used to achieve this goal [Kochan *et al.* 1981]:

- Rather than probing one pin at a time, simultaneously probe all the pins of a device using an IC clip. If several inputs have errors, then follow the one with the earliest error (this also reduces the amount of probing for tracing feedback loops).

- Skip probing the output of a suspected device and directly probe its inputs. For the example in Figure 12.13, after observing an error at i, one would directly probe the inputs of m. The output j would be probed only if no errors are detected at the inputs, to distinguish between the faulty device m and an open between j and i.

- Probe only those device inputs that can affect the output with errors (see Figure 12.14).

- Among the inputs that can influence the output with errors, probe first the control lines; if no errors are detected at the control lines, then probe only those data inputs enabled by the values of the control lines. For the multiplexor shown in Figure 12.15, if no errors are found at the select inputs S_0 and S_1, the next line probed is D_3 (because it is the input selected by $(S_0,S_1) = 11$).

- Use a fault dictionary to provide a starting point for probing.

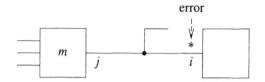

Figure 12.13

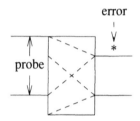

Figure 12.14

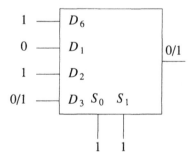

Figure 12.15

Techniques similar to those used in guided-probe testing are used in electron-beam testing [Tamama and Kuji 1986].

12.4 Diagnosis by UUT Reduction

The *C*ontrollability, *O*bservability and *M*aintenance *E*ngineering *T*echnique (COMET) developed by Chang and Heimbigner [1974] is based on a design for diagnosability technique which requires selective disabling of portions of the UUT, coupled with the ability of testing only the enabled part. With these provisions, the testing and diagnosis process follows the diagram shown in Figure 12.16.

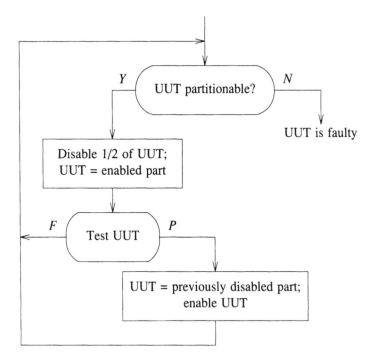

Figure 12.16 Testing and diagnosis process in COMET

Initially, the UUT is the entire circuit and the process starts when its test fails. While the failing UUT can be partitioned, half of the UUT is disabled and the remaining half is tested. If the test passes, the fault must be in the disabled part, which then becomes the UUT.

COMET has been applied to processor-level testing, where every board in the processor can be independently disabled, either by physically disconnecting it or by setting its outputs to a logically inactive state.

Example 12.7: Consider a processor composed of four boards, referred to as A, B, C, and D. Assume that B is the faulty board. After the fault-detection tests for the entire processor fail, C and D are disabled and fault-detection tests for A and B are run. These tests fail and then B is disabled and the tests for A are run. These tests pass and implicate B as the faulty unit.   □

Note that COMET does not require fault-location tests. Applying fault-detection tests to the enabled part of the UUT, and the ability to reduce the UUT progressively, are sufficient to locate faults.

12.5 Fault Diagnosis for Combinational Circuits

The fault-location methods discussed in this section are applicable to combinational circuits and also to sequential circuits that are transformed into combinational ones during testing using one of the scan-design techniques presented in Chapter 9.

Structural Analysis

If we assume a single fault that does not create new connections, then there exists a path from the site of the fault to each of the outputs where errors have been detected. Hence the fault site belongs to the intersection of the cones of all the failing outputs (Figure 12.17).

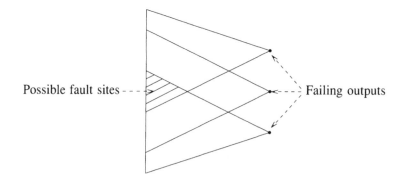

Figure 12.17   Cone intersection

(This observation is also true in a sequential circuit, but its application is more difficult because there a propagation path may span several time frames). Unlike fault dictionaries, which rely only on the first error, this simple structural analysis tries to find faults that can explain all the observed errors, and it often yields acceptable diagnostic resolution [Hsu *et al.* 1981].

If we restrict ourselves to the SSF model, the set of suspected faults can be further reduced by taking into account the inversion parities of the paths between the fault site and the failing POs. Let l be a possible fault site. The fault l s-a-v is a *plausible* fault only if for every error value v_o obtained at an output o, there exists a path with inversion parity $v \oplus v_o$ between l and o.

Example 12.8: When testing the circuit shown in Figure 12.18, assume that 1/0 errors are observed at E and F. Then the fault site must be in the subcircuit whose output is D (note that this does not include $C2$). The fault C s-a-0 is excluded because the unique path between C and E has odd inversion. The plausible faults are C s-a-1, B s-a-1, $C1$ s-a-1, and D s-a-0.   □

Fault Simulation

Once a set of plausible faults has been identified by structural analysis, these faults can then be simulated to select the ones that generate a response identical to the actual response of the UUT [Arzoumanian and Waicukauski 1981]. In this simulation, one compares the output errors produced by every fault with the errors observed during the

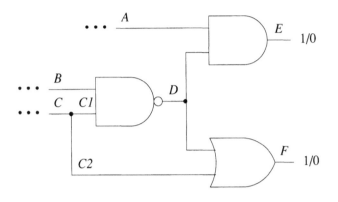

Figure 12.18

testing. The comparison is done vector by vector, and every fault that mismatches is dropped from the set of simulated faults. The faults remaining after all vectors have been simulated are consistent with the observed symptom, and they are equivalent under the applied test. The actual fault belongs to this equivalence class. Compared to building a fault dictionary, this fault simulation has the advantage of processing a much smaller set of faults. Also the resulting fault resolution is generally higher because now faults are required to account for all the observed errors.

Example 12.9: Consider again the circuit and the test given in Figure 12.2. Assume that tests t_1 and t_4 failed. Because the error value is 1 in t_1 and 0 in t_4, there must exist two propagation paths with different inversion parities from the fault site to the output m. Hence the fault site is one of the lines $\{a,b,c,f\}$ and the set of plausible faults is $\{a_0,a_1,b_1,c_1,f_0,f_1\}$. Only a_1 and f_0 make t_1 fail, so the other faults are dropped after simulating t_1. Then only f_0 produces an error in t_4, and we conclude that f_0 is the actual fault. By contrast, the result obtained using a fault dictionary built with fault dropping on the first error would be the class $\{a_1,d_1,f_0,i_1,k_1\}$.   □

Note that it is possible that none of the plausible faults can reproduce the response obtained from the UUT. Such a result shows that the actual fault contradicts the initial assumptions made about its nature (single fault that does not create new connections). For example, the actual fault could be a multiple stuck fault or a bridging fault that is not equivalent under the applied test with any SSF in the circuit.

12.6 Expert Systems for Diagnosis

Expert systems are programs applying Artificial Intelligence techniques in an attempt to emulate the problem-solving ability of a human expert in a specialized field [Nau 1983]. An expert system relies on a *knowledge base*, which is a data base representing the knowledge applied by a human expert, coupled with an *inference engine* which searches the knowledge base to retrieve and apply the information pertinent to the problem to be solved.

Expert systems for diagnosis [Hartley 1984, Laffey *et al.* 1986, Mullis 1984, Wilkinson 1984] are built by encoding the techniques followed by expert troubleshooters to locate a fault, given certain error symptoms. Because these diagnosis techniques are specific to one target system, such expert systems have been developed mainly for high-priced, complex digital systems, where they can provide substantial economic benefits accumulated during the lifetime of the system. An important advantage of an expert system is that it makes the expert's knowledge instantly available in the field, thus allowing diagnosis to be performed by less skilled personnel. In general, the use of an expert system leads to a faster and more accurate fault-location process, and hence it minimizes the downtime of the target system.

Most of the knowledge of an expert system is a collection of *if-then rules* (such systems are also referred to as *rule-based* systems). The general form of a rule is

if (condition) **then** action(s)

In the following example, the condition is based on some of the observed symptoms and the action suggests specific RUs as suspects (likely to be faulty):

if (*ALU-test* failed **and** *register-test* passed) **then** (suspect board *A* or board *B*)

Usually the tests are grouped according to the part of the system they exercise (i.e., *ALU-test*) and the symptoms relate to the overall results of groups of tests, rather than individual vectors. Faults are located only to large field-RUs (typically boards).

The conditions can also relate to the current state of the diagnosis process (that is, which RUs are currently suspected). Other actions may rule out a previously suspected RU, or they may request the application of additional tests or the replacement of a suspected RU with a fault-free one and rerunning of some tests. For example:

if (board *A* is suspect **and** board *B* is suspect)
    then (replace board *A*; apply *ALU-test*)
if (*Data-Path-test* passed) **then** (assume board *C* is fault-free).

Additional knowledge may be included in the knowledge base to represent

- structural information about the target system (its RUs and their interconnections);

- relations between RUs and tests (which RUs are checked by which tests);

- general assumptions about the nature of the fault (no more than one faulty RU, etc.)

Figure 12.19 illustrates the general structure of an expert system for diagnosis. The diagnosis data consist of the known facts (symptoms obtained so far, currently suspected RUs). Based on these facts, the inference engine repeatedly searches the knowledge base to find the applicable rules, which are then used to establish new facts or to control the testing experiment to obtain new facts. The expert system controls the ATE directly and/or by interacting with an operator.

The inference engine may reason using *forward chaining* — reaching conclusions by finding and applying the rules whose conditions are true in the current state — or *backward chaining* — starting with a hypothesis and searching for supporting facts to confirm it. Some systems can combine the two strategies.

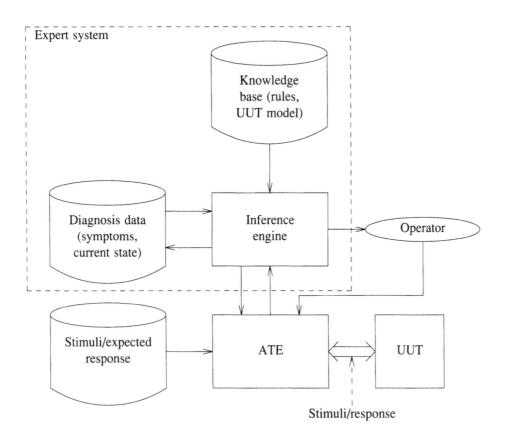

Figure 12.19   Expert system for diagnosis

A limitation of rule-based expert systems for diagnosis is their use of *shallow knowledge* — a collection of empirical associations that are not directly related to the intended behavior of the target system. This is why such expert systems cannot successfully process a new symptom that does not match the ones stored in the knowledge base.

12.7 Effect-Cause Analysis

The *effect-cause analysis* [Abramovici and Breuer 1980] processes the response obtained from the UUT to determine the possible stuck-at faults that can generate that response, based on *deducing internal values* in the UUT (Figure 12.20).

Any line for which both 0 and 1 values are deduced can be neither *s-a-*1 nor *s-a-*0, and it is identified as *normal* (fault-free). Faults are located on some of the lines that could not be proved normal. The ensemble of normal and faulty lines is referred to as a *fault situation*. The effect-cause analysis employs an *implicit multiple stuck-fault model*, and it does not require fault enumeration or precomputing faulty responses.

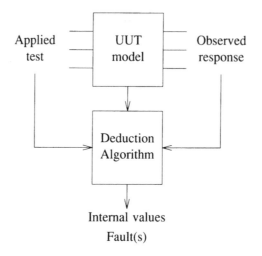

Figure 12.20   Effect-cause analysis

Internal values are computed by the *Deduction Algorithm*, which implements a line-justification process whose primary goal is to justify all the values obtained at the POs, given the tests applied at the PIs. The values of a normal PI are the values set by the applied tests, and the values of every other normal line must be justified by values of its predecessors. The following example introduces the basic concepts of the Deduction Algorithm.

Example 12.10:   Consider the circuit and the test shown in Figure 12.21.

Assume that the observed response is $D=101$. Because D has both 0 and 1 values, D is normal. Then its values must be justified by those of B and C. First $D=1$ in t_1 implies $B=C=1$ in t_1. The value 1 deduced for B shows that B is not s-a-0; therefore in every test that applies 1 to the PI B, B will have value 1. Hence $B=1$ in t_1 implies $B=1$ in t_2. This, in turn, implies $C=0$ in t_2 (because $D=0$ in t_2 and D is normal). At this point, both 0 and 1 values have been deduced for C, hence C is a normal line. Then all the known values of C (in t_1 and t_2) are complemented and assigned to A. Because A is a PI, $A=0$ in t_1 shows that A will have value 0 in every test that sets $A=0$; hence $A=0$ in t_3. Then $C=1$ in t_3. From $D=1$ in t_3 we determine $B=1$ in t_3. Because t_3 applies a 0 value to B, this shows that B is s-a-1.   □

In this simple example, all internal values have been deduced via a chain of implications. The implications involving values of different lines in the same test — called *horizontal implications* — are similar to the backward and the forward implications described in Chapter 6, except that they apply only to lines identified as normal. The example also illustrates a new type of implications — called *vertical implications* — which involve values of the same line in different tests. Vertical implications are based on the concept of *forced value* (FVs).

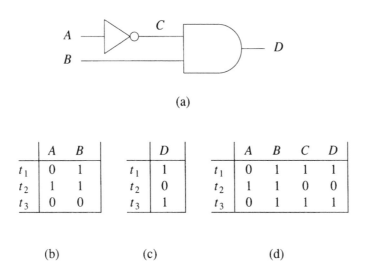

(a)

	A	B
t_1	0	1
t_2	1	1
t_3	0	0

	D
t_1	1
t_2	0
t_3	1

	A	B	C	D
t_1	0	1	1	1
t_2	1	1	0	0
t_3	0	1	1	1

(b)   (c)   (d)

Figure 12.21   (a) Circuit (b) Applied test (c) Obtained response (d) Deduced values

Definition 12.1: Let v be the expected value of line l in test t. We say that v is a *forced value*, if for every possible fault situation, either l has value v in t, or else l has value $\bar{v}$ in every test.

Clearly, all the expected values of a PI are FVs. In the previous example, either $A=0$ in t_1 (if A is normal or s-a-0), or else $A=1$ in every test (if A is s-a-1). The FVs are not restricted only to PIs. If C is normal, C propagates the values from A; hence either $C=1$ in t_1, or else $C=0$ in every test. The statement "$C=1$ in t_1, or else $C=0$ in every test" is also true when C is s-a-1 or s-a-0 and therefore is true for any fault situation involving A and C. Since the fault situation of the other lines does not affect C, the statement is true for any fault situation in the circuit; hence C has FV 1 in t_1. The same reasoning can be applied for t_2 or t_3; thus we can conclude that the FVs of the output of an inverter are the complements of the FVs of its input. Because a normal fanout branch takes all the values of its stem, it is easy to show that a fanout branch has the same FVs as its stem. The propagation of FVs through multi-input gates is based on the following theorem.

Theorem 12.1: If all inputs of a gate G have the same FV in a test t, than the corresponding output value of G is a FV in t.

Proof: Without loss of generality, assume that G is an AND gate. First assume that all inputs of G have FV 1 in t. If G is normal, its value is determined by its inputs; then either $G=1$ in t (if all its inputs are 1), or else $G=0$ in every test (if at least one input is 0 in every test). The same is true if G is s-a-1 or s-a-0, hence G has FV 1 in t. Now assume that all inputs of G have FV 0 in t. Then either at least one input is 0 in t, or else all inputs are 1 in every test. If G is normal, then either $G=0$ in t, or else $G=1$ in every test. The same is true if G is s-a-0 or s-a-1, hence G has FV 0 in t. Therefore, if all inputs of G have the same FV in t, the corresponding output value of G is also a FV in t.   □

The following example illustrates the computation and the use of FVs by the
Deduction Algorithm.

Example 12.11: Consider the circuit and tests given in Figure 12.22. Figure 12.22(c)
shows the FVs computed as a preprocessing step, starting with the FVs of the PIs
(ignore now the values in parentheses).

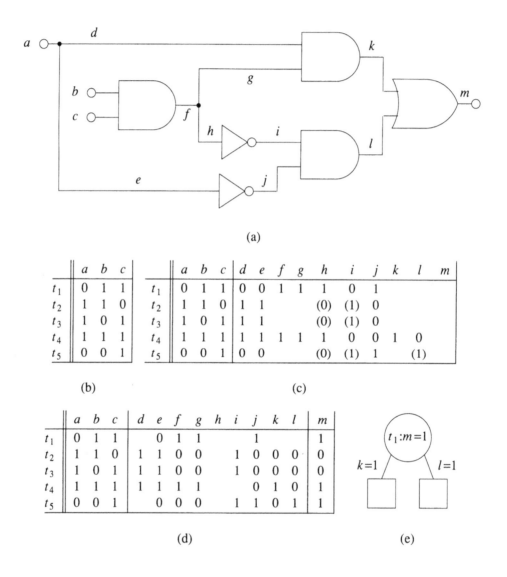

(a)

(b)

	a	b	c
t_1	0	1	1
t_2	1	1	0
t_3	1	0	1
t_4	1	1	1
t_5	0	0	1

(c)

	a	b	c	d	e	f	g	h	i	j	k	l	m
t_1	0	1	1	0	0	1	1	1	0		1		
t_2	1	1	0	1	1			(0)	(1)	0			
t_3	1	0	1	1	1			(0)	(1)	0			
t_4	1	1	1	1	1	1	1	1	1	0	0	1	0
t_5	0	0	1	0	0			(0)	(1)	1		(1)	

(d)

	a	b	c	d	e	f	g	h	i	j	k	l	m
t_1	0	1	1		0	1	1		1				1
t_2	1	1	0	1	1	0	0		1	0	0	0	0
t_3	1	0	1	1	1	0	0		1	0	0	0	0
t_4	1	1	1	1	1	1	1			0	1	0	1
t_5	0	0	1		0	0	0		1	1	0	1	1

(e)

$t_1 : m = 1$

$k = 1$ $l = 1$

Figure 12.22 (a) Circuit (b) Applied tests (c) Forced values (d) Obtained response
and initial implications (e) Decision tree

Assume that the obtained response is 10011. Figure 12.22(d) shows the implications
obtained by processing this response. (A good way to follow this example is to mark
values as they are deduced.) First we determine that m is normal. In t_2 and t_3, $m=0$

implies $k=l=0$. The FV 0 of l in t_4 means that either $l=0$ in t_4, or else $l=1$ in every test. Since 0 values have been deduced for l, it follows that $l=0$ in t_4 (this is an instance of vertical implication). This in turn implies $k=1$ in t_4. Hence k is normal. From $k=1$ in t_4, we deduce $d=g=1$, both of which generate vertical implications, namely $d=1$ in t_2 and t_3, and $g=1$ in t_1. In t_2 and t_3, $d=1$ implies $g=0$ (to justify $k=0$). Then g becomes normal, and all its known values are assigned to its stem f, which is also identified as normal. Next, $f=1$ in t_1 implies $b=c=1$, which, by vertical implications, show that $b=1$ in t_2 and t_4, and $c=1$ in t_3, t_4 and t_5. Then $f=0$ in t_2 is justified by $c=0$, and $f=0$ in t_3 by $b=0$. The latter implies $b=0$ in t_5, which propagates forward to make $f=g=k=0$. From $k=0$ and $m=1$ in t_5 we deduce $l=1$, which shows that l is normal. Then $l=1$ implies $i=j=1$ in t_5, and $j=1$ in t_5 implies $j=1$ in t_1.

Known values of a stem generate new FVs on its fanout branches. For example, $f=0$ in t_2 shows that either $h=0$ in t_2 or else $h=1$ in every test; hence h has FV 0 in t_2. Then the deduced values of f in t_2, t_3 and t_5 propagate as FVs on h and further on i. (The new FVs — enclosed in parentheses — are added to those in Figure 12.22(c)). Because $i=1$ in t_5, i must also have 1 values in all tests where it has FV 1; hence $i=1$ in t_2 and t_3. From these we deduce $j=0$ in t_2 and t_3; then $j=0$ in t_4 is obtained by vertical implication. Now j becomes normal, and its values are propagated to e and to a.

Now all the values given in Figure 12.22(d) have been deduced and no more implications can be made. The only value not yet justified is $m=1$ in t_1. First we try $k=1$ (see the decision tree in Figure 12.22(e)); this implies $d=1$ in t_1, which results in $d=1$ in t_5 (because d has FV 0 in t_1). At this point we have obtained a *solution*, i.e., a consistent set of values where all known values of the normal lines are justified. The only lines not reported as normal are d, i, and h. Because $d=1$ in every test, d is identified as *s-a-1*. (Moreover, because a is normal, we can locate an *open* between a and d). In addition, i may also be *s-a-1*, as only 1 values have been deduced for i. A second solution is obtained by justifying $m=1$ in t_1 by $l=1$; this identifies i *s-a-1* as another fault that can generate the analyzed response.   □

Unlike conventional diagnosis methods, *the effect-cause analysis does not use the concept of error*; in the above examples, faults were located without knowing the expected output values. The effect-cause analysis has also been extended to sequential circuits [Abramovici and Breuer 1982], where, because it does not require predictable errors, it can diagnose even faults that prevent initialization.

The Deduction Algorithm tries to justify simultaneously all the values obtained at the POs. Like the line justification algorithms described in Chapter 6, it alternates between implications and decisions. Backtracking is used either to recover from incorrect decisions, or to generate all solutions. Every solution identifies a possible fault situation in the circuit under test. When the Deduction Algorithm fails to obtain any solution, the processed response cannot be generated by any (single or multiple) stuck-at fault. Here the effect-cause analysis recognizes the presence of a fault not consistent with the assumed stuck-fault fault model, such as a bridging fault or a functional fault.

If the processed response happens to be the expected one, then one solution identifies every line as normal. If the Deduction Algorithm generates additional solutions, they *identify faults not detected by the applied tests*.

The effect-cause analysis can also help guided-probe testing [Abramovici 1981]. This application uses only the implication part of the Deduction Algorithm and replaces the decision process (and its associated backtracking) with probing. In other words, instead of making decisions about values on the inputs of a gate whose value is not justified, we probe them (if possible). This guided-probe testing algorithm achieves a higher diagnostic resolution than conventional methods and requires less probing points.

12.8 Diagnostic Reasoning Based on Structure and Behavior

Like the effect-cause analysis, the diagnosis technique described by Davis [1984] analyzes the observed response of the UUT, but it is more general since it is not limited to a particular fault model. Unlike a rule-based expert system using shallow knowledge about a specific digital system, this method is general and it uses *deep knowledge* based on the structure and the behavior of the circuit under test. Such a diagnosis approach relying on *model-based reasoning* is also referred to as a *first principles* approach.

The behavior of the components in the structural model of the UUT is described by *simulation rules* and *inference rules*. Simulation rules represent input-to-output value mapping, while inference rules correspond to backward implication. The behavior of a component is considered as introducing a *constraint* on the values of its inputs and outputs, and the entire circuit is regarded as a *constraint network*.

The initial analysis makes the following assumptions.

- The interconnections are fault-free and only the components can be faulty.

- No more than one module is faulty.

- Faulty components do not increase their number of states.

(These simplifying assumptions are explicit and they can be relaxed later.)

The first step is to determine an initial set of suspected components, that is, components whose incorrect behavior may explain the output errors. Based on the single faulty-module assumption, this step is done by intersecting the sets of components affecting the outputs with errors.

The next step is checking the validity of every suspected component by a technique called *constraint suppression*. It consists of temporarily suspending all the behavior rules of a suspected component and checking whether the resulting behavior of the circuit is consistent with the observed response.

Example 12.12: Consider the circuit in Figure 12.23, composed of three multipliers (M1, M2, and M3) and two adders (A1 and A2) interconnected by busses. The values shown are the decimal representations of the bus values.

Assume that instead of the expected value $F = 12$, in the first test we obtain the erroneous response $F = 10$ (Figure 12.23(a)). The suspected components affecting the output F are A1, M2, and M1. First we check whether A1 can be the faulty component by suspending the constraints introduced by A1 (i.e., we disable the

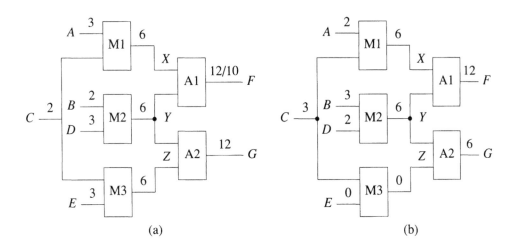

Figure 12.23

relation $F = X + Y$). Since the other components work correctly, the symptom describing the misbehavior of A1 is $\{X=6, Y=6, F=10\}$. In the second test (Figure 2.23(b)), A1 receives the same inputs ($X=6$, $Y=6$), and based on its behavior observed in the first test, we expect $F = 10$. The observed value, however, is the correct one ($F=12$). Based on the assumption that the faulty A1 is still a combinational component, we reached a contradiction which proves that A1 cannot be faulty.

Next, we restore the correct rules of A1 and check whether M2 can be faulty. Since M1 is fault-free, we deduce $X = 6$. Since A1 is fault-free, from $X = 6$ and $F = 10$ we infer that $Y = F - X = 4$. But $Y = 4$, $Z = 6$, and A2 fault-free imply $G = 10$, which contradicts the observed value $G = 12$. Therefore M2 cannot be faulty.

The last suspected component is M1. Since M2 is fault-free, we have $Y = 6$ in the first test. From $F = 10$, $Y = 6$, and A1 fault-free we infer $X = F - Y = 4$. Thus the symptom describing the misbehavior of M1 is $\{A=3, C=2, X=4\}$. (In the second test, however, M1 works correctly). Thus we conclude that M1 is faulty. If we have a hierarchical structural model, the analysis can continue inside M1 using a lower-level structural model of the multiplier.   □

If no consistent explanation of the observed response can be found, the reasoning system is capable of relaxing the initial assumptions and repeat the analysis. For example, if no single faulty component can generate the observed response, the system may assume that two components are simultaneously faulty, and repeat the constraint suppression analyzing pairs of suspected components.

Other systems performing model-based diagnosis are described in [Genesereth 1984] and [Purcell 1988].

REFERENCES

[Abramovici 1981] M. Abramovici, "A Maximal Resolution Guided-Probe Testing Algorithm," *Proc. 18th Design Automation Conf.*, pp. 189-195, June, 1981.

[Abramovici 1982] M. Abramovici, "A Hierarchical, Path-Oriented Approach to Fault Diagnosis in Modular Combinational Circuits," *IEEE Trans. on Computers*, Vol. C-31, No. 7, pp. 672-677, July, 1982.

[Abramovici and Breuer 1980] M. Abramovici and M. A. Breuer, "Multiple Fault Diagnosis in Combinational Circuits Based on an Effect-Cause Analysis," *IEEE Trans. on Computers*, Vol. C-29, No. 6, pp. 451-460, June, 1980.

[Abramovici and Breuer 1982] M. Abramovici and M. A. Breuer, "Fault Diagnosis in Synchronous Sequential Circuits Based on an Effect-Cause Analysis," *IEEE Trans. on Computers*, Vol. C-31, No. 12, pp. 1165-1172, December, 1982.

[Arzoumanian and Waicukauski 1981] Y. Arzoumanian and J. Waicukauski, "Fault Diagnosis in an LSSD Environment," *Digest of Papers 1981 Intn'l. Test Conf.*, pp. 86-88, October, 1981.

[Chang and Heimbigner 1974] H. Y. Chang and G. W. Heimbigner, "LAMP: Controllability, Observability, and Maintenance Engineering Technique (COMET)," *Bell System Technical Journal*, Vol. 53, No. 8, pp. 1505-1534, October, 1974.

[Davis 1984] R. Davis, "Diagnostic Reasoning Based on Structure and Behavior," *Artificial Intelligence*, Vol. 24, pp. 347-410, December, 1984.

[Genesereth 1984] M. R. Genesereth, "The Use of Design Descriptions in Automated Diagnosis," *Artificial Intelligence*, Vol. 24, pp. 411-436, December, 1984.

[Groves 1979] W. A. Groves, "Rapid Digital Fault Isolation with FASTRACE," *Hewlett-Packard Journal*, pp. 8-13, March, 1979.

[Hartley 1984] R. T. Hartley, "CRIB: Computer Fault-Finding Through Knowledge Engineering," *Computer*, Vol. 17, No. 3, pp. 76-83, March, 1984.

[Havlicsek 1986] B. L. Havlicsek, "A Knowledge Based Diagnostic System for Automatic Test Equipment," *Proc. Intn'l. Test Conf.*, pp. 930-938, September, 1986.

[Hsu *et al.* 1981] F. Hsu, P. Solecky, and R. Beaudoin, "Structured Trace Diagnosis for LSSD Board Testing — An Alternative to Full Fault Simulated Diagnosis," *Proc. 18th Design Automation Conf.*, pp. 891-897, June, 1981.

[Kime 1979] C. R. Kime, "An Abstract Model for Digital System Fault Diagnosis," *IEEE Trans. on Computers*, Vol. C-28, No. 10, pp. 754-767, October, 1979.

[Kochan *et al.* 1981] S. Kochan, N. Landis, and D. Monson, "Computer-Guided Probing Techniques," *Digest of Papers 1981 Intn'l. Test Conf.*, pp. 253-268, October, 1981.

[Koren and Kohavi 1977] I. Koren and Z. Kohavi, "Sequential Fault Diagnosis in Combinational Networks," *IEEE Trans. on Computers*, Vol. C-26, No. 4, pp. 334-342, April, 1977.

[Laffey *et al.* 1986] T. J. Laffey, W. A. Perkins, and T. A. Nguyen, "Reasoning About Fault Diagnosis with LES," *IEEE Expert*, pp. 13-20, Spring, 1986.

[Mullis 1984] R. Mullis, "An Expert System for VLSI Tester Diagnostics," *Proc. Intn'l. Test Conf.*, pp. 196-199, October, 1984.

[Nau 1983] D. S. Nau, "Expert Computer Systems," *Computer*, Vol. 16, No. 2, pp. 63-85, February, 1983.

[Purcell 1988] E. T. Purcell, "Fault Diagnosis Assistant," *IEEE Circuits and Devices Magazine*, Vol. 4, No. 1, pp. 47-59, January, 1988.

[Rajski 1988] J. Rajski, "GEMINI — A Logic System for Fault Diagnosis Based on Set Functions," *Digest of Papers 18th Intn'l. Symp. on Fault-Tolerant Computing*, pp. 292-297, June, 1988.

[Richman and Bowden 1985] J. Richman and K. R. Bowden, "The Modern Fault Dictionary," *Proc. Intn'l. Test Conf.*, pp. 696-702, November, 1985.

[Savir and Roth 1982] J. Savir and J. P. Roth, "Testing for, and Distinguishing Between Failures," *Digest of Papers 12th Annual Intn'l. Symp. on Fault-Tolerant Computing*, pp. 165-172, June, 1982.

[Shirley and Davis 1983] M. Shirley and R. Davis, "Generating Distinguishing Tests Based on Hierarchical Models and Symptom Information," *Proc. Intn'l. Conf. on Computer Design*, pp. 455-458, October, 1983.

[Tamama and Kuji 1986] T. Tamama and N. Kuji, "Integrating an Electron-Beam System into VLSI Fault Diagnosis," *IEEE Design & Test of Computers*, Vol. 3, No. 4, pp. 23-29, August, 1986.

[Tendolkar and Swann 1982] N. N. Tendolkar and R. L. Swann, "Automated Diagnostic Methodology for the IBM 3081 Processor Complex," *IBM Journal of Research and Development*, Vol. 26, No. 1, pp. 78-88, January, 1982.

[Tulloss 1978] R. E. Tulloss, "Size Optimization of Fault Dictionaries," *Digest of Papers 1978 Semiconductor Test Symp.*, pp. 264-265, October, 1978.

[Tulloss 1980] R. E. Tulloss, "Fault Dictionary Compression: Recognizing When a Fault May Be Unambiguously Represented by a Single Failure Detection," *Digest of Papers 1980 Test Conf.*, pp. 368-370, November, 1980.

[Wilkinson 1984] A. J. Wilkinson, "A Method for Test System Diagnostics Based on the Principles of Artificial Intelligence," *Proc. Intn'l. Test Conf.*, pp. 188-195, October, 1984.

[Yau 1987] C. W. Yau, "ILIAD: A Computer-Aided Diagnosis and Repair System," *Proc. Intn'l. Test Conf.*, pp. 890-898, September, 1987.

PROBLEMS

12.1 For the circuit shown in Figure 12.2(a), show that the faults d s-a-1 and i s-a-1 are indistinguishable.

12.2 For the circuit and the test set shown in Figure 12.2, determine the fault resolution of a fault dictionary built

a. without fault dropping;

b. with fault dropping after the first detection;

c. with fault dropping after two detections.

12.3 Construct a diagnosis tree for the circuit shown in Figure 12.2(a), assuming that the tests given in Figure 12.2(b) are applied in reverse order (i.e., t_5, ..., t_1).

12.4 For the circuit of Figure 12.2(a), try to generate a test that distinguishes between the faults (a) g s-a-1 and h s-a-1 (b) g s-a-1 and j s-a-1.

12.5 For the circuit and the faulty responses shown in Figure 12.2, find the closest match for the response produced by the bridging fault (a,f).

12.6 For the circuit and the test shown in Figure 12.21, try to deduce the internal values corresponding to the response $D = 010$.

13. SELF-CHECKING DESIGN

About This Chapter

In previous chapters we have considered the problems associated with detection of faults by observation of responses to tests. In this chapter we will consider some design procedures that simplify fault diagnosis or detection: specifically design of *self-checking* systems in which faults can be automatically detected by a subcircuit called a *checker*. Such circuits imply the use of coded inputs. After a brief introduction to a few fundamental concepts of coding theory, specific kinds of codes, including parity-check codes, Berger codes, and residue codes, are derived, and self-checking designs of checkers for these codes are presented.

13.1 Basic Concepts

In some cases it may be possible to determine from the outputs of a circuit C whether a certain fault f exists within the circuit without knowing the value of the expected response. This type of testing, which can be performed "on-line" is based on checking some invariant properties of the response. In this case, it is unnecessary to test explicitly for f, and the circuit is said to be self-checking for f. Another circuit, called a *checker*, can be designed to generate an error signal whenever the outputs of C indicate the presence of a fault within C. It is desirable to design circuits, including checkers, to be self-checking to as great an extent as possible (i.e., for as many faults as possible).

For an arbitrary combinational circuit with p inputs and q outputs, all 2^p input combinations can occur, as can all 2^q possible output combinations. If all possible output combinations can occur, it is impossible to determine whether a fault is present by just observing the outputs of the circuit, assuming no knowledge of the corresponding inputs. However, if only $k < 2^q$ output configurations can occur during normal operation, the occurrence of any of the $2^q - k$ other configurations indicates a malfunction (regardless of the corresponding input). Thus, faults that result in such a an "illegal" output can be detected by a hardware checker.

Example 13.1: Consider a circuit that realizes the combinational functions $f_1(x_1,x_2)$ and $f_2(x_1,x_2)$ described by the truth table of Figure 13.1(a). Note that the output configuration $f_1 = f_2 = 1$ never occurs. Any fault that leads to this configuration can be automatically detected by the checker shown in Figure 13.1(b), which generates a 1-output, indicating an error, if and only if $f_1 = f_2 = 1$. Note that this checker will fail to detect faults that cause an incorrect but "legal" output configuration. □

In a circuit that has output configurations that do not occur during fault-free (i.e., normal) operation, the outputs that do occur are called *(valid) code words,* and the nonoccurring configurations are called *invalid,* or *noncode words.* We will speak of both input codes and output codes in this manner. Considerable work has been done on specifying codes that are useful for detecting and correcting errors.

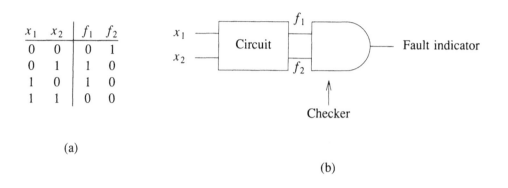

x_1	x_2	f_1	f_2
0	0	0	1
0	1	1	0
1	0	1	0
1	1	0	0

(a)

(b)

Figure 13.1   Automatic fault checking for Example 13.1

13.2 Application of Error-Detecting and Error-Correcting Codes

Codes are commonly classified in terms of their ability to detect or correct classes of errors that affect some fixed number of bits in a word. Thus a code is *e-error detecting* if it can detect any error affecting at most e bits. This implies that any such error does not transform one code word into another code word. Similarly a code is *e-error correcting* if it can correct any error affecting at most e bits. This implies that any two such errors e_1, e_2, affecting words w_1 and w_2, respectively, do not result in the same word.

The *Hamming distance d* of a code is the minimum number of bits in which any two code words differ. The error-detecting and error-correcting capability of a code can be expressed in terms of d as shown in the table of Figure 13.2. (The proof of these results can be found in virtually any book on coding theory).

In general, additional outputs, called *check bits*, must be generated by a circuit in order that its output word constitute a code with useful error capabilities. The *parity-check code* is the simplest such code. For this code, $d = 2$, and the number of check bits is one (independent of the number of output bits in the original circuit). There are two types of parity-check code, even and odd. For an even code the check bit is defined so that the total number of 1-bits is always even; for an odd code, the number is always odd. Consequently any error affecting a single bit causes the output word to have an incorrect number of 1-bits and hence can be detected. Note that in an arbitrary circuit a single fault may cause an error in more than one output bit due to the presence of fanout. Hence care must be taken in designing a circuit in which detection of errors is based on a code with limited error-detecting capabilities.

Example 13.2:   For the functions f_1, f_2 of Example 13.1, the check bit y_e for an even parity-check code is defined so that for any input the total number of 1-output bits among f_1, f_2, and y_e is even (Figure 13.3). Thus if $x_1 = 0$, $x_2 = 1$, since $f_1 = 1$ and $f_2 = 0$, then y_e must have value 1 so that the total number of 1-bits among f_1, f_2, and y_e will be even. For an odd parity-check code, the check bit y_o would be as shown in Figure 13.3. Note that y_e and y_o are always complements of each other.   □

d	Capability
1	none
2	1-error detection, 0-error correction
3	2 -error detection, 1-error correction
.	
.	
.	
$e + 1$	e-error detection, $\lceil \dfrac{e}{2} \rceil$ -error correction
.	
.	
.	
$2e + 1$	$2e$-error detection, e-error correction

Figure 13.2   Capability of a code with distance d

x_1	x_2	f_1	f_2	y_e	y_o
0	0	0	1	1	0
0	1	1	0	1	0
1	0	1	0	1	0
1	1	0	0	0	1

Figure 13.3

The check bits required in a code can be thought of as constituting redundancy, since they are only required for error detection. The other bits are called *information bits*. A generalized class of parity-check codes that have greater error-detecting and/or error-correcting capability can be defined. A single-error-correcting code for q information bits requires c check bits where $2^c \geq q + c + 1$. The value of c for various values of q is shown in Figure 13.4.

The c check bits and q information bits form a word with $(c+q)$ bits, $b_{c+q} \cdots b_2 b_1$. In the conventional Hamming code the check bits occur in bits b_{2^i}, $0 \leq i \leq c-1$. The values of these check bits are defined by c parity-check equations. Let p_j be the set of integers whose binary representation has a 1-value in position b_j, (i.e., $p_j = \{I \mid b_j(I) = 1\}$, where $b_j(n)$ denotes the value of the j-th bit (from the right) of a binary integer n. Then the values of the check bits are defined by the c parity-check equations of the form

$$\sum_{k \in P_i} b_k = 0 \qquad i = 1, ..., c$$

q	c
1	2
4	3
11	4
26	5
57	6
120	7

Figure 13.4   Values of q and c for single-error-correcting parity-check codes

where the sum is modulo 2. An error in bit b_j will result in incorrect parity for exactly those equations for which j is in p_i. Thus the erroneous bit can be computed from these c parity-check equations.

Example 13.3:   Let $q = 4$. Then $2^c \geq q + c + 1 = 5 + c$, and hence $c = 3$. The single-error-correcting code defines a 7-bit word with check bits b_1, b_2, and b_4. The value of the check bits for any given value of the information bits can be computed from the three parity-check equations.

$$b_1 \oplus b_3 \oplus b_5 \oplus b_7 = 0 \quad \text{defined by } p_1 \tag{1}$$

$$b_2 \oplus b_3 \oplus b_6 \oplus b_7 = 0 \quad \text{defined by } p_2 \tag{2}$$

$$b_4 \oplus b_5 \oplus b_6 \oplus b_7 = 0 \quad \text{defined by } p_3 \tag{3}$$

Thus if the information bits have the values $b_3 = 1$, $b_5 = 0$, $b_6 = 0$, and $b_7 = 1$, then the check bits will have the values $b_1 = 0$ (defined by substitution in equation 1 above), $b_2 = 0$ (from equation 2) and $b_4 = 1$ (from equation 3), and thus the encoded word is 1 0 0 1 1 0 0. If an error occurs in bit position b_4, the word becomes 1 0 0 0 1 0 0. If we recompute the three parity-check equations from this word, we derive the values 1 0 0 from equations 3, 2, and 1 respectively. The binary number formed from these three equations is 100 (binary 4) thus indicating an error in bit b_4. Thus the single-bit error can be corrected.   □

The use of codes enables hardware to be designed so that certain classes of errors in the circuit can be automatically detected or corrected through the use of a hardware checker, as shown in Figure 13.5. Note that errors in the checker may not be detected. Many codes have been developed that can be used in the design of such self-checking circuits. The type of code to be used may vary depending on the type of circuit. For data-transmission busses, a parity-check code may be adequate. For other types of functions, however, we may wish to use a code for which the check bits of the result can be determined from the check bits of the operands.

Consider the design of an arithmetic adder using an even parity-check code. For the two additions illustrated in Figure 13.6, in both cases the operand check bits are 0 and 1 respectively, but the check bit of the sum $A + B_1$ is 0 and the check bit of the sum

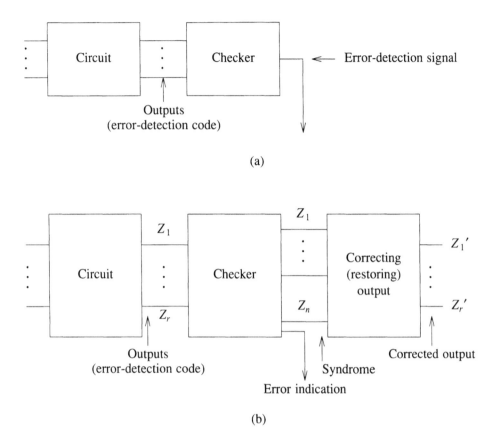

Figure 13.5   (a) Error-detecting circuit (b) Error-correcting circuit

$A + B_2$ is 1. Therefore it would be necessary to recompute the value of the check bit after each addition, and errors in the addition itself would not be detected.

Another class of codes, called *residue codes*, has the desirable property that for the arithmetic operations of addition, subtraction, and multiplication, the check bits of the result can be determined directly from the check bits of the operands. This property is called *independent checking*. Several different types of residue code have been formulated. We will consider only one such class. In this code the rightmost p bits are the check bits. The check bits define a binary number C, and the information bits define another number N. The values of the check bits are defined so that $C = (N)$ modulo m, where m is a parameter called the *residue* of the code, and the number of check bits is $p = \lceil \log_2 m \rceil$.

Example 13.4:   Consider the derivation of the check bits for a residue code with three information bits I_2, I_1, I_0 and $m = 3$. Since $\lceil \log_2 m \rceil = 2$, check bits C_1 and C_0 are

$$A \quad = 0\ 0\ 0\ 1 \qquad\qquad A \quad = 0\ 0\ 0\ 1$$
$$\underline{B_1 = 0\ 1\ 0\ 1} \qquad\qquad \underline{B_2 = 0\ 0\ 1\ 1}$$
$$A + B_1 = 0\ 1\ 1\ 0 \qquad\qquad A + B_2 = 0\ 1\ 0\ 0$$

$$C(A) = 1,\ C(B_1) = C(B_2) = 0,\ C(A + B_1) = 0,\ C(A + B_2) = 1$$
$$C(X) \text{ is the check bit of } X$$

Figure 13.6   Parity not preserved by addition

required. Their values are defined so that the binary number $C = C_1 C_0 =$ (N) modulo 3, where N is the binary number $I_2 I_1 I_0$, as shown in Figure 13.7.   □

I_2	I_1	I_0	N	C	C_1	C_0
0	0	0	0	0	0	0
0	0	1	1	1	0	1
0	1	0	2	2	1	0
0	1	1	3	0	0	0
1	0	0	4	1	0	1
1	0	1	5	2	1	0
1	1	0	6	0	0	0
1	1	1	7	1	0	1

Figure 13.7   A 3-bit residue code with $m = 3$

We will now prove that for this type of residue code, the check bits of the sum (product) of a set of operands is equal to the sum (product) of the check bits of the operands.

Theorem 13.1:   Let $\{a_i\}$ be a set of operands. The check bits of the sum are given by $(\sum a_i) \bmod m$, and the check bits of the product by $(\Pi\ a_i) \bmod m$. Then

a. The check bits of the sum are equal to the sum of the check bits of the operands modulo m.

b. The check bits of the product are equal to the product of the check bits of the operands modulo m.

Proof

a. Let $a_i = k_{i1} m + k_{i2}$ where $0 \le k_{i2} < m$. Then $(\sum a_i) \bmod m = (\sum (k_{i1}\ m + k_{i2}))$ $\bmod m = (\sum k_{i2}) \bmod m = (\sum (a_i) \bmod m) \bmod m$ (since k_{i2} is the residue of a_i).

b. The proof of part b is similar to part a and is left as an exercise.   □

Example 13.5

a. Consider the addition shown in Figure 13.8(a) using residue codes with $m = 3$. The information bits of the sum represent the number 6 and (6) mod 3 = 0. The sum of the check bits mod 3 is also 0. Thus the sum of the check bits modulo m is equal to the check bits of the sum.

b. Consider the multiplication shown in Figure 13.8(b) using residue codes with $m = 3$. The product of the check bits modulo m is equal to 2. The information bits of the product represent the number 8, and (8) mod 3 = 2. Thus the product of the check bits modulo m is equal to the check bits of the product.   □

Information bits			Check bits			Information bits			Check bits		
0	0	1	0	1	0	0	0	1	0	1	0
0	1	0	0	0	1	0	1	0	0	0	1
0	1	1	0	1	1	1	0	0	0	1	0

(a)	(b)

Figure 13.8

The use of residue codes to check addition is illustrated by the system of Figure 13.9, which computes the check bits, $C(A+B)$, of the sum $A + B$ and compares the result with the modulo m sum of the check bits, $(C(A)+C(B))$ mod m. This circuit will detect an error that causes these two computations to be unequal. Let us now consider the class of errors that can be detected by this class of residue codes.

If a residue code defines code words with $s = p + q$ bits, then an error pattern E can be defined as an s-bit binary vector in which $e_i = 1$ if bit i is in error and $e_i = 0$ if bit i is correct. For a number N with check bits C where $C = (N)$ mod m, such an error may change N to N' and/or C to C'. Such an error will be detected, provided $C' \neq N'$ mod m. Let us first consider single-bit errors.

Theorem 13.2: In a residue code with m odd, all single-bit errors are detected.

Proof: A single-bit error can affect either the value of C or N but not both. We therefore consider two cases.

Case 1: Assume the single erroneous bit is an information bit. If bit i of the information segment is in error, then $N' = N \pm 2^i$ and $C' = C$. Then N'mod $m = (N \pm 2^i)$ mod $m = N$ mod $m \pm (2^i)$ mod $m = C \pm (2^i)$ mod m. For m odd, (2^i) mod $m \neq 0$ and hence the error is detected.

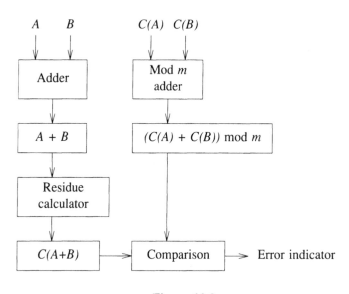

Figure 13.9

Case 2: Assume the single erroneous bit is a check bit. If bit i of the check segment is in error, then $C' = (C \pm 2^i) \bmod m \neq C$ for m odd. Hence all single-bit errors are detected.   □

If m is odd, some single-bit errors may be indistinguishable. If m is even, some single-bit errors may not even be detected.

Example 13.6

a. Let $m = 3$. Consider the code word 11000 in which the rightmost two bits are check bits. For the single-bit error affecting bit 4 (from the right), the erroneous word is 10000, which is not a code word, since $N = 4$, $C = 0$, and $N \bmod 3 = 1 \neq 0$. Therefore this error will be detected. However, the same erroneous word could result from the code word 00000 due to a single-bit error pattern affecting bit 5 (from the right). Thus this code can not distinguish these two single-bit errors and hence the code is not single-error correcting.

b. Let $m = 2$. Consider the code word 1100 with the rightmost bit as the only check bit. A single-bit error affecting bit 3 results in the word 1000. Since $I' = 4$ and $C' = 0$ and $I' \bmod 2 = 0 = C'$, this is a valid code word, and the single-bit error is not detected.   □

Thus the parameter m must be odd for the code to detect single errors. As m increases, the number of check bits required increases, and the error-detecting and error-correcting capabilities of the code vary with the specific value of m in a complex manner that will not be considered herein [Sellers *et al.* 1968].

13.3 Multiple-Bit Errors

The error-detecting capabilities of codes with respect to multiple-bit errors are also of interest. For a parity-check code, all errors that affect an odd number of bits will be detected, whereas even bit errors will not be. It should be noted, however, that in many technologies the most probable multiple-bit errors are not random but have some special properties associated with them such as *unidirectional errors* (all erroneous bits have the same value) and *adjacent-bit errors* (all bits affected by an error are contiguous). Some common codes are useful for detecting such multiple-bit errors.

The k/n (k-out-of-n) code consists of n-bit words in which each code word has exactly k 1-bits. These codes will detect all unidirectional errors, since a unidirectional error will either result in an increase or a decrease in the number of 1-bits in a word. This code is a *nonseparable code*, in which it is not possible to classify bits as check bits or information bits. When such codes are used, the circuitry involved must perform computations on the entire word rather than just the information bits of a word.

The *Berger codes* are separable codes. For I information bits, the number of check bits required is $C = \lceil \log_2(I + 1) \rceil$, forming a word of $n = I + C$ bits. The C check bits define a binary number corresponding to the Boolean complement of the number of information bits with value 1. Thus with three information bits, two check bits are required. If the information bits are 110, the check bits define the Boolean complement of 2 (since there are two information bits with value 1) and hence the check bits are 01. For $I = 3$ and $C = 2$, the complete code is shown in Figure 13.10.

I_1	I_2	I_3	C_1	C_0
0	0	0	1	1
0	0	1	1	0
0	1	0	1	0
0	1	1	0	1
1	0	0	1	0
1	0	1	0	1
1	1	0	0	1
1	1	1	0	0

Figure 13.10   Berger code for $I=3$ and $C=2$

Berger codes detect all unidirectional errors (Problem 13.5). Among separable codes that detect all unidirectional errors, Berger codes are optimal in requiring fewer check bits for I information bits. For large values of r, however, m-out-of-n codes require fewer bits to achieve r valid code words.

In the *modified residue code*, $m - 1$ check bits are defined so that the total number of 1-bits in a code word is a multiple of the parameter m. This code, which is a generalization of the parity-check code, detects all unidirectional errors affecting fewer than m bits. For $m = 3$, if the information bits have the value 110, the check bits have the value 01 or 10.

Given a set of kn bits arranged as k n-bit words, adding a parity-check bit to each word results in k $(n + 1)$-bit words in which any single-bit error can be detected as well as any multiple-bit error so long as no word has more than one erroneous bit. Alternatively, we can define one n-bit check word C, where for all i, $1 \leq i \leq n$, the i-th bit of C is a parity check over the i-th bit of all k words. This results in $(k + 1)$ n-bit words in which all single-bit errors are detected, all multiple-bit errors within one word are detected, and all multiple-bit errors affecting different bits of different words are detected. This technique can also be used with error-correcting codes to obtain multiple-bit error correction.

13.4 Checking Circuits and Self-Checking

The use of on-line testing (i.e., checking circuits in conjunction with coded outputs) to detect faults has the following advantages over off-line testing.

1. Intermittent faults are detected.

2. Output errors are detected immediately upon occurrence, thus preventing possible corruption of data.

3. The distribution of checkers throughout a digital system provides a good measure of the location of a fault by the location of the checker at which the error is detected.

4. The software diagnostic program is eliminated (or at least simplified substantially).

Checking by means of hardware can be combined with a general reliability strategy called *rollback*, in which the status of a system is periodically saved, and upon detection of an error the system is reconfigured to its most recent previous valid saved condition and the subsequent sequence of inputs is repeated. Repeated failures cause an interrupt. (This strategy is effective for intermittent failures of short duration).

There are also several disadvantages associated with on-line testing by means of hardware (i.e., checking circuits).

1. More hardware is required, including a hardware checker.

2. The additional hardware must be checked or tested (*checking the checker problem*). In general, some faults cannot be automatically detected.

Thus the use of hardware for testing raises the problem of how to handle faults in the checking unit. This has led to the study of *totally self-checking circuits* (*and checkers*) [Carter and Schneider 1968]. A circuit is *fault secure* for a set of faults F, if for any fault in F, and any valid (code) input, the output is a noncode word or the correct code word, never an invalid code word. A circuit is *self-testing* for a set of faults F if for any fault f in F, there exists an valid (code) input that detects f. A *totally self-checking circuit* is both fault secure and self-testing for all faults (in the set of faults of interest). Fault secureness insures that the circuit is operating properly if the output is a code word. Self-testing insures that it is possible for any fault to be detected during normal operation. Hence a totally self-checking circuit need not have a special diagnostic program but can be completely tested by means of hardware*. One intrinsic difficulty associated with such a method is the possibility of faults within the checker. For the system to be totally self-checking, both the circuit and the checker must also be self-testing and fault secure. The part of the circuit containing faults that cannot be detected in this manner must be tested by other means and is sometimes referred to as *hardcore*. It is desirable to design the checking circuit so that the hardcore is localized and/or minimized.

13.5 Self-Checking Checkers

Most of the research of self-checking has been related to the design of self-checking checkers. A checker (for a specific code) is usually defined as a single-output circuit which assumes the value 0 for any input corresponding to a code word and assumes the value 1 for any input corresponding to a noncode word. This output is used as an indication of error. Since the output is 0 for all code inputs, a checker cannot be designed in this manner so as to be self-testing for a s-a-0 fault on its output, since this fault can only be detected by applying an input corresponding to a noncode word, and this will never occur during normal operation. Signals that have only one possible value during normal operation are called *passive*. One approach to the design of self-checking checkers is to replace all passive signals by a pair of signals each of which, during normal operation, could assume both logical values, 0 and 1. Thus a checker could be designed, as shown in Figure 13.11, where the two outputs of C_1 take the values (0,1) or (1,0) for code inputs and (0,0) or (1,1) for noncode inputs. If we assume that a single signal is necessary to indicate error, additional logic is required to generate this single signal z from the two nonpassive signals. In Figure 13.11 the subcircuit C_2 generates z as the exclusive-OR of the two signals generated by C_1. However, since z is a passive signal, C_2 is always non-self-testing. Thus the hardcore (i.e., the logic that is not self-testable) has actually not been eliminated but has been localized and minimized.

The concept of realizing a passive function as a pair of nonpassive functions can be formalized. Consider a binary function f with values 0,1, which are mapped into two binary functions z_1, z_2 in which $(z_1,z_2) = (0,0)$ or $(1,1)$ corresponds to $f = 1$ and $(z_1,z_2) = (0,1)$ or $(1,0)$ corresponds to $f = 0$. The set of functions (z_1,z_2) is called a *morphic function* corresponding to f. It is possible to design some morphic functions so as to be totally self-checking.

---

* If a circuit is totally self-testing for a set of faults F, then it is possible for any fault in F to be detected during normal operation by hardware checking. Of significance is the probability that the fault will be detected within t units of time from its occurrence. The expected value of t has been referred to as the *error latency* of the circuit [Shedletsky and McCluskey 1975]. If F is the set of single stuck faults, hardware detection is only useful if the error latency of the circuit is much less than the expected time for a second fault to occur so that a multiple fault does not exist.

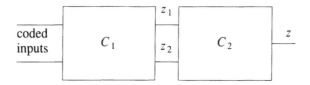

Figure 13.11

13.6 Parity-Check Function

Consider the 3-bit parity-check function represented by the Karnaugh map of
Figure 13.12. The corresponding morphic function is shown in Figure 13.13, where
each a_i can be 0 or 1.

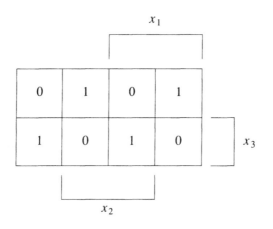

Figure 13.12   Karnaugh map of 3-input parity-check function

In general for an n-variable function there are 2^n variables a_i defined in the
corresponding morphic function. The question remains as to whether these values of
a_i can be selected so that both z_1 and z_2 are self-testing and hence self-checking. For
the parity-check function of Figure 13.13, the choice $a_i = 0$, $i \le 3$, $a_i = 1$, $i \ge 4$ leads
to the circuit shown in Figure 13.14(a), which can be verified to be totally
self-checking, where the code inputs correspond to all inputs with an even number of
1-bits. This type of realization can be generalized for an n-input parity-check function.
In the totally self-checking circuit, the set of n variables is partitioned into two disjoint
sets A_i, B_i, each with at least one variable, and parity-check functions of these two
variable sets are realized. The resultant circuit, shown in Figure 13.14(b), is totally
self-checking. (An inverter on one of the outputs is required to produce outputs of
(0,1) or (1,0) for even parity inputs.) This morphic function realization can be shown

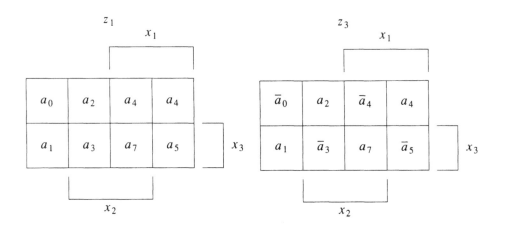

Figure 13.13 Karnaugh maps representing morphic function corresponding to 3-input parity-check function

to be a minor modification of a normal checker realization shown in Figure 13.15. Thus the hardcore of the normal circuit has now been placed with the decision logic (which interprets the error-identification signals $(0,0)$ and $(1,1)$ and produces a passive error signal). The hardcore has been *localized* rather than eliminated.

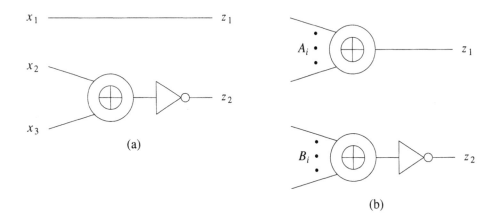

Figure 13.14 (a) Self-checking 3-bit parity checker (b) General self-checking parity checker

13.7 Totally Self-Checking m/n Code Checkers

A totally self-checking $k/2k$ code checker can also be designed [Anderson and Metze 1973]. The checker has two outputs, f and g. The $2k$ inputs are partitioned into two

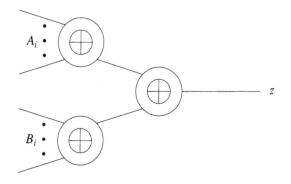

Figure 13.15 Parity-check circuit

disjoint sets of k inputs each, x_A and x_B. The function f is defined to have the value 1 if and only if i or more of the variables in x_A have the value 1 and $k - i$ or more of the variables in x_B have the value 1, for i odd. Similarly g is defined to have the value 1 if i or more variables in x_A and $k - i$ or more variables in x_B have the value 1, for i even. For code inputs (i.e., exactly k of the $2k$ variables have the value 1) then $f = 1$ and $g = 0$ or $g = 1$ and $f = 0$, while for noncode inputs with fewer than k 1-bits, $f = 0$ and $g = 0$, and for noncode inputs with more than k 1-bits, $f = g = 1$. The general form of the circuit is as shown in Figure 13.16, where T_{Aodd} has outputs for each odd value of i, and the ith output assumes the value 1 if i or more inputs assume the value 1. The subcircuits T_{Aeven}, T_{Bodd}, and T_{Beven} are similarly defined. It can be shown that this circuit is both self-checking and fault secure, and hence totally self-checking, for all single stuck faults. Other realizations of this checker have a large number of faults that cannot be tested by code word inputs.

The self-checking checker for $k/2k$ codes can also be used to realize a general k/n checker where $n \neq 2k$ [Marouf and Friedman 1978]. The general form of the circuit realization is shown in Figure 13.17. The AND array consists of a single level of k-input AND gates, one for each subset of k variables, which generate all possible products of k of the n variables. Thus, for a code word input, the output of this array will have exactly one 1-signal, while for noncode inputs with more than k 1-bits, two or more of the output signals will have the value 1, and for noncode inputs with fewer than k 1-bits, none of the output signals will have the value 1. The OR array consists of a single level of OR gates, which converts the $1/\binom{n}{k}$ code on the z signals to a $p/2p$ code, where p must satisfy the constraint

$$2p \leq \binom{n}{k} \leq \binom{2p}{p}.$$

The k/n checker design represented in Figure 13.17 might be considered inefficient, since it requires a great amount of hardware and *all* valid code words as tests in order

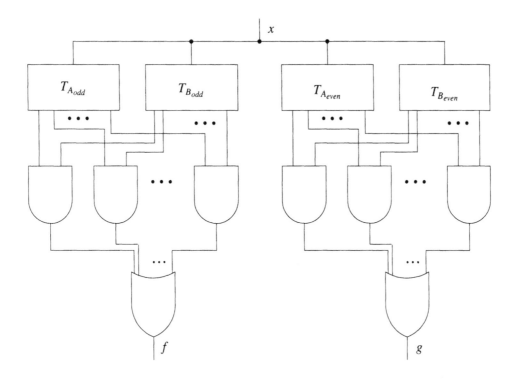

Figure 13.16   Totally self-checking $k/2k$ checker

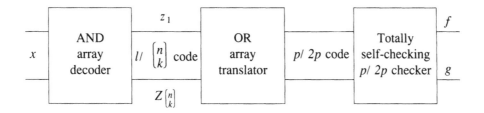

Figure 13.17   A k/n checker design

to detect all single stuck faults. Subsequent research was related to attempts to finding improved realizations that were totally self-checking and required either a simpler circuit design and/or fewer tests to detect all single stuck faults. A more efficient checker can be designed, as shown in Figure 13.18, consisting of three subcircuits, C_1, C_2, and C_3. The circuit C_1 has n inputs and Z outputs where $Z = 4$ for $m/(2m+1)$ codes, $Z = 5$ for $2/n$ codes, and $Z = 6$ for any other m/n code. In normal operation, C_1 receives m/n code words as inputs and produces a $1/Z$ code on its outputs. The circuit

C_2 translates the 1/Z code on its inputs to a 2/4 code on its outputs, and C_3 is a 2/4 code checker. All three of the subcircuits are designed as totally self-checking circuits. This type of realization can be shown in general to require a much smaller set of tests to detect all single stuck faults compared to the design shown in Figure 13.16.

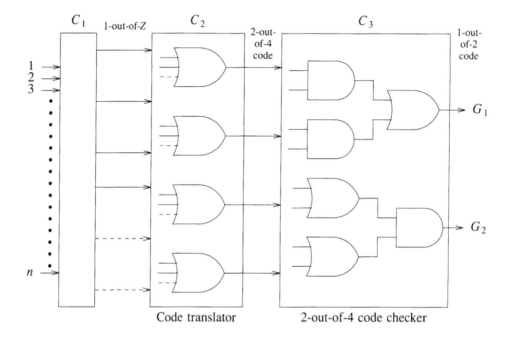

Figure 13.18 m-out-of-n code checker

13.8 Totally Self-Checking Equality Checkers

Totally self-checking equality checkers have also been designed [Anderson 1971]. These circuits test two k-bit words to determine if they are identical. For $k = 2$, the circuit of Figure 13.19 is a totally self-checking equality checker for the two words (a_1, a_2) and (b_1, b_2). This can be generalized for arbitrary values of k. Such an equality checker can be used as the basis for checkers for various operations and codes as illustrated previously in the design of checkers for addition using residue codes.

13.9 Self-Checking Berger Code Checkers

A general structure for a totally self-checking checker for separable codes is shown in Figure 13.20 [Marouf and Friedman 1978b]. The circuit N_1 generates check bits from the information bits. The equality checker, N_2, compares the check bits of the input word with the check bits generated by N_1.

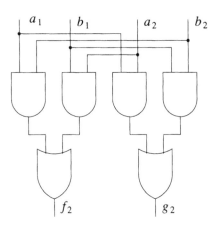

Figure 13.19   Totally self-checking equality checker

For the Berger code, the generator circuit N_1 can be constructed from full-adder modules, as shown in Figure 13.21, for the case $I = 7$, $C = 3$. A set of four test vectors is shown.

This type of design is easily generalized for the case where I, the number of information bits, is $2^k - 1$. The generator circuit, N_1, is easily tested with eight inputs sufficient to detect all single stuck faults within N_1 as well as all multiple faults occurring within a single full-adder module within N_1.

For the case where $I = 2^k - 1$, all of the 2^k possible combinations of check-bit values occur in some code words. Therefore a two-level realization of the equality checker, N_2, will be totally self-checking. However, such a realization is very inefficient both in terms of the number of gates required and the number of tests required to detect all single stuck faults. A much more efficient realization is a tree circuit formed as an interconnection of one-bit comparator modules. Although somewhat more complex, relatively efficient designs of both N_1 and N_2 can be derived for the cases where $I \neq 2^k - 1$ [Marouf and Friedman 1978b].

A significant amount of research has been directed toward the development of more efficient totally self-checking circuits for many of the more common codes, some of which have been considered herein [Jha 1989, Gastanis and Halatsis 1983, Nikolos *et al.* 1988], as well as to specific technologies such as PLAs [Mak *et al.* 1982]. Little has been done, however, in the way of a general theory for the design of totally self-checking circuits for arbitrary combinational or sequential functions.

13.10   Toward a General Theory of Self-Checking Combinational Circuits

We will now consider the general problem of self-checking circuit design for arbitrary functions. It is easily shown that the inputs of a self-checking circuit must be coded in a distance d code where d is at least 2 in order for the circuit to be fault secure with

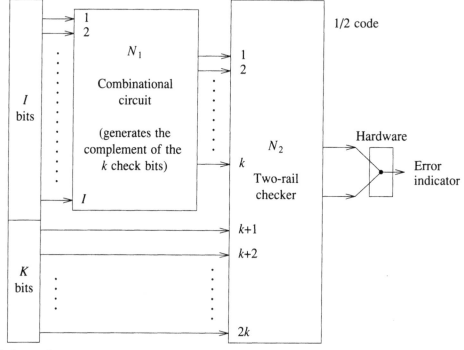

Figure 13.20 Totally self-checking checker for separable codes

respect to single stuck input faults. The outputs must be similarly coded to be fault secure with respect to single stuck output faults. If it is assumed that there are no input faults and the inputs are uncoded, then self-testing is automatically achieved and the circuit must only be designed to be fault secure. If the outputs of the circuit are defined to be of even parity, they define a distance 2 code. However, it is possible for a single fault to affect two output bits, as illustrated by the fault a s-a-0 with input $(x_1, x_2) = (0,1)$ in the circuit of Figure 13.22(b). In this case the circuit is not fault secure, since the normal output $(f_1, f_2, f_3) = (1,1,0)$ and the faulty output $(f_1, f_2, f_3) = (0,0,0)$ are both valid possible circuit outputs for some input value. The circuit can be redesigned to prevent such an error pattern, as shown in Figure 13.22(c), wherein duplicated signals have been generated to ensure odd parity errors. Note that each of the gates G_1 and G_2 of Figure 13.22(b) have been replaced by two gates, and it must be further assumed that no input fault can affect both of these gates. Thus if no input faults are considered, it is possible to design fault-secure realizations of arbitrary combinational functions.

As we have seen previously, if we assume that a single passive signal is required for error indication design of totally self-checking circuits cannot completely eliminate the hardcore. Referring to Figure 13.11, if C_1 is to be totally self-checking and C_2 is to represent the hardcore, the question arises whether the logic in C_1 can be substantially

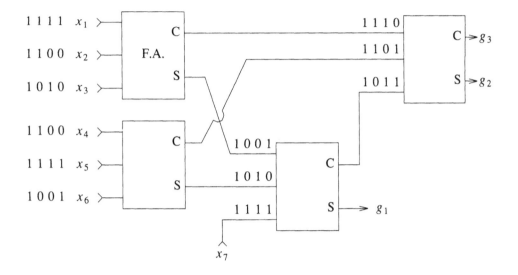

Figure 13.21   Self-checking Berger code checker, *I*=7, *C*=3

simplified if C_2 is made somewhat more complex. It appears likely that a general theory of self-checking circuits would not be restricted to two-output morphic functions and would explicitly recognize the trade-off between the hardcore and the degree of self-checking. In this connection a class of circuits that are totally fault secure but only partially self-testing, and a class of circuits that are totally self-testing but are only fault secure for a subset of all possible inputs may be considered. However, no general design procedures for such circuits have yet been developed.

13.11 Self-Checking Sequential Circuits

The use of codes and the concepts of self-checking design are also applicable to the design of self-checking sequential circuits. Consider the sequential-circuit model of Figure 13.23, in which the outputs and state variables are coded. Self-checking can be obtained by designing the combinational logic so that

1. For any fault internal to C^* and for any input, either the output and the next state are correct or the output and/or the next state is a noncode word.

2. For any state corresponding to a noncode word resulting from a fault f in C, and for any input, the next state generated by C with fault f is a noncode word and the output is also a noncode word.

---

*   If input faults are also to be considered, the inputs must be coded.

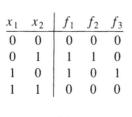

x_1	x_2	f_1	f_2	f_3
0	0	0	0	0
0	1	1	1	0
1	0	1	0	1
1	1	0	0	0

(a)

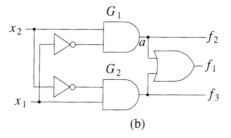

(b)

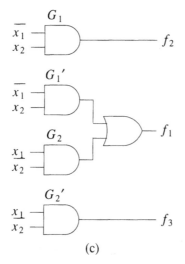

(c)

Figure 13.22   Redesign of a circuit to restrict effect of single stuck faults

The net result from these conditions is that C serves as a checker on the coded state variables.

A conceptually similar design procedure is shown in Figure 13.24. The outputs of the sequential circuit C_1 are encoded in a k/n code, which is input to a k/n checker. The checker generates a signal R_s, which is required for the next clock pulse (or in the case of asynchronous circuits, for the next input signal) to be generated. Thus for many faults the system will stop as soon as an erroneous output is generated.

A considerable amount of research has been focused on the application of the basic concepts of self-checking to the design of more complex systems including microprogrammed control units and microprocessors [Nicolaidis 1987, Paschalis *et al.* 1987].

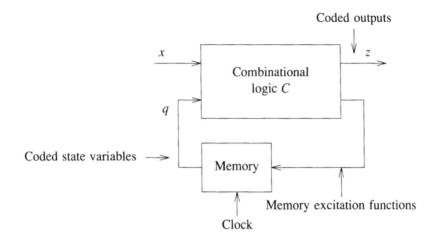

Figure 13.23 Sequential circuit model

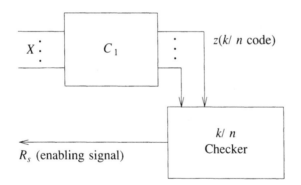

Figure 13.24 Use of codes for self-checking in sequential circuits

REFERENCES

[Anderson 1971] D. A. Anderson, "Design of Self-Checking Digital Networks Using Coding Technique," Univ. of Illinois CSL Report R-527, September, 1972.

[Anderson and Metze 1978] D. A. Anderson and G. Metze, "Design of Totally Self-Checking Check Circuits for *m*-out-of-*n* Codes," *IEEE Trans. on Computers*, Vol. C-22, No. 3, pp. 263-269, March, 1973.

[Carter and Schneider 1968] W. C. Carter and P. R. Schneider, "Design of Dynamically Checked Computers," *IFIP Proceedings*, Vol. 2, pp. 873-883, 1968.

[Gastanis and Halatsis 1983] N. Gastanis and C. Halatsis, "A New Design Method for *m-out-of-n* TSC Checkers," *IEEE Trans. on Computers*, Vol. C-32, No. 3, pp. 273-283, March, 1983.

[Jha 1989] N. K. Jha, "A Totally Self-Checking Checker for Borden's Code," *IEEE Trans. on Computer-Aided Design*, Vol. 8, No. 7, pp. 731-736, July, 1989.

[Mak *et al.* 1982] G. P. Mak, J. A. Abraham, and E. S. Davidson, "The Design of PLAs with Concurrent Error Detection," *Proc. 12th Annual Intn'l. Symp. Fault-Tolerant Computing*, pp. 300-310, June, 1982.

[Marouf and Friedman 1978a] M. Marouf and A. D. Friedman, "Efficient Design of Self-Checking Checkers for Any *m*-out-of-*n* Code," *IEEE Trans. on Computers*, Vol. C-27, No. 6, pp. 482-490, June, 1978.

[Marouf and Friedman 1978b] M. Marouf and A. D. Friedman, "Design of Self-Checking Checkers for Berger Codes," *Digest of Papers 8th Annual Intn'l. Conf. on Fault-Tolerant Computing*, pp. 179-184, June, 1978.

[Nicolaidis 1987] M. Nicolaidis, "Evaluation of a Self-Checking Version of the MC 68000 Microprocessor," *Microprocessing and Microprogramming*, Vol. 20, pp. 235-247, 1987.

[Nikolos *et al.* 1988] D. Nikolos, A. M. Paschalis, and G. Philokyprou, "Efficient Design of Totally Self-Checking Checkers For All Low-Cost Arithmetic Codes," *IEEE Trans. on Computers*, Vol. 37, pp. 807-814, July, 1988.

[Paschalis *et al.* 1987] A. M. Paschalis, C. Halatsis, and G. Philokyprou, "Concurrently Totally Self-Checking Microprogram Control Unit with Duplication of Microprogram Sequencer," *Microprocessing and Microprogramming*, Vol. 20, pp. 271-281, 1987.

[Sellers *et al.* 1968] F. F. Sellers, M. Y. Hsiao, and L. W. Bearnson, *Error Detecting Logic for Digital Computers*, McGraw-Hill, New York, New York, 1968.

[Shedletsky and McCluskey 1975] J. J. Shedletsky and E. J. McCluskey, "The Error Latency of a Fault in a Combinational Digital Circuit," *Digest of Papers 1975 Intn'l. Symp. on Fault-Tolerant Computing*, pp. 210-214, June, 1975.

PROBLEMS

13.1.

a. Consider a 6-bit residue code in which $m = 3$ with the rightmost two bits being check bits. For each of the following, assuming at most a single-bit error, determine if such an error is present and if so which bits might be erroneous.

$$010110, \; 011110, \; 011001$$

b. Consider a 7-bit Hamming single-error correction code. For each of the following, assuming at most a single-bit error, determine the erroneous bit if any.

$$0101100, \; 0101101, \; 0111101$$

13.2. Consider a residue code and an error that results in the interchange of two successive bits. Prove that all such errors are detected, or present a counterexample to this conjecture.

13.3. Prove that the $k/2k$ checker of Figure 13.16 is totally self-checking for all single stuck faults.

13.4. Consider a residue code and a burst error that results in unidirectional errors in a sequence of k successive bits. Will all such errors be detected?

13.5. Prove that Berger codes detect all unidirectional errors.

13.6. Design a 2/4 code checker in the form of Figure 13.16, and prove that the circuit is totally self-checking for all single stuck faults.

13.7. Consider the following design of a k-bit equality checker ($k \geq 3$) to determine the equivalence of two words $(a_1, a_2, ..., a_k)$ and $(a_1', a_2', ..., a_k')$. The circuit has two outputs $\int_k$ and g_k defined by the recursive equations

$$\int_k = f_{k-1} b_k + g_{k-1} a_k$$

$$g_k = \int_{k-1} a_k + g_{k-1} b_k$$

where $b_k = \bar{a}_k'$ and $\int_2$ and g_2 are as defined in Figure 13.19. Verify that this circuit is totally self-checking for all single stuck faults.

14. PLA TESTING

About This Chapter

Since programmable logic arrays (PLAs) can implement any Boolean function, they have become a popular device in the realization of both combinational and sequential logic circuits and are used extensively in VLSI designs and as LSI devices on printed circuit boards. Testing of PLAs to detect hardware faults has been studied by numerous researchers. A partial taxonomy of PLA test methods is shown in Figure 14.1. In this chapter we survey many of these methods. First we discuss fault models and traditional methods of test generation. Then we briefly describe test generation algorithms designed specially for PLAs. In Section 14.4 we present several semi-built-in and fully built-in test methodologies for PLAs. In Section 14.5 we compare these various techniques in terms of such measures as area overhead, test time, and fault coverage.

14.1 Introduction

A PLA consists of three parts: an input decoder, an AND array, and an OR array. Figure 14.2 shows the basic structure of a PLA. The input decoder partitions the n primary inputs into $2n$ bit lines. There are two common forms of input decoders: 1-bit decoders that transform each input x_i into x_i and its complement, and 2-bit decoders that transform a pair of inputs into the four minterms of the two variables.

Depending on the technology, a PLA implements two-level logic of various forms, such as NAND-NAND or NOR-NOR. For simplicity we assume the logic is of the form AND-OR, with the AND array forming the product (conjunctive) terms and the OR array the disjunctive terms. The intersection of a bit or output line with a product line is called a *crosspoint*. The proper logical terms are created by either making or not making a connection at each crosspoint. Usually a transistor exists at each crosspoint, and if the connection is not wanted, the associated transistor is disabled by one of several techniques, such as cutting a line.

A PLA can be described by a matrix $P = (A, O)$, where A is an $m \times n$ matrix representing the AND array, O is an $m \times k$ matrix representing the OR array, and n, m, and k are the number of inputs, product terms, and outputs, respectively. This matrix is called the PLA's *personality*. The function realized by a PLA can also be represented by a set of *cubes*, each cube corresponding to a row in the personality matrix. A simple example is given in Figure 14.3. Here a 1(0) in the AND array corresponds to a connection to x_i ($\overline{x}_i$); an x denotes no connection to x_i or $\overline{x}_i$. A 1(0) in the OR array represents a connection (no connection). The PLA shown in Figure 14.3(a) implements the functions

$$f_1 = \overline{x}_1 + x_3 x_4 + \overline{x}_2 \overline{x}_3$$

$$f_2 = x_3 x_4 + \overline{x}_1 \overline{x}_2 \overline{x}_3 \overline{x}_4 + x_2 x_3 + x_1 x_2$$

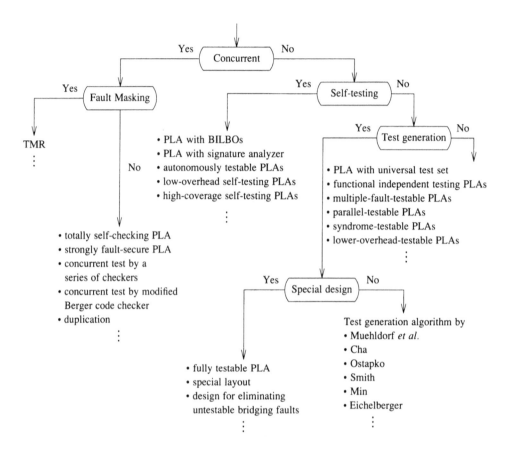

Figure 14.1 A taxonomy of PLA test methodologies

14.2 PLA Testing Problems

The widespread application of PLAs makes PLA testing an important issue. Though PLAs offer many advantages, they also present new testing problems.

14.2.1 Fault Models

In testing digital circuits, the most commonly considered fault model is the stuck-at fault. A s-a-1 or s-a-0 fault may occur on any wire within a PLA, including input/output lines, bit lines, and product lines. Stuck-at faults alone cannot adequately model all physical defects in PLAs. Because of the PLA's array structure, a new class of faults, known as *crosspoint faults*, often occur. A crosspoint fault is either an extra or a missing connection at a crosspoint in the AND or OR array of a PLA. There are four kinds of crosspoint faults:

1. **Shrinkage fault** — an extra connection between a bit line and a product line in the AND array that causes the implicant to shrink because it includes one additional input variable;

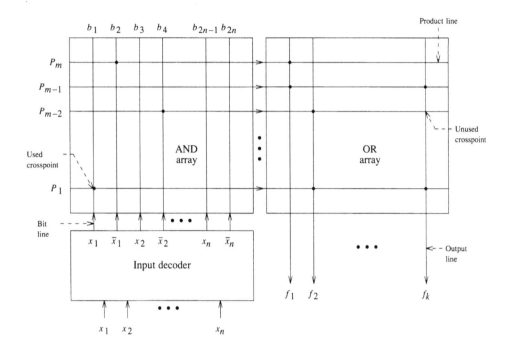

Figure 14.2 A general PLA structure with 1-bit input decoders

2. **Growth fault** — a missing connection between a bit line and a product line in the AND array that causes the implicant to grow because it becomes independent of an additional input variable;

3. **Appearance fault** — an extra connection between a product line and an output line in the OR array that causes the corresponding implicant to appear in the output function;

4. **Disappearance fault** — a missing connection between a product line and an output line in the OR array that causes an implicant to disappear from the corresponding output function.

Note that missing crosspoints are equivalent to some stuck-at faults, but extra connections cannot be modeled by stuck faults. To illustrate these faults, refer to Figure 14.3(a). If an extra connection (shrinkage fault) exists between P_1 and b_5, then f_1 contains the term $\bar{x}_1 x_3$ rather than just $\bar{x}_1$, and the function f_1 loses the two minterms covered by $\bar{x}_1 x_2 \bar{x}_3$. If the connection between P_4 and b_2 is missing (growth fault), then the product term on line P_4 becomes $\bar{x}_2 \bar{x}_3 \bar{x}_4$, and f_2 inherits the extra minterm $x_1 \bar{x}_2 \bar{x}_3 \bar{x}_4$. Note that there are $(2n + k)m$ possible single crosspoint faults and $2^{(2n+k)m} - 1$ different single and multiple crosspoint faults. For large PLAs, explicit consideration of single crosspoint faults is difficult; for multiple faults it is totally impractical.

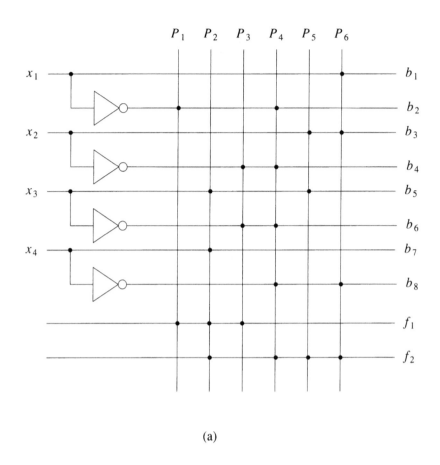

(a)

$$P = \begin{array}{c} \begin{array}{cccccc} x_1 & x_2 & x_3 & x_4 & & f_1 & f_2 \end{array} \\ \left[\begin{array}{cccccc} 0 & x & x & x & \quad 1 & 0 \\ x & x & 1 & 1 & \quad 1 & 1 \\ x & 0 & 0 & x & \quad 1 & 0 \\ 0 & 0 & 0 & 0 & \quad 0 & 1 \\ x & 1 & 1 & x & \quad 0 & 1 \\ 1 & 1 & x & 0 & \quad 0 & 1 \end{array} \right] \end{array} = [A, O]$$

(b)

Figure 14.3   A PLA example (a) A PLA implementation (b) A personality matrix

Because of the way PLAs are made, multiple faults are common. Multiple stuck faults, multiple crosspoint faults, or combinations of these faults may occur in newly manufactured circuits. The compact array structure of PLAs makes shorts between two adjacent lines more likely to occur. The effect of shorts between a pair of lines is determined by the technology used to manufacture the PLA. Generally, either a "high" or "low" voltage will dominate, so these shorts can be modeled as OR or AND bridging faults.

The various fault modes that may occur in a PLA can make test generation a complex process. However, analysis of the relation between different kinds of faults reduces the complexity of the problem. It has been shown that a complete test set for single crosspoint faults also covers most single stuck faults in input decoders and output lines, as well as many shorts and a large portion of multiple faults [Cha 1978, Ostapko and Hong 1979, Smith 1979]. Min [1983a] presented a unified fault model and proved that any stuck-at fault or bridging fault (of AND type) is equivalent to a multiple crosspoint fault. It has been verified that 98 percent of all multiple crosspoint faults of size 8 and less are inherently covered by every complete single crosspoint fault test set in a PLA [Agarwal 1980]. These results suggest that single crosspoint faults should be of primary concern in testing. If other classes of faults are considered significant, special effort must be made to ensure for their high fault coverage.

14.2.2 Problems with Traditional Test Generation Methods

A PLA corresponds to a two-level sum-of-product circuit with input inverters, although in nMOS technology it is often implemented by two levels of NOR-NOR gates with output inverters. One way to generate tests for a PLA is first to convert the PLA into a two-level gate circuit and then to find tests for stuck faults in the "equivalent" gate circuit. Many algorithms exist for generating tests for such circuits (see Chapter 6). However, these methods share two serious problems. First, although the two-level circuit is logically equivalent to the PLA, as far as fault behavior is concerned, they are not equivalent. Some faults in the PLA, such as an extra crosspoint fault in the AND array, cannot be modeled as a stuck fault in the gate circuit. Therefore a high fault coverage is not guaranteed. Second, traditional test generation algorithms are not always effective for PLAs because PLAs have high fanin, reconvergent fanout, and redundancy. Although exhaustive testing is not affected by these factors, it becomes less applicable as the size of the PLA increases.

In general PLAs are not efficiently tested by random test vectors due primarily to the large number of *crosspoints used* in the AND array. Suppose a product line p is connected to j bit lines ($j \leq n$). To test for a missing crosspoint fault at $A(p, b_i)$, one needs to place a 0 on the i-th bit line and 1's at all the other $(j-1)$ connected bit lines. Since there are 2^j possible patterns on these j bit lines and only one pattern can test this fault, the probability of detecting such a missing connection with a random pattern is approximately $1/2^j$. Since j is frequently 10 or larger, many random patterns may be needed to achieve a high fault coverage.

14.3 Test Generation Algorithms for PLAs

Since conventional test generation methods are not suitable for PLAs, several ad hoc test generation approaches have been developed [Bose and Abraham 1982, Cha 1978, Eichelberger and Lindbloom 1980, Min 1983b, Muehldorf and Williams 1977, Ostapko and Hong 1979, Smith 1979, Wei and Sangiovanni-Vincentelli 1985]. It has been shown that a PLA's regular structure leads to more efficient test generation and fault simulation algorithms than for random logic. We will only briefly review these concepts.

14.3.1 Deterministic Test Generation

The basic idea behind most PLA test generation algorithms is path sensitization, namely, to select a product line and then sensitize it through one of the output lines. We will show that a test must match or almost match (only one bit differs from) a product term in the AND array. If a PLA's personality is known, tests of this nature can easily be found.

The *sharp* operation, denoted by "#"[1], is especially useful for finding such a test. For example, let ci be the cube representation of product line P_i. Suppose the connection between P_i and the j-th input variable is missing, thus creating a growth fault. The cube ci' representing the faulty P_i would be the same as ci except the j-th bit changes to x. To detect this missing crosspoint fault, a test t must be covered by $ci' \# ci$. Let $z(i,k)$ be a set of cubes representing product lines that are connected to output line k, except ci. To propagate the error through output k, the test t must also cause all product lines represented by $z(i,k)$ to be 0. That is to say, $t \in (ci' \# ci) \# z(i, k)$.

If the result is empty, another k should be tried until a test t is found. If no test can be computed, the fault is undetectable. Formulas for generating tests for other crosspoint faults can be similarly defined [Ostapko and Hong 1979, Smith 1979].

Example 14.1: Consider the PLA shown in Figure 14.3. Assume the crosspoint between lines P_2 and b_7 is missing; hence the product term $x_3 x_4$ becomes x_3. Then $(ci' \# ci)$ is equivalent to $x_3(\overline{x_3 x_4}) = x_3 \overline{x}_4$. Choosing $k = 1$, $z(i,k)$ is equivalent to $\overline{x}_1 + \overline{x}_2 \overline{x}_3$, and $(ci' \# ci) \# z(i, k)$ is equivalent to $x_3 \overline{x}_4 (\overline{x}_1 + \overline{x}_2 \overline{x}_3)' = x_1 x_3 \overline{x}_4$. Hence the set of tests is represented by $1x10$.   □

To determine fault coverage, after a test is generated, test analysis is carried out instead of conventional fault simulation. A PLA's regularity makes it possible to determine directly what faults are detected by a given test. For example, if a test t results in only product line P_j being set to 1, and there is an output line q that does not connect to P_j, then an appearance fault at position (j,q) in the OR array can be detected by t. Such rules of fault analysis can be easily formulated and work much faster than fault simulation for random logic. Details of test analysis for PLAs are given in [Ostapko and Hong 1979, Smith 1979].

Under the single-fault assumption, most test generation algorithms can achieve a high fault coverage for detectable stuck faults and crosspoint faults. AND-type bridging faults can also be covered by some algorithms [Cha 1978, Ostapko and Hong 1979]. Min [1983b] proposed a test generation algorithm for irredundant PLAs that covers multiple crosspoint faults. The tests produced by this algorithm also detect stuck-at faults and bridging faults. In general, the size of the test set produced by these algorithms is bounded by the number of crosspoints in a PLA, namely $m(2n + k)$.

---

1. By definition, $a \# b = a\overline{b}$.

14.3.2 Semirandom Test Generation

Deterministic test generation for PLAs is feasible but laborious. Random patterns can be easily generated but are often ineffective. A PLA's regular structure allows deterministic and random test generation methods to be combined into an effective and inexpensive *semirandom test generation* technique [Eichelberger and Lindbloom 1980]. A PLA has a direct logic correspondence with a two-level AND-OR combinational circuit with all input variables and their complements supplied. It is well-known that the circuit shown in Figure 14.4(a) can be tested, under the single stuck-fault assumption, by applying a set of critical cubes to each AND gate (see Figure 14.4(b)) and sensitizing its output through the OR gate. The critical cubes are hard to generate randomly. However, it is probable that a random pattern on G_2 and G_3 will result in a sensitized path for G_1, since for an AND gate with i inputs, $2^i - 1$ out of 2^i possible input combinations lead to an output of 0.

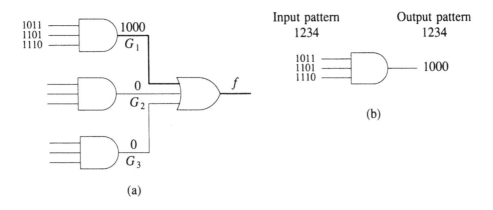

(a)

(b)

Figure 14.4   (a) Critical patterns for an AND gate together with a sensitized test path (b) Critical cubes for an AND gate

Based on these observations, a semirandom test pattern generation procedure can be developed that assigns critical cubes deterministically and assigns unspecified bits in the corresponding input vector arbitrarily. Such fully specified patterns are candidate tests. They are subject to *pattern evaluation* for determining if the random assignments provide for a sensitized path.

This heuristic test generation method is fast, but it has a limitation in that it only deals with missing crosspoint faults. For 31 PLAs studied, an average fault coverage of 98.44 percent of the modeled faults, which did not include extra connections, was achieved. All undetected crosspoint faults were found to be redundant. Thus semirandom tests can be used for PLAs. Such tests can be easily and quickly generated. However, the resulting test sets are usually large, and hence fault coverage is difficult to determine. To circumvent this problem, a strategy that mixes biased random patterns and deterministic tests can be used. The biased random-pattern method is first used to generate tests for most of the faults. Then the uncovered faults are processed using a deterministic method. The resulting test set can be substantially reduced using fault simulation. An example of such a system is PLATYPUS [Wei and

Sangiovanni-Vincentelli 1985], which employs several heuristics and algorithms used in the PLA logic optimizer ESPRESSO-II [Brayton *et al.* 1984].

The test generation methods discussed provide a software solution to the PLA testing problem. They do not require any hardware changes to the PLA. Since test generation and the application of large test sets are expensive, several alternative methods have been developed for testing PLAs.

14.4 Testable PLA Designs

As PLAs increase in size, more test patterns have to be generated and stored. Sophisticated automatic test equipment is needed to execute the test process. Hence stored-pattern testing becomes a time-consuming and expensive task. To alleviate this problem, several hardware-oriented approaches have been developed that add extra built-in test (BIT) circuitry to the original PLA such that the modified PLA can be easily tested. Most techniques reported to date fall into one of four categories, namely, special coding, parity checking, divide and conquer, and signature analysis. In the following, we will summarize basic principles of testable PLA design methodologies and give examples for each category.

14.4.1 Concurrent Testable PLAs with Special Coding

A PLA's regular memory-like structure suggests the application of special coding for either concurrent or off-line fault detection. The most popular code for PLA testing is the parity code; these techniques will be discussed in a separate section.

To test a PLA concurrently, i.e., during normal operation, requires that during fault-free operation only one product line can be activated by any input vector. Not every PLA has this property. However, simultaneous activation of product lines can be detected and removed, which, unfortunately, increases PLA size [Wang and Avizienis 1979]. In this section we assume that at most one product line is activated by any input vector.

14.4.1.1 PLA with Concurrent Error Detection by a Series of Checkers

A technique proposed by Khakbaz and McCluskey [1981] makes use of the following facts about a PLA:

- The bit lines in the AND array naturally form a two-rail code, i.e., x_i and $\bar{x}_i$.

- During normal operation the signals on the m product lines form a 1-out-of-m code. (This condition is sometimes referred to as a *nonconcurrent* PLA.)

- The fault-free output patterns are determined by the PLA's personality matrix. They can be coded into some error-detection code by adding extra output lines to the OR array.

The proposed testable PLA, shown in Figure 14.5, has three checkers. C_1 is a totally self-checking (TSC) 1-out-of-m checker on all product lines and detects any fault that destroys the nonconcurrent property, such as a product line stuck at 1(0), or any missing and/or extra crosspoint in the AND array. C_2 is a TSC two-rail checker that tests all single stuck-at faults on the bit lines and input decoders. C_3 is an output-code checker. Its complexity depends on how the outputs are coded. The coding is implemented in that portion of the OR array denoted by D. The simplest code makes all output patterns have even (odd) parity. Here, only one extra output line needs to be added, and C_3 would be a parity checker. In general, C_3 is not a TSC checker and may not be fully tested during normal operation, since the inputs to C_3 are basically the PLA's outputs that are not directly controllable.

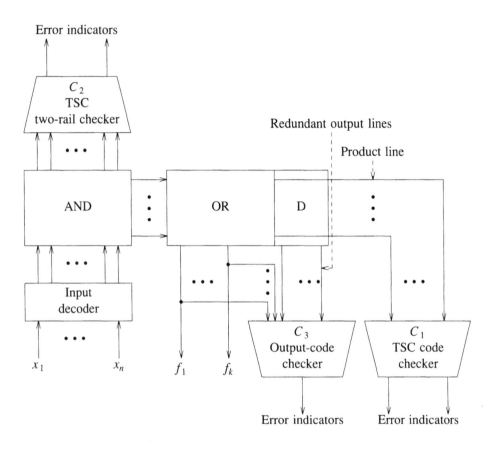

Figure 14.5   A concurrent testable PLA design (CONC1)

Testing occurs concurrently with normal operation. Most errors are caught by one of the three checkers. However, the circuit is not totally self-checking. Therefore off-line testing is still needed to ensure a high fault coverage. This technique combines concurrent error detection with off-line testing by using the same added circuits for both modes of testing. A simple test generation procedure that produces a complete test set has been developed, which is partially function-dependent, and covers faults in both the original PLA and the additional hardware [Khakbaz and McCluskey 1981].

14.4.1.2 Concurrent Testable PLAs Using Modified Berger Code

Output errors of a non-concurrent PLA caused by single stuck-at and crosspoint faults and some multiple faults inside a PLA, except input stuck-at faults, are unidirectional

[Dong and McCluskey 1981]. Consequently, any codes that detect unidirectional errors, such as n-out-of-m codes [Wakerly 1978] and Berger codes [Berger 1961], may be used for the concurrent testing of PLAs. One design that uses these concepts is shown in Figure 14.6. The inputs are associated with a parity bit x_{n+1}. A duplicate parity checker (parity tree) and a totally self-checking two rail checker C_1 detect faults on input lines, bit lines, and in the input decoder. An on-line generator C_2 produces check symbols C^* for each output pattern, which the checker C_1 compares with check symbol C which is generated by the OR array. These check symbols are used to detect output errors.

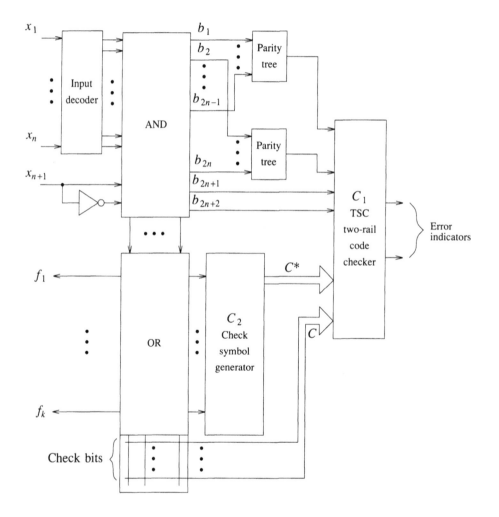

Figure 14.6 A PLA design for concurrent testing (CONC2)

Modified Berger (MB) codes [Dong 1982] are separable codes in which each code word consists of a data part D and a check symbol C. Let $O(D)$ be the number of 1s in D. Then a check symbol C for a MB codeword $[D\ C]$ can be obtained by first setting $C' = O(D) \bmod M$, and then encoding C' into a codeword C that detects unidirectional errors. Here M is a predetermined integer and all unidirectional errors of size not equal to multiples of M can be detected by MB codes. By proper selection of M, all single faults in the AND and OR array can be detected. Suppose the number of inputs is odd, and the inputs are associated with a parity bit x_{n+1}. Faults on input lines, on bit lines, and in the input decoders are detected by a pair of parity checkers (trees) in which one tree generates the parity of all x_i's and the other generates the parity of all $\bar{x}_i$'s. During normal operation, the outputs of the two trees are always complement values, which together with the parity bit, are monitored by a TSC two-rail checker.

Since all other faults cause unidirectional errors on the outputs, the PLA's outputs are checked using a MB code. A check symbol is attached to every output pattern. During normal operation, an on-line generator produces a check symbol C^* for each output pattern, which is compared with the attached check symbol C using a TSC two-rail checker. Any disagreement between these two signals indicates some fault in either the PLA or the checking circuits.

This scheme is suitable for PLAs with a large number of outputs and product lines and in situations where each product line is shared by a small number of outputs, since the number of check bits required only depends on M. If $O(D)$ is small, a small M can be chosen and hence area overhead is not too large. Also with a small number of code bits, the likelihood of every codeword being applied to the PLA increases, so that the totally self-checking property of the two-rail checker can be ensured.

14.4.2 Parity Testable PLAs

Since PLAs have a regular array structure, it is possible to design a PLA so that it can be tested by a small set of deterministic tests that are function-independent, i.e., independent of the personality matrix. This is possible because of two important concepts.

1. Let N_i be the number of used crosspoints on bit line b_i. One can add an extra product line and make connections to it such that every bit line has an even (odd) number of connections with product lines. Then any single crosspoint fault on b_i changes the parity of N_i. The same is true for output lines. Therefore single crosspoint faults can be detected by parity checking on these lines.

2. To test a PLA easily, it must be possible to control individually each bit and product line, and sensitize each product line through the OR array.

14.4.2.1 PLA with Universal Test Set

One of the first easily testable PLA designs employing a universal test set that is thus independent of the PLA's personality is shown in Figure 14.7 [Fujiwara and Kinoshita 1981]. We refer to this design as UTS.

One column and one row are appended to the AND array and the OR array, respectively, so that each bit line in the AND array can have an odd number of connections and the portion of each product line in the OR array can have an even number of connections. Two parity checkers consisting of cascades of XOR gates are used to examine the parities of the two arrays during testing.

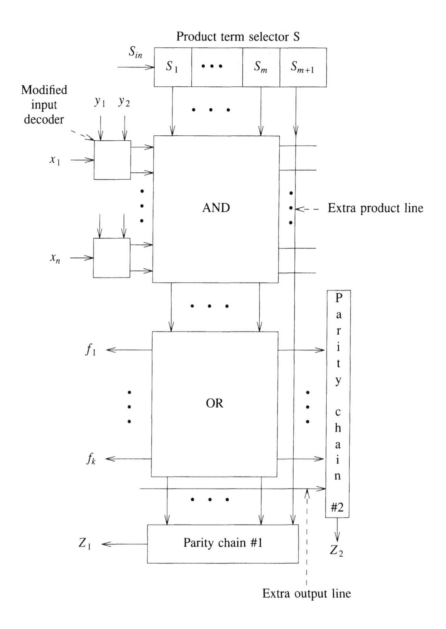

Figure 14.7   PLA with universal test set (UTS)

A selection circuit is used to activate the columns and rows in the PLA so that every crosspoint can be individually selected and tested. In UTS, a product line is selected by adding a shift register S as shown in Figure 14.7. Each bit S_i of S is connected to a product line P_i to create a new product line $P_i' = P_iS_i$. Thus if $S_i=1$ and $S_j=0$ for all $j \neq i$, product line P_i' is selected and sensitized to the output, since all other product lines are 0.

Bit lines are selected by modifying the input decoder as shown in Figure 14.8. By properly assigning values to y_1, y_2, and the x_is, all but one input line can be set to 0. Thus that line is selected and tested. During normal operation $y_1 = y_2 = 0$.

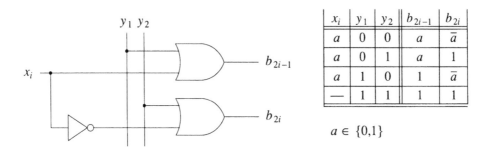

x_i	y_1	y_2	b_{2i-1}	b_{2i}
a	0	0	a	$\bar{a}$
a	0	1	a	1
a	1	0	1	$\bar{a}$
—	1	1	1	1

$a \in \{0,1\}$

Figure 14.8   Augmented input decoder and its truth table

Based on this augmented PLA design, a universal test set can be derived (see Figure 14.9), which is applicable to all PLAs regardless of their function. This test set attempts to sensitize each product line and bit line, one at a time. For example, when the test I_{j0} or I_{j1} is applied, all bit lines and the j-th product line are set to 1, and all other product lines are set to 0. Then the crosspoints along the j-th product line in the OR array are tested by the parity chain connected to the output lines, and the result can be observed from Z_2. Similarly, the AND array is tested by applying the J_{i0}s and J_{i1}s and observing the output Z_1. All single stuck-at and crosspoint faults are detected by the parity checkers. This test set can also detect stuck-at faults in the input decoder and the added hardware. The size of the universal test set is linear with n and m. The test I_1 is used to check stuck-at-1 faults in the product term selector.

Hong and Ostapko [1980] independently proposed another design for function-independent testable PLAs (FIT), which employs the same philosophy as UTS. The test set for this design is similar to that of UTS.

14.4.2.2 Autonomously Testable PLAs

For UTS- or FIT-type PLAs, the test vectors must be either stored or generated externally, and the test results have to be monitored during the test process. Yajima and Aramaki [1981] have augmented these schemes so that the resulting PLA circuitry can generate a universal test using a built-in feedback-value generator, a product-term selector, and an input shift register (see Figure 14.10). The feedback-value generator produces test data based on its current state and the outputs of various parity-tree generators. A feedback shift register C_1 is used as both a product-term selector and simultaneously as a signature analyzer. Test results are applied to the product lines using the product-term selector, and evaluated when the test operation is completed by the signature analyzer and detector circuit. Any single fault in the original PLA, and in most of the additional circuits, changes the final signature and therefore is detected. However, some faults in the feedback-value generator and the signature detector are not covered; hence these circuits need to be duplicated to achieve a high fault coverage.

	x_1 $\cdots$ x_i $\cdots$ x_n	y_1 y_2	S_1 $\cdots$ S_j $\cdots$ S_m	Z_1 Z_2
I_1	— $\cdots$ — $\cdots$ —	— —	0 $\cdots$ 0 $\cdots$ 0	0 0

For $j = 1, ..., m$

	x_1 $\cdots$ x_i $\cdots$ x_n	y_1 y_2	S_1 $\cdots$ S_j $\cdots$ S_m	Z_1 Z_2
I_{j0}	0 $\cdots$ 0 0 0 $\cdots$ 0	1 0	0 $\cdots$ 0 1 0 $\cdots$ 0	1 1
I_{j1}	1 $\cdots$ 1 1 1 $\cdots$ 1	0 1	0 $\cdots$ 0 1 0 $\cdots$ 0	1 1

For $i = 1, ..., n$

	x_1 $\cdots$ x_i $\cdots$ x_n	y_1 y_2	S_1 $\cdots$ S_j $\cdots$ S_m	Z_1 Z_2
J_{i0}	1 $\cdots$ 1 0 1 $\cdots$ 1	0 1	1 $\cdots$ 1 1 1 $\cdots$ 1	e_m —
J_{i1}	0 $\cdots$ 0 1 0 $\cdots$ 0	1 0	1 $\cdots$ 1 1 1 $\cdots$ 1	e_m —

$$e_m = \begin{cases} 0, & \text{if } m \text{ is odd} \\ 1, & \text{if } m \text{ is even} \end{cases}$$

"—" represents a don't-care condition

Figure 14.9   Universal test set for PLAs

The advantages of autonomously testable PLAs are that test patterns need not be generated *a priori* nor stored, hence field testing becomes easy. Since the tests are function-independent, any PLA can be modified in a uniform way.

14.4.2.3 A Built-In Self-Testable PLA Design with Cumulative Parity Comparison

To obtain more efficient testable PLA designs, some researchers have focused on reducing area overhead and/or increasing fault coverage. This can be done using the idea of parity compression [Fujiwara 1984, Treuer *et al.* 1985]. In the testable designs discussed previously [Fujiwara and Kinoshita 1981, Hong and Ostapko 1980], two parity checkers are used to monitor the parity of the two arrays. These checkers can be replaced by a *cumulative parity comparison* method, i.e., by accumulating parity signals in a flip-flop and comparing its value with expected values only at specific times. This scheme is illustrated in Figure 14.11. Two control lines C_1 and C_2 are added to the AND array to disable all x_i's and $\bar{x}_i$'s, respectively. C_1, C_2, and the primary inputs can be used together to select each bit line. As before, a shift register is added to select product lines. One or two product lines are appended to the AND array so that every bit line has (1) an odd number of used crosspoints, and (2) an odd number of unused crosspoints. The same is done for the OR array. Area is saved by eliminating the parity-checking circuit for the product lines. Only one parity chain is employed at the PLA's outputs, and cumulative parity comparison is used to detect errors.

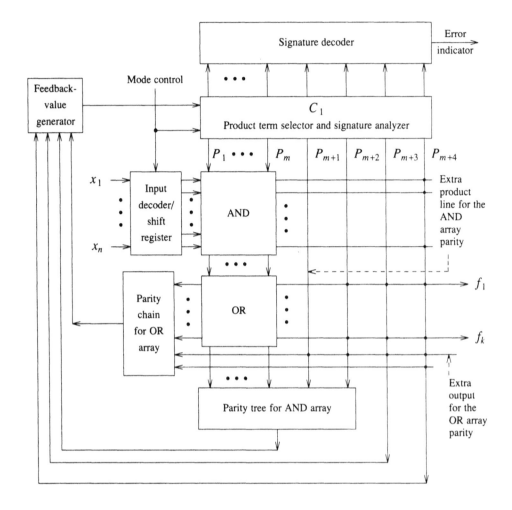

Figure 14.10 An autonomously testable PLA (AUTO)

In the test mode, a universal test set of length $2m(1 + n) + 1$ is applied to the inputs of the PLA. Faults in the OR array can be detected by the parity checker on the outputs. The AND array is tested as follows. Suppose we select one bit line b_i and activate product lines one at a time. When P_j is activated, if there is a device connecting b_i and P_j, P_j is forced to 0 and the output pattern should be all 0s; thus the parity value of the output is 0. If there is no device connecting b_i and P_j, P_j is 1 and the output should have an odd parity. Since there are an odd number of used and unused crosspoints on each bit line, an odd number of 1s and 0s are produced from the output of the parity tree. This information is cumulated in a parity counter that produces the parity of the parity sequence. An interesting property of this scheme is that the sequence of cumulated parity bits at $2n+2m+1$ selected check points is simply a sequence of alternating 0s and 1s. Hence it is easy to generate the expected value on-line. The comparator is shown at the bottom of Figure 14.11. The proven fault

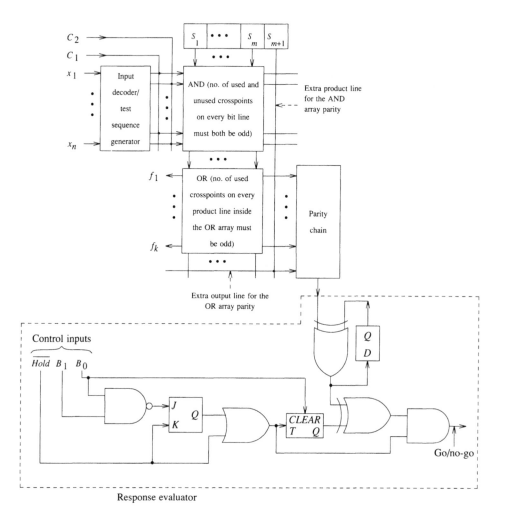

Figure 14.11 A testable PLA with cumulative parity comparison (CPC)

coverage of this testable design is high; all single and $(1 - 2^{-(m+2n)}) \times 100$ percent of all multiple crosspoint faults and all stuck-at and bridging faults are covered.

14.4.3 Signature-Testable PLAs

Signature analysis is a simple and effective way for testing digital systems, and several self-testing PLA designs using this concept have been proposed. In these approaches, a set of input patterns is applied and the results are compressed to generate a signature, which, when compared with a known correct value, determines whether the PLA is faulty.

14.4.3.1 PLA with Multiple Signature Analyzers

One approach proposed by Hassan and McCluskey [1983] uses at least four linear feedback shift registers, $L1$, $L2$, G, and LS, the first three having the same length and characteristic polynomial (see Figure 14.12). We refer to this design as SIG. Here G is used as a maximum-length sequence generator for exhaustive input-pattern generation and the others for signature analysis. The complemented bit lines ($\overline{b}_i$) are fed into $L1$ and the true bit lines (b_i) into $L2$. These two signature analyzers are used to detect all single and multiple stuck-at faults on the input and the bit lines. In the absence of faults at the input and bit lines, the signature analyzer LS can detect faults in the AND and OR array.

In the test mode, all possible input patterns are generated by G. Responses from the PLA are compacted in the three signature analyzers. The final signatures are shifted out for inspection. Using this scheme, all multiple bit-line stuck-at faults, all output faults, most product-line stuck-at faults, and most crosspoint faults are detected. This method is only practical for PLAs with a small number of inputs/outputs and a large number of product lines. Because of the lack of parity checkers, the delay per test pattern is very small.

14.4.3.2 Self-Testable PLAs with Single Signature Analyzer

Grassl and Pfleiderer [1983] proposed a simple function-independent self-testable (FIST) design for PLAs that attempts to reduce the number of test patterns and additional hardware and also takes into account the important aspect of PLA layout. It appears to be applicable to large PLAs. Figure 14.13 shows the block diagram for this design. The input decoder is modified to be a test pattern generator during test mode. It shifts a single "1" across the bit lines. Product lines are selected by the shift register SHR and the selector SEL. The SHR selects a pair of adjacent product lines at a time, so its length is only half of the number of product terms. The SEL will in turn connect product line 1, 3, 5, ... or 2, 4, 6, ... to ground, thus resulting in a unique selection of each product line. During testing, every crosspoint in the AND array is addressed using the two shift registers. The results are analyzed in the multiple-input signature register (MISR) that acts as a conventional latch register during normal operation.

Splitting the product line selector into the SHR and SEL allows the added circuit to fit physically into the narrow pitch of a PLA array, substantially reducing area overhead. A detailed implementation of such a two-stage m-bit shift register has been given in [Hua *et al.* 1984]. It consists of a $\lceil m/2 \rceil$-bit shift register and $\lceil m/2 \rceil$ 2-to-2 switch boxes, shown in Figure 14.13(b). A compact parity-checker design is also given. The layout for these BIT structures can fit into the PLA's normal pitch. The connection cost between the BIT hardware and the original PLA is almost zero. This design and layout technique has been adopted by many designers.

The SHR and SEL circuitry operate as follows. Consider the switch box B_i driven by S_i. If $S_i = 0$, then $P_{2i} = P_{2i-1} = 0$; and if $S_i = 1$, then $P_{2i} = 1$ if $Q = 1$, and $P_{2i-1} = 1$ if $Q = 0$. During the test mode a single "1" is shifted through the register consisting of S_1, S_2, ..., $S_{m/2}$, which is clocked at one half the normal clock rate. When $Q = 0$, the odd-numbered product terms are selected in turn, while when $Q = 1$ the even-numbered product terms are selected. When $T_2 = 0$ the circuit is in the non-test

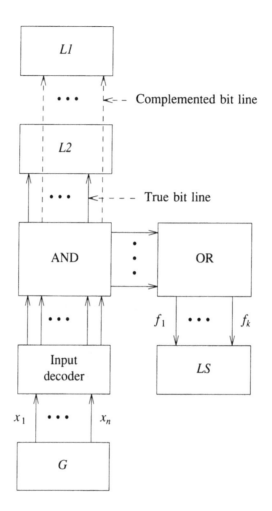

Figure 14.12   Self-testable PLA using signature analysis (SIG)

mode and the connections to the Ps are cut, leaving all the Ps in their high impedance mode. In the test mode, $T_2 = 1$ and the values of the Ps are determined by the state of the two-stage shift register. Now flip-flop A toggles every clock time. Combining this operation with the "1" propagating up the shift register produces the desired selection of the P_j lines.

14.4.4 Partitioning and Testing of PLAs

Testing becomes more complex as the size of the PLA increases. A common strategy for dealing with large problems is that of divide and conquer. This principle has also been applied to the design of testable PLAs.

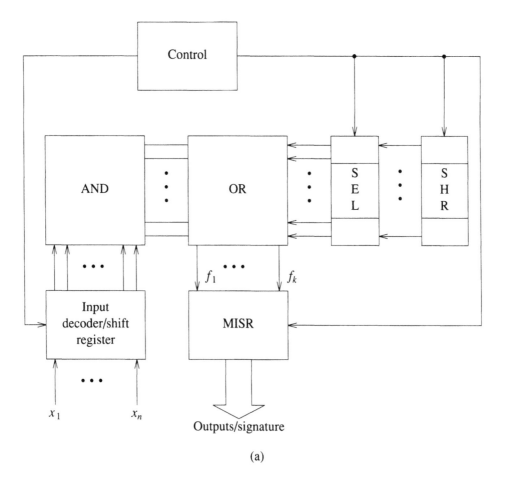

(a)

Figure 14.13   Block diagram of a self-testing PLA (FIST)

14.4.4.1 PLA with BILBOs

Daehn and Mucha [1981] suggested a partitioning approach for self-testable PLAs that partitions a PLA into three parts — an input decoder, an AND array, and an OR array — and then inserts three BILBOs between these parts as shown in Figure 14.14. Since BILBOs can be used for both test pattern generation and response evaluation, in the test mode the partitioned blocks can be tested separately by properly controlling the BILBOs. Each block is tested by letting the BILBO at its input operate as a test generator and the BILBO at its output operate as a signature analyzer. The final signature is shifted out for inspection. The three parts of the PLA are tested one by one.

After partitioning, the AND array and the OR array are just arrays of NOR gates. All inputs are controllable and outputs are observable. Testing now becomes a simple

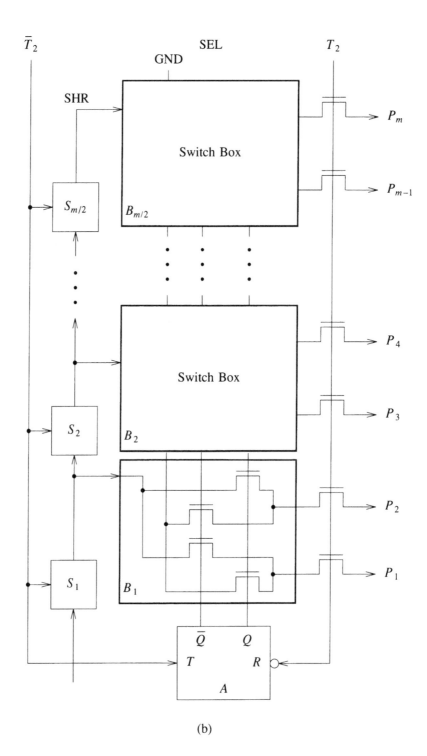

(b)

Figure 14.13   (Continued)

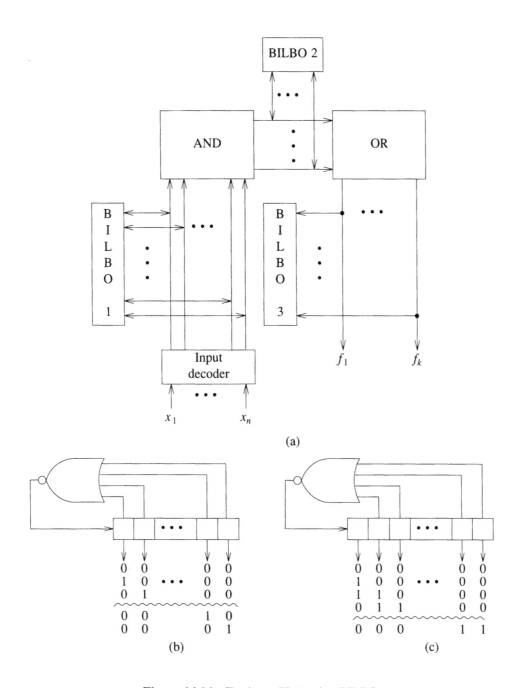

Figure 14.14 Testing a PLA using BILBOs

task. It is well known that a NOR gate can be fully tested by the following simple sequence:

$$\begin{matrix} 0 & 0 & ... & 0 \\ 1 & 0 & ... & 0 \\ 0 & 1 & ... & 0 \\ & & ... & \\ 0 & 0 & ... & 1 \end{matrix}$$

A NOR gate array can be tested by the same patterns. Hence the test generator need not be a pseudorandom pattern generator producing all input combinations, but can rather be a nonlinear feedback shift register, as shown in Figure 14.14(b), producing the patterns given above. In this way, the number of test patterns is greatly reduced. This sequence can detect all single stuck faults, crosspoint faults, and bridge faults in the AND or OR array. The faults in the input decoder are tested by a similar sequence that can be produced by the nonlinear feedback shift register shown in Figure 14.14(c).

14.4.4.2 Parallel-Testable PLAs

In most testable designs discussed so far, the product lines are tested individually one at a time. This simplifies testing and leads to a high fault coverage. However, an m-bit shift register is required and a long test sequence of the order of $O(nm)$ may be necessary. By exploiting possible parallelism in testing product lines, high fault coverage can be maintained, while area overhead and the number of test vectors are reduced [Boswell *et al.* 1985].

The conditions for an entire PLA to be testable in parallel (*parallel testable*) are too stringent to be satisfied by general PLAs. It is possible, however, to partition the product lines into groups such that each group is tested in parallel. The procedure for forming these partitions is complex and will not be presented here. A shift register R_1 is added to the circuit, each of its cells controlling one partitioned block of product lines. An output Z_1 is added that connects to all bit lines. Finally a $2n$-bit shift register and an input-control circuit R are inserted for controlling each bit line individually. An example of a parallel-testable PLA (PTPLA) is shown in Figure 14.15.

In the test mode, some patterns are first applied to the entire PLA; then the groups of product lines are tested in sequence. Within each group, product lines are tested in parallel. The lengths of the shift register used for selecting product lines is reduced from m to the number of groups, which is usually less than $m/2$. The number of test patterns is reduced as well because the tests applied to each group are basically the same as those applied previously to each individual line. The test sequence is simple and universal, but the response is function-dependent. It has been shown that all single and multiple stuck faults and crosspoint faults are detected using this technique.

14.4.4.3 Divide-and-Conquer Strategy for Testable PLA Design

The optimum partitioning of a PLA into parallel-testable blocks in a divide-and-conquer (DAC) strategy is a difficult problem. Partitioning a PLA structurally into three parts and inserting BILBOs is easy but requires a substantial amount of area overhead. Figure 14.16(a) illustrates a partitioning technique that has low area overhead and can be easily applied to any PLA [Saluja and Upadhyaya 1985]. The product lines are partitioned into J groups, and a J-bit shift register R_1 is used to control each group. Within each group, there are at most 2^I product lines, where $I = \lceil log_2(m/J) \rceil$; an individual line is selected by a decoder in the decoder-parity AND

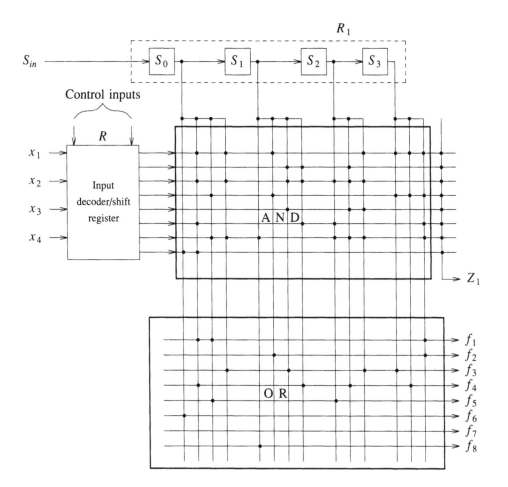

Figure 14.15   A parallel-testable PLA structure (PTPLA)

array (DPAA) [Upadhyaya and Saluja 1988], shown in Figure 14.16(b). During testing, groups of product lines are tested one by one. Within a group, product lines are also tested sequentially. This design does not reduce the number of test patterns, but it reduces the number of shift register cells for selecting product lines. This leads to a simpler layout. It has been proven that for this scheme, a PLA can be tested for all faults by a universal test set of length $m(3+2n+\lceil log(m/J)\rceil)+c$, where c is a constant.

14.4.5 Fully-Testable PLA Designs

Because of redundancy and concurrency in PLAs and the diversity of fault models, it is difficult to test a PLA and achieve a high fault coverage without changing the PLA design. The testable design methods discussed so far solve this problem by inserting a considerable amount of built-in test hardware. To design inherently testable PLAs, several methods have been proposed that simplify the test generation process or

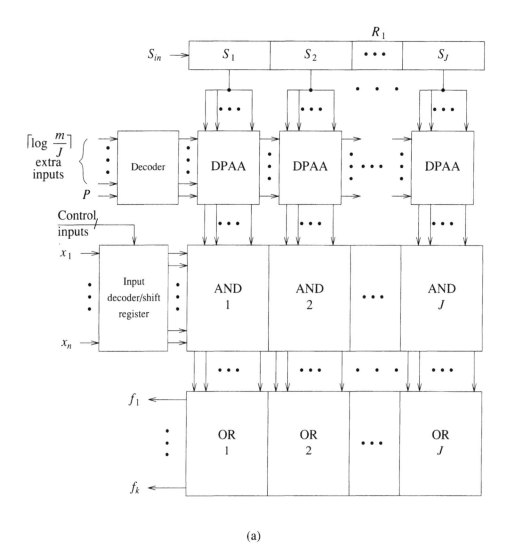

(a)

Figure 14.16   A testable PLA design with DAC partitioning (a) Design-for-test
architecture (b) A decoder-parity AND array (DPAA)

improve the fault coverage of existing test generation algorithms by employing little or
no BIT circuitry.

A Design of Testable PLAs by Specialized Layout

An example of this type has been proposed by Son and Pradhan [1980]. The PLA is
modified to be nonconcurrent in the sense that the AND array only consists of
mutually disjoint implicates of the function being implemented. The product lines of
the PLA are arranged such that the crosspoint connection patterns on adjacent output
lines are distinct. If this cannot be done, an extra output (test point) should be added

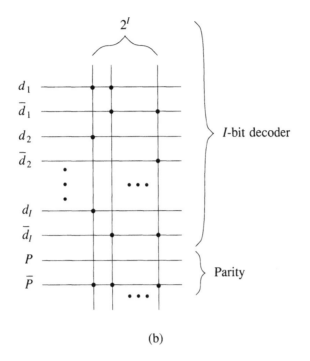

(b)

Figure 14.16   (Continued)

to make each pair of neighboring output lines differ from the others. It has been shown that PLAs designed in such a way have the property that the test set for crosspoint faults also covers stuck-at faults and bridging faults. A simple algebraic test generation algorithm exists that calculates the tests for all crosspoint faults.

A PLA Design for Testing Single Bridging Faults

Pradhan and Son [1980] have also shown that there are some undetectable bridging faults in PLAs that can *invalidate* a test set for crosspoint faults. Thus removal of undetectable faults should be considered to ensure a high fault coverage. Such undetectable bridging faults only occur between adjacent bit lines and adjacent product lines. The authors have developed a simple testable design for eliminating untestable bridging faults. For example, if the bridging faults produce AND functions, all bridging faults can be made testable by adding one product line and one output line.

A Design for Complete Testability of PLAs

To achieve a completely testable design, a PLA must be converted into a *crosspoint-irredundant PLA*, namely, a PLA in which all the crosspoint faults are detectable. Any PLA can be made crosspoint-irredundant by adding control inputs. This design also has the advantage that a test set for single crosspoint faults is sufficient to test all single stuck-at faults and bridging faults as well [Ramanatha and

Biswas 1982, 1983]. The complexity of test generation to achieve a high fault coverage is thus reduced.

Low-Overhead Design of Testable PLAs

Most testable designs of PLAs use a shift register to select individual product lines. An m-bit shift register takes up a significant amount of silicon area. A shift register is used to avoid concurrency that is inherent in most PLAs, i.e., an input pattern may activate two or more product lines. Each product term can be uniquely selected by increasing the Hamming distance among product terms [Bozorgui-Nesbat and McCluskey 1984]. Clearly if the Hamming distance between any pair of product terms is at least 1, any input patterns that activate product term P_j will not activate any other P_i for $i \neq j$. For this technique, extra inputs are added such that the Hamming distance between any two product terms is at least 2. Then a main test pattern and n auxiliary test patterns are applied to each product line. For P_i, the main pattern is a completely specified input that activates P_i, and the n auxiliary patterns are those inputs that have exactly one bit different from the main pattern. All main test patterns are found directly from the PLA's personality. Auxiliary patterns can be easily generated. Since an auxiliary pattern has only one bit different from a main pattern, the Hamming distance between any auxiliary pattern for P_i and a main pattern for other P_j is at least 1. Therefore, when testing P_i, no fault-free P_j $(j \neq i)$ can be activated; i.e., each product line can be individually selected and tested in the test mode. All single stuck-at faults, missing crosspoint faults, or extra crosspoint faults can be detected using this technique. Since no extra circuitry except for control input lines is added, area overhead is low. However, the problem of finding the minimal number of extra inputs and assigning connections to product lines is NP-complete. A heuristic procedure for this problem is described in [Bozorgui-Nesbat and McCluskey 1984], but it also requires extensive computation.

Fully testable (FT) PLAs are alternatives to BIT PLAs. They offer a high degree of testability by changing the logic design of the PLA and adding check points. Since little or no change is made to the PLA's physical structure, they do not create any layout problem. In general, converting a given PLA into a FT PLA is a complex task, and software tools for logic modification and test generation are required. The amount of overhead involved in constructing a fully testable PLA is function-dependent and may be high.

14.5 Evaluation of PLA Test Methodologies

14.5.1 Measures of TDMs

PLAs have many testing and testable design methods. Each method has its advantages and disadvantages. There are several common measures that help in characterizing and evaluating DFT techniques. These measures can be classified into four categories, namely (1) testability characteristics, (2) resulting effect on the original design, (3) requirements for test environment, and (4) design costs.

Testability characteristics specify the degree of testability a test method can achieve, for example, the type of faults detected and the fault coverage; if it supports fault masking, concurrent testing, or self-testing; and whether the tests are function-dependent. A circuit is usually made testable by changing the design. This may result in additional delay under normal operation, area overhead, and extra I/O pins. Most methods cannot operate without additional support and control logic. Requirements for the application environment specify the hardware needed to support a method in a real-time test process. Finally, design costs relate to the difficulty of implementing a test method, e.g., how complicated it is to modify the logic and the layout. In the following, we will briefly discuss some of these measures and then present a table indicating the values of these measures for numerous test methods.

14.5.1.1 Resulting Effect on the Original Design

Area overhead: BIT circuitry requires extra area, referred to as BIT area. The area overhead for BIT is

$$\text{area overhead} = \frac{BIT \ area}{original \ area} \cdot$$

Different CAD systems use different design rules and generate different PLA layouts. To estimate area overhead, the floor plan of a PLA given in [Treuer *et al.* 1985] will be used, from which it follows that

$$\text{area of original PLA} = 130nm + 65mk + 900n + 300m + 550k + 2200 \ (\lambda^2)$$

where n, m, and k are the number of inputs, product lines, and outputs of the PLA, respectively, and λ is a scale factor that is a measure of the resolution of the manufacturing process. For simplicity, one can take the value of λ to be 1 micron. Since BIT circuits consist mainly of standard cells, such as shift register cells and parity checker cells, we will measure the additional area as follows:

$$\text{BIT area} = [\ \Sigma_i \ \# \ \text{of} \ type_i \ \text{cells} \times \text{area of a} \ type_i \ \text{cell}] \times \text{connection cost.}$$

Here connection cost is 1 if a BIT cell can be connected to the original PLA without requiring extra area for wiring. Otherwise, the connection cost is greater than 1. The dominant factor in area overhead is the part of BIT area that grows with PLA size. As no accurate figure for this area can be obtained without carrying out the chip layout, the estimates to be presented emphasize the major part of area overhead, although constant components are also considered. The estimates tend to give a lower bound on area overhead.

Extra delay: Adding BIT hardware may have some side effect on system performance, such as additional delay encountered during normal circuit operation. This delay may affect the system's normal operation or change the system's critical path. Extra delay is measured in terms of gate delays.

Delay per test: Delay per test refers to the maximal number of gate delays in each test cycle, i.e., the maximal time for the effect of one test pattern to propagate through the PLA. It is assumed that the original PLA has three gate delays that are attributed to the input decoder, the AND array, and the OR array. Delay per test partially determines the test application rate and may affect the system clock cycle.

14.5.1.2 Requirements for Test Environment

Test application time: Test application time specifies a lower bound for the time required to complete the entire test process. Suppose (1) there is no overlapping of tests between successive test cycles, (2) each test cycle finishes in one system clock cycle, and (3) all clock cycles are of the same length. Then

test application time = delay per test × length of the shortest test sequence
that contains the entire test set in the right order

Note that a test sequence may be different from a sequence of patterns in a test set, because some transition sequences may be necessary in order to apply a given set of tests.

Means for test generation: The way in which tests are generated partially determines the applicability of a test method. To produce the required test inputs and responses before testing, some software tools may be necessary. There are several cases.

Case 1: No tools are required. For example, for the self-testable design shown in Figure 14.11, the tests [Treuer *et al.* 1985] are generated by the BIT hardware and responses are evaluated by the BIT circuit. The logic modification procedure is function-independent, and there is no need for *a priori* test generation.

Case 2: The test patterns are function-independent, but the responses are function-dependent, so simulation is required in order to determine the correct response. For example, in the method using signature analysis [Hassan 1982], the tests are pseudo-random sequences and are generated on-line. However, the expected signature is determined by the function realized by the PLA and has to be determined before testing.

Case 3: The test patterns are function-dependent but can be generated by a simple program.

Case 4: The test patterns are function-dependent and can only be generated by a complex program.

14.5.2 Evaluation of PLA Test Techniques

Various criteria can be used to compare different PLA test techniques. Based on this information, an evaluation matrix has been constructed that contains attribute values for most known PLA test techniques. This matrix is given in Figure 14.18. The abbreviations for the test techniques used in the evaluation matrix are listed in Figure 14.17. None of these techniques masks faults, and only CONC1 and CONC2 support concurrent testing.

Several of the characteristics of these test techniques vary radically with a PLA's parameters. A more complete study of these measures can be found in [Zhu and Breuer 1986a]. In general, area overhead tends to decrease as PLA size increases, where the size of a PLA is defined to be $m(2n + k)$. However, different test techniques are more sensitive to one parameter than another. For example, the area overhead for SIG varies much differently with large values of m than do UTS, BIST2, and FIST. Because of the complexity in evaluating and selecting a test technique for a particular PLA, an expert system has been constructed to aid designers in this task [Zhu and Breuer 1986b]. The techniques presented are applicable to normal PLAs. It is possible to reduce the area of a PLA by carrying out a process referred to as *PLA folding* [Hachtel *et al.* 1982]. Here more than one input and/or output variable can share the same column. Folding introduces a new fault mode, called a *cut point* fault. Ways of detecting this fault as well as how to extend the DFT and BIST techniques presented here to folded PLAs have been developed by Breuer and Saheban [1987].

EXH	Exhaustive test
CONC1	PLA with concurrent error detection by a series of checkers [Khakbaz and McCluskey 1981] — see Figure 14.5
CONC2	Concurrent testable PLA using error-detection code [Dong and McCluskey 1982] — see Figure 14.6
UTS	PLA with universal test set [Fujiwara and Kinoshita 1981] — see Figure 14.7
FIT	Function-independent-testable PLA [Hong and Ostapko 1980]
AUTO	Autonomously testable PLA [Yajima and Aramaki 1981] — see Figure 14.10
BILBO	PLA with BILBOs [Daehn and Mucha 1981] — see Figure 14.14
SIG	PLA with multiple signature analyzers [Hassan and McCluskey 1983] — see Figure 14.12
FIST	Function-independent self-testable PLA [Grassl and Pfleiderer 1983] — see Figure 14.13
TLO	Testable PLA with low overhead and high fault coverage [Khakbaz 1984]
BIT	Built-in tests for VLSI finite-state machines [Hua et al. 1984]
CPC	PLA design with cumulative parity comparison [Fujiwara 1984] — see Figure 14.11
LOD	Low-overhead design of testable PLA [Bozorgui-Nesbat and McCluskey 1984]
PTPLA	Parallel-testable PLA [Boswell et al. 1985] — see Figure 14.15
BIST2	Built-in self-testable PLA [Treuer et al. 1985]
HFC	High-fault-coverage built-in self-testable PLA [Upadhyaya and Saluja 1988]
DAC	Divide-and-conquer strategy for testable PLA [Saluja and Upadhyaya 1985] — see Figure 14.16
TSBF	PLA design for testing single bridging faults [Pradhan and Son 1980]
TBSL	Design of testable PLA by specialized layout [Son and Pradhan 1980]
DCT	Design for complete testability of PLA [Ramanatha and Biswas 1982]
EL	Test generation algorithm by Eichelberger and Lindbloom [1980]
SMITH	Test generation algorithm by Smith [1979]

Figure 14.17 Abbreviations of test techniques used in Figure 14.18

TDM	Single stuck	Single crosspoint	Single bridge	Multiple stuck-at	Multiple crosspoint Unidirectional	Multiple crosspoint Others	Multiple bridge
EXH	100	100	100	100	100	100	100
CONC1	99*	99*	99*				
CONC2	100	100			100		
UTS	100	100					
FIT	100	100					
AUTO	99*	100					
BILBO	99*	99*	99*				
SIG	98*	98*		98*			
FIST	99*	99*	99*	99*	99*	99*	99*
TLO	100	100		100	100	100	
BIT	100	100	100	100	100	100	
CPC	100	100	100	100	100	100	
LOD	100	100	100	100	100	100	
PTPLA	100	100		100	100		
BIST2	100	100	100	$100(1-2^{-(m+2n)})$	$100(1-2^{(m+2n)})$		$100(1-2^{-m})$
HFC	100	100			100	99.9*	
DAC	100	100		100	100	100	
TSBF			100				
TBSL	100	100	100				
DCT	100	100	100				
EL			¯ 97 (This figure is for missing crosspoint faults only.)				
SMITH	100	100			100		

The table is headed **FAULT COVERAGE (%)**

Note: Blank entries represent unknown data. * = estimated value.

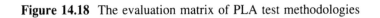

Figure 14.18 The evaluation matrix of PLA test methodologies

TDM	Self testing	Function dependency		Test generation	Extra IO pins	Extra delay	Delay per test	Number of tests
		Tests	Responses					
EXH	no	no	yes	no	0	0	3	2^n
CONC1	no	yes	no	easy	5	0	12	$2m+D+8$
CONC2	no	no	no	no	2	0	3	0
UTS	no	no	no	no	5	1	$3+2logm$	$2(n+m)+1$
FIT	no	no	no	no	$6+log(n+1)$	1	$4+2logm$	$5+2.5n+2m$
AUTO	yes	no	no	no	2	1	$6+2(k+1)$	$2(n+m)+9$
BILBO	yes	no	yes	no	$3 \sim 9$	0	4	$2n+m+k+3$
SIG	yes	no	yes	no	4	0	6	2^n
FIST	yes	no	yes	no	1	3	7	$2nm$
TLO	no	yes	yes	easy	3	0	3	$m(7+n)+2$
BIT	yes	no	no	no	5	1	$m+4$	$2n+m+2$
CPC	no	yes	no	no	4	0	3	$2n(m+1)+m+3$
LOD	no	yes	yes	easy	F.D.	0	3	$(n+1)m$
PTPLA	no	no	yes	no	4	1	4	$2nJ+4n+4$
BIST2	yes	no	no	no	4	1	$1.25k+3$	$2m(n+1)+1$
HFC	no	no	yes	no	3	1	6	$m(2n+k+3)+k+2$
DAC	no	no	yes	no	4	1	4	$m(3+2n+logm/J)+4$
TSBF	no	yes	no	med.	2	0	3	$m+1$
TBSL	no	yes	yes	hard	1	0	3	F.D.
DCT	no	yes	yes	hard	F.D.	0	3	F.D.
EL	no	yes	yes	med.	0	0	3	random
SMITH	no	yes	yes	hard	0	0	3	F.D.

Note: F. D. stands for function dependent. J is the number of product line groups.

Figure 14.18 (Continued)

TDM	Test application time	Test storage	Area overhead
EXH	$(3)2^n$	0	0
CONC1	$12(2m+D+8)$	$(2m+D+8)(3+n)$	$15680(n/2-1)+APC(k+1)+AMC$
CONC2	0	0	F.D.
UTS	$(3+2logm)(2n+3m)$	$5(5+n)$	$(m+1)ASR+(m+k)APC+64(6n+m+k+15)$
FIT	$(4+2logm)(5+2.5n+2m)$	$8(5+2.5n+2m)$	$5397+977n+2113m+1021k$
AUTO	$(8+2k)(2n+2m+9)$	0	$2215m+1764n+1420k+11752$
BILBO	$4(2n+2m+max(n,k)+1)$	$2n+2m+max(n,k)$	$1.3(2n+m+max(n,k))AB$
SIG	$(6)2^n$	$n(2+\lceil k/n \rceil)$	$1.3(3n+\lceil k/n \rceil)ASG$
FIST	$14nm$	k	$(2n+m)ASR+1.1k(ASG)$
TLO	$3m(7+n)+6$	$(7m+nm+2)(n+k+2)$	$m(64+ASR)+400$
BIT	$(m+4)(2n+m+2)$	0	$(2n+m+1)ASR+(m+k+2)APC+2AF+128(m+2n+k)+896$
CPC	$6n(m+1)+3m+9$	$(k+1)(2n(m+1)+m+3)$	$(m+1)ASR+384n+64m+64k+944$
LOD	$3m(n+1)$	$m(n+nk+k)$	$C(128m+640)$
PTPLA	$8nJ+24n+12$	F.D.	$64m+128n+696+ASR(2n+J)$
BIST2	$(1.25k+3)(2nm+2m+1)$	0	$1360m+2160n+760k$
HFC	$6m(2n+k+1)+6k$	0	$(2n+m+k+2)ASR+(2AC+5)logm+128m+1040$
DAC	$4m(2+2n+logm)+16$	0	$(1+2n)ASR+(1+\lceil log(m/J) \rceil)AC$
TSBF	$3m+3$	$(m+1)(n+2)$	$64(2n+m+k)+816$
TBSL	F.D.	F.D.	F.D.
DCT	F.D.	F.D.	$16n+8k+256m+1989$
EL	random	random	0
SMITH	F.D.	F.D.	0

Note: C is an integer that is function-dependent and usually small. D is the number of used crosspoints. ASR=area of 1-bit shift register. APC=area of 1-bit parity checker. AMC=area of a 1-out-of-m checker. ASG=area of a signature analyzer cell. AB=area of a BILBO register cell. AC=area of 1-bit counter. AF=area of a flip-flop.

Figure 14.18 (Continued)

TDM	EXTRA TEST SUPPORT MEANS REQUIRED		
	Test set generation	Test application	Response evaluation
EXH	response simulation	counter | LFSR	ATE
CONC1	simple TG program	ATE | ROM	monitor error indication
CONC2	none	none	none
UTS	none	ATE | on-line PG	monitor two extra outputs
FIT	none	ATE | on-line PG	ATE | ROM & comparator
AUTO	none	none	none
BILBO	obtain signature	control BILBO function	S.R. & comparator
SIG	obtain signature	none	ROM & comparator
FIST	obtain signature	none	comparator | signature decoder
TLO	simple TG program	ATE | ROM & on-line PG	ATE
BIT	none	none	monitor 3 extra outputs
CPC	response simulation	ATE | on-line PG	ATE
LOD	simple TG program	ATE | ROM & on-line PG	ATE
PTPLA	response simulation	ROM & simple control	ATE
BIST2	none	none	none
HFC	count crosspoints	ROM & simple control	none
DAC	response simulation	none	ATE
TSBF	special TG program	ATE | ROM	ATE | ROM & comparator
TBSL	special TG program	ATE	ATE
DCT	special TG program	ATE	ATE
EL	special TG program	ATE	ATE
SMITH	special TG program	ATE	ATE

Figure 14.18 (Continued)

TDM	Extra lines			Extra	Assumptions & remarks
	Bit	Product	Output	transistors	
EXH	0	0	0	0	
CONC1	0	0	1	$10n+5k+logm(m+5)$	2-rail checker and 2 EOR trees
CONC2	2	0	C	F.D.	C = the length of check symbols
UTS	0	1	1	$4n+11m+5k+6$	column rank of OR array is k
FIT	0	3	1	$11m+5k+0.5nlogn+29$	
AUTO	0	4	2	$12n+12m+5k+50$	feedback generator is not duplicated
BILBO	0	0	0	$16(2n+m+k)$	
SIG	0	0	0	$6(3n+\lceil k/n \rceil)$	
FIST	0	0	0	$12n+6m+8k$	
TLO	0	0	1	$6m$	
BIT	0	2	2	$12n+11m+5k$	PLA can be a sequential circuit
CPC	0	3	1	$6(m+1)$	
LOD	$2C$	0	0	0	C is an integer and needs to be calculated
PTPLA	$2log(m/J)$	0	0	$6J+2log(mJ)$	J is no. of product line groups
BIST2	0	3	1	$11n+6m+3.5k+30$	for n odd; otherwise 2 product lines are added
HFC	0	0	2	$12n+6m+6k+19logm$	
DAC	$2+2J$	0	0	$12n+6J+10log(m/J)$	J is no. of product line groups
TSBF	0	1	1	0	bridging faults are all AND and detectable
TBSL	0	0	1	0	AND array consists of disjoint product terms
DCT	2	2	2	0	PLA is crosspoint irredundant; bridging is AND
EL	0	0	0	0	only missing crosspoint faults are considered
SMITH	0	0	0	0	size of test set is bounded by $n(2m+k)$

Figure 14.18 (Continued)

REFERENCES

[Agarwal 1980] V. K. Agarwal, "Multiple Fault Detection in Programmable Logic Arrays," *IEEE Trans. on Computers*, Vol. C-29, No. 6, pp. 518-522, June, 1980.

[Berger 1961] J. M. Berger, "A Note on Error Detection Codes for Asymmetric Channels," *Inform. Control*, Vol. 4, No. 1, pp. 68-73, March, 1961.

[Bose and Abraham 1982] P. Bose and J. A. Abraham, "Test Generation for Programmable Logic Arrays," *Proc. 19th Design Automation Conf.*, pp. 574-580, June, 1982.

[Boswell *et al.* 1985] C. Boswell, K. Saluja and K. Kinoshita, "A Design of Programmable Logic Arrays for Parallel Testing," *J. Computer Systems Science and Engineering*, Vol. 1, pp. 5-16, October, 1985.

[Bozorgui-Nesbat and McCluskey 1984] S. Bozorgui-Nesbat and E. J. McCluskey, "Lower Overhead Design for Testability for PLAs," *Proc. Intn'l. Test Conf.*, pp. 856-865, October, 1984.

[Brayton *et al.* 1984] R. K. Brayton, G. D. Hachtel, C. T. McMullen, and A. L. Sangiovanni-Vincentelli, *Logic Minimization Algorithms for VLSI Synthesis*, Kluwer Academic Publishers, Norwell, MA, 1984.

[Breuer and Saheban 1987] M. A. Breuer and F. Saheban, "Built-In Test for Folded Programmable Logic Arrays," *Microprocessors and Microsystems*, Vol. 11, No. 6, pp. 319-329, July/August, 1987.

[Cha 1978] C. W. Cha, "A Testing Strategy for PLAs," *Proc. 15th Design Automation Conf.*, pp. 326-331, June, 1978.

[Daehn and Mucha 1981] W. Daehn and J. Mucha, "A Hardware Approach to Self-Testing of Large Programmable Logic Arrays," *IEEE Trans. on Computers*, Vol. C-30, No. 11, pp. 829-833, November, 1981.

[Dong 1982] H. Dong, "Modified Berger Codes for Detection of Unidirectional Errors," *Digest of Papers 12th Annual Symp. on Fault-Tolerant Computing*, pp. 317-320, June, 1982.

[Dong and McCluskey 1981] H. Dong and E. J. McCluskey, "Matrix Representation of PLA's and an Application to Characterizing Errors," CRC Technical Report 81-11, Stanford University, September, 1981.

[Dong and McCluskey 1982] H. Dong and E. J. McCluskey, "Concurrent Testing of Programmable Logic Arrays," CRC Technical Report 82-11, Stanford University, June, 1982.

[Eichelberger and Lindbloom 1980] E. B. Eichelberger and E. Lindbloom, "A Heuristic Test-Pattern Generator for Programmable Logic Array," *IBM Journal of Research and Development*, Vol. 24, No. 1, pp. 15-22, January, 1980.

[Fujiwara and Kinoshita 1981] H. Fujiwara and K. Kinoshita, "A Design of Programmable Logic Arrays with Universal Tests," *IEEE Trans. on Computers*, Vol. C-30, No. 11, pp. 823-828, November, 1981.

[Fujiwara 1984] H. Fujiwara, "A New PLA Design for Universal Testability," *IEEE Trans. on Computers*, Vol. C-33, No. 8, pp. 745-750, August, 1984.

[Grassl and Pfleiderer 1983] G. Grassl and H-J. Pfleiderer, "A Function-Independent Self-Test for Large Programmable Logic Arrays," *Integration, the VLSI Magazine*, Vol. 1, pp. 71-80, 1983.

[Hassan and McCluskey 1983] S. Z. Hassan and E. J. McCluskey, "Testing PLAs Using Multiple Parallel Signature Analyzers," *Digest of Papers 13th Annual Intn'l. Symp. on Fault-Tolerant Computing*, pp. 422-425, June, 1983.

[Hachtel *et al.* 1982] G. D. Hachtel, A. R. Newton, and A. L. Sangiovanni-Vincentelli, "An Algorithm for Optimal PLA Folding," *IEEE Trans. on Computer-Aided Design*, Vol. CAD-1, No. 2, pp. 63-76, April, 1982.

[Hong and Ostapko 1980] S. J. Hong and D. L. Ostapko, "FITPLA: A Programmable Logic Array for Functional Independent Testing," *Digest of Papers 10th Intn'l. Symp. on Fault-Tolerant Computing*, pp. 131-136, October, 1980.

[Hua *et al.* 1984] K. A. Hua, J. Y. Jou, and J. A. Abraham, "Built-In Tests for VLSI Finite-State Machines," *Digest of Papers 14th Intn'l. Symp. on Fault-Tolerant Computing*, pp. 292-297, June, 1984.

[Khakbaz and McCluskey 1981] J. Khakbaz and E. J. McCluskey, "Concurrent Error Detection and Testing for Large PLAs," *IEEE Trans. on Electron Devices*, Vol. ED-29, pp. 756-764, April, 1982.

[Khakbaz 1984] J. Khakbaz, "A Testable PLA Design with Low Overhead and High Fault Coverage," *IEEE Trans. on Computers*, Vol. C-33, No. 8, pp. 743-745, August, 1984.

[Min 1983a] Y. Min, "A Unified Fault Model for Programmable Logic Arrays," CRC Technical Report 83-5, Stanford University, May, 1983.

[Min 1983b] Y. Min, "Generating a Complete Test Set for Programmable Logic Arrays," CRC Technical Report 83-4, Stanford University, May, 1983.

[Muehldorf and Williams 1977] E. I. Muehldorf and T. W. Williams, "Optimized Stuck Fault Test Pattern Generation for PLA Macros," *Digest of Papers 1977 Semiconductor Test Symp.*, pp. 89-101, October, 1977.

[Ostapko and Hong 1979] D. L. Ostapko and S. J. Hong, "Fault Analysis and Test Generation for Programmable Logic Arrays (PLA's)," *IEEE Trans. on Computers*, Vol. C-28, No. 9, pp. 617-627, September, 1979.

[Pradhan and Son 1980] D. K. Pradhan and K. Son, "The Effect of Undetectable Faults in PLAs and a Design for Testability," *Digest of Papers 1980 Test Conf.*, pp. 359-367, November, 1980.

[Ramanatha and Biswas 1982] K. S. Ramanatha and N. N. Biswas, "A Design for Complete Testability of Programmable Logic Arrays," *Digest of Papers 1982 Intn'l. Test Conf.*, pp. 67-73, November, 1982.

[Ramanatha and Biswas 1983] K. S. Ramanatha and N. N. Biswas, "A Design for Testability of Undetectable Crosspoint Faults in Programmable Logic Arrays," *IEEE Trans. on Computers*, Vol. C-32, No. 6, pp. 551-557, June, 1983.

[Reddy and Ha 1987] S. M. Reddy and D. S. Ha, "A New Approach to the Design of Testable PLA's," *IEEE Trans. on Computers*, Vol. C-36, No. 2, pp. 201-211, February, 1987.

[Saluja and Upadhyaya 1985] K. K. Saluja and J. S. Upadhyaya, "Divide and Conquer Strategy for Testable Design of Programmable Logic Arrays," *Proc. 4th Australian Microelectronics Conf.*, May, 1985.

[Saluja *et al*. 1983] K. K. Saluja, K. Kinoshita and H. Fujiwara, "An Easily Testable Design of Programmable Logic Arrays for Multiple Faults," *IEEE Trans. on Computers*, Vol. C-32, No. 11, pp. 1038-1046, November, 1983.

[Saluja *et al*. 1985] K. K. Saluja, H. Fujiwara, and K. Kinoshita, "Testable Design of Programmable Logic Arrays with Universal Control and Minimal Overhead," *Proc. Intn'l. Test Conf.*, pp. 574-582, 1985. Also in *Intn'l. Journal of Computers and Mathematics with Applications*, Vol. 13, No. 5/6, pp. 503-517, February, 1987.

[Smith 1979] J. E. Smith, "Detection of Faults in Programmable Logic Arrays," *IEEE Trans. on Computers*, Vol. C-28, No. 11, pp. 848-853, November, 1979.

[Somenzi *et al*. 1984] F. Somenzi, S. Gai, M. Mezzalamo, and P. Prinetto, "PART: Programmable Array Testing Based on a Partitioning Algorithm," *IEEE Trans. on Computer-Aided Design*, Vol. CAD-3, No. 2, pp. 142-149, April, 1984.

[Son and Pradhan 1980] K. Son and D. K. Pradhan, "Design of Programmable Logic Arrays for Testability," *Digest of Papers 1980 Test Conf.*, pp. 163-166, November, 1980.

[Treuer *et al*. 1985] R. Treuer, H. Fujiwara, and V. K. Agrawal, "Implementing a Built-In Self-Test PLA Design," *IEEE Design & Test of Computers*, Vol. 2, No. 2, pp. 37-48, April, 1985.

[Upadhyaya and Saluja 1988] J. S. Upadhyaya and K. K. Saluja, "A New Approach to the Design of Built-In Self-Testing PLAs for High Fault Coverage," *IEEE Trans. on Computer-Aided Design*, Vol. 7, No. 1, pp. 60-67, January, 1988.

[Wakerly 1978] J. Wakerly, *Error Detecting Codes, Self Checking Circuits and Applications*, American-Elsevier, New York, 1978.

[Wang and Avizienis 1979] S. L. Wang and A. Avizienis, "The Design of Totally Self-Checking Circuits Using Programmable Logic Arrays," *Digest of Papers 9th Annual Intn'l. Symp. on Fault-Tolerant Computing*, pp. 173-180, June, 1979.

[Wei and Sangiovanni-Vincentelli 1985] R-S. Wei and A. Sangiovanni-Vincentelli, "PLATYPUS: A PLA Test Pattern Generation Tool," *Proc. 22nd Design Automation Conf.*, pp. 197-203, June, 1985.

[Yajima and Aramaki 1981] S. Yajima and T. Aramaki, "Autonomously Testable Programmable Logic Arrays," *Digest of Papers 11th Annual Intn'l. Symp. on Fault-Tolerant Computing*, pp. 41-43, June, 1981.

[Zhu and Breuer 1986a] X. Zhu and M. A. Breuer, "A Survey of Testable PLA Designs," *IEEE Design & Test of Computers*, Vol. 5, No. 4, pp. 14-28, August, 1988.

[Zhu and Breuer 1986b] X. Zhu and M. A. Breuer, "A Knowledge Based TDM Selection System," *Proc. 1986 Fall Joint Computer Conf.*, pp. 854-863, November, 1986.

PROBLEMS

14.1 Draw a PLA representation for the sum and carry functions over three variables.

14.2 For the PLA shown in Figure 14.3, indicate an extra crosspoint fault that cannot be modeled as a s-a fault in the sum of product representation of f_1 or f_2.

14.3 Prove that a missing crosspoint fault is equivalent to some stuck-at fault in the sum-of-product representation of a PLA.

14.4 Describe a test generation procedure, similar to the one presented in this chapter for missing crosspoints, for detecting shrinkage, appearance, and disappearance crosspoint faults.

14.5 Modify the PLA of Example 1 so that no two product lines are activated by any input vector.

14.6 Construct a testable PLA for the PLA of Example 1 using the concurrent error detection by a series of checkers technique discussed in Section 14.4.1.1. For every possible checkpoint fault that can occur to product line P_3, construct an input test that will detect the fault and indicate the values on the error-indicators. Show the output error-detection code used.

14.7 Complete the design of the PLA of Example 1 using the universal test set method described in Section 14.4.2.1. Construct a table of all the test vectors in the universal test set, and indicate next to each vector which faults are detected. Use the same format for the test vectors shown in the text.

14.8 For the UTS technique listed in Figure 14.17 compute the information shown in Figure 14.18. Assume $n = 32$, $m = 64$, and $k = 16$.

14.9 Consider the PLA design shown in Figure 14.3(a). Construct the Karnaugh graph for f_1 and f_2 for each of the following situations, and indicate the product terms implemented by the PLA.

    (a) The original PLA;

    (b) A shrinkage fault caused by a connection between bit line b_1 and product line P_1;

    (c) A shrinkage fault caused by a connection between bit line b_6 and P_1;

(d)   A growth fault caused by the missing connection between bit line b_1 and P_6;

(e)   An appearance fault caused by an extra connection between P_4 and f_1;

(f)   A disappearance fault caused by the missing connection between P_6 and f_2.

14.10   Explain how a shrinkage fault can be functionally equivalent to a disappearance fault.

14.11   Let $f_1 = x_1 x_2$ and $f_2 = x_1 x_2 x_3$. If these functions were implemented using a PLA, the requirement for concurrent testability is violated because both product terms can be activated at the same time. Show how to reimplement f_1 or f_2 so that only one product term is activated at a time. Estimate the area penalty for this modification.

15. SYSTEM-LEVEL DIAGNOSIS

About This Chapter

The advent of large, parallel computing systems containing hundreds or thousands of processing elements means that the problems associated with the testing and diagnosis of such systems are especially challenging and important. In this chapter we consider a formal model of system-level diagnosis in which one processing element is used to test and diagnose other processing elements. Using this model, several different measures of system diagnosability are defined and evaluated. Necessary and sufficient conditions on system structure to obtain defined levels of diagnosability are derived for the basic model and its various generalizations.

15.1 A Simple Model of System-Level Diagnosis

With the advent of VLSI, there has been much research and development related to large computing systems containing many processing elements connected in a network. In such systems an interesting possibility is their diagnosis using one subsystem (i.e., processing element) to diagnose other subsystems. A formal model for such diagnosis developed by Preparata, Metze, and Chien [Preparata *et al.* 1967], called the PMC model, has led to a large amount of research. This model is based upon the following three assumptions:

1. A system can be partitioned into units (called *modules*), and a single unit can individually test another unit.

2. On the basis of the responses to the test that is applied by a unit to another unit, the test outcome has a binary classification, "pass" or "fail" (i.e., the testing unit evaluates the tested unit as either fault-free or faulty).

3. The test outcome and evaluation are always accurate (i.e., a fault-free unit will be diagnosed as fault-free and a faulty unit will be diagnosed as faulty) if the testing unit is fault-free, but the test outcome and evaluation may be inaccurate if the testing unit itself is faulty.

With these assumptions a diagnostic system may be represented by a diagnostic graph in which each vertex (node) v_i corresponds to a unit in the system, and a branch (arc) from v_i to v_j (denoted by v_{ij}) corresponds to the existence of a test by which unit v_i evaluates unit v_j. The test outcome a_{ij} associated with v_{ij} is assumed to be as follows:

$a_{ij} = 0$ if v_i and v_j are both fault-free;

$a_{ij} = 1$ if v_i is fault-free and v_j is faulty;

$a_{ij} = \times$ (unspecified and indeterminate) if v_i is faulty regardless of the status of v_j (i.e., $\times$ can be 0 or 1).

The various test outcome situations are depicted in Figure 15.1.

Two implicit assumptions required by this assumed test outcome are that faults are permanent and that the tests applied by unit v_i to unit v_j can detect all possible faults in v_j.

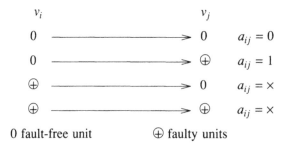

Figure 15.1   Assumed test outcomes in the Preparata-Metze-Chien model

In the PMC model, faulty units can be identified by decoding the set of test outcomes, referred to as the *syndrome*, of the system. A fault in unit v_i is *distinguishable* from a fault in unit v_j if the syndromes associated with these two faults are different. The two faults are *indistinguishable* if their syndromes could be identical. These definitions can be directly extended to define distinguishable and indistinguishable multiple faults (i.e., sets of faulty units), sometimes referred to as *fault patterns*.

Example 15.1: Figure 15.2 shows a system with five units $v_1, ..., v_5$. The test syndromes shown in lines (a) and (b) correspond to the single faulty units v_1 and v_2 respectively. Since these two test syndromes have opposite values of a_{51}, these two faults are distinguishable. Line (c) shows the test syndrome associated with the multiple fault pattern $\{v_1,v_2\}$ (i.e., both v_1 and v_2 are faulty). This fault pattern is distinguishable from v_2 since lines (b) and (c) have opposite values of a_{51}. However, since the test syndromes in lines (a) and (c) may not be different, the single fault v_1 is indistinguishable from the multiple fault $\{v_1,v_2\}$. □

Two measures of diagnosability, *one-step diagnosability* and *sequential diagnosability*, were originally defined [Preparata *et al.* 1967]. A system of n units is one-step t-fault diagnosable if all faulty units in the system can be identified without replacement, provided the number of faulty units does not exceed t. A system of n units is sequentially t-fault diagnosable if at least one faulty unit can be identified without replacement, provided the number of faulty units does not exceed t.

Sequential diagnosability implies a multistep diagnostic procedure for the identification of all faulty units. In the first iteration one or more faulty units are identified and replaced by other units, which are assumed to be fault-free. After this replacement the test is rerun and additional faulty units may be identified. The process is repeated until all faulty units are identified and replaced, requiring at most t iterations (steps). As previously stated, it is assumed that all replacement units are fault-free and that no faults occur during the testing process.

The following theorem presents some general *necessary* properties of the diagnostic graph required for one-step diagnosability.

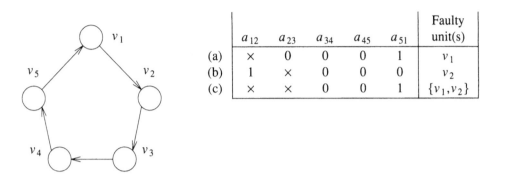

	a_{12}	a_{23}	a_{34}	a_{45}	a_{51}	Faulty unit(s)
(a)	×	0	0	0	1	v_1
(b)	1	×	0	0	0	v_2
(c)	×	×	0	0	1	$\{v_1, v_2\}$

Figure 15.2   A system and associated test outcomes

Theorem 15.1:   In a one-step t-fault diagnosable system

a.   There must be at least $2t+1$ units.

b.   Each unit must be diagnosed by at least t other units.

Proof

a.   Suppose there are $n \leq 2t$ units. Then the vertices can be partitioned into two disjoint sets A and B each containing at most t vertices. The diagnostic graph can then be represented as shown in Figure 15.3, where a_{AA} is the set of connections within A, a_{AB} is the set of connections from A to B, a_{BA} is the set of connections from B to A, and a_{BB} is the set of connections within B. Figure 15.3(b) shows the value of the test syndromes for the two disjoint fault patterns consisting of (1) all units in A being faulty and (2) all units in B being faulty.

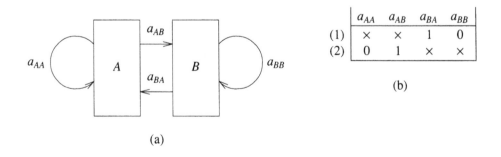

	a_{AA}	a_{AB}	a_{BA}	a_{BB}
(1)	×	×	1	0
(2)	0	1	×	×

(b)

(a)

Figure 15.3   Partition of system into two subsystems

Since no test outcome must have a different value for these two fault patterns, they are indistinguishable and hence the system is not one-step t-fault diagnosable.

b. Suppose some module v_i is tested by $k < t$ other modules $v_1, v_2, ..., v_k$. Consider the following two sets of faulty units:

$$A = \{v_1, v_2, ..., v_k\}$$

$$B = \{v_1, v_2, ..., v_k, v_i\}.$$

These fault conditions will produce indistinguishable test syndromes, and since both fault patterns contain at most t faults, the system is not one-step t-fault diagnosable. □

The conditions of Theorem 15.1 are necessary but are not sufficient, as can be seen for the system of Figure 15.4(a). Since the test syndromes for fault patterns $\{v_1\}$ and $\{v_2\}$ may be identical, this system is not one-step one-fault diagnosable, although it does satisfy both of the necessary conditions of Theorem 15.1 for $t = 1$.

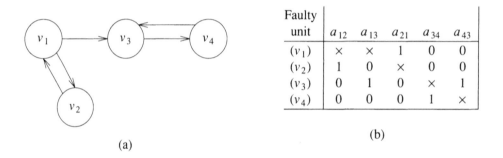

Faulty unit	a_{12}	a_{13}	a_{21}	a_{34}	a_{43}
(v_1)	×	×	1	0	0
(v_2)	1	0	×	0	0
(v_3)	0	1	0	×	1
(v_4)	0	0	0	1	×

(a)

(b)

Figure 15.4 A system and single fault test syndromes

However, the conditions of Theorem 15.1 can be used to show that a class of systems for which the number of vertices is $n = 2t + 1$ and each unit is tested by exactly t other units is optimal (i.e., has the minimal number of testing links) for any one-step t-fault diagnosable system. Such a system has $m = nt$ testing links, which is minimal. One such optimal class of system is called a $D_{\delta t}$ system. In such a system there is a testing link from v_i to v_j if and only if $j = (i + \delta m)$ mod n, where n is the number of vertices, δ is an integer, and $m = 1, 2, ..., t$. Figure 15.5 shows such a system with $\delta = t = 2$ (i.e., a D_{22} system). $D_{\delta t}$ systems in which the values of δ and $n = 2t + 1$ are relatively prime are optimal with respect to the number of testing links, for one-step t-fault diagnosability.

Some general classes of system that are sequentially t-fault diagnosable have also been developed. One such class of systems has $m = n + 2t - 2$ testing links and is illustrated by the system shown in Figure 15.6 for the case $t = 6$ and $n = 14$.

A *single-loop system* is a system in which each unit is tested by and tests exactly one other unit, and all units of the system are contained within one testing loop. For such

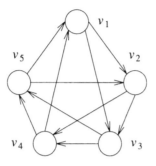

Figure 15.5 An optimal 2-fault diagnosable system

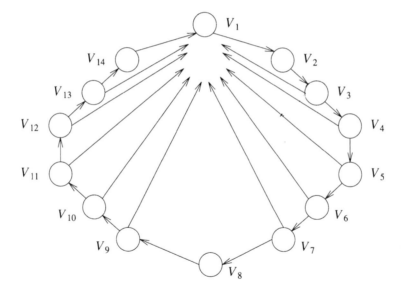

Figure 15.6 A sequentially 6-fault diagnosable system

systems having n units, if $t = 2b + c$, and $c = 0$ or 1, a necessary and sufficient condition for sequential t-fault diagnosability is

$$n \geq 1 + (b+1)^2 + c(b+1).$$

In order for a system to be sequentially t-fault diagnosable for any set of fault patterns $(F_1, F_2, ..., F_r)$ that are not distinguishable, all fault patterns in this set must have a common element (i.e., $F_1 \cap F_2 \cap ... \cap F_r \neq \emptyset$). Since the system in the proof of

Theorem 15.1(a) does not satisfy this condition, any system with fewer than $2t + 1$ units is not sequentially t-fault diagnosable.

Example 15.2: We will examine the diagnostic capabilities of the system of Figure 15.7. Since $n = 5$, from Theorem 15.1(a) we know that the system is at most 2-fault diagnosable (either sequentially or in one step). Furthermore, since vertex 3 is only tested by vertex 2, it follows from Theorem 15.1(b) that the system is at most one-fault diagnosable in one step. The tables of Figure 15.7 show the test outcomes for all single and double fault patterns.

Since for each pair of fault patterns containing a single fault the fault syndromes are distinguishable (i.e., at least one test outcome has to have a different binary value for the two different fault patterns), this system is one-step one-fault diagnosable. To determine if the system is sequentially 2-fault diagnosable, we must examine and compare the test syndromes for all fault patterns containing two faults to determine distinguishable test syndromes. The set of fault patterns that do not satisfy the one-step distinguishability criteria are {(2,3),(2,4)}, which have a common element 2, and {(3,4),(3,5)}, which have a common element 3. Therefore the system is sequentially 2-fault diagnosable, since for any test outcome at least one module can definitely be determined as faulty and replaced.   □

The PMC model of diagnosable systems is restricted to systems in which an individual unit is capable of testing another unit, and the measures of diagnosability are restricted to worst-case measures and to repair strategies in which only faulty units are replaced. It has not been possible to apply this simple model of system-level diagnosis directly to actual systems. However, it has laid the groundwork for subsequent research that has attempted to generalize the model and add more realistic constraints associated with actual systems. Because of the extensive research that has been done in this area and the lack of current practical applications, we will present a relatively broad overview of the general direction of this work and omit a detailed presentation of the mathematics involved.

15.2 Generalizations of the PMC Model

15.2.1 Generalizations of the System Diagnostic Graph

In the PMC model a test is applied by a single unit, which if faulty invalidates the test. This assumption can be generalized to include the possibility that application of a test requires the combined operation of more than one unit [Russell and Kime 1975a], as well as the possibility that a unit is known to be fault-free at the beginning of the diagnosis [Russell and Kime 1975b]. For example, the diagnosis of the IBM System/360 Model 50 has been described with a generalized diagnostic graph (GDG) [Hackl and Shirk 1965], shown in Figure 15.8. Here the units are represented as follows: v_1 is the main storage, v_2 is the ROM control, v_3 is the ALU, v_4 is the local storage, and v_5 is the channel. Each unit v_i has associated with it a corresponding fault condition f_i. The test t_1 for the fault f_1 associated with unit v_1 will be valid even in the presence of other faulty units. Therefore the GDG has no arc labeled t_1. The single arc labeled t_2 from f_1 to f_2 indicates (as in the basic PMC model) that unit v_1 (previously verified by t_1) is sufficient to test and verify unit v_2. The two arcs labeled t_3 from f_1 to f_3 and from f_2 to f_3 respectively indicate that if either v_1 or v_2

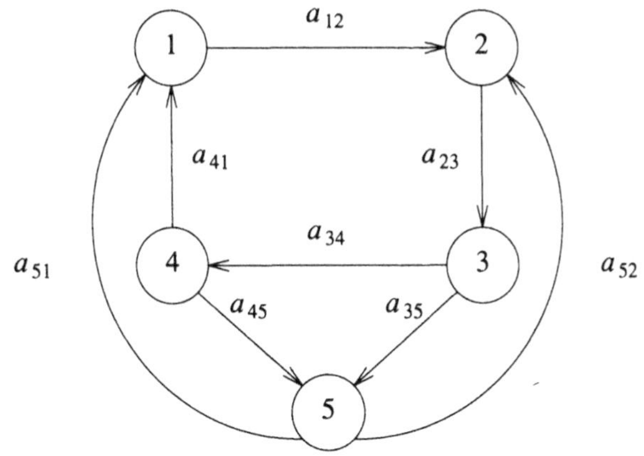

Faulty units	a_{12}	a_{23}	a_{34}	a_{35}	a_{41}	a_{45}	a_{51}	a_{52}
1	×	0	0	0	1	0	1	0
2	1	×	0	0	0	0	0	1
3	0	1	×	×	0	0	0	0
4	0	0	1	0	×	×	0	0
5	0	0	0	1	0	1	×	×

Faulty units	a_{12}	a_{23}	a_{34}	a_{35}	a_{41}	a_{45}	a_{51}	a_{52}
(1,2)	×	×	0	0	1	0	1	1
(1,3)	×	1	×	×	1	0	1	0
(1,4)	×	0	1	0	×	×	1	0
(2,3)	1	×	×	×	0	0	0	1
(2,4)	1	×	1	0	×	×	0	1
(2,5)	1	×	0	1	0	1	×	×
(3,4)	0	1	×	×	×	×	0	0
(3,5)	0	1	×	×	0	1	×	×
(4,5)	0	0	1	1	×	×	×	×

Figure 15.7

or both are faulty, then t_3 will be invalidated. Similarly tests t_4 and t_5 require v_1, v_2, and v_3 for them all to be fault-free.

A very general approach consists of using algebraic expressions to represent conditions under which a test is invalidated as well as to represent fault conditions detected by a test [Adham and Friedman 1977]. A set of fault patterns is described by a Boolean expression in the variables $\{f_1, f_2, ..., f_n\}$, where f_i is 1 if v_i is faulty and f_i is 0 if v_i is fault-free. Associated with any test t_k are two Boolean functions, the invalidation function $I(t_k)$, which has the value 1 for any fault pattern that invalidates the test t_k,

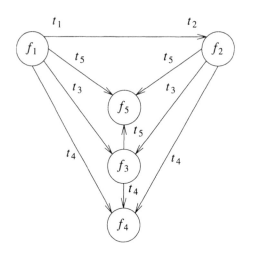

Figure 15.8   Diagnostic graph of IBM System/360 Model

and the detection function $D(t_k)$, which has the value 1 for any fault pattern that is detected by t_k. Although this model can handle many cases that cannot be handled by graph-based models, it is too complex to be useful for large systems.

The complexities inherent in the algebraic model can be reflected in a much simpler, but less accurate, probabilistic model in which a test detects a fault f_i with probability p_i where $1 > p_i > 0$ [Fujiwara and Kinoshita 1978]. We shall now consider several similar types of generalization of the PMC model.

15.2.2 Generalization of Possible Test Outcomes

In the PMC model the outcome of a test applied to unit v_j by unit v_i is defined as shown in Figure 15.1. This assumption of the PMC model can be generalized in many ways, as shown in Figure 15.9 [Kavianpour 1978].

The different models represented in this table range from a "perfect" tester, A_∞, in which the test outcome always corresponds to a perfect diagnosis of faulty units even when the testing unit is itself faulty, to the "0-information" tester, A_0, which never provides an assured-reliable test outcome even when the testing unit is fault-free. Many variants of the original model of Figure 15.1 as summarized in Figure 15.9 have been considered by various researchers.

The PMC model is represented by the column labeled A_p. The column labeled A_0 represents "zero information" since the test outcome is always unpredictable regardless of the condition of v_j (the unit being tested) and v_i (the unit testing v_j). The column labeled A_∞ represents "total information" since the test outcome always represents accurately the condition of v_j. In column A_{pT} the test outcome is unpredictable when v_i is not faulty and v_j is faulty. Therefore v_a cannot detect all faults in v_j. This represents "incomplete" or "partial testing". In all of the seven models other than A_0 and A_{pT} the test outcome always accurately reflects the condition of v_j when v_i is not

a_{ij}	A_∞	A_W	A_B	A_y	A_μ	A_λ	A_p	A_{pT}	A_0
$0 \to 0$	0	0	0	0	0	0	0	0	×
$0 \to \oplus$	1	1	1	1	1	1	1	×	×
$\oplus \to 0$	0	1	×	0	0	1	×	×	×
$\oplus \to \oplus$	1	1	1	0	×	×	×	×	×

Figure 15.9　Different models of system diagnosis

faulty with differing assumptions when v_i is faulty such as no information (the PMC model A_p), always indicating a fault (A_w), and never indicating a fault (A_y).

15.2.3 Generalization of Diagnosability Measures

Originally only two measures of system-level diagnosability, one-step t-fault diagnosability and sequential t-fault diagnosability, were proposed. Both these measures imply a very conservative philosophy of system maintenance in which only faulty units are permitted to be replaced. In actuality there exists a trade-off between the necessity of repeating tests (in order to identify all the faulty units in a sequentially diagnosable system) and the necessity of replacing fault-free units. Consider, for example, the n-unit single-loop system shown in Figure 15.10, with the assumed test outcome defined by $a_{12} = a_{23} = 1$ and all other $a_{ij} = 0$. Assuming $n > t$, we can conclude that unit v_2 is faulty and units v_i, v_{i+1}, ..., v_n, v_1 are all fault-free where $i = t + 2$. Units v_3, ..., v_{i-1} may be either faulty or fault-free. To determine the status of these units we could replace unit v_2 and then repeat the tests (possibly t times in all) and eventually determine the status of all units while ensuring that no fault-free unit has ever been replaced. Alternatively, without repeating the tests we could replace units v_2, ..., v_{i-1}, ensuring that all faulty units have been replaced but perhaps replacing some fault-free units as well. Thus the trade-off between repeating tests (i.e., speed of diagnosis) and diagnostic accuracy (i.e., the number of fault-free units that are replaced) becomes apparent. To reflect this situation another measure of system-level diagnosis has been defined [Friedman 1975]. A system is k-step t/s (read t-out-of-s) diagnosable if by k applications of the diagnostic test sequence any set of $f \leq t$ faulty units can be diagnosed and repaired by replacing at most s units. Clearly $s \geq t$ and $n \geq s$. Measures of the diagnosability of the system are the average or expected value of $s - f$ as well as the maximum value of $s - f$. These new measures have been considered for regular systems such as $D_{\delta t}$ systems, of which single-loop systems are a special case.

An interesting special case of t/s diagnosability is when $s = t$. In this case all fault patterns consisting of t faulty units are exactly diagnosed, but patterns consisting of $f < t$ faulty units may be inexactly diagnosed (requiring the replacement of at most $t - f$ fault-free units) [Kavianpour and Friedman 1978]. Consider the system of Figure 15.11. Since each unit is only tested by two other units, the system is not one-step 3-fault diagnosable. This is verified by the test outcome shown in

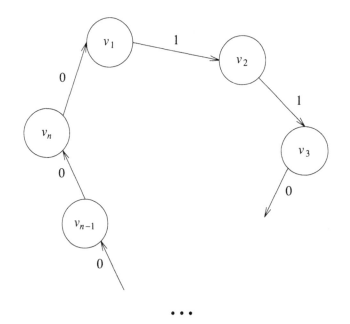

Figure 15.10   A single-loop system with n units

Figure 15.11(a). Assuming $t = 3$, we can deduce that unit v_7 is not faulty, since if it were faulty then v_6, v_5, and v_4 would also be faulty (as a consequence of the test outcomes ($a_{67}=a_{56}=a_{45}=0$), and hence there would be four faulty units, thus violating the upper bound $t = 3$. Since v_7 is fault-free, then v_1 and v_2 must be faulty (as a consequence of the test outcomes $a_{71} = a_{72} = 1$). In a similar manner we can then deduce that v_6, v_5, and v_4 are not faulty. Since v_3 is only tested by v_1 and v_2, both of which are faulty, we cannot determine whether v_3 is faulty or not. Thus this system is not one-step 3-fault diagnosable.

It can be shown, however, that the system is one-step 3/3 diagnosable. This implies that any fault pattern consisting of three faulty units can be exactly diagnosed, but some fault patterns of $f < 3$ faulty units can only be diagnosed to within three units. Consider the test outcome shown in Figure 15.11(b) produced by the set of three faulty units $\{v_1,v_3,v_4\}$. We can deduce that v_2 is not faulty, since if v_2 were faulty then v_7, v_6, and v_5 would also be faulty (as a consequence of the test outcomes $a_{72} = a_{67} = a_{56} = 0$), thus violating the upper bound $t = 3$. Since v_2 is not faulty, v_3 and v_4 are faulty (as a consequence of the test outcomes $a_{23} = a_{24} = 1$). Similarly we then deduce that v_7 is not faulty and consequently v_1 is faulty. We have thus identified the set of three faulty units. It can be shown that this can always be done for any set of three faulty units for the system shown. It is thus apparent that t/t diagnosability necessitates fewer testing branches than t–fault diagnosability, thus again

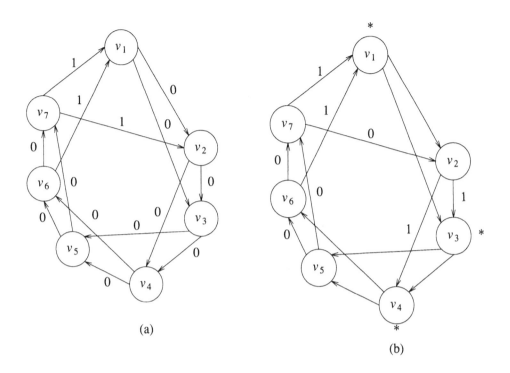

Figure 15.11   A system demonstrating t/t diagnosability

demonstrating the inherent trade-off in system-level diagnosis between accuracy of diagnosis and system complexity.

Many other generalizations of the PMC model have been studied. Such generalizations allow constraints encountered in actual systems to be more closely modeled. However, additional extensions and modifications are necessary to make the model applicable to actual systems [Friedman and Simoncini 1980].

The assumption that a unit can test other units requires complete access to the units being tested. This may necessitate a complex interconnection network. Testing links represent logical interconnections, and the directed graph is a logical representation of the diagnostic capabilities of a system. Consequently, a directed graph is not necessarily a simplified representation of a system's data-flow structure. It must still be determined what types of internal system organizations can most efficiently support the diagnostic procedures implied by a directed graph. Modeling of the diagnostic phase as well as normal operation of actual systems requires additional attention. An integrated approach to the modeling of fault-tolerant computing systems should consider both normal and diagnostic operations as well as reconfiguration. Thus, despite a considerable amount of work in this research area, practical applications still seem quite remote.

REFERENCES

[Adham and Friedman 1977] M. Adham and A. D. Friedman, "Digital System Fault Diagnosis," *Journal of Design Automation & Fault-Tolerant Computing*, Vol. 1, No. 2, pp. 115-132, February, 1977.

[Barsi *et al.* 1976] F. Barsi, F. Grandoni, and P. Maestrini, "A Theory of Diagnosability of Digital Systems," *IEEE Trans. on Computers*, Vol. C-25, No. 6, pp. 885-893, June, 1976.

[Blount 1977] M. Blount, "Probabilistic Treatment of Diagnosis in Digital Systems," *Proc. 7th Annual Intn'l. Conf. on Fault-Tolerant Computing*, pp. 72-77, June, 1977.

[Dahbura and Masson 1984] "An $O(n^{2.5})$ Fault Identification Algorithm for Diagnosable Systems," *IEEE Trans. on Computers*, Vol. C-33, No. 6, pp. 486-492, June, 1984.

[Friedman 1975] A. D. Friedman, "A New Measure of Digital System Diagnosis," *Digest of Papers 1975 Intn'l. Symp. on Fault-Tolerant Computing*, pp. 167-169, June, 1975.

[Friedman and Simoncini 1980] A. D. Friedman and L. Simoncini, "System-Level Fault Diagnosis," *Computer*, pp. 47-53, March, 1980.

[Fujiwara and Kinoshita 1978] H. Fujiwara and K. Kinoshita, "Connection Assignment for Probabilistically Diagnosable Systems," *IEEE Trans. on Computers*, Vol. C-27, No. 3, pp. 280-283, March, 1978.

[Hackl and Shirk 1965] F. J. Hackl and R. W. Shirk, "An Integrated Approach to Automated Computer Maintenance," *IEEE Conf. on Switching Theory and Logical Design*, pp. 298-302, October, 1965.

[Hakimi and Nakajima 1984] S. L. Hakimi and K. Nakajima, "On Adaptive Systems Diagnosis," *IEEE Trans. on Computers*, Vol. C-33, No. 3, pp. 234-240, March, 1984.

[Kavianpour 1978] A. Kavianpour, "Diagnosis of Digital System Using t/s Measure," Ph.D. Thesis, University of Southern California, June, 1978.

[Kavianpour and Friedman 1978] A. Kavianpour and A. D. Friedman, "Efficient Design of Easily Diagnosable Systems," *3rd USA-Japan Computer Conf.*, pp. 14.1-14.17, 1978.

[Kreutzer and Hakimi 1987] "System-Level Fault Diagnosis: A Survey," *Microprocessing and Microprogramming*, Vol. 20, pp. 323-330, 1987.

[Mallela and Masson 1978] S. Mallela and G. M. Masson, "Diagnosable Systems for Intermittent Faults," *IEEE Trans. on Computers*, Vol. C-27, No. 6, pp. 560-566, June, 1978.

[Preparata *et al.* 1967] F. P. Preparata, G. Metze, and R. T. Chien, "On the Connection Assignment Problem of Diagnosable Systems," *IEEE Trans. on Electronic Computers*, Vol. EC-16, No. 6, pp. 848-854. December, 1967.

[Russell and Kime 1975a] J. D. Russell and C. R. Kime, "System Fault Diagnosis: Closure and Diagnosability With Repair," *IEEE Trans. on Computers*, Vol. C-24, No. 11, pp. 1078-1088, November, 1975.

[Russell and Kime 1975b] J. D. Russell and C. R. Kime, "System Fault Diagnosis: Masking, Exposure, and Diagnosability Without Repair," *IEEE Trans. on Computers*, Vol. C-24, No. 12, pp. 1155-1161, December, 1975.

PROBLEMS

15.1 Consider a diagnostic graph consisting of a loop of n modules.

a. Prove that such a system is one-step t-fault diagnosable only for $t < 2$.

b. Prove that such a system is one-step 2/3 fault diagnosable if $n > 6$ by showing that for any fault pattern produced by two or fewer faults, at least $n - 3$ modules can be ascertained to be properly functioning under the assumption that at most two modules can be faulty.

15.2 Consider a system whose diagnostic graph has five nodes $\{0,1,2,3,4\}$ and an edge from i to $(i + 1)$ mod 5 and from i to $(i + 2)$ mod 5 for all i.

a. Prove that such a system is one-step two-fault diagnosable and sequentially two-fault diagnosable.

b. What is the maximum number of edges that can be removed from this graph so that it is still one-step two-fault diagnosable, or so that it is still sequentially two-fault diagnosable?

15.3

a. Prove that the system shown in Figure 15.12, is one-step 3/3 diagnosable.

b. Prove that if any edge is removed the system is not one-step 3/3 diagnosable.

c. For the test outcome shown in Figure 15.12, assuming $t = 3$, determine the set of faulty modules to within $s = 3$ units.

15.4 For the system shown in Figure 15.13 and for each of the models of Figure 6.9

a. Determine the maximum value of t for which the system is one-step t fault diagnosable.

b. Determine the maximum value of t^1 for which the system is sequentially t^1-fault diagnosable.

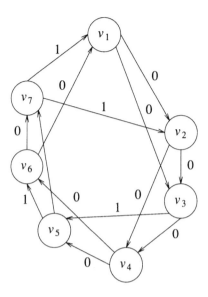

Figure 15.12   Problem 15.3

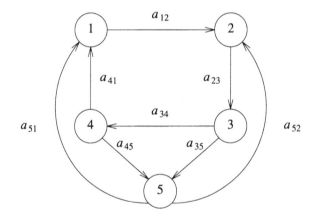

Figure 15.13   Problem 15.4

CPSIA information can be obtained
at www.ICGtesting.com
Printed in the USA
BVHW01*1918070718
520825BV00013B/148/P

9 780780 310629